Molecular Electronic Devices II

Forrest L. Carter

Chemistry Division
Naval Research Laboratory
Washington, D.C.

Marcel Dekker, Inc.   *New York and Basel*

0 2604710x

CHEMISTRY

Library of Congress Cataloging-in-Publication Data

Molecular electronic devices II.

   Includes index.
   1. Molecular electronics--Congresses. I. Carter,
Forrest L.
TK7874.M53272 1987   621.381'71   87-9174
ISBN 0-8247-7562-7

MARCEL DEKKER, INC.

270 Madison Avenue, New York, New York 10016

Current printing (last digit):
10 9 8 7 6 5 4 3 2 1

PRINTED IN THE UNITED STATES OF AMERICA

Preface

Molecular Electronic Devices II carries forth the exciting concepts, theories, and experiments needed to further stimulate multinational and interdisciplinary scientists to future achievements in molecular electronics.

Theoretical advances in several areas provide support for a variety of molecular-size switches, including electron tunnel switches, optical switches, and two-soliton switches. One of these, the soliton valve, has been formulated into AND and OR gates, an inverter, and even a four-bit fan memory. The beauty of molecular electronics is that molecular-size components can be formulated to produce non-Boolean devices with great economy. For example, a molecular analog of the CASE statement in the PASCAL language has been formulated by a high school student. Another, more elaborate, assembly of mechanical "molecular" devices has been proposed as the molecular analog of a mechanical computer.

A bridge must be built to span the difference between macro- and molecular technology. The imaginative sculpturing of Don Kendall to

produce both advanced integrated circuits and novel mechanical silicon devices has provided the kind of inspiration needed. The three-
dimensional chemical structures assembled by Hans Kuhn from two-
dimensional layers of Langmuir-Blodgett films provide insight into
the fabrication of simple molecular electronic devices. Interest in
L-B films has also led to studies in self-organization, an area of
extreme interest if molecular electronics is to become a reality.

The interconnection of topics in the book is fascinating. For
example, the need to develop and test concepts of neural networks in
membranes follows for the small size and vulnerability to damage of
conjugated switches such as soliton valves. Other related topics
include a possible link between Davydov solitons, the α-helix structure and anesthesia, grooves in silicon and biological structures,
silicon-L-B film-resists-phthalocyanines, and optical signals and
switches.

Entirely new concepts of computer architecture will evolve from
the concepts of cellular automata. These small computing elements,
such as finite-state machines, are linked only by nearest neighbors
in forming one-, two-, or three-dimensional arrays. The simplest
and smallest three-state finite-state machine is the soliton valve
consisting of approximately 20 atoms. Multiple-state finite-state
machines are readily constructed from cyclic configurations of soliton valves. John Barker showed that two-dimensional cellular automata can be used to accept two-dimensional data and then multiply and
transmit them to eight different process arrays. Other two-dimensional
arrays from Conway's Game of "Life" can be considered as flowing
switches. Together, these two uses of cellular automata can be seen
as a new form of computer floating in a "sea of cellular automata."

A major challenge of molecular electronics is fabrication at the
molecular level. Rather than etching and sculpturing in blocks of
silicon the approach suggested here is to build up structures chemically from the molecular level. For example, a modular synthetic
method inspired by the Merrifield approach has been suggested for a
$(SN)_x$ "molecular wire." Similarly, "molecular construction" has

been suggested for the formation of two-dimensional arrays on semi-conductor substrates using biological monolayers as etchable resists and a molecular lithography. The storehouse of molecular fabrication techniques is becoming very rich indeed.

Forrest L. Carter

Contents

Contents

Contributors

Augustine A. Abia Department of Chemistry, University of Nevada, Reno, Nevada

Aleksander Balter School of Medicine, University of Maryland, Baltimore, Maryland

John R. Barker Department of Physics, University of Warwick, Coventry, England

William A. Barlow Imperial Chemical Industries plc, Runcorn, Cheshire, United Kingdom

T. W. Barrett Chemistry Division, Naval Research Laboratory, Washington, D. C.

Gilbert Baumann Duke University, Durham, North Carolina

Richard C. Benson Applied Physics Laboratory, Milton S. Eisenhower Research Center, Johns Hopkins University, Laurel, Maryland

Richard D. Burkhart Department of Chemistry, University of Nevada, Reno, Nevada

David K. Campbell Center for Nonlinear Studies and Theoretical Division, Los Alamos National Laboratory, Los Alamos, New Mexico

Forrest L. Carter Chemistry Division, Naval Research Laboratory, Washington, D. C.

Gary M. Carter GTE Laboratories, Inc., Waltham, Massachusetts

Y. J. Chen GTE Laboratories, Inc., Waltham, Massachusetts

Henryk Cherek Department of Biological Chemistry, School of Medicine, University of Maryland, Baltimore, Maryland

Timothy Coffey Director of Research, Naval Research Laboratory, Washington, D. C.

J. Comas Naval Research Laboratory, Washington, D. C.

K. L. Davis Naval Research Laboratory, Washington, D. C.

K. Eric Drexler M.I.T. Space Systems Laboratory, Cambridge, Massachusetts

D. Duckworth Naval Research Laboratory, Washington, D. C.

George Easton Duke University, Durham, North Carolina

Barbara L. Eyres Imperial Chemical Industries plc, Runcorn, Cheshire, United Kingdom

G. Fariss Department of Macromolecular Science, Case Institute of Technology, Case Western Reserve University, Cleveland, Ohio

D. K. Ferry Department of Electrical Engineering, Colorado State University, Fort Collins, Colorado

Donald J. Freed Imperial Chemical Industries Americas, Wilmington, Delaware

B. L. Giammara Microelectronics Center of North Carolina, Research Triangle Park, North Carolina

R. O. Grondin Department of Electrical Engineering, Colorado State University, Fort Collins, Colorado

Michael P. Groves Department of Computing Science, University of Adelaide, Adelaide, South Australia

J. S. Hanker Dental Research Center, University of North Carolina at Chapel Hill, Chapel Hill, North Carolina

Richard A. Hann Imperial Chemical Industries plc, Runcorn, Cheshire, United Kingdom

Dagmar Higelin Physikalisches Institut, Universitat Stuttgart, Stuttgart, West Germany

Robert C. Hoffman Applied Physics Laboratory, Milton S. Eisenhower Research Center, Johns Hopkins University, Laurel, Maryland

Richard N. Johnson University of North Carolina at Chapel Hill, Chapel Hill, North Carolina

David C. Joy Bell Laboratories, Murray Hill, New Jersey

Don L. Kendall Institut Nacional de Astrofisica, Optica, y Electronica, Pueblo, Mexico

S. W. Kirchoefer Naval Research Laboratory, Washington, D. C.

Hans Kuhn Department for Development of Molecular Systems, Max Planck Institute for Biophysical Chemistry, Karl Friedrich Bonhoeffer Institute, Gottingen-Nikolausberg, Federal Republic of Germany

Joseph R. Lakowicz Department of Biological Chemistry, School of Medicine, University of Maryland, Baltimore, Maryland

Jerome B. Lando Department of Macromolecular Science, Case Institute of Technology, Case Western Reserve University, Cleveland, Ohio

Aldrich N. K. Lau Department of Chemistry, University of Minnesota, Minneapolis, Minnesota

Jean-Pierre Launay Laboratoire de Chimie des Metaux de Transition, Universite Pierre et Marie Curie, Paris, France

Albert F. Lawrence Design Development Laboratory, Hughes Aircraft Company, Long Beach, California

Scott P. Layne Center for Nonlinear Studies and Theoretical Division, Los Alamos National Laboratory, Los Alamos, New Mexico

C. M. Loeffler Department of Electrical Engineering, Colorado State University, Fort Collins, Colorado

R. Magno Naval Research Laboratory, Washington, D. C.

James H. McAlear Gentronix Laboratories, Inc., Rockville, Maryland

Robert M. Metzger Department of Chemistry, University of Mississippi, University, Mississippi

Larry L. Miller Department of Chemistry, University of Minnesota, Minneapolis, Minnesota

D. J. Nagel Naval Research Laboratory, Washington, D. C.

Dale E. Newbury Center for Analytical Chemistry, National Bureau of Standards, Washington, D. C.

Charles A. Panetta Department of Chemistry, University of Mississippi, University, Mississippi

Theodore O. Poehler Applied Physics Laboratory, Milton S. Eisenhower Research Center, Johns Hopkins University, Laurel, Maryland

W. Porod Department of Electrical Engineering, Colorado State University, Fort Collins, Colorado

Richard S. Potember Applied Physics Laboratory, Milton S. Eisenhower Research Center, Johns Hopkins University, Laurel, Maryland

R. H. Propst Biomedical Engineering and Mathematics Curriculum, University of North Carolina at Chapel Hill, Chapel Hill, North Carolina

Stephen R. Quint Department of Biomedical Engineering, University of North Carolina at Chapel Hill, Chapel Hill, North Carolina

P. E. Rapp Department of Physiology and Biochemistry, The Medical College of Pennsylvania, Philadelphia, Pennsylvania

Scott E. Rickert Department of Macromolecular Science, Case Institute of Technology, Case Western Reserve University, Cleveland, Ohio

Gareth G. Roberts Department of Applied Physics and Electronics, The University at Durham, Durham, United Kingdom

G. A. Rozgonyi Microelectronics Center of North Carolina, Research Triangle Park, North Carolina

Lynne A. Samuelson GTE Laboratories, Inc., Waltham, Massachusetts

Daniel J. Sandman GTE Laboratories, Inc., Waltham, Massachusetts

R. L. Schmidt Naval Research Laboratory, Washington, D. C.

A. Schultz Naval Research Laboratory, Washington, D. C.

B. Simic-Glavaski The Chemistry Department and Case Center for Electrochemical Sciences, Case Western Reserve University, Cleveland, Ohio

Hans Sixl Physikalisches Institut, Universitat Stuttgart, Stuttgart, West Germany

A. Snow Chemistry Division, Naval Research Laboratory, Washington, D. C.

Kenneth R. Speck Applied Physics Laboratory, Milton S. Eisenhower Research Center, Johns Hopkins University, Laurel, Maryland

James H. Steven Imperial Chemical Industries plc, Runcorn, Cheshire, United Kingdom

Mrinal K. Thakur Department of Macromolecular Science, Case Institute of Technology, Case Western Reserve University, Cleveland, Ohio

Richard B. Thompson Department of Biological Chemistry, School of Medicine, University of Maryland, Baltimore, Maryland

Sukant Tripathy GTE Laboratories, Inc., Waltham, Massachusetts

Martyn V. Twigg Imperial Chemical Industries plc, Billingham, Cleveland, United Kingdom

Kevin M. Ulmer Director of Exploratory Research, Genex Corporation, Gaithersburg, Maryland

John M. Wehrung Gentronix Laboratories, Inc., Rockville, Maryland

Hank Wohltjen Chemistry Division, Naval Research Laboratory, Washington, D. C.

Bernard A. Zempolich Naval Air Systems Command Headquarters, Washington, D. C.

I
Molecular Switching and Materials

Opening Remarks

Dr. Timothy Coffey / Director of Research, Naval Research
Laboratory, Washington, D.C.

Good morning, ladies and gentlemen. I am pleased to be here to
welcome you to the second International Workshop on Molecular
Electronic Devices.

I think the promise of the field of science and engineering
which you are beginning to develop is truly immense. If successful, its scientific and sociologic consequences will be awesome.
Successful development of practical molecular electronic devices
will have an impact equal to that of any of the major scientific
breakthroughs of the Twentieth Century.

Moreover, practical sources will substantially accelerate the
already rapid pace of technology advancement. One can imagine,
for example, super computers which one could hold in the palm of
one's hand. One can imagine the development of devices with
exceedingly complete and sensitive sensors. These would be

sight, sound, smell, touch, and the ability to make reasoned use
of these senses. Such achievements could have great profound
benefits to mankind.

Like all great advances, however, they also present problems
which needn't impede the pace of progress but which, the history
of Twentieth Century technology tells us, we probably should
begin to worry about early in the development phase. For
instance, the military application which will, and should occur,
could be destabilizing. I think our political leaders and
military planners have to worry about that.

Perhaps a more serious problem relates to whether or not
society itself has the resilience to adapt without major dislocations to such potentially rapid and profound changes in
technology. We need to consider the current economic dislocations which are occurring as a result of, in light of what might
occur here, the relatively modest advances in computer technology and robotics. Development of practical molecular electronic devices will far outstrip the latter accomplishments.

I think in some ways society, through its educational
system, must find a way to better prepare our young people to
cope with technological advances which will lead to radically
different employment opportunities from those which might exist
when a child enters school. Each of us as citizens bear some
responsibility to address this problem. Those of us who are
responsible for the technology push bear more responsibility than
most.

Now I have been speaking as though the scientific and
technological problems here are in hand. This, of course, is
hardly the case. The problems which you face are truly for-

midable. Their solution will take many years, will require the finest minds, and an almost dogged persistence.

The vision, however, is clear and it captures the imagination. We at NRL are proud of the contributions that NRL personnel have already made and will encourage our people, along with all of you, to excel in this important undertaking.

The Naval Research Laboratory, therefore, wishes you success at this workshop and success in the exciting years ahead. I thank you very much for the opportunity to address you.

1

Toward Organic Rectifiers

Robert M. Metzger and Charles A. Panetta/ Department of
Chemistry, The University of Mississippi, University, MS

I. INTRODUCTION

This is an up-to-the-minute report on our current efforts to
test the tantalizing proposal by Aviram, Ratner, Seiden and others
in 1974 [1-4] that a unimolecular organic "rectifier" or
"switch" D-σ-A may be constructed from a potent organic electron
donor molecule, D, connected through a covalent "sigma" bridge,
σ, to a strong organic electron acceptor molecule, A.

This proposal has been reviewed tangentially by Aviram [5]
and Ratner [6] at the previous NRL Molecular Electronic Devices
Workshop in March 1981. Our own results up to November 1982 have
been summarized elsewhere [7].

We report below on the successful synthesis of such a D-σ-A
molecule, where D is tetrathiafulvalene (TTF, 1) A is 7,7,8,8-
tetracyanoquinodimethan (TCNQ, 2) and σ is a carbamate or urethan
linkage [7,8]. However, chemical purification of this product has

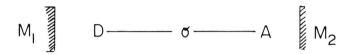

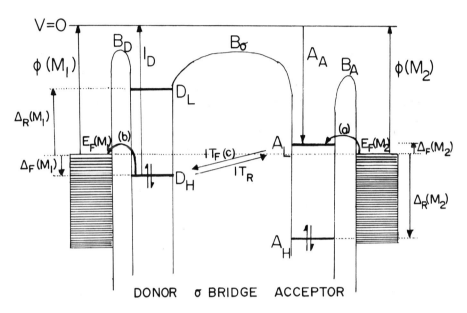

FIGURE 1.   Energy levels versus distance for $M_1|D-\sigma-A|M_2$ rectifier
device, adapted from Aviram and Ratner, Ref. [1].

the separate D and A molecules are minimally perturbed when D-σ-A
is synthesized. Also shown in Fig. 1 are the workfunctions $\phi(M_1)$
and $\phi(M_2)$ of the two metal films, the ionization potential I_D of
the donor part of the molecule, and the electron affinity A_A of
the acceptor part. For gas-phase organic D molecules:

$$D(g) \longrightarrow D^+(g) + e^-(g) \qquad I_D \qquad (1)$$

I_D is of the order of 6 eV (good donor) to 9 eV (poor donor); for
gas-phase organic A molecules:

$$A(g) + e^-(g) \longrightarrow A^-(g) \qquad -A_A \qquad (2)$$

A_A is of the order of 1 eV (poor acceptor) to 3 eV (good

been frustratingly slow. Thus we cannot yet present a final
answer on the feasibility of the Aviram–Ratner proposal.

In Section II we review the essentials of the Aviram–Ratner
proposal; in Section III we discuss its potential utility; in
Section IV we detail the six practical obstacles to its
realization. Section V presents our synthesis of the first D–σ–A
molecule, and Section VI briefly outlines related synthetic
projects being carried out elsewhere.

II. THE AVIRAM–RATNER PROPOSAL

Aviram and Ratner suggested [1] that a unimolecular oriented
film of organic molecules D–σ–A [where D = <u>strong</u> electron donor
(but poor electron acceptor) e.g. TTF, σ = saturated covalent
"sigma" bridge, A = <u>strong</u> electron acceptor (but poor electron
donor) e.g. TCNQ] sandwiched between two metal films M_1, M_2 could
be a unimolecular electrical rectifier.

Their argument is based on the relative positions of several
energy levels (Fig. 1): the Fermi energies $E_F(M_1)$ and $E_F(M_2)$ of
metal films M_1, M_2, the highest occupied molecular orbitals (HOMO)
of donor (D_H) and acceptor (A_H), and the lowest occupied molecular
orbitals (LUMO) of donor (D_L) and acceptor (A_L), and on their
shifts under forward and reverse electrical bias. Also shown on
Fig. 1 are the chemisorptive potential energy barriers B_D and B_A
for the M_1–D and A–M_2 interfaces, and the wider barrier B_σ
presented by the σ bridge, whose role is to keep the D and A parts
of D–σ–A physically far apart, so that through–bond (TB) [9]
electronic interactions are strongly favored over through space
(TS) [9] interactions, and so that the molecular energy levels of

acceptor). In a condensed phase I_D and A_A are considerably
shifted [10].

The essential details of the Aviram-Ratner proposal are as
follows: under moderate forward bias $E_F(M_1)$ and D_H will be
brought to the same energy, and $E_F(M_2)$ and A_L will also be at the
same energy; the molecular energy levels will be shifted by the
electric field perturbation, but so that A_L will remain above D_H.
When $E_F(M_2)$ matches A_L, resonant (elastic) electron tunneling (a)
occurs from M_2 (cathode) to A; when $E_F(M_1)$ matches D_H, resonant
electron tunneling (b) occurs from D to M_1 (anode); the resultant
$D^+-\sigma-A^-$ then returns, (c), to $D-\sigma-A$ by inelastic TB internal
tunneling (IT_F) from the lowest vibrational-rotational sublevel
of A_L to an excited sublevel of D_H, and hence by radiationless
transitions to the lowest sublevel of D_H: this IT_F is clearly
irreversible. Thus under forward bias the following process is
expected:

$$\text{FORWARD BIAS: GO} \qquad M_1 \Big| D-\sigma-A \Big| M_2 \xrightarrow[\text{(b) } D \longrightarrow M_1]{\text{(a) } M_2 \longrightarrow A} M_1^- \Big| D^+-\sigma-A^- \Big| M_2^+ \qquad (3)$$

$$\xrightarrow{\text{(c) } IT_F} M_1^- \Big| D-\sigma-A \Big| M_2^+$$

If steps (a) and (b) are to occur readily under forward bias
voltage V, the electron affinity of A, A_A, must be close to the
work function of M_2, $\phi(M_2)$, and the ionization potential of D, I_D,
must be close to $V + \phi(M_1)$. Now transition metal work functions ϕ
are in the range 3.1 eV (Gd) to 5.5 eV (Pt) [11]; for TTF $I_D =$
6.85 eV [12] and for TCNQ $A_A = 2.8$ eV [13]. However one might
hope that in the solid state I_D (and A_A) would be depressed
(raised) from their gas-phase values. Under reverse bias the

voltages $\Delta_R(M_1)$ and $\Delta_R(M_2)$ needed to match Fermi energies with D_L and A_D are too large; thus the process:

REVERSE
BIAS: $M_1 \| D-\sigma-A \| M_2 \xrightarrow{\quad\times\quad} M^+ \| D^--\sigma-A^+ \| M_2^-$ (4)
NO GO

is energetically not favored since D_L and A_H are far from the Fermi energies $E_F(M_1)$ and $E_F(M_2)$ respectively. What distinguishes process (3) from process (4) is that the energy level differences $\Delta F(M_1)$, $\Delta F(M_2)$ are much smaller than $\Delta_R(M_1)$, $\Delta_R(M_2)$.

The gas-phase analog of the electron transfer process (3):

$$D(g) + A(g) \longrightarrow D^+(g) + A^-(g) \qquad (5)$$

requires an energy $I_D-A_A = 6.8 - 2.8 = 4.0$ eV for D = TTF and A = TCNQ; the gas-phase analog of the reverse process (4):

$$D(g) + A(g) \longrightarrow D^-(g) + A^+(g) \qquad (6)$$

requires an estimated energy [14,15] of about 9.6 eV. Under reverse bias another electron transfer mechanism can also occur: internal tunneling IT_R from D_H to A_L becomes possible if and only if the external electric field is strong enough to raise D_H to match or be higher than A_L. This should occur only at high reverse bias. Then the process:

REVERSE
BIAS: $M_1 \| D-\sigma-A \| M_2 \xrightarrow{\;IT_R\;} M_1 \| D^+-\sigma-A^- \| M_2 \longrightarrow M_1^+ \| D-\sigma-A \| M_2^-$ (7)

becomes possible. The model molecule proposed by Aviram and Ratner [1], based on TTF and TCNQ, is molecule $\underline{3}$. For this mole-

$\underline{3}$

cule, using estimates I_D = 5.3 eV, A_A = 5.0 eV, $\phi(M_1)$ = $\phi(M_2)$ =
5.1 eV the I–V curve of Fig. 2 was computed where I was in A m^{-2}
(<u>not</u> A cm^{-2}) and V was in volts. The rectification of the model
molecule was <u>conceptually</u> demonstrated.

III. POSSIBLE PRACTICAL ADVANTAGE OF AN ORGANIC RECTIFIER

The obvious foreseeable practical advantage of an organic
rectifier is its size: the dimensions of molecule $\underline{3}$ are
(roughly) 20Å (length) by 10Å (height) by 5Å (thickness) assuming

a geometry as planar as is allowed by the fairly bulky [2.2.2] –
bicyclooctane group in the σ bridge. Thus, if super thin metal
films, or doped polyacetylene conductors are used as M_1 and M_2,
then a rectifier of thickness 50Å (5nm) can be imagined. If this
device becomes technically feasible, and if other molecular
analogs of transistors also can be devised, then a great
reduction in scale of electrical circuit elements can be
imagined.

From the current pn junction rectifiers of typical thickness
100 to 1000 nm, such an organic rectifier would present a size
reduction of a factor of 20 to 200.

If the Aviram–Ratner proposal can be extended to three
rectifiers in series, with the middle molecule acting as a gate,
one can dream about molecular transistors.

IV. PRACTICAL OBSTACLES IN BUILDING ORGANIC RECTIFIERS:

There are at least six practical obstacles that must be
overcome before an organic rectifier becomes a reality.

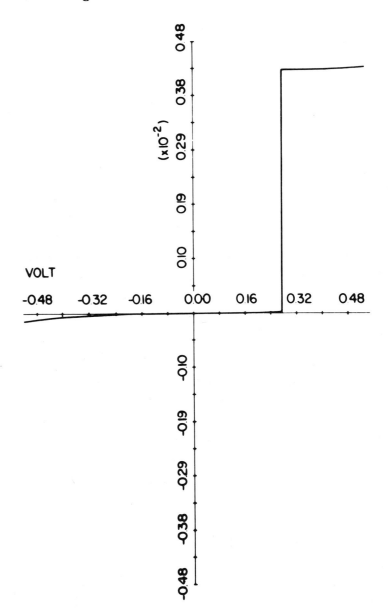

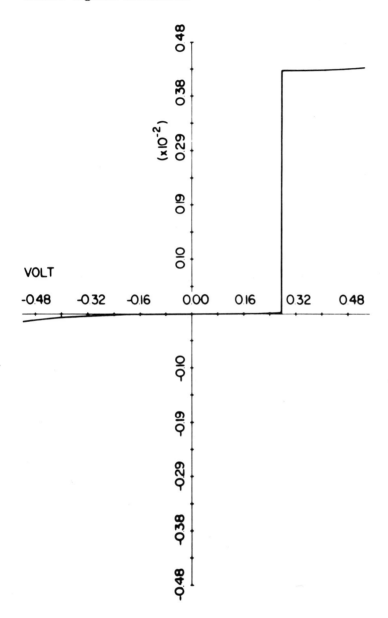

FIGURE 2.   I-V characteristics predicted by Aviram and Ratner for
            molecule $\underline{3}$, from Ref. [1].   I is in A/m^2, V in volts,
            I_D = 5.3 eV, A_A = 5.0 eV, $\phi(M_1)$ = $\phi(M_2)$ = 5.1 eV;
            processes (3), (6), and (7) were included in the model
            calculations.

1) The <u>first</u> obstacle is a synthetic one: to build D–σ–A
one must find a chemical coupling reaction between functionalized
D and A derivatives, D–X and Y–A:

$$D-X + Y-A \longrightarrow D-\sigma-A \tag{8}$$

which is <u>so fast</u> that the competing charge transfer complex
formation:

$$D-X + Y-A \longrightarrow (D-X)^+(Y-A)^- \tag{9}$$

is inhibited. Normally, D–X and Y–A are such strong reducing and
oxidizing agents, respectively, that most coupling attempts
follow eq. (9) and not eq. (8). In fact, underivatized TTF and
TCNQ form the partially charge–transferred quasi–one dimensional
metallic conductor $TTF^{+0.59} TCNQ^{-0.59}$ [16–20]. However, as is
reported below, this first obstacle has been <u>removed</u>: a urethan
linkage between TTF and TCNQ has been built [7,8].

2) The <u>second</u> obstacle (actually, a design criterion) is
that σ must be long enough, and the D and A ends must be far
enough, that the molecular energy levels of D and A are not
grossly perturbed. The σ must <u>not</u> be flexible to the extent that
the D and A moieties are bent over to form a U–shaped molecule
with strong through–space interaction between D and A. Several
such insulating bridges have been built recently [21–25] because
of the wide–spread research interests in separating π systems and
chromophores by long σ bridges in order to study through–bond
energy electron transfer processes and hopefully, to inhibit
back–charge transfer (as Mother Nature does so well in
photosynthesis).

3) The <u>third</u> obstacle is in forming a dense, highly packed
monolayer <u>4</u> of oriented D–σ–A molecules between two metal films:

$$M_1 \quad \begin{array}{c} D-\sigma-A \\ D-\sigma-A \\ D-\sigma-A \end{array} \quad M_2$$

$$\underline{4}$$

Two techniques suggest themselves for the assembly of such films:
the Langmuir-Blodgett (LB) film technique [24,25] using water as
a support medium in a film balance: this technique has been
perfected recently by Kuhn and co-workers [28-34]; another
technique involves attaching monolayers covalently to derivatized
metal surfaces using silanizing reagents [35-38]. Usable LB
films of the neutral D-σ-A molecules will be obtained only if
uniform dense films of type 5 or 6 are formed but not
of type 7:

	D-σ-A				A-σ-D				A-σ-D	
H$_2$O	D-σ-A	Air		H$_2$O	A-σ-D	Air		H$_2$O	D-σ-A	Air
	D-σ-A				A-σ-D				A-σ-A	
	5				6				7	

Most LB films studied by Kuhn and Möbius are mainly of
amphiphilic molecules. A prototype of these is cadmium
arichidate, $Cd^{++}(CH_3-(CH_2)_{18}COO^-)_2$ which can be represented as 
∿∿● , where ∿∿∿ is the fat-soluble (hydrophobic) 19-
carbon "tail" and ● is the ionic, water-soluble (hydrophilic) -
COO⁻ "head". Thus arrays of type 8 can be formed:

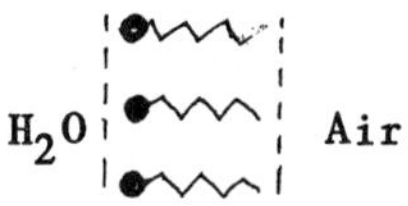

$$H_2O \quad \begin{array}{c} ●∿∿ \\ ●∿∿∿ \\ ●∿∿∿ \end{array} \quad Air$$

8

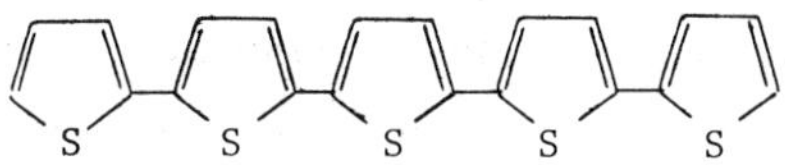

9

COOH

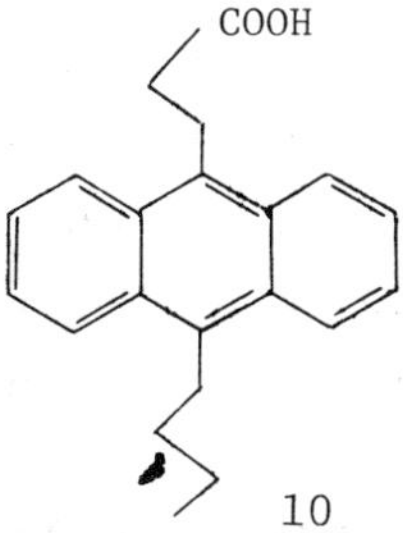

10

However, quinquethienyl <u>9</u> [31,34] and a lightly substituted anthracene <u>10</u> [39] are the only known non-amphiphilic molecules which do form monolayers, so there is some hope that <u>5</u> and <u>6</u> can be formed. Either the D or the A of D–σ–A must show selectivity towards the water interface (<u>5</u>, <u>6</u>) and thus defeat an otherwise natural tendency to cancel molecular dipole moments, as in <u>7</u>. If the D–σ–A molecule is obtained in its zwitterionic form, D^+–σ–A^-, then instead of using water in the film balance, mercury could be used, [40-42] and a LB film <u>11</u> could be prepared.

$$
\text{Hg} \quad
\begin{array}{|c|}
\hline
D^+\text{–}\sigma\text{–}A^- \\
D^+\text{–}\sigma\text{–}A^- \\
D^+\text{–}\sigma\text{–}A^- \\
\hline
\end{array}
\quad \text{Air}
$$

<u>11</u>

The technique of attaching D–σ–A to a suitable derivatized metal surface [35-38] is quite attractive, but its surface coverage is usually, (but not always [36]) less dense than obtainable in LB films. Also, the size of the usual chemical coupling reagents (silanes) would create a larger metal-to-D–σ–A distance than may be desirable.

Of course, a very dense well-packed monolayer coverage of metal film M_1 is essential, or else the deposition of a metallic film M_2 on top of the monolayer will make electrical short circuits to M_1.

A brief LB experiment involving a presumed neutral form of D–σ–A is described at the end of the next section.

4) The <u>fourth</u> obstacle is that the thickness d of D–σ–A monolayers between metal films M_1 and M_2 must make TS tunneling between M_1 and M_2 negligibly small compared to TB tunneling. For either form of tunneling one can use Bethe's expression [43] for the conductivity σ_t [32,33]:

$$\sigma_t = (e^2 \sqrt{2m\phi}/h^2) \exp[-4\pi d \sqrt{2m\phi}/h] \qquad (10)$$

where e is the charge and m the mass of the electron, h is
Planck's constant, ϕ is the work function from metal to
insulator,

$$\phi = \varphi - A_I \qquad (11)$$

φ is the work function from metal to vacuum, and A_I is the
electron affinity of the insulator (for TB tunneling). For TS

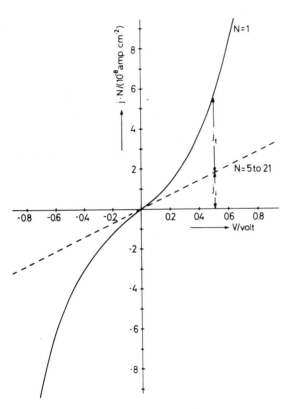

FIGURE 3. IV characteristics observed by Mann and Kuhn, Ref.
[32] for a single LB monolayer (N = 1) of cadmium
arachidate Cd^{++} $[CH_3(CH_2)_{18}COO^-]_2$, thickness of
about 28Å, sandwiched between Al and Hg. Solid
line is for N = 1 monolayer (Al electrode
positively biased). Dashed line is for N = 5 to
21 monolayers (background impurity current j_i); j_t
is the through-bond thickness-dependent tunneling
current, and $j = j_t + j_i$.

tunneling $A_I = 0$ is used. Kuhn and coworkers, in an elegant series of experiments, showed that eqns (10,11) are applicable to TB tunneling across one to four monolayers of cadmium arichidate [32,33] (see Fig. 3). If σ_t is in $\Omega^{-1}cm^{-1}$, d is in Å, and ϕ is in eV:

$$\sigma_t = 446 \, \phi^{1/2} \exp(-1.079d \, \phi^{1/2}). \qquad (12)$$

The crucial difference between TS and TB tunneling is that in the former A_I is set equal to zero. Using $\phi = 5$ eV, $A_I = 2$ eV [33] one obtains from eq. (12) theoretical ratios $\sigma_t(TB)/\sigma_t(TS)$ of 1.332, 178, 4.1×10^4, 9.4×10^6 for d = 1, 10, 20, and 30 Å respectively. Thus, a D–σ–A film thickness d > 15 Å should easily eliminate the competition from through–space tunneling processes.

5) The fifth obstacle to a practical D–σ–A rectifier is that process (3) must be as fast as competing processes in commercial silicon–based pn junctions. A reassuring recent result is that intramolecular electron transfer rates (half–lives) $t_{1/2} < 5 \times 10^{-10}$s have been measured [44] for electron transfer from the negative ion (D⁻) of the biphenyl group of 12 to the cinnamoyl (A) group of 12.

D

12

A

6) The sixth obstacle, or rather, practical limitation, is that excessive voltages across the film may chemically damage it. The forward and reverse bias voltages between M_1 and M_2 should not, in all probability, exceed ±1 volt. Also, heating of D–σ–A films during deposition of the second metal film must be carefully avoided.

V. SYNTHESIS AND PRELIMINARY CHARACTERIZATION OF FIRST D-σ-A MOLECULE

The molecule 3 suggested by Aviram and Ratner was judged difficult to synthesize, and probably not sufficiently planar for good monolayer packing. There are two conceptual approaches to synthesizing a D-σ-A molecule. The first is to "build" the whole molecular framework P-σ-Q, and then transforming P and Q in careful steps by protecting-group techniques into D and A respectively. The second approach is to add functional groups X to D, Y to A, and find a reaction that will yield D-σ-A (eq. (8)). This second approach was followed. Several efforts were made to build ethylene or ester linkages between TTF and TCNQ; they all failed [8] because the charge-transfer reaction (eq. 9) overwhelmed the bridge formation. The esterification reaction under controlled electrochemical potential, to discourage formation of TTF radical cations and TCNQ radical anions, failed, in part because of interference from the supporting electrolyte, NBu_4PF_6. Some years ago Hertler [45] synthesized the insulating TTF-TCNQ copolymer 15 using a urethan linkage:

Evidently the urethan coupling reaction is very rapid.

Accordingly, at the University of Mississippi the singly
substituted 2-isocyanatotetrathiafulvalene 18 was synthesized from
the known 2-carboxytetrathiafulvalene [46]:

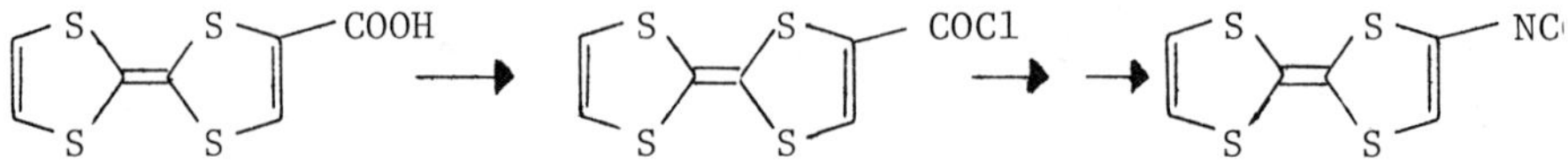

16   17   18
                     m.p. 130-135°C dec   m.p.75-80°C
                     dark purple needles

Then 18 was treated with the known 2-(2-hydroxyethoxy)-5-bromo-
TCNQ 19 [46] in the presence of dibutyltin dilaurate catalyst

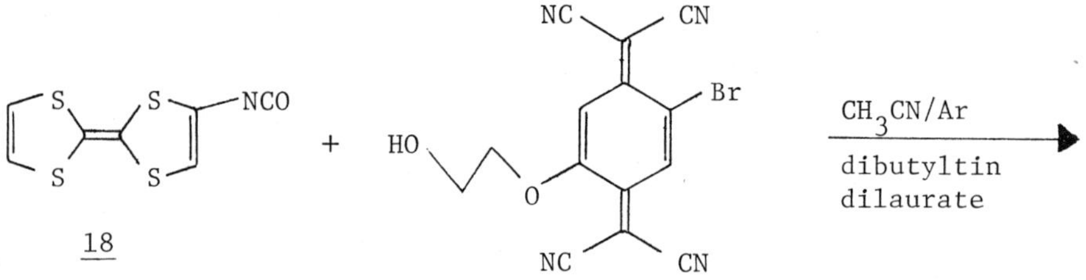

18   19

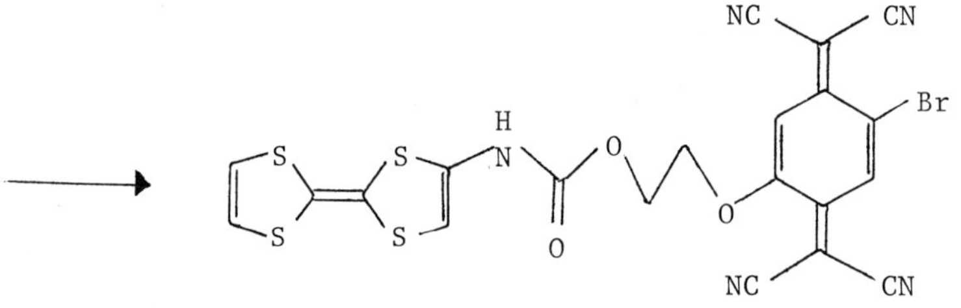

20 A m.p. 145-150°C dec
    B m.p. 105-108°C

[47] to yield the TTF-σ-TCNQ urethan, 20 [8]. Two products were
isolated in this reaction [8]. Product A, extracted with aceto-
nitrile or dichloromethane, was a brown-purplish powder, m.p. 145-

150°C dec, IR (KBr) 3500-3400, 3080, 2180 (broad: negatively charged C≡N), 1740-1700, and 1600 (carbamate), 1240-1220 cm^{-1}, VUV (CH_3CN) 853 nm ($log_{10}\varepsilon$ 4.57), 763 (4.24), 443 (s, 4.45), 416 (4.63), 297 (2.47), soluble in H_2O. Elemental analysis for $C_{21}H_{10}BrN_5O_3S_4$: calc (found) C 42.86(43.26), N 11.90(11.33), Br 13.58(13.72), S 21.80(17.40). An intense EPR powder spectrum with large g-factor anisotropy is observed. A cyclic voltammogram of 20 A shows good reversible oxidation waves, but the reduction waves are not what one would expect. Nevertheless, product A seems to be the zwitterionic (D^+-σ-A^-) form of 20. In a Lauda film balance 20 A dissolves in water, and forms no films. Film balance work with Hg instead of H_2O [40-42] is planned.

Product B, extracted with $CHCl_3$, m.p. 105-108°C, IR (CH_3CN) 3500-3400, 2210 (sharp, neutral -C≡N), 1735, 1600, 1525, 745 cm^{-1}, VUV (CH_3CN) 412 ($log_{10}\varepsilon$ 4.53), 401(4.52), 291(4.08) forms micro-crystals [x-ray lines at d = 7.225Å (w^{-5}), 6.981 (w^{-4}), 6.230 (w^-), 4.974 (w^{-2}), 4.405 (w^{-5}), 4.282 (w^{-3}), 3.897 (w^{-3}), 3.593 (w^{-4}), 3.515 (w), 3.437 (w^{-5}), 2.926 (w^{-5}), 2.317 (w^{-6}), which can be indexed for five possible triclinic lattices of reasonable cell dimensions]. The cyclic voltammogram of 20 B shows reasonable reversible oxidation and reduction waves, but of different intensities (presence of impurities?). Satisfactory elemental analyses of 20 B could not be obtained. It is possible that product B is the neutral (D-σ-A) form of 20, but this is far from certain. In a Lauda film balance product B formed a unimolecular film at 10°C, surface tension 12.7 mN/m. The molecular area was determined as 134 ± 50 Å/molecule (provided that B has indeed the molecular weight of 20). A Langmuir-Blodgett film supported on a Pt-coated glass slide gave a resistance of about 20 Ω (measured as Pt/film/Hg droplet). A cyclic voltammogram of this Pt/film/Hg droplet (swept from −0.3 V to +0.3 V) gave a spurious "rectification": a similar "recti-

fication" (rather: oxidation-reduction of Hg versus Pt!!) was
seen when no LB film was interposed!!   For molecule 20 a fairly
planar configuration is reasonable, and one can estimate a molecular size of 20Å (length) by 10Å (height) by 3.5Å (thickness).
Then, the measured LB molecular area implies a film made of D-σ-A
molecules laid down, slanted, or almost flat with thickness not
over 6Å; for such a thin film rectification may not be expected
to occur.   Sofar, products 20 A and 20 B could only be "purified" by repeated solvent washings.   Standard chromatographic
separations using silica gels have failed, since 20 seems to
react with the ionic impurities on the silica gel.   Sublimation
of 18, 19, and gentler chromatographic separations (cellulose,
sephadex, reverse-phase TLC plates) are planned.

Nevertheless, the first synthetic goal of covalently linking
TTF to TCNQ has been accomplished.   Spectroscopic, crystallographic, and electrochemical studies, as well as further
Langmuir-Blodgett film studies will be vigorously pursued as soon
as better D-σ-A samples become available.

VI.   RELATED SYNTHETIC EFFORTS BY OTHER RESEARCH GROUPS

Other groups are engaged in synthetic programs similar to
assembling TTF-σ-TCNQ molecules.   For several years Staab and
coworkers have worked towards synthesizing a para-cyclophane (21)
analog of TTF TCNQ: this would be TTF⌐σ⌐TCNQ, 22,

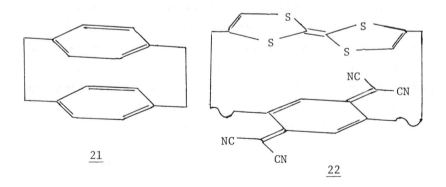

<u>21</u>

<u>22</u>

but such a molecule has not yet been prepared. However, [2.2]-
tetrathiafulvalenophane, <u>23</u> and [3.3]-tetrathiafulvalenophane, <u>24</u>

have been prepared [48,49] as well as <u>25</u> [50] and, most recently,
a neutral [2.2] phane of TMPD (N,N,N',N'-tetramethylparaphenylene-
diamine, <u>26</u>) with TCNQ [51].

Also, a group in Israel is studying D-σ-A-σ-D interactions
[52] and has recently synthesized the TCNQ derivative <u>27</u> [53] as a

prototype for such D-σ-A-σ-D systems.

VII. CONCLUSION

It has been shown that a unimolecular electronic device based on Aviram and Ratner's proposal is possible if certain stringent criteria are met. A strong electron donor (TTF) has been linked through a covalent urethan linkage to a strong electron acceptor (TCNQ): thus the synthetic obstacle to testing the Aviram-Ratner proposal has been overcome. Chemical purification, monolayer assembly, and device testing are the next logical steps in our efforts.

VIII. ACKNOWLEDGEMENT

We wish to thank Drs. Jamil Baghdadchi and Frank Yeh for their diligence and perseverance, Dr. Sukant Tripathy of GTE, Inc. for collaborative work on the Langmuir-Blodgett films, the National Science Foundation (Grant DMR 80-15658, 82-41625) for financial support and the Naval Research Laboratory and the Office of Naval Research for their kind hospitality during this meeting.

REFERENCES

1. A. Aviram and M. A. Ratner, Chem. Phys. Lett., _29_, 277 (1974).

2. A. Aviram and M. A. Ratner, Bull. Am. Phys. Soc., _19_, 341 (1974).

3. A. Aviram and M. A. Ratner, IBM Research Report RC 5419 (#23668) 15 May 1975 (114 pages).

4. A. Aviram, M. J. Freiser, P. E. Seiden, and W. R. Young, U.S. Patent 3,953,874, (27 April 1976).

5. A. Aviram, P. E. Seiden, and M. A. Ratner, in Proceedings of the First Molecular Electronic Devices Workshop, F. L. Carter, Editor, NRL Memorandum Report 4662, (22 October 1981); also in Molecular Electronics Devices, F.L. Carter, Ed., Marcel Dekker, New York, 1982, page 5.

6. R. Kozloff and M. A. Ratner, in Proceedings of the First Molecular Electronic Devices Workshop, F. L. Carter, Editor, NRL Memorandum Report 4662, 22 October 1981; also in Molecular Electronics Devices, F. L. Carter, Ed., Marcel Dekker, New York, 1982, page 31.

7. R. M. Metzger and C. A. Panetta in Proceedings of
    International CNRS Colloquium on Low-Dimensionl Organic
    Conductors, Dec. 1982, J. Physique (Orsay, Fr.) Colloque, in
    press.

8. J. Baghdadchi, Ph.D. Dissertation, University of Mississippi,
    Dec. 1982.

9. R. Hoffmann, A. Imamura, and W. J. Hehre, J. Am. Chem. Soc.,
    90, 1499 (1968).

10. P. Nielsen, A. J. Epstein, and D. J. Sandman, Solid State
    Commun., 15, 53 (1974).

11. Amer. Institute of Physics Handbook, D. E. Gray, Ed.,
    III Edition, McGraw-Hill, New York 1972, pages 9-173.

12. R. Gleiter, E. Schmidt, D. O. Cowan, and J. P. Ferraris, J.
    Electron Spectrosc., 2, 207 (1973).

13. R. N. Compton and C. D. Cooper, J. Chem. Phys., 66, 4325
    (1977).

14. Z. G. Soos, Chem. Phys. Letters, 63, 179 (1979).

15. F. Herman and I. P. Batra, Phys. Rev. Letters, 33, 94 (1974).

16. J. Ferraris, D. O. Cowan, V. Walatka, Jr., and J. H.
    Perlstein, J. Am. Chem. Soc., 95, 948 (1973).

17. G. A. Thomas, D. E. Schafer, F. Wudl, P. M. Horn, D. Rimai,
    J. W. Cook, D. A. Glocker, M. J. Skove, C. W. Chiu, R. P.
    Groff, J. L. Gillson, R. C. Wheland, L. R. Melby, M. G.
    Salamon, R. A. Craven, G. DePasquali, A. N. Bloch, D. O.
    Cowan, V. V. Walatka, Jr., R. E. Pyle, R. Gemmer, T. O.
    Poehler, G. R. Johnson, M. G. Miles, J. A. Wilson, J. P.
    Ferraris, T. F. Finnegan, R. J. Warmack, V. F. Raaen, and D.
    Jerome, Phys. Rev., B13, 5105 (1976).

18. M. J. Cohen, L. B. Coleman, A. F. Garito, and A. J. Heeger,
    Phys. Rev., B13, 5111 (1976).

19. T. J. Kistenmacher, T. E. Phillips, and D. O. Cowan, Acta
    Crystallogr. Sect. B, 30, 763 (1974).

20. F. Denoyer, R. Comès, A. F. Garito, and A. J. Heeger, Phys.
    Rev. Lett., 35, 445 (1975).

21. R. S. Davidson, R. Bonneau, J. Joussot-Dubien and K. J.
    Toyne, Chem. Phys. Lett., 63, 269 (1979).

22. P. Pasman, F. Rob, and J. W. Verhoeven, J. Am. Chem. Soc.,
     104, 5127 (1982).

23. J. W. Verhoeven, Rec. Trav. Chim. Pays-Bas, 99, 369 (1980).

24. T.-F. Ho, A. R. McIntosh, and J. R. Bolton, Nature 286, 254
     (1980).

25. S. Nishitani, N. Kurata, Y. Sakata, S. Misumi, M. Migita, T.
     Okada, and N. Mataga, Tetrahedron Lett., 22, 2099 (1981).

26. K. B. Blodgett, J. Am. Chem. Soc., 57, 1007 (1935).

27. K. B. Blodgett and I. Langmuir, Phys. Rev. B51, 964 (1937).

28. H. Kuhn, D. Möbius, and H. Bücher in Physical Methods of
     Chemistry, A. Weissberger and B. W. Rossiter, Eds., Vol. I
     Part IIIB, Wiley, New York 1972, Chapter 7 page 577.

29. D. Möbius, Acc. Chem. Research, 14, 63 (1981).

30. H. Kuhn, Pure Appl. Chem., 51, 341 (1981).

31. H. Kuhn, Pure Appl. Chem., 53, 2105 (1981).

32. B. Mann and H. Kuhn, J. Appl. Phys., 42, 4398 (1971).

33. B. Mann, H. Kuhn, and L. V. Szentpály, Chem. Phys. Letters,
     8, 82 (1971).

34. U. Schoeler, K. H. Tews, and H. Kuhn, J. Chem. Phys., 61,
     5009 (1974).

35. R. W. Murray, Acc. Chem. Res., 13, 135 (1980).

36. H. Abruna, T. J. Meyer, and R. W. Murray, Inorg. Chem., 18,
     3233 (1979).

37. E. E. Polymeropoulos and J. Sagiv, J. Chem. Phys., 69, 1836
     (1978).

38. L. Netzer and J. Sagiv, J. Am. Chem. Soc., 105, 674 (1983).

39. P. S. Vincett, W. A. Barlow, F. T. Boyle, J. A. Finney, and
     G. G. Roberts, Thin Solid Films, 60, 265 (1979).

40. A. H. Ellison, J. Phys. Chem., 66, 1867 (1962).

41. T. Smith, J. Colloid Interf. Sci., 26, 509 (1968).

42.   B. J. Kinzig in Proceedings of the First Molecular Electronic
      Devices Workshop, F. L. Carter, Editor, NRL Memorandum Report
      4662, (22 October 1981), page 360; also in <u>Molecular
      Electronics Devices</u>, F. L. Carter, Ed., Marcel Dekker,
      New York, 1982, page 223.

43.   A. Sommerfeld and H. Bethe in 'Handbuch der Physik' Vol.
      24, Geiger and Scheel, Eds., Springer, Berlin 1933, p. 450.

44.   L. T. Calcaterra, G. L. Closs, and J. R. Miller, J. Am. Chem.
      Soc., <u>105</u>, 670 (1983).

45.   W. R. Hertler, J. Org. Chem., <u>41</u>, 1412 (1976).

46.   D. C. Green, J. Org. Chem., <u>44</u>, 1476 (1979).

47.   T. Francis and M. P. Thorne, Can. J. Chem., <u>54</u>, 24 (1976).

48.   H. A. Staab, J. Ippen, C. Tao-pen, C. Krieger, and B.
      Starker, Angew. Chem. Int. Ed. Engl., <u>19</u>, 66 (1980).

49.   J. Ippen, C. Tao-pen, B. Starker, D. Schweitzer, and H. A.
      Staab, Angew. Chem. Int. Ed. Engl., <u>19</u>, 67 (1980).

50.   H. A. Staab and G. H. Knaus, Tetrahedron Lett., 4261 (1979).

51.   H. A. Staab, private communication.

52.   S. Bittner, private communication.

53.   J. Y. Becker, J. Bernstein, S. Bittner, N. Levi and S. Shaik,
      submitted to J. Am. Chem. Soc.

2

Photochromic N-Salicylideneaniline:
Spectroscopic Properties and Possible Applications

Hans Sixl and Dagmar Higelin/Physikalisches Institut, Universität
Stuttgart, Teil 3, D-7000 Stuttgart 80, West Germany

I. INTRODUCTION

The recent interest in photochromic organic materials is based on
their possible applications in photochromic or phototropic glasses,
in holographic memories and in conventional computer technology as
optical information storage elements (1). With respect to molecular
electronic devices future applications may be expected in computer
technology (2), some of which will be discussed in the second part
of this paper. Due to their wide-spread applications, there has
been a growing interest in solid state reactions, their mechanisms
and some of the relevant factors, which influence them, have been
reviewed very recently (3).

Among the different photochromic organic molecules some anils of
salicylaldehyde have attracted the interest of chemists and physicists, due to their reversible photoreactivity in solutions, rigid
glasses, and crystals. The best investigated representant is the

N-salicylideneaniline. The stable ground state configuration is
described by the colorless enol E configuration with an intramolecular hydrogen bridge between the oxygen and the nitrogen.

Upon irradiation with UV-light of energy $h\nu_1$ the enol E configuration changes to the trans-keto QC configuration as shown in Scheme 1.
In the solid state this gives rise to a typical red coloration,
which may be reversibly bleached either photochemically using
visible light of different energy $h\nu_2$ or thermally upon annealing
of the sample.

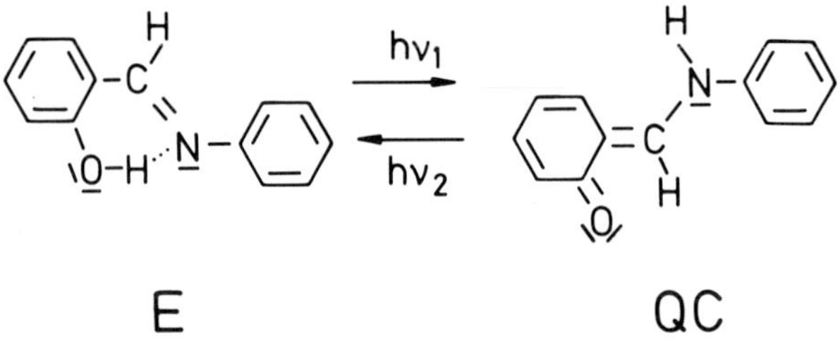

$$E \qquad\qquad\qquad QC$$

SCHEME 1.   Photochromism of N-salicylideneaniline.

In the course of the photochromic reaction at least two quinoid
intermediates QA and QB have been identified by optical absorption
and emission spectroscopy (4-7). This is a consequence of the fact
that the reaction is not restricted to proton migration from the
oxygen to the nitrogen, which is characterized by a very fast back
reaction corresponding to the keto-enol tautomerism of these
systems as shown in Scheme 2.

In thermochromic anils the equilibrium of Scheme 2 may be shifted
by heat from the E towards the QB configuration. In contrast to the
thermochromic reaction photochromism involves geometrical framework changes, which disrupt the hydrogen bond. In salicylideneaniline this change is given by a "rotation" around the C_1C_7 double
bond. Thus, the QC photoproduct configuration of Scheme 1 is stable
against the proton back reaction to the enol E configuration.

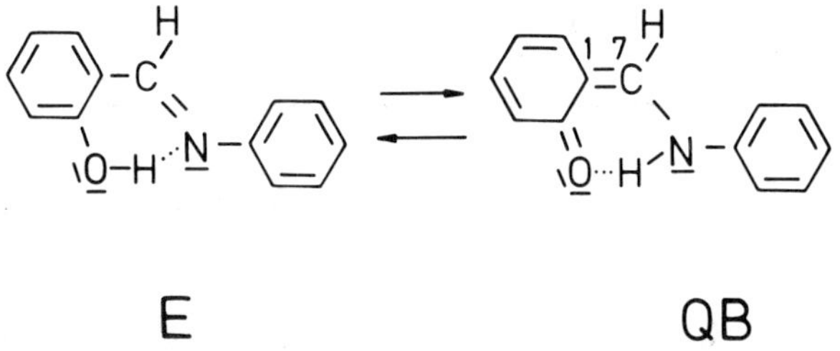

SCHEME 2.   Keto-enol tautomeric equilibrium.

In this contribution we first present the spectroscopic properties
of salicylideneaniline in a crystalline matrix. These comprise the
low-temperature optical absorption and emission of salicylideneaniline in dibenzyl single crystal matrix. From the temperature
dependence of the photochemical reactions the energy barriers between the reaction products are determined. In a second part we are
speculating upon the possibilities of future applications of photochromic salicylideneaniline for molecular electronic devices.

II.   LOW TEMPERATURE OPTICAL SPECTRA

Optical absorption and emission spectroscopy of the salicylideneaniline doped dibenzyl crystals was performed using a variable
temperature cryostat. The low-temperature absorption spectrum of
the original transparent crystal of Fig. 1 a is characterized by a
broad absorption of the enol E configuration in the ultraviolet
spectral region. After UV-irradiation of the crystal at 10 K the
structured spectrum of Fig. 1 b is obtained, which is responsible
for the pink coloration of the QC photoproduct. Simultaneously the
intensity of the broad UV-absorption decreases. Optical bleaching
of the photoproduct absorption is performed upon irradiation into
the photoproduct absorption bands with visible light above 150 K.
Both, the forward and back reactions are linearly dependent on the
intensity of the irradiation.

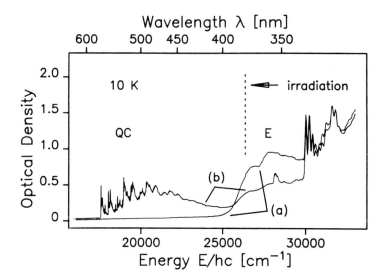

FIGURE 1.   Optical absorption spectrum of salicylideneaniline in a
            dibenzyl host crystal (a) uncolored (b) colored crystal.
            Spectrum (b) is obtained from (a) by 60 min irradiation
            of the crystal with UV-light of wavelengths 280 nm < λ <
            380 nm.

The low-temperature emission spectra of the salicylideneaniline
molecules in the dibenzyl host crystals are shown in Fig. 2 a and
b together with the corresponding absorption spectra. In Fig. 2 a
there is an unusually large shift of about 5500 cm^{-1} between the
E absorption and the QA and QB emission of the uncolored species.
This shift has been attributed to the proton transfer OH•••N → O•••HN
within the salicylideneaniline molecule following immediately the
excitation of the E configuration. The emission therefore arises
from the quinoid configurations QA and QB (7) with essentially
different lifetimes (5,6). In Fig. 2 b there is no shift between
the zero-phonon bands of the QC photoproduct absorption and
emission. Therefore the absorbing and emitting species are identical. A close look at the spectra confirms the mirror symmetry of
the absorption and emission. The sharp spectra indicate the disruption of the original hydrogen bonds of the E and QB structures,

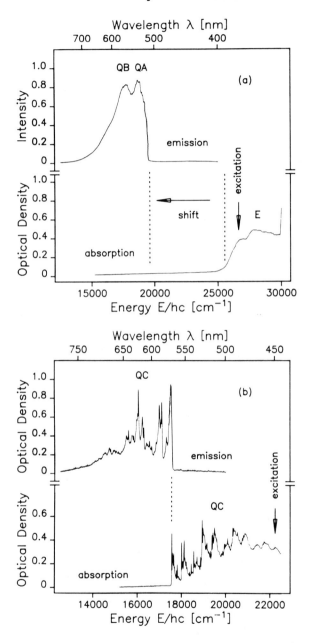

FIGURE 2.   Low-temperature emission and absorption spectra of the
            salicylideneaniline-dibenzyl mixed crystal.
            (a) uncolored crystal (b) colored crystal.

according to the assignment of the photoproduct to the QC trans-
keto configuration (8,9).

III. REACTION KINETICS

The photochemical reactions of the salicylideneaniline molecules in
the dibenzyl host crystals are thermally activated. The activation
energies of the forward and back reactions are deduced from the
time dependencies of the integral changes of the photoproduct absorption spectra during photocoloration and optical bleaching as
shown in Fig. 3 a and b. The monoexponential built up and decay
functions are described by rate constants of the type $k = k_0 \cdot$
$\exp(-E_a/kT)$. From the slopes of the corresponding Arrhenius plots
the activation energies $E_a^{(1)} = (130 \pm 30)$ meV, $E_a^{(2)} = (290 \pm 30)$ meV

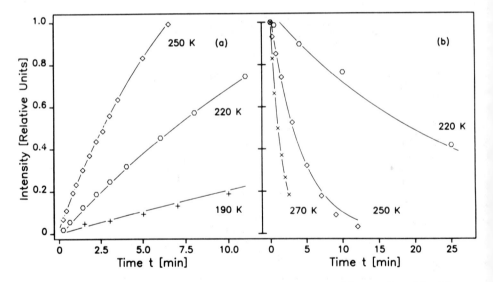

FIGURE 3. Time dependence of the photochromic effect of salicyli-
           deneaniline in dibenzyl host crystals. The calculated
           curves are monoexponential.
           (a) Increase of the intensity of the photoproduct
               absorption during photocoloration with UV-light,
           (b) decay of the intensity during optical bleaching with
               visible light.

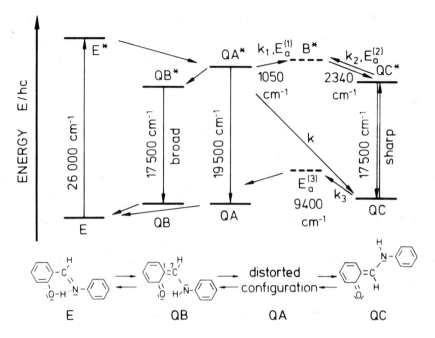

FIGURE 4. Energy level scheme of the forward photoreaction and of
the optical and thermal back reactions of salicylideneaniline in dibenzyl host crystals.

and $E_a^{(3)}$ = (1200 ± 100)meV of the optical forward (1) and of the
optical (2) and thermal (3) back reactions are obtained.

IV. REACTION PATHWAYS

The energy level scheme of Figure 4 summarizes the essential optical and thermal pathways of the forward and back reactions as deduced from our experiments on salicylideneaniline in dibenzyl host
crystals. Excitation of the enol E results in an extremely Stokes
shifted QA and QB emission. No emission from the E species has been
observed. Therefore the energy is converted within the excited
states from the E* to the QA* and QB* species. According to previous
interpretations (5,6) QA* is the precursor of both, the QB tautomer and the QC photoproduct.

The forward photoreaction is characterized by two pathways with
rate constants k and k_1. The activated process (k_1) is determined
by a potential barrier B* between the excited state QA* and QC*.
The non-activated pathway is supposed to lead directly to the QC
ground state.

The back reaction is also characterized by two pathways k_2 and k_3.
The purely thermal pathway k_3, which is dominant above room temperature, is characterized by a ground state barrier B. Upon irradiation into the QC absorption an intense QC emission is observed,
which is in competition with the thermally activated optical back
reaction. The potential barrier B* of this pathway is assumed to
be identical to that of the forward reaction. All the data given
in the figure have been determined in our experiments.

V.   CONCEPTION OF MOLECULAR ELECTRONIC DEVICES

Photochromic and thermochromic materials are of great interest for
the design of molecular electronic devices, due to their ability
of detecting, generating and switching of solitons (2). This will
be demonstrated below by the enol E and the QC photoproduct configurations of the salicylideneaniline molecules combined with a
trans-polyacetylene chain commonly introduced for information and
charge transfer in organic polymers.

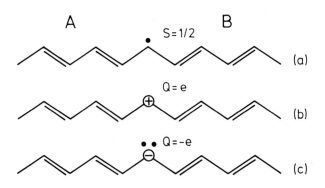

SCHEME 3.   Neutral and charged solitons (10).

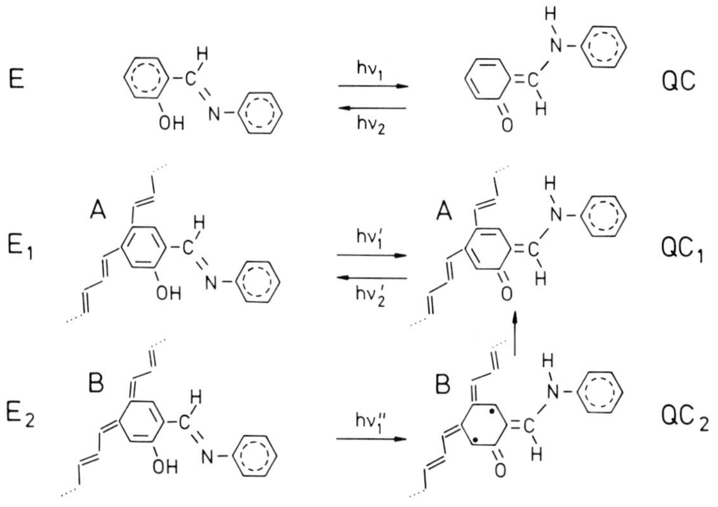

SCHEME 4. Design of molecular electronic devices using salicylideneaniline molecules and trans-polyacetylene chains.

In this paper we will be concerned with neutral or charged solitons (10) as shown in Scheme 3, involving a structural change or
phase boundary between the A and B structures of the polyacetylene
chain. These mobile bond alternation defect structures have been
introduced originally by Pople and Walmsley (11).

The radical electron of the neutral soliton (a) has no charge but
a spin of $S = 1/2$. If the electron is removed (b) a positively
charged soliton is obtained. Upon addition of an electron (c) a
negatively charged soliton is obtained.

Applying the above three types of solitons for the transfer of
charges or informations in a molecular electronic device we may be
able to generate, to switch and to detect them optically using
combined structures of photochromic salicylideneaniline molecules
and trans-polyacetylene chains. This is shown in Scheme 4 by the
different E and QC structures. Due to their different conjugation
and chemical structures all states are expected of having different

absorption and emission spectra as demonstrated in this paper by
the different spectra of the pure E and QC configurations. Because
of the extension of the conjugated photochromic system an appreciable red shift of the E_1, E_2, QC_1 and QC_2 spectra is expected.

By the introduction of the polyacetylene chain, the cyclic conjugated benzene ring of the E configuration is restricted to a
specific bond alternation structure in E_1. By changing the structure of the polyacetylene chain from A to B the cyclic conjugation
of the benzene ring is lost in the E_2 configuration. In the QC_1
situation the bond alternation is lost in the acetylene chain at
the salicylideneaniline molecules. The QC_2 configuration is only
possible with a radical electron pair and therefore is expected
to be very unstable.

VI. SOLITON DETECTION, GENERATION AND SWITCHING

If the salicylideneaniline molecule is part of an extended trans-
polyacetylene chain, the solitons of the chain are transmitted in
the enol E configuration as shown in Scheme 4 by the different
structures A and B of the polyacetylene chain in the E_1 and E_2
states. Therefore the E configuration allows passage of a soliton.
By the passage the structures change from E_1 to E_2 and vice versa.
The corresponding changes in the optical spectra may be utilized
to detect the soliton passage.

Soliton passage is not possible in the photoproduct QC_1 and QC_2
configuration, where either the conjugation of the polyacetylene
chain is lost (QC_1) or an unstable radical pair is formed corresponding to the disruption of one π bond of the salicylideneaniline
molecule. The QC_2 species therefore is expected to decay into the
QC_1 state via emission of two solitons in opposite directions of
the trans polyacetylene chains. This reaction may be initiated by
the photochemical forward reactions

$$E_2 \xrightarrow{h\nu_1''} QC_2 \to QC_1 + 2 \text{ solitons.} \tag{1}$$

Thus a soliton pair may be generated photochemically. In the reaction

$$E_1 \xrightarrow{h\nu'_1} QC_1 \qquad (2)$$

no solitons are generated. This is also valid for the corresponding optical and thermal back reactions

$$QC_1 \xrightarrow{h\nu'_2} E_1 \text{ and } QC_1 \xrightarrow{kT} E_1. \qquad (3)$$

By the forward photoreactions of eqs. (1) and (2) soliton transmission along the trans polyacetylene chain is switched off. By the back reactions soliton transmission is switched on. In this way we have designed an optical soliton switch, provided that the photochromic properties of the salicylideneaniline molecules may be maintained in the combined configurations. This is presumably the case as shown by a series of experiments on differently substituted salicylideneaniline molecules (8), which show that substitution of the keto group affects neither the photochromic nor the thermochromic behavior of the anils of salicylaldehyde. This remarkable insensitivity to the host structure is based on its purely molecular type of photoreaction involving structural changes and distortions of only part of the molecule.

At the present stage the conclusions of this very last section are speculative, because they need experimental verification. However, they may serve as a stimulation to increase our efforts in the tayloring and design of molecular units, which may be useful in some future photosensitive molecular electronic devices.

ACKNOWLEDGEMENTS

This work has been supported by the Stiftung Volkswagenwerk.

REFERENCES

1. J. Rajchmann, J. Appl. Phys. 41, 1376 (1970), A. Szabo,
    US Patent No. 3, 896, 420 (1975), G. Casto, D. Haarer,
    R. M. Macfarlane and H. P. Trommsdorff, US Patent No. 4,
    101, 976 (1978).

2. F. L. Carter in Molecular Electronic Devices, Proceedings of
    an International Workshop, Washington ed. by F. L. Carter,
    Marcel Dekker, New York (1982), Chapter V.

3. J. M. Thomas, Phil. Trans. Roy. Soc. A 277, 251 (1974),
    G. M. J. Schmid et al. Solid State Photochemistry ed. by
    D. Ginsburg, Verlag Chemie, Weinheim (1976).

4. M. Ottolenghi and D. S. McClure, J. Chem. Phys. 46, 4620 (1967),
    R. Potashnik and M. Ottolenghi, J. Chem. Phys. 51, 3671 (1969).

5. P. F. Barbara, P. M. Rentzepis and L. E. Brus, J. Am. Chem.
    Soc. 102, 2786 (1980).

6. R. Nakagaki, R. Kobayashi, J. Nakamura and S. Nagakura, Bull.
    Chem. Soc. Japan 50, 1909 (1977).

7. D. Higelin and H. Sixl, Chem. Phys. (1983) in press.

8. M. D. Cohen and G. M. J. Schmidt, J. Chem. Phys. 66, 2442
    (1962), M. D. Cohen, Y. Hirshberg and G. M. J. Schmidt, J. Chem.
    Soc. 2051 and 2060 (1964), M. D. Cohen and S. Flavian, J. Chem.
    Soc. B 117 and 334 (1967).

9. E. Hadjoudis, J. of Photochem. 17, 355 (1981).

10. W. P. Su, J. R. Schrieffer and A. J. Heeger, Phys. Rev. Lett.
    42, 1698 (1978), Phys. Rev. B 22, 2099 (1980), A. G. MacDiarmid
    and A. J. Heeger in Molecular Electronic Devices see Ref. 2,
    Chapter XX.

11. J. A. Pople and S. H. Walmley, Mol. Phys. 5, 15 (1962).

3

Electron Transfer in Mixed Valence Compounds and Their Possible Use as Molecular Electronic Devices

Jean-Pierre Launay/Laboratoire de Chimie des Métaux de Transition, Université Pierre et Marie Curie, 4, Place Jussieu, 75230 Paris Cedex 05, France

I. INTRODUCTION

In the search for possible molecular electronic devices, it is interesting to consider systems for which the electronic and / or molecular structure is able to rearrange as a function of time. Thus for example, hydrogen bonded molecules(1), fluxional systems, spin-crossovers, dynamical Jahn-Teller effects could be envisaged. However, in this broad class of compounds, mixed valence systems occupy a privileged position. Indeed, due to the presence of several metal ions in different formal oxidation states, they are built to present intramolecular electron exchange. The dynamical effects associated with this electron motion have been the subject of much theoretical as well as experimental work. In particular, it has been shown that two limiting electronic structures can occur : one in which the electrons are completely delocalized over the different sites (class III systems in Robin and Day's classification (2)), and one in which the electrons are at least partly localized on definite

sites (class II in the same classification). This distinction is very
analogous to the problem of protonic motion in hydrogen bonded systems, for which the potential energy curve can present either a single
or a double minimum(1).

The case of mixed valence systems present however some special features which seem advantages for use in molecular devices. First the
dynamical process is an electron exchange, so that a mixed valence
molecule can be considered as a prototype fragment of a semi-conductor. Thus it can be conceived to connect it by molecular wires to the
external world so that the information would be processed directly as
an electrical current. Secondly, a number of theoretical descriptions
are available, which have been devised either for the mixed valence
case (3-8), or for the related case of electron transfers in redox
reactions (9). Generally speaking, the theoretical descriptions do
not require extensive ab initio calculations. They use rather phenomenological parameters such as electron phonon coupling constant and
electronic interaction matrix elements which can be intuitively or
empirically related to the chemical structure. Finally, besides the
thermal activation process for electron transfer, there also exists
an optical process. This provides a second possibility to couple the
system to external influences and the relevant excited state surfaces
can be readily calculated.

II. THE TIME SCALE PROBLEM

Before going further on, we shall briefly discuss the important time
scale problem since it also applies to other dynamical systems such
as hydrogen bonding or soliton propagation. For simplicity we consider a binuclear system for which the localized electronic states are
ϕ_a and ϕ_b. If the electronic interaction is large enough, then delocalized ground and excited states occur, the wave functions of which
are given by

$$\psi+ = \frac{1}{\sqrt{2}} (\phi_a + \phi_b) \cdot \chi_i$$

$$\psi- = \frac{1}{\sqrt{2}} (\phi_a - \phi_b) \cdot \chi_i$$

in which χ_i is a common vibrational wave function describing a symmetrical disposition of the nuclei (8). There are no vibronic coupling effects, so that the system is correctly described by a Born-Oppenheimer product. The ground state is intrinsically delocalized so that no charge motion actually occurs between the two sites. Of course, if the system is prepared in one of the localized states, and then allowed to evolve, it is mathematically possible to describe a charge oscillation. This is easily performed by writing the time dependent wave function as

$$\psi = \psi_+ \exp\left(-\frac{i\,E_+ t}{\hbar}\right) + \psi_- \exp\left(-\frac{i\,E_- t}{\hbar}\right)$$

This describes a non stationary state oscillating between sites a and b with a frequency $\dfrac{E_+ - E_-}{h}$. One must however consider that this effect cannot be used in a molecular electronic device. Indeed the energy splitting is such that the associated frequency falls into the UV-visible domain (time scale 10^{-14} sec). Thus this very fast motion cannot couple with any chemical process (10). Furthermore due to relaxation effects, the system will not stay in this high-energy non stationary state, but rather rapidly relax to the ground state. In fact, the above process describes the resonance case for which the limiting formula have no physical existence. An example of such situation is provided by the mixed valence cyclic system $\left[W_2^{VI} W_2^{V} O_8 Cl_8 (H_2O)_4\right]^{2-}$ which has been studied by the Molecular Orbital and the Valence Bond Methods (11). Strongly coupled systems of these kind are not interesting for MED purposes (12).

If, on the contrary, the electronic interaction decreases markedly, the oscillation frequency can be low enough so that electronic and nuclear motions can couple together. This in turn is an additional cause of slowing down the charge transfer motion since the electron must carry a lattice distorsion with him (polaron). Mathematically, there is a coupling of vibronic functions corresponding to different electronic wave functions ; thus for example a function such as $\dfrac{1}{\sqrt{2}}(\phi_a + \phi_b)\chi_i$ can mix with $\dfrac{1}{\sqrt{2}}(\phi_a - \phi_b)\chi_j$. In this complicated

situation, the nice Born-Oppenheimer separation of electronic and
nuclear motions is lost. Furthermore, there are several thermally
accessible levels corresponding to different electronic wave functions.
Thus, in this last case, the different resonant formula acquire some
physical reality for 2 reasons = (i) non stationary states can be
reached by thermal activation and (ii) the electron exchange becomes
slow enough to be monitored by physical methods, for instance
Mossbauer (13), ESR (14), flash photolysis (15) and so on ... The
time evolution of such systems allows applications for MEDs.

It thus appears that the most promising systems are the weakly coupled
ones. Reducing the electronic interaction can be a priori achieved
by separating the metal centers by more than a single bridge or by
realizing a suitable orientation of the concerned orbitals so that a
zero overlap would occur.

Of the two most exciting perspectives, i.e. the molecular memory
and the switching device, the former seems to be the most difficult
to conceive starting from a mixed valence molecule. This is because
most systems investigated to date have been chosen such that a nota-
ble electronic interaction exists. This would lead to a poor stability as a memory element. In contradistinction, a switching device,i.e.
a system in which the intramolecular "conductivity" could be dramatically altered, appears feasible in view of recent works on tricentric systems (25).

III. ELECTRON TRANSFER IN A TRICENTRIC SYSTEM

In the following, we shall try to foresee how could it be possible to
control the electron transfer between two parts of a molecule. We
consider a tricentric system A B A $^-$ where A are two chemically equivalent localization sites and B is a central unit which can temporarily
localize the exchanging electron. The energy levels are such that in
the ground state configuration, the electron is localized on one end
of the molecule, i.e. we have either A$^-$ B A or A B A$^-$. The case where
A are metal sites and B an organic bridging ligand has been considered by a number of authors (16-19). However, in most cases, the

energy levels of the bridging ligand are high, so that the configuration A B⁻ A is virtual state. In the present treatment, we consider rather the case where B is a third metal site of the same chemical nature as A, so that the configuration A B⁻ A is not very high in energy. Representative examples are given on figure 1. In the case of the pyrazine bridged ruthenium trimers prepared by Taube et al (22) (Compound III), it has been shown that some end-to-end communication occurs, and this is assigned to the low energy of the Ru^{II}-Ru^{III}-Ru^{II} configuration.

Some physical insight on the electron exchange process can be gained by the calculation of the adiabatic potential energy surfaces. Although the semi-classical concept of a system evolving along a single potential energy surface can be criticized (8), it can give an overall view on the basic process and retains heuristic properties.

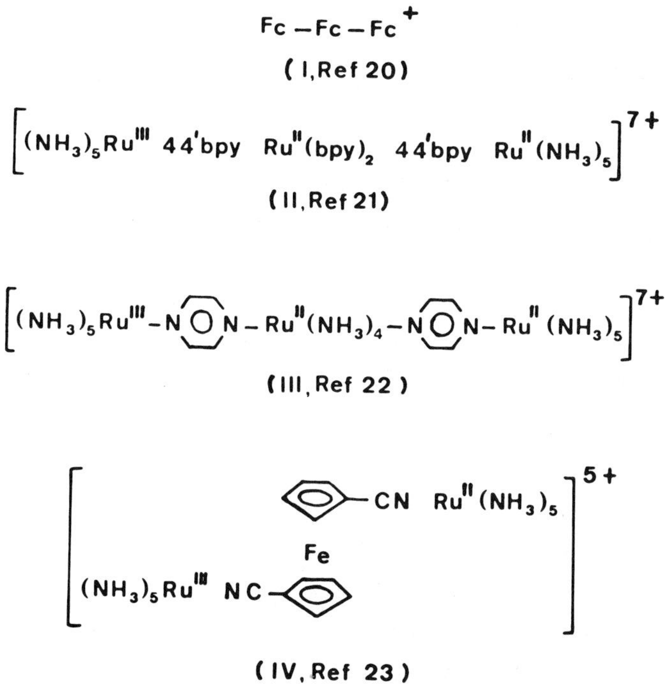

$$Fc-Fc-Fc^{+}$$

(I, Ref 20)

$$\left[(NH_3)_5Ru^{III}\ 44'bpy\ Ru^{II}(bpy)_2\ 44'bpy\ Ru^{II}(NH_3)_5\right]^{7+}$$

(II, Ref 21)

$$\left[(NH_3)_5Ru^{III}-N\bigcirc N-Ru^{II}(NH_3)_4-N\bigcirc N-Ru^{II}(NH_3)_5\right]^{7+}$$

(III, Ref 22)

$$\left[\begin{array}{c}\bigcirc-CN\ Ru^{II}(NH_3)_5\\ Fe\\ (NH_3)_5Ru^{III}\ NC-\bigcirc\end{array}\right]^{5+}$$

(IV, Ref 23)

Figure 1. Examples of A B A systems (bibliographic reference numbers in parentheses)

In particular it allows some predictions based on chemical intuition.
Finally, if necessary, it can be completed by taking into account
the effects of non-adiabaticity and nuclear tunneling, which can be
introduced as corrections (24).

The procedure we use here is a modification of the treatment of a
symmetrical (i.e. with 3-fold symmetry) trinuclear array (25). Recently two analogous calculations have been reported (26-27).

We first consider the three basis states $|\psi_a\rangle$ $|\psi_b\rangle$ and $|\psi_c\rangle$ obtained
by allocating the extra electron (or hole) successively on sites A,
B and C. In the absence of electronic interaction, the corresponding
potential energies are written as (25):

$$E_a = \frac{1}{2} k\, Q_A^2 + 1Q_A + \frac{1}{2} kQ_B^2 + \frac{1}{2} kQ_C^2$$

$$E_b = \frac{1}{2} kQ_A^2 + \frac{1}{2} kQ_B^2 + 1Q_B + \frac{1}{2} kQ_C^2 + \Delta$$

$$E_c = \frac{1}{2} kQ_A^2 + \frac{1}{2} kQ_B^2 + \frac{1}{2} kQ_C^2 + 1Q_C$$

In these expressions, k is the force constant, assumed to be the
same for all sites and for the two oxidation states, 1 is the electron-phonon coupling constant, Q_A, Q_B and Q_C are coordinates for
symmetrical (breathing) modes, Δ is the difference in zero-point
energies such that $|\psi_b\rangle$ is destabilized with respect to the two
others.

If now we allow the electronic interaction to occur between A and B
on one hand, B and C on the other hand, the energies are obtained by
diagonalizing the following potential energy matrix :

$$\begin{bmatrix} E_a & W & 0 \\ W & E_b & W \\ 0 & W & E_c \end{bmatrix}$$

where W is the electronic coupling parameter. In the following, W is
considered as a phenomenological parameter integrating all effects
of through space and through bond interactions. Thus W can be either
negative or positive according to which combination (in-phase or out-
of-phase) is the most stable (25).

The nuclear coordinates are now transformed in the following way :

$$Q_1 = 3^{-1/2} (Q_A + Q_B + Q_C)$$

$$Q_2 = 6^{-1/2} (2Q_B - Q_A - Q_C)$$

$$Q_3 = 2^{-1/2} (Q_A - Q_C)$$

Actually, these combinations were set up for the problem of three equivalent sites (25). In the present case Q_1 and Q_2 are of the same symmetry and may interact, so that the above combinations do not necessarily correspond to pure vibrational modes. However we keep these expressions since it can be shown that the Q_1 mode does not intervene in electron transfer. In effect the three diagonal terms now take the form

$$E_A = \frac{1}{2} k \sum_{1,3} Q_i^2 + 3^{-1/2} 1Q_1 - 6^{-1/2} 1Q_2 + 2^{-1/2} 1Q_3$$

$$E_B = \frac{1}{2} k \sum_{1,3} Q_i^2 + 3^{-1/2} 1Q_1 + (2/3)^{1/2} 1Q_2 + \Delta$$

$$E_C = \frac{1}{2} k \sum_{1,3} Q_i^2 + 3^{-1/2} 1Q_1 - 6^{-1/2} 1Q_2 - 2^{-1/2} 1Q_3$$

Since they all present the same parabolic dependence in Q_1, we can expect that the system will evolve at constant Q_1, and drop the Q_1 terms (25).

After adimensionalization by putting :

$$q_i = Q_i / (-1/k) \ , \ e_a = E_a / (1^2/k) \text{ etc..}$$

$$w = W/(1^2/k) \qquad \delta = \Delta/(1^2/k)$$

we obtain

$$e_a = \frac{1}{2} (q_2^2 + q_3^2) + 6^{-1/2} q_2 - 2^{-1/2} q_3$$

$$e_b = \frac{1}{2} (q_2^2 + q_3^2) - (2/3)^{1/2} q_2 + \delta$$

$$e_c = \frac{1}{2} (q_2^2 + q_3^2) + 6^{-1/2} q_2 + 2^{-1/2} q_3$$

Potential energy curves e = f (q_2 , q_3) now depend on two independent adimensional parameters w and δ . The first one is a measure of electronic interaction and the second one a measure of the destabilization of the A B$^-$ A configuration, both parameters being expressed in units of $(1^2/k)$ which is the electron-phonon coupling term. Numerical computation of these energy curves was readily performed on a PDP 11-23 computer, using the general solution for a cubic secular equation.

IV. RESULTS

For representative values of the parameters, i.e. a low $|w|$ (weak interaction between sites) and as small positive δ (corresponding to a slight destabilization of the A B$^-$ A configuration), three minima are observed on the lowest potential energy surface (see fig. 2). They are connected by saddle points such that two distinct paths

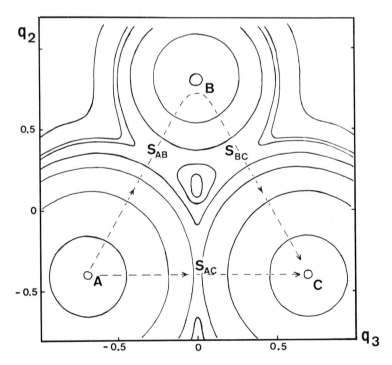

Figure 2. A typical lowest energy surface e = $f(q_2, q_3)$.w = 0.05
δ = 0.2

are possible for going from one $\bar{A}$ B A state to the equivalent other. Thus one clearly sees the two different mechanisms for electron transfer :

(i) a direct electron transfer in which only the two coordination spheres of both ends distort in a concerted way, while the coordination sphere around B does not change (trajectory at constant q_2). At the saddle point (S_{AC} fig. 2), the electronic interaction is possible only through mixing with the ψ_b wave function, since there is no direct overlap between ψ_a and ψ_c. Thus, for instance the symmetrical combination ($\psi_a + \psi_c$) can mix with ψ_b while the antisymmetric one ($\psi_a - \psi_c$) cannot. This effect has been described by several authors considering the case where B is an organic bridgind ligand (6,16,19). Usually, the influence of the third configuration is refered to as "superexchange".

(i i) an indirect electron transfer with temporary charge localization on site B (path A S_{AB} B S_{BC} C on figure 2). Here the coordination sphere around B plays a role in the process. This case can be viewed, as first approximation, as two consecutive electron transfers. It corresponds to the "chemical mechanism" of Halpern and Orgel (16), or the "radical intermediate mechanism" of Taube et al (28).Strictly speaking one   can talk of a chemical mechanism only if in the intermediate state, the whole system has time to relax to its equilibrium configuration (17). We shall not discuss here if this condition is fulfilled. In any case, the activation energy for this process will be mainly determined by the first energy barrier.

Although these two kinds of processes have been already considered (17,18), the relevant potential energy surfaces have not been computed until now. Furthermore, the possible coexistence of the two mechanisms in the same system was generally not envisaged. For instance, in a recent paper, Root and Ondrechen (27) performed a similar calculation. However, in their model system, there was no electron-phonon coupling with the central site, so that only one electron exchange mechanism (the direct one) emerged. As we   shall see below, the coexistence of the two mechanisms has important consequences from the point of view of realizing a switching device.

We have explored the consequences of variations of the two parameters
w and δ . Increasing $|w|$ at constant δ tends to wipe the central hill
and to make the different saddle point merge together. Thus for high
$|w|$ values, the distinction between the two mechanisms vanishes. Con-
versely, if we keep $|w|$ rather low and constant (0.05) and vary δ , the
two mechanisms retain their individuality, but they are modified in
a different way. Thus we have plotted on fig. 3 the heights of the
two activation barriers as functions of δ . It can be seen that the
activation barrier for direct electron-transfer does not change with δ.
This is because this barrier represents the work necessary for adjusting the two coordination spheres of the two ends to a common radius.
On the contrary, and as could be expected, the energy barrier for the
indirect electron transfer increases almost linearly with δ . For
low values of δ , the indirect electron transfer appears easier
because the effect of the electronic interaction parameter w is
greater on saddle points such as S_{AB} or S_{BC} than S_{AC}. Consequently
there exists a given w value for which the two energy barriers have
equal heights (fig. 3).

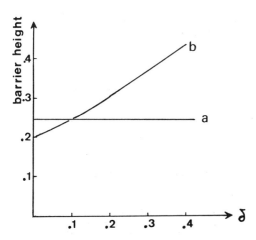

Figure 3.   Heights of the energy barriers as functions of δ, for
            $w = -0.05$ a = direct electron transfer   b = indirect
            electron transfer

At first sight, if we want to control the electron transfer by acting on the δ parameter, the existence of the direct electron transfer process which is δ independent is a major drawback. Fortunately the rate of electron transfer does not depend on the barrier height alone, but also on the splitting of the energy surfaces in the saddle point regions.

Inspection of the first excited state energy surface shows that this splitting is much greater in the vicinity of saddle points S_{AB} and S_{BC} than near S_{AC}. This is clearly shown on fig. 4, where cross sections of the energy surfaces have been made through the direct and indirect paths. Near S_{AB} or S_{BC}, the energy splitting is of the order of $2 |w|$, while near S_{AC}, simple perturbation theory shows the splitting to be of the order of $4w^2/(1 + 2\delta)$. Thus one expects a non adiabatic behavior for the direct electron transfer.

The simplest way of taking this effects into account is to use the Landau Zener equation (29) :

$$P = 1 - \exp (-8\pi H_{ij}^2/v|S_1 - S_2|)$$

where P is the probability for electron transfer, $2H_{ij}$ is the gap between the energy surfaces, v the velocity of the nuclei and $S_1 - S_2$ the difference in slope at the crossing point between the surfaces without interaction.

It is generally considered that if $2H_{ij}$ is lower than ca 0.02 eV, the electron transfer will present a non adiabatic behaviour (16,30).

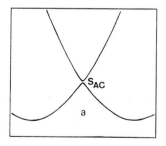 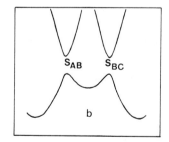

Figure 4.   Cross sections through the direct (a) and the indirect (b) paths. Same conditions as Figure 4.

In the example given fig. 2, one has splitting of 0.1 and 0.008
respectively for the indirect and direct electron transfer. If the
energy unit l^2/k is taken to be of the order of 1.0 ev, which seems a
reasonable value, then the absolute splittings are 0.1 and 0.008 eV,
showing that the direct electron transfer will effectively exhibit
a non adiabatic behaviour. Thus one can expect the overall rate to
be dominated by the other process, even if the activation barrier is
higher.

One should also notice that the behaviour near saddle point S_{AC} is
very sensitive to the presence of a small amount of direct interac-
tion between states $|\psi_a\rangle$ and $|\psi_c\rangle$. This direct interaction can
either reinforce or hinder the energy splitting due to the interac-
tion with $|\Psi_b\rangle$, since, as explained above, the electronic coupling
terms can be either negative or positive. Thus, for w = - 0.05 and
δ = 0.2, inclusion of a direct positive interaction w' = 0.004 leads
to a zero splitting for the direct electron transfer. For such
systems, the only reaction path would be by the indirect electron
exchange.

Finally the shapes of the two excited state surfaces have been investigated. The first one exhibits minima for q_2 and q_3 values corresponding to the three saddle points of the lowest surface. The second
excited state surface exhibits only one minimum for $q_2 = q_3 = 0$. In
principle, optical excitation on these surfaces, followed by vibrational relaxation sould lead to a change of state, i.e. starting
from A⁻ B A, one could obtain A B⁻ A or A B A⁻.

V. POSSIBILITIES FOR CONTROLLING THE ELECTRON TRANSFER

According to the above treatment, the rate of electron transfer between both ends could be modulated by a rapid modification of the δ
parameter. In principle any physical process affecting the energy
levels of the central unit could be envisaged, for instance switching
of hydrogen bonds, ferroelectric distorsions, soliton propagation,
influence of an electric field (this last case would require a bent
A B A system with the electric field perpendicular to the A---A line).

The existence of the excited state surfaces provides also the possibility of light induced electron transfer. This has been already
observed. For example, irradiation of the following complex :

$$(NH_3)_5 \; Ru^{II} \longrightarrow N\bigcirc N \; -- \; Ru^{III} \; (edta)^+$$

on a Ru – pyrazine charge transfer band generated a $Ru^{III} \; pyz^- Ru^{III}$
charge transfer state. This was followed by a fast reaction yielding
the unstable isomer with Ru^{III} coordinated to NH_3 and Ru^{II} coordinated to edta. Eventually, this isomer gave back the original material
(15).

A more striking example is provided by the following system :

$$(NH_3)_5 \; Co^{III}OOC \diagdown \overset{\overset{\textstyle Cu^I}{|}}{\underset{HC \; = \; CH_2}{}}$$

in which the thermodynamically allowed electron transfer between
Co^{III} and Cu^I does not occur at a measurable rate. However irradiation on a Cu^I —— ligand charge transfer band could trigger the
impeded reaction (31). This result shows the power of the photochemical process. Moreover, the possibility of preparing a system in
which the thermal electron transfer is blocked deserves attention,
in relation with memory applications.

There are still other possibilities for controlling the electron
transfer. Instead of changing the energy levels of a central unit
(δ) we could envisage to change the electronic interaction between
two sites (w). Thus for instance, the binuclear system $\left[V_2O_3(pmida)_2\right]^-$
exhibits a strong electronic interaction which is ascribed to a
coplanar orientation of the d_{xy} orbitals of vanadium sites. Consequently one can predict that in a "twisted" configuration, the electronic
interaction should vanish. This system is not itself convenient for
such an effect since it does not exhibit free rotation in solution.
However the synthesis of analogous systems with some fluxional character is conceivable. There is in particular a broad class of molecules exhibiting twisted intramolecular change transfer states (34)
which could probably be used as bridging ligands.

Finally, there have been some reports of an influence of protonation state on the electronic interaction. Thus deprotonation of the bridging ligand NCCRHCN leads to a large increase in metal-metal interaction in $\left[Ru^{II}(NH_3)_5NCCRN\ Ru^{III}(NH_3)_5\right]^{4+}$ (32). A more fully characterized example can be found in the mixed valence substituted heteropolyanion $\left[P_2W_{15}V_2^{V}V^{IV}O_{62}\right]^{10-}$, which exhibits ESR spectra characteristic of an electron rapidly hopping at room temperature between three chemically equivalent vanadium sites. However, upon protonation of one of the V-O-V bridges, the electronic interaction appears to be "shut off" (33). This is probably the result of a dramatic change in the electron exchange parameter (w) between these two sites. These results are promising since protonic motions are fast and a switching effect becomes conceivable.

VI. CONCLUSION

Although there are no present examples of a molecular device based on mixed valence systems, some attractive possibilities begin to emerge. Clearly the study of fluxional or hydrogen bonded mixed valence systems would be worthwhile. However the major lack of results occurs in the field of weakly coupled systems. It would be also desirable to expand the range of experimental methods in order to obtain direct measurements of electron transfer rates.

REFERENCES

1. A. Aviram, P.E. Seiden and M.A. Ratner. "Molecular Electronic Devices", F.L. Carter Ed., M. Dekker, New York (1982), p 5. R.C. Haddon and F.H. Stillinger. ibid p 19.

2. M.B. Robin and P. Day. Adv. Inorg. Chem. Radiochem. 10, 247, (1967).

3. N.S. Hush, Progr. Inorg. Chem. 8, 391, (1967).

4. N.S. Hush, Chem. Phys. 10, 361, (1975)

5. B. Mayoh and P. Day J. Am. Chem. Soc. 94, 2885, (1972)

6. B. Mayoh and P. Day. Inorg. Chem. 13, 2273, (1974)

7. S.B. Piepho, E.R. Krausz and P.N. Schatz. J. Am. Chem. Soc. 100, 2996, (1978)

8. K.Y. Wong and P.N. Schatz, Prog. Inorg. 28, 369, (1981)

9. See for instance R.D. Cannon. "Electron transfer reactions". Butterworths, London (1980) and references therein.

10. S.Y. Chu and S.L. Lee. Chem. Phys. Letters, 71, 363, (1980)

11. J.J. Girerd and J.P. Launay. Chem. Phys. 74, 217, (1983)

12. A. Aviram and M.A. Ratner. Chem. Phys. Letters, 29, 277, (1974)

13. C.T. Dziobkowski, J.T. Wrobleski and D.B. Brown. Inorg. Chem. 20, 679, (1981)

14. C. Sanchez, J. Livage, J.P. Launay, M. Fournier and Y. Jeannin. J. Am. Chem. Soc. 104, 3194, (1982)

15. C. Creutz, P. Kroger, T. Matsubara, T.L. Netzel and N.Sutin. J. Am. Chem. Soc. 101, 5442, (1979)

16. J. Halpern and L.E. Orgel. Disc. Farad. Soc. 29, 32, (1960)

17. Y.I. Kharkats, A. K. Madumarov and M.A. Vorotyntsev. J. Chem. Soc. Farad. II, 70, 1578, (1974)

18. A.M. Kuznetsov and J. Ulstrup. J. Chem. Phys. 75, 2047, (1981)

19. S. Larson. J. Am. Chem. Soc. 103, 4034, (1981)

20. G.M. Brown, T.J. Meyer, D.O. Cowan, C. le Vanda, F. Kaufman, P.V. Roling and M.D. Rausch. Inorg. Chem. 14, 506, (1975)

21. M.J. Powers, R.W. Callahan, D.J. Salmon and T.J. Meyer. Inorg. Chem. 15, 894, (1976)

22. A. Von Kameke, G.M. Tom and H. Taube. Inorg. Chem. 17, 1790, (1978)

23. N. Dowling and P.M. Henry. Inorg. Chem. 21, 4088, (1982)

24. B.S. Brunschwig, J. Logan, M.D. Newton and N. Sutin. J. Am. Chem. Soc. 102, 5799, (1980)

25. J.P. Launay and F. Babonneau. Chem. Phys. 67, 295, (1982)

26. S.A. Borshch, I.N. Kotov and I.B. Bersuker. Chem. Phys. Letters, 89, 381, (1982)

27. L.J. Root and M.J. Ondrechen, Chem. Phys. Letters, 93, 421, (1982)

28. F. Nordmeyer and H. Taube. J. Am. Chem. Soc. 94, 6403, (1972)

29. L.D. Landau, Phys. Z. 1, 88, (1932) ; 2, 46, (1932)

30. N.S. Hush. Trans. Farad. Soc. 57, 557, (1961)

31. J.K. Hurst and R.H. Lane, J. Am. Chem. Soc. 95,1703, (1973) ; J.K. Farr, L.G. Hulett, R.H. Lane and J.K. Hurst. J. Am. Soc. 97, 2654 (1975)

32. H. Krentzien and H. Taube. J. Am. Chem. Soc. 98, 6379, (1976)

33. S.P. Harmalker and M.T. Pope. J. Am. Chem. Soc. 103, 7381, (1981)

34. Z.R. Grabowski, K. Rotkiewicz, A. Siemiarczuk, D.L. Cowley and W. Baumann. Nouv. J. Chim. 3, 443, (1979)

4

Molecular Switching in Neuronal Membranes

Gilbert Baumann and George Easton /
Duke University, Durham, North Carolina

Stephen R. Quint and Richard N. Johnson /
University of North Carolina, Chapel
Hill, North Carolina

I. INTRODUCTION

Computers are sometimes referred to as thinking machines or electronic brains. To some extent such a comparison between computer
and brain is justified. After all, both the computer and the brain
process information using electrical signals. But there are fundamental differences between computers and brains. For example, in
the computer the signal is produced by electrons in the solid state,
while in the brain this is achieved with cations and anions diffusing in the fluid state. Consequently, information travels with the
speed of light in computers, while in nerve the velocity of signal
propagation is typically in the order of one to 100 meters per
second. That is, computer and brain differ in signal speed by a
factor of three million to 300 million. Yet, in spite of this
limitation the brain is capable of recognizing and analyzing complex patterns within a fraction of a second. Such brain functions
cannot be performed with even the fastest and largest modern computers. The computing elements in the brain--the neurons--are

interconnected in circuits that have evolved over millions of years
to perform highly specific tasks in a very efficient way.

We do not know all the principles underlying the brain circuits. If we did, and if we could apply them to computers, then
surely computers could be made to perform some brain functions with
incredible speed. How the neuron works, however, is now quite well
understood. All specific functions--signal propagation, signal
integration, and signal transmission from neuron to neuron--involve
special proteins that switch ionic currents on and off. In other
words, the elementary switching operations involved in information
processing in the neuron take place at the molecular level. With
continuing miniaturization in computer technology we may eventually
be able to approach the molecular level and we may then apply some
of the insights gained from the neuron to designing molecular
switches for computers.

II.   BACKGROUND AND DEFINITIONS

All electrical switching takes place in the neuron's cell membrane.
The structural element of the biological membrane is the lipid bi-
layer (Fig. 1). Membrane lipids have dual solubility preference.
The polar head group (represented in Fig. 1 by a hatched disk) is

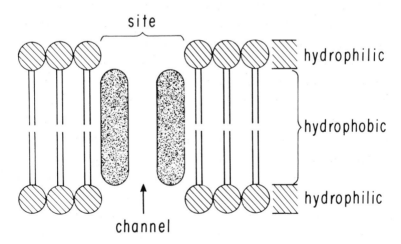

FIGURE 1. Lipid bilayer with a protein site forming a channel for
ion current.

hydrophilic (i.e., readily dissolves in water) and the two apolar fatty-acid chains (represented in Fig. 1 by two lines) are hydrophobic (i.e., are not water-soluble). Under water, lipids form bilayers with an apolar, hydrophobic (fatty or oily) layer sandwiched between two polar, hydrophilic surface layers. The hydrophobic layer is a dielectric and, therefore, almost impermeable to any electrically charged particles such as ions. That is, the lipid bilayer is an excellent electric insulator and ions can cross a membrane only at special locations called transport sites, or sites for short.

One type of site--the sodium pump--is responsible for keeping the membrane charged. The sodium pump is driven by ATP. It moves sodium ions out of the neuron and potassium ions into the neuron by way of active transport. This results in transmembrane gradients for these ions.

The neuron uses the energy stored in these ion gradients across its outer membrane to drive other processes such as the generation and propagation of nerve signals. It does this with the help of sites for passive transport. One particular type of site for passive transport functions by forming channels. A channel is a hole or pore formed by membrane protein through which ions can more or less freely flow down their concentration gradient across the membrane.

The ease with which ions can pass through a channel is expressed in its electrical conductance. Because channels are formed by protein which is in thermal motion at body temperature, they randomly open and close. Thus, the single-channel conductance continually fluctuates between on and off. It was originally thought that such conductance fluctuations were due to some kind of molecular flaps or gates that are part of a permanently formed channel. Thus, electrophysiologists called these switching sites in membranes gated sites, and referred to the process of opening and closing as gating.

Some sites are chemically gated. That is to say, their tendency to open depends on the presence of special transmitter

molecules. Other sites are electrically gated and their tendency
to open depends on the voltage across the membrane. Some sites are
both chemically and electrically gated.

Signal propagation in the neuron results from the interaction
of two types of sites whose conductances are voltage-dependent. One
type forms channels that are selective for potassium ions, and the
other forms channels that are selective for sodium ions. They are
thus called potassium channels and sodium channels. A membrane containing such electrically gated sites is called electrically excitable. Since the tendency for the formation of these channels depends on the voltage which, in turn, is affected by the currents
flowing through the channels, conductance and voltage are functions
of each other. One can study the kinetic aspects of electrically
excitable

membranes by using the voltage-clamp technique, whereby the voltage
across the membrane is made to change in controlled voltage steps
and the resulting current is measured.

III. EXPERIMENTAL OBSERVATIONS

It has recently become feasible to study the kinetics of a single
gating site. Examples of current-versus-time curves (1,2) recorded
from single sites are shown in Fig. 2. The potassium currents
(Fig. 2, left) were recorded at steady state. The sodium currents
(Fig. 2, right) were recorded during an imposed voltage pulse of
10 mV. Nine sample records and the average of 300 records of sodium current are shown. Looking at any given record in Fig. 2, one
finds that sites switch between an open and a closed state. They
do this at random time intervals. If the direction of the time
axis in a given conductance-versus-time curve from a single site is
reversed, the new curve is just as plausible as the original. That
is, at the level of single sites, the kinetics are reversible. All
types of sites studied so far display, essentially, these same
kinetics.

The current records obtained from a single site repeatedly subjected to the same voltage-pulse protocol can be summed. This was
done in the case of the sodium-conducting site on the right. The

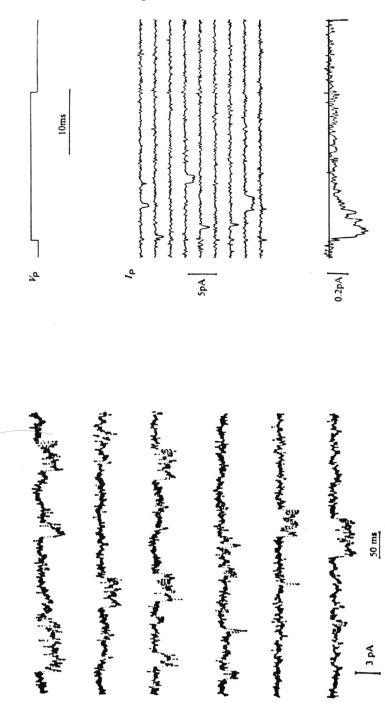

FIGURE 2. Current-versus-time curves recorded from a potassium gating site (left) and a sodium gating site (right). (Reproduced with permission from Nature 285:142, 1980 and 287:448, 1980. Copyright by Macmillan Journals Ltd.)

record of the summed sodium currents shows a dip. Since the current-voltage relation in membrane channels is usually ohmic, the
current records can directly be converted into conductance records
by flipping them upside down and rescaling. The record of the
summed conductance obtained in this way from the record of the
summed currents would thus show a hump instead of a dip. This
characteristic hump indicates that the probability of the random
openings of sodium channels changes with time in a very particular
way. Surprisingly, the summed kinetics are not reversible.

An experiment equivalent to voltage clamping the same site
many times and summing the resulting records is to voltage clamp
a membrane patch containing many electrically gated sites, say
1,000 or more. An ensemble of such sites in a membrane is called
a conductance system. Fig. 3 shows how conductance systems typically behave in response to a voltage step (3). The summed kinetics
of many potassium conductance sites are clearly different from the
summed kinetics of many sodium conductance sites. The responses
of the potassium conductance system to three voltage steps are
shown on the left in Fig. 3. The system arrives at the new steady
state by way of monophasic kinetics featuring an initial delay.

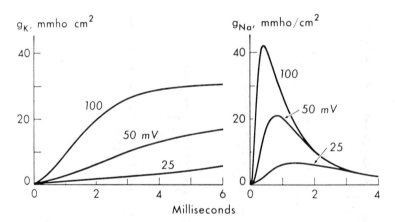

FIGURE 3. Conductance-versus-time plots from the potassium conductance system (left) and the sodium conductance system (right) in
response to three voltage-clamp steps of 25, 50, and 100 mV. (Reproduced with permission from K.S. Cole, <u>Membranes, Ions and
Impulses</u>, p. 272, University of California Press, Berkeley, 1968).

The responses of the sodium conductance system to the same three voltage steps are shown on the right in Fig. 3. It arrives at the new steady state by way of biphasic kinetics resulting in the characteristic hump mentioned earlier. The falling phase in the biphasic kinetics is called inactivation. It is remarkable that for all the electrically gated systems ever studied, only mono-phasic and biphasic kinetics have been found.

The conductance at the new steady state can be plotted as a function of the clamped voltage. In such plots the function is typically very steep. That is, a small voltage change usually re-sults in a large conductance change.

The fact that gating is voltage-dependent indicates that the electric field across the membrane exerts a force on the electric charge or charges associated with the molecular structure responsi-ble for gating. The only possible molecular response is charge movement or dipole reorientation. That is, the gating process it-self should produce a current. This current was predicted (4) and discovered (5-7). The charge movement during the gating process causes a displacement current referred to as gating current. The gating current is small in comparison to the ionic current. It is therefore difficult to observe gating current in the presence of ionic current. The ionic current, however, can be suppressed, for example, by replacing in the solution those ions that can pass through channels by impermeant ions. Yet, even in the absence of ionic current it is not feasible to measure gating current directly, because the displacement current includes a capacitive current that originates from movement of charges in the membrane unrelated to the gating process. Gating current can be measured indirectly, how-ever, as the difference between the capacitive current in response, for example, to two voltage steps of equal size in opposite direct-ions.

Fig. 4 shows asymmetric displacement current or asymmetry cur-rent recorded from the sodium conductance system in response to three voltage pulses of different duration (8). Notice that for the short voltage-pulse duration, the on- and the off-asymmetry cur-

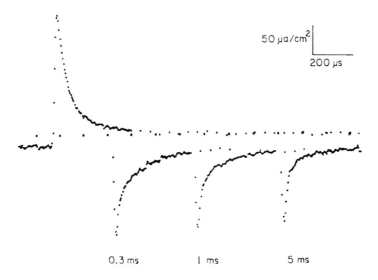

$50 \, \mu a/cm^2$

$200 \, \mu s$

0.3 ms   1 ms   5 ms

FIGURE 4. Displacement currents owing to electric charge associated with the molecular switching mechanism in response to three on-
off voltage pulses of different on-pulse duration. (Reproduced
with permission from Cold Spring Harbor Symp. Quant. Biol. XL: 301
(1976). Copyright by The Cold Spring Harbor Laboratory).

rents are similar in amplitude but opposite in sign. For longer
voltage-pulse duration, however, the amplitudes are increasingly
different. The usual interpretation of this is that the charge associated with gating becomes more and more immobilized with increasing voltage-pulse duration. The phenomenon is, thus, called charge
immobilization. Charge immobilization always occurs together with
inactivation. The two phenomena also disappear together in the
neuron under certain experimental conditions (9).

    The question arises as to how two kinds of sites having simi-
lar microscopic kinetics can have such different macroscopic
kinetics. Furthermore, how can random and reversible opening and
closing of sites give predictable and irreversible macroscopic
kinetics? This implies that nature can make reliable systems from
unreliable components. These questions can only be answered if we
understand how gating works at the molecular level. That is, how
do gating proteins physically switch the ionic currents?

An answer to this question came from experiments with artificial lipid bilayers made excitable by special molecules called
excitability-inducing molecules. Such man-made excitable membranes
have the same basic electrical properties as nerve cell membranes
(10). In particular, just like nerve cell membranes, they produce
conductance-versus-time curves that are either monophasic or biphasic. The clue that led to the discovery of their mode of operation was the finding that all excitability-inducing molecules have
a strong tendency to aggregate (11,12).

IV. AGGREGATION GATING

Aggregation is defined as the interaction of subunits in a step-by-
step manner (reactions r, s, and t in Fig. 5a). A monomer (i.e., a

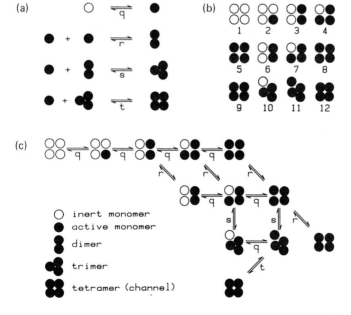

FIGURE 5. Aggregation gating (switching). a) The eight elementary
reactions taking place in the proposed gating site. b) The 12 site
configurations of an aggregation gating site consisting of four
monomers. c) Transition scheme of the aggregation gating site.
(Reproduced with permission from Structure and Function in Excitable
Cells (D.C. Chang, I. Tasaki, W.J. Adelman, Jr., and H.R. Leuchtag,
eds.), Plenum Publishing Corp., New York, 1983).

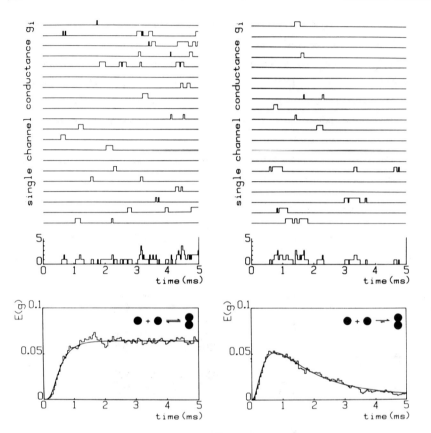

FIGURE 6. Relation between Monte Carlo simulations and Markov-process computations of a single aggregation-gating site for low (left) and high dimer stability (right). (Reproduced with permission from the same source as Fig. 5. Parts of this figure reproduced with permission from the J. theor. Biol., $\underline{93}$: 798 (1981). Copyright by Academic Press Inc., (London) Ltd.).

single subunit) can reversibly interact with another monomer to form a dimer. A dimer can reversibly interact with a monomer to form a trimer. A trimer can reversibly interact with a monomer to form a tetramer. Now assume that the monomers can be in an inert and an active state. Only active monomers can aggregate and are supplied or withdrawn by a voltage-dependent reversible reaction (reaction q in Fig. 5a). Further assume that the tetramer forms the hole through the membrane that acts as ion channel (13,14).

If these four reversible reactions can take place in a site consisting of four monomers, then the site can exist in 12 configurations (Fig. 5b) that are connected by 14 reversible transitions (Fig. 5c). Notice that each of the 14 reversible transitions involves one elementary reaction.

An extensive study of the kinetics of the channel-forming tetramer in response to voltage-clamp steps applied to this transition scheme gave the following results (13,14):

1. One can derive all basic kinetics of potassium gating if one assumes that the dimer is just as stable as two single monomers.

2. Also, one can derive all basic kinetics of sodium gating if one assumes that the dimer is more stable than two single monomers.

3. Significantly, too, one cannot derive basic kinetics from this scheme that have not been observed experimentally.

Fig. 6 shows Monte Carlo simulations of the single-site conductance and the expected conductance calculated with a Markov process algorithm (14) assuming low dimer stability on the left and high dimer stability on the right in response to a voltage step from −80 to 0 mV under voltage clamp. At the top of Fig. 6 the 20 sample simulations obtained assuming low dimer stability (left) resemble the 20 sample simulations obtained assuming high dimer stability (right). That is, they are both random and reversible. This corresponds well with the experimental findings in nerve. But, by summing as few as 20 traces for both assumptions, the collective kinetics already become clearly different. By summing the simulated kinetics of some 5,000 sites for each assumption, the resulting two jagged curves approach the calculated smooth curves of expected conductances which show no inactivation on the left and inactivation on the right. Again, this corresponds well with the experimental findings in nerve.

If the conductance at the new steady state obtained from the model is plotted as a function of the clamped voltage, one obtains a steep relation. That is, just like in the experimental data, a small voltage change results in a large conductance change.

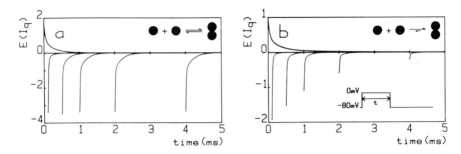

FIGURE 7. Kinetics of expected gating currents derived from a
Markov-process characterization of a single aggregation-gating site
for low dimer stability (left) and high dimer stability (right).
(Reproduced with permission from the J. theor. Biol., 99: 254
(1982). Copyright by Academic Press Inc. (London) Ltd.).

Fig. 7 shows expected gating currents calculated with a Markov
process algorithm (15) for the same two assumptions--low dimer stability and high dimer stability--of Fig. 6. Notice the phenomenon
of charge immobilization is absent on the left and present on the
right. Again, this corresponds well with the experimental findings
in nerve.

V. MOLECULAR INTERPRETATION

All molecules of known chemical structure that induce switching
phenomena in lipid bilayers have common properties (13). They all
are cigar-shaped and are about as long as the hydrophobic layer of
the membrane is thick. They all can be sectioned longitudinally
such that one face is hydrophilic and the remaining face is hydrophobic. In other words, they are amphiphilic or have dual solubility nature. Finally, all excitability-inducing molecules are dipoles.

How does a molecule with these properties interact with a
lipid bilayer? A hypothesis was proposed (13) and its essential
features are outlined here. The hypothesis states that the amphiphilic excitability-inducing molecules tend to remain at the hydro-
philic/hydrophobic interface of the lipid bilayer. There, the
place in nerve, then instead of dipole reorientation there may be

a conformational change, and instead of aggregation of protein subunits there may be aggregation between domains in a single protein.

The proposed molecular switching mechanism is unlike anything
we can experience in everyday life and is therefore often misinter-
preted. A frequent misunderstanding concerns the explanation provided by the model for the phenomenon of inactivation. It is sometimes suggested that inactivation in aggregation gating occurs because, in response to a voltage step, activated monomers diffuse towards each other, form channels, and then diffuse away from each
other, thereby producing the hump in the conductance that is experimentally observable. As Fig. 5c indicates, a site consisting of
four protein subunits or four domains in a single protein can undergo configurational changes that involve almost no lateral movement.

In reality, inactivation in this model occurs for another reason: Consider a single site consisting of four subunits (or domains)
and assume that in response to an appropriate voltage step they happen to go into their active conformation and are now ready to aggregate. First, two monomers may interact to form a dimer. The
two remaining monomers may either interact with each other or one
of them may interact with the dimer to form a trimer which may then
interact with the last monomer to form a conducting tetramer. In
the first case the site tends to remain in the dimer-dimer configuration because of the stability of the dimers. Thus, a site
trapped in this configuration is inactivated. In the second case
the site does not tend to remain in the conducting tetramer configuration because both trimer and tetramer formation are readily reversible. That is, while the site can easily become nonconducting
once it conducts, it cannot easily become reactivated once it is
inactivated. Thus, the longer the voltage-clamp pulse, the higher
the probability of finding the site inactivated.

Similarly, the phenomenon of charge immobilization can be given
a molecular interpretation. In an ensemble of aggregation-gating
sites with high dimer stability, the longer the voltage-clamp pulse,
hydrophilic face of the molecules is in contact with the hydrophilic layer of the membrane and their hydrophobic face interacts with

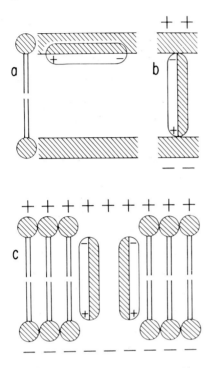

FIGURE 8. Proposed gating (switching) mechanism. a) Amphiphilic excitability-inducing molecule at hydrophilic-hydrophobic interface of the lipid bilayer. b) Dipole reorientation owing to external electric field. c) Aggregation causes configurational changes that lead to the formation of a watery ion channel.

the hydrophobic membrane layer (Fig. 8a). In response to an appropriate externally applied voltage across the membrane, the molecules--owing to their dipolar nature--tend to align parallel to the electric field. In this position their hydrophilic face is exposed to the hydrophobic layer of the membrane (Fig. 8b). Because this is an energetically unfavorable arrangement, molecules aligned in this way tend to aggregate. Such a configuration is energetically favorable because the hydrophilic faces are exposed to each other, while their hydrophobic faces are exposed to the hydrophobic layer of the membrane. By doing this they form a watery channel that allows ions to flow down their concentration gradient across the electrically insulating membrane.

The flow of ionic current locally changes the voltage across the membrane. The voltage change affects the availability of gating molecules for channel formation and, thus, changes the ionic current. This interdependence between ionic current and trans-membrane voltage is a prerequisite for the generation of nerve signals.

While experimental evidence appears to confirm this mechanism for a number of excitability-inducing molecules (16), at present, the molecular structure of the proteins that provide the potassium and the sodium channels in nerve is not known. However, the similarity of the basic steady-state and kinetic data from the neuron and from artificial excitable membranes suggests a common mechanism, and the proposed model is a good candidate to provide this mechanism. The model has already been useful in predicting novel kinetic features in the ionic currents of the sodium conductance system (17) which were subsequently confirmed experimentally (18). But while it appears likely that the underlying principles in the molecular switching mechanism of all excitable membranes are essentially the same, molecular details may well be different for different systems. For example, if indeed aggregation gating took the more sites become inactivated and, thus, in less and less sites can active monomers readily change to inert monomers and, in doing so, produce gating current. That is, in the inactivated site, gating charge is immobilized.

VI. CONCLUSION

We can now answer the questions raised earlier. The fundamental switching component in the neuron may be made up of a small number of protein subunits or domains that aggregate to form an ion channel. Small molecular systems are unreliable because of the inherent randomness at the microscopic scale. But while the state of the individual switching site is decided by chance, we may now know the rules that govern the site. These rules may be the rules of aggregation. If it were aggregation, then we also know the difference in these molecular rules for the conductance systems involved in

signal propagation: Either low dimer stability or high dimer stability prevails. This molecular difference would give rise to the different macroscopic kinetics.

In conclusion, in order to overcome the problem of inherent
randomness, nature does not rely on single switching sites, but on
the average behavior of many such sites. This device of redundancy
is used over and over in biological systems. A similar approach may
have to be considered for the design of molecular computers.

REFERENCES

1. F. Conti and E. Neher, Nature, 285, 140 (1980).

2. C.F. Sigworth and E. Neher, Nature, 287, 447 (1980).

3. K.S. Cole, Membranes, Ions, and Impulses, University of Cali-
    fornia Press, Berkeley, 1968.

4. A.L. Hodgkin and A.F. Huxley, J. Physiol., 117, 500 (1952).

5. M.F. Schneider and W.K. Chandler, Nature, 242, 244 (1973).

6. C.M. Armstrong and F. Bezanilla, Nature, 242, 459 (1973).

7. R.D. Keynes and E. Rojas, J. Physiol., 239, 393 (1974).

8. F. Bezanilla and C.M. Armstrong, Cold Spring Harbor Symp.
    Quant. Biol., 40, 297 (1976).

9. W. Almers, Rev. Physiol. Biochem. Pharmacol., 82, 96 (1978).

10. P. Mueller, Electrical excitability in bilayers and cell mem-
    branes, in: Energy Transducing Mechanisms, (E. Racker, ed.),
    MTP Intern. Rev. Sci., Biochem. Ser. 1, Vol. 3, p. 75, Butter-
    worth, London.

11. A.I. McMullen and J.A. Stirrup, Biochim. Biophys. Acta, 241,
    807 (1971).

12. J. Bukovsky, J. Biol. Chem., 252, 8884 (1977).

13. G. Baumann and P. Mueller, J. supramolec. Struct., 2, 538
    (1974).

14. G. Baumann and G.S. Easton, J. theor. Biol. 93, 784 (1981).

15. G. Baumann and G.S. Easton, J. theor. Biol.,99, 249 (1982).

16. J.E. Hall, in: Membrane Transport (S.L. Bonting and J.J.H.
    Depont, eds.), Elsevier/North Holland Biomedical Press, 1981, p.
    107.

17. G. Baumann, Biophys. J., 35, 699 (1981).

18. C.L. Schauf and G. Baumann, Biophys. J., 35, 707 (1981).

5

Analysis of Excited State Reactions by Fluorescence Phase Shift and Modulation Spectroscopy

Joseph R. Lakowicz and Aleksandr Baltr /
Department of Biological Chemistry,
University of Maryland School of Medicine,
Baltimore, Maryland

ABSTRACT

The chemical and physical properties of organic molecules are frequently altered in excited electronic states. Such changes may be the basis for the future use of organic molecules as molecular electronic devices. In the following sections we describe the use of fluorescence phase-shift and demodulation spectroscopy to measure the spectral characteristics of molecules in ground and excited states, and to measure the kinetic constants which govern the conversion between the different states formed from the initially excited state. We chose simple examples which illustrate the characteristic features of phase-modulation data for fluorophores which undergo excited state reactions. Using these examples we demonstrate that phase-modulation data allows the following: (1) An excited state process can be distinguished from ground state heterogeneity. (2) The emission spectra of the individual species can be calculated, or recorded directly. (3) A reversible excited state process can be distinguished from a irreversible process, and the reverse reaction rate can be measured. Finally (4), time-dependent spectral shifts due to multiple solvent-fluorophore interactions can be distinguished from a one-step process. Hence, phase-modulation fluorescence spectroscopy can provide detailed information concerning the nature and kinetics of excited state reactions.

INTRODUCTION

The functional properties of molecular electronic devices are likely to depend upon the interaction of these devices with light, and upon the different chemical and electronic properties of these molecules in the ground and the first singlet excited state. The altered properties of organic molecules in the excited state are frequently evident from their altered reactivity. Electronic excitation often results in a modified distribution of electrons, which in turn alters the chemical properties of the excited molecule. Well known examples of this phenomenon include excimer and exciplex formation, the gain or loss of protons, solvent relaxation around fluorophores whose emission spectra are sensitive to solvent polarity, and geometric rearrangements of the absorbing molecule which are triggered by light absorption. These and other excited state reactions result in substantial shifts of the fluorescence emission spectra to lower energies or longer wavelengths. Such altered spectral and/or chemical properties may ultimately serve to indicate the arrival of a impulse (photon) at the receiver of a molecular circuit. Of course, considerable technical difficulties need to be overcome to realize this objective.

The excited state interactions of a fluorescent molecule (fluorophore) with other substances in its immediate environment are generally studied by both steady state and time-resolved fluorescence spectroscopy. In the steady state method one examines the fluorescence yields and emission spectra using a variety of solution conditions. For example, the fluorescence emission spectra of anthracene may be examined in the presence of varying quantities of diethylaniline (DEA).

In the excited state anthracene forms a charge transfer complex with DEA. Formation of this complex is evident from the decreased fluorescence intensity of anthracene, and from the appearance of a new emission at longer wavelengths. Of course, this is the emission from the charge transfer complex (exciplex). The information available from the steady-state data includes the fluorescence intensities and spectral distributions. Generally, one attempts to interpret these data in terms of the time-dependent interactions of the fluorophore with the added substance. Often, estimates of the fluorescence lifetimes and of the diffusion coefficients are used to aid interpretation of the steady-state data.

Of course, an excited state reaction is a time-dependent process, and the instantaneous absorption of light initiates the reaction. Consequently, time-resolved methods are frequently used to investigate the excited state processes. In the time-resolved method the sample is excited with a brief pulse of light, typically at least several-fold shorter than the shortest decay time of the sample. Then, the fluorescence intensity and/or photon output is measured at various times following excitation [1]. These time-dependent data can yield considerable detail about the system under study, such as the lifetimes of the individual species, the emission spectra of these species, and the kinetic constants which govern the interchange among the energy levels. Depending upon the nature of the reaction (reversible, irreversible, one-step or continuous), these parameters may be obtained with varying degrees of accuracy and/or confidence. To date, the time-resolved methods have been widely used to investigate a variety of excited state processes [1].

In this paper we describe an alternative method which is experimentally simple to implement, and which also reveals the desired spectral and kinetic constants for the excited state reaction. This method is phase-modulation fluorescence spectroscopy. In this technique the sample is excited with light whose amplitude is modulated sinusoidally at a frequency comparable to the reciprocal decay times. The desired information about the sample is contained in the phase delay of the emission relative to the excitation (ϕ) and in the extent to which the emission is demodulated relative to the modulated excitation (m). Phase-modulation fluorescence spectroscopy permits the forward and reverse rates of excited state reactions to the measured, allows direct recording of emission spectra from individual species, and can distinguish stepwise and continuous processes. In the following sections we present examples which illustrate each of these potential applications.

THEORY

For phase-modulation fluorometry the intensity of the excitation is sinusoidally modulated. The emission is a forced response to the excitation, and is therefore modulated at the same circular frequency as the excitation ($\omega = 2\pi$ x frequency in Hz). The duration of the excited state causes the emission to be delayed in phase by an angle ϕ relative to the excitation. Also, the emission is less modulated (demodulated) relative to the excitation. These phenomena are illustrated schematically in Figure 1. The relative amplitude of the modulated emission (B/A) is smaller than the relative amplitude of the modulated excitation (b/a). The phase angle (ϕ) and the demodulation factor (m) are both measured and used to calculate the phase (τ^p) and modulation (τ^m) lifetimes using

$$\tan \phi = \omega \tau^p \qquad\qquad \tau^p = \omega^{-1} \tan \phi \qquad\qquad (1)$$

$$m = \left[1 + \omega^2 (\tau^m)^2 \right]^{-1/2} \qquad\qquad \tau^m = \omega^{-1} \left[m^{-2} - 1 \right]^{1/2} \qquad\qquad (2)$$

For example, assume a modulation frequency of 30 MHz and a lifetime of 9 nsec. One can readily calculate a phase shift of 59.5° and a demodulation factor of 0.5. We stress that interpretation of the observed values (ϕ and m) in terms of the lifetimes (τ^p and τ^m) is only correct if the decay of intensity is a single exponential. More complex equations are required for mixtures of fluorophores (ground state heterogeneity) or for an excited state process.

Consider now the more complex case of an excited state reaction, as is illustrated schematically in Figure 2. In this model the initially excited Franck-Condon state (F) is assumed to decay to the ground state with a rate constant Γ_F and by relaxation (k_1) to a relaxed state (R). The R state can decay either by emission (Γ_R) or by the reverse reaction (k_2). For simplicity we have included the radiative (λ_F, λ_R) and the non-radiative (k_F, k_R) decay rates in the Γ values ($\Gamma_i = \lambda_i + k_i$). Since emission from the R state is lower in energy one can generally observe F and R separately, to some extent, by selection of the wavelength. We will now briefly summarize the equations needed to interpret the phase and modulation values observed for the F and R states [2,3].

The lifetime of the F state is given by $\tau_F = (\Gamma_F + k_1)^{-1}$. Under favorable circumstances emission from the F state may be separately observable on the blue side of the emission. This lifetime may be obtained from either the phase angle (ϕ_F) or the demodulation factor (m_F) of its emission,

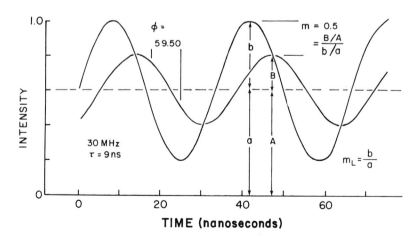

Figure   1.   Intuitive description of phase and modulation lifetime
                      measurements [1].

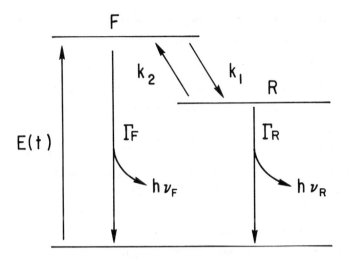

Figure   2.   Jablonski   diagram   for   a   two-state   (one-step)
                      reversible excited state reaction [2].

$$\tan \phi_F = \omega / (\Gamma_F + k_1) = \omega \tau_F \qquad (3)$$

$$m_F = \frac{\Gamma_F + k_1}{\sqrt{(\Gamma_F + k_1)^2 + \omega^2}} = \frac{1}{\sqrt{1 + \omega^2 \tau_F^2}} \qquad (4)$$

The phase angle (ϕ_R) and the demodulation factor (m_R) of the
reacted state, when measured relative to the exciting light, do not
directly yield the lifetimes by the usual expressions [4], which are
similar to eqs. 1 and 2.   These measured parameters are instead
dependent upon the kinetic constants of both the F and R states,

$$\tan \phi_R = \frac{\omega (\Gamma_R + \Gamma_F + k_1)}{\Gamma_R (\Gamma_F + k_1) - \omega^2} \qquad (5)$$

$$m_R = m_F \frac{\Gamma_R}{\sqrt{\Gamma_R^2 + \omega^2}} \qquad (6)$$

It is important to notice that the intrinsic properties of the R state can be measured directly by phase-modulation fluorometry. Wavelengths must be chosen which allow each state to be observed independently. Then, the phase difference between the R and F states ($\Delta\phi = \phi_R - \phi_F$) and the demodulation factor between these states (m_R/m_F) can be used to calculate the intrinsic lifetime of the R state τ_R. By "intrinsic" we mean the lifetime of the R state if this state could be excited directly.

$$\tan \Delta\phi = \omega / \Gamma_R = \omega\tau_R \tag{7}$$

$$\frac{m_R}{m_F} = \frac{\Gamma_R}{\sqrt{\Gamma_R^2 + \omega^2}} = \frac{1}{\sqrt{1 + \omega^2\tau_R^2}} \tag{8}$$

We note that similar expressions, which contain only the kinetic constants of the R state, apply for reversible excited-state reactions [2,5].

Wavelength-dependent phase ($\phi(\lambda)$) and modulation ($m(\lambda)$) values permit calculation of the emission spectrum of the F and R states. Assume these states can be observed independently on the blue and red sides of the emission, respectively. Then the fractional intensity of each state can be calculated from

$$\alpha_F(\lambda) = \frac{m(\lambda) \cos \phi(\lambda) - m_R \cos \phi_R}{m_F \cos \phi_F - m_R \cos \phi_R} \tag{9}$$

$$\alpha_R(\lambda) = \frac{m(\lambda) \cos \phi(\lambda) - m_F \cos \phi_F}{m_R \cos \phi_R - m_F \cos \phi_F} \tag{10}$$

where $m(\lambda)$ and $\phi(\lambda)$ are the wavelength-dependent modulations and phase angles, and m_F, m_R, and ϕ_F, ϕ_R are the modulation and phase values observed on the blue and red sides of the emission. These fractional intensities, along with the steady-state emission spectrum, can be used to calculate the spectra of the unreacted ($I_F(\lambda)$) and the reacted ($I_R(\lambda)$) states using

$$I_F(\lambda) = \alpha_F(\lambda)I(\lambda) \tag{11}$$

$$I_R(\lambda) = \alpha_R(\lambda)I(\lambda) \tag{12}$$

where $I(\lambda)$ is the steady-state emission spectrum.

Phase-modulation methods can also be used to distinguish between an excited-state process and ground-state heterogeneity. If the emission at a given wavelength results from an excited-state process then the apparent phase and modulation lifetimes increase with increased modulation frequency, phase angles can exceed 90°, and the ratio $m/\cos \phi$ can exceed unity. Observation of any of these effects proves that the observed emission did not arise from direct excitation. These unique features are a result of the phase angles of each emitting species being additive and the total demodulation being the product of the individual demodulation factors. Use of trigonometric relations allows one to demonstrate from eqs. 3 and 5 that

$$\phi_R = \phi_F + \phi_R' \tag{13}$$

where ϕ_R' is the intrinsic phase angle of the R state, $\tan \phi_R' = \omega\tau_R$. Furthermore, eq. 6 yields

$$m_R = m_F m_R' \tag{14}$$

where

$$m_R' = \frac{1}{\sqrt{1 + \omega^2 \tau_R^2}} \tag{15}$$

Using eqs. 13 and 14 one may readily show that the ratio $m_R/\cos \phi_R$ is larger than unity for nonzero values of ϕ_F and ϕ_R'. Ratios larger than unity are equivalent to apparent phase lifetimes (τ^P) which are larger than the apparent modulation lifetimes (τ^m). In addition, the additive nature of the individual phase angles results in the increase in apparent lifetimes with modulation frequency and the possibility of phase angles in excess of 90°. Eqs. 3-15 were used in our analysis of the two-state reactions, i.e., protonation of acridine and exciplex formation by anthracene.

RESULTS

Excited State Protonation of Acridine

For the first example of the analysis of excited state reactions by phase-modulation fluorometry we chose the excited state protonation of neutral acridine (Ac) by ammonium ion. The fluorescence from the acridinium cation (AcH^+) is shifted to longer wavelengths relative to the emission from acridine (Figure 3). Increasing concentrations of ammonium ion, at pH 8.3, yield a progressive quenching of the short wavelength emission with concomitant appearance of the emission from the acridinium ion (Figure 3). These spectral shifts are the result of protonation of the excited neutral acridine molecule by ammonium. Protonation occurs because excitation results in an increase in the pK_a of acridine to 10.7 [6].

The spectral properties of the acridine-acridinium system nicely illustrate the theory described above. It is a two-state process (Figure 2) which is essentially irreversible ($k_2 = 0$) [6]. Hence, the excited-state protonation of acridine provides an ideal model system to verify the simplest case of our theory [2]. In addition, a moderate degree of complexity is provided by the spectral overlap at wavelengths larger than 410 nm. Examination of the emission spectra (Figure 3) reveals a region where only the neutral species emits (390-410 nm), and a region of moderate overlap where the spectrum is dominated by the AcH^+ emission (> 500 nm).

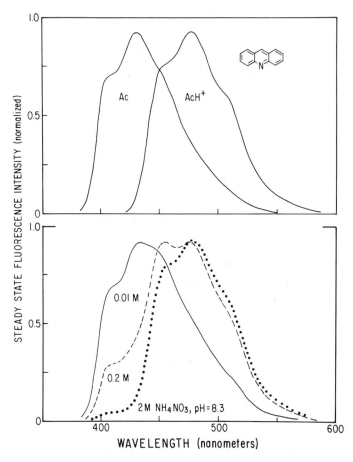

Figure   3.   Fluorescence emission spectra of acridine [3].

Apparent Phase Lifetimes for Acridine-Acridinium

The apparent phase lifetimes (τ^P) for acridine in 0.05 N NaOH, 0.1 N H_2SO_4 and 2 M NH_4NO_3 are shown in Figure 4. In the acidic and basic solutions, where only one species is present in both the ground and excited states, the lifetimes (or phase angles) are independent of emission wavelength. By contrast, in 2 M NH_4NO_3 the lifetimes are highly dependent upon wavelength because of protonation of acridine subsequent to excitation. At short wavelengths (410 nm) where neutral

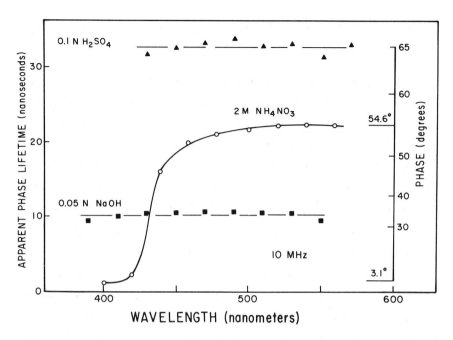

Figure   4.   Apparent phase lifetimes of acridine [3].

acridine emits, the lifetimes decrease according to the Stern-Volmer equation with a dynamic quenching constant for ammonium ion of 0.78 x 10^9 $M^{-1}s^{-1}$. At longer wavelengths the apparent lifetime increases until a nearly constant value is reached for wavelengths longer than 500 nm. In this wavelength range the emission is dominated by the acridinium ion. The constant lifetime, or more correctly phase angle, on the red and blue sides of the emission may be regarded as evidence for the two-state model. In contrast, if the overall emission were shifting to longer wavelengths according to the continuous Bakhshiev model, such constant regions are not expected [7,8]. Overall, the phase data for acridine may be regarded as typical for a two-state reaction with moderate spectral overlap. These characteristics are a decrease in lifetime on the blue side of the emission, an increase in lifetime with emission wavelength and nearly constant lifetimes on the blue and red sides of the emission.

Direct Measurement of the Lifetime of the Product of an Excited State
Reaction

A valuable property of phase fluorometry is its ability to measure
directly the intrinsic lifetime of the product of an excited state
reaction. If information is to be stored in the energy levels of
molecules, then the duration molecules remain in each level must be
known. For reaction products such lifetime determinations are complex
because the reacted state is not populated by direct excitation, but
rather by the excited state population of the initially excited state.

The intrinsic lifetime of the reacted species is revealed by the
phase difference ($\Delta\phi$) between the blue and red sides of the emission
($\Delta\phi = \phi_R - \phi_F$). By intrinsic we mean the lifetime which would be
observed if this species were formed by direct excitation rather than by
an excited-state reaction. For example, consider the phase difference
between 400 and 560 nm shown on Figure 4. This phase angle difference
($\Delta\phi = 51.5°$) yields the lifetime of the acridine cation (τ (AcH^+))
according to

$$\tan \Delta\phi = \omega\tau(AcH^+) \tag{13}$$

One may understand this simple result by realizing that the excited
neutral acridine population is in fact the origin of the excited acridinium
molecules, and thus the excitation pumping function. The calculated
lifetime in 2 M NH_4NO_3 is 20 ns, which is considerably shorter than that
of acridinium in 0.1 N H_2SO_4 (30 ns). In this instance the shorter value
is a result of quenching of the emission by NH_4NO_3, and not spectral
overlap of the emission of acridine and acridinium.

Proof of an Excited State Reaction by Comparison of Phase and
Modulation Data

Phase measurements alone, at a single modulation frequency,
cannot be used to calculate the individual lifetimes of the components of

a sample which contains more than one fluorophore. Hence, the increase in phase angle at long wavelengths shown in Figure 4 could also be attributed to a second directly excited fluorophore with a longer lifetime. Of course, the decrease in the phase lifetime at short wavelengths indicates a quenching process. A rigorous proof of the presence of an excited-state process is obtainable by comparison of the phase shift and demodulation data from the sample. Figure 5 shows apparent phase and modulation lifetimes for acridine in 0.2 NH_4NO_3, and the ratio m/cos ϕ. On the blue edge $\tau^P \simeq \tau^m$ and m/cos $\phi \simeq 1$, indicative of a single-exponential decay and thus, an irreversible reaction [9]. In the central overlap region, $\tau^P < \tau^m$ and m/cos $\phi < 1$, which is indicative of heterogeneous emission [4]. The most interesting and informative results were observed for wavelengths longer than 500 nm. At these longer wavelengths $\tau^P > \tau^m$ and m/cos $\phi > 1$. Such results are impossible for an exponential decay or for any degree of

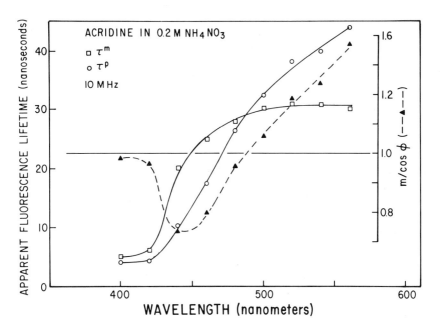

Figure   5.   Apparent phase and modulation lifetimes of acridine [3].

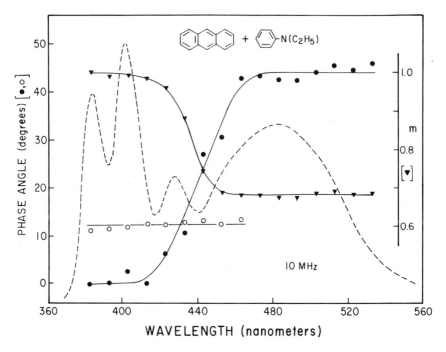

Figure   6.   Phase   angles   and   demodulation   factors   for   the
                         anthracene-diethylaniline exciplex [3].

heterogeneity. This observation proves that the emission at these longer
wavelengths was not a result of direct excitation, but rather a result of
an excited-state process [2,7,10].

Resolution of the Emission Spectra of Anthracene and its Exciplex with Diethylaniline

In the analysis of an excited state process it may be desirable to
resolve the emission spectra of the individual species. In the study of
the excited-state protonation of acridine, we had available control
samples which displayed the emission spectra of each species. This is
not always the case, as is illustrated by the formation of a charge-
transfer complex between anthracene and diethylaniline (Figure 6). In
this instance, the long wavelength exciplex emission cannot be observed

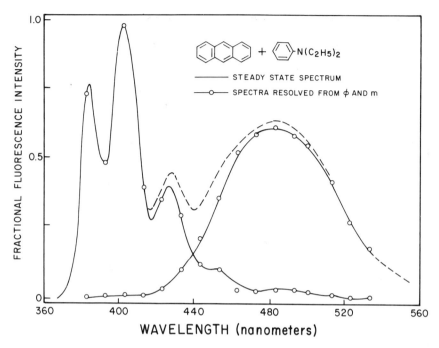

Figure   7.   Emission spectra of anthracene and its exciplex from
the wavelength-dependent phase and modulation data
[3].

in the absence of the shorter wavelength anthracene emission.
Nonetheless, the individual spectra can be calculated from the phase-
modulation data measured across the emission spectra. It is apparent
that the values of ϕ and m are constant on the blue and red sides of the
emission. These data are adequate to calculate the emission spectra of
both the monomeric and complexed forms of anthracene (Figure 7).
Equations 9-12 were used to calculate these resolved spectra. These
calculated spectra agree precisely with those obtained previously using
phase-sensitive detection of fluorescence [11]. The ability to decompose
the steady-state spectra into two spectra which sum to match the
steady-state spectrum demonstrates that the two-state model is
adequate. If additional species were present, the decomposition would
not be successful. We note that this experiment was not designed to
detect minor components.

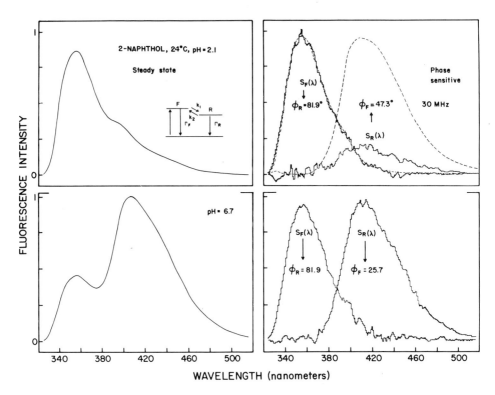

Figure   8.   Steady state and phase sensitive spectra of naphthol
                    [5]. The dashed lines in the upper right panel
                    represent the steady state spectra of naphthol (pH =
                    0.5) and naphtholate (pH - 12.0). Unless indicated
                    otherwise, the pH was 6.7.

Reversible Excited State Dissociation of 2-Naphthol

The excited state dissociation of 2-naphthol is an informative
model because the proton loss is reversible. In this section we show how
phase sensitive detection of fluorescence can be used to determine the
reverse reaction rate constant for the naphtholate anion.

The spectral properties of a reacting fluorophore become
considerably more complex if the excited state process is reversible.
However, irrespective of the complexity of the decay law of each
species (naphthol and naphtholate), the emission of each species displays

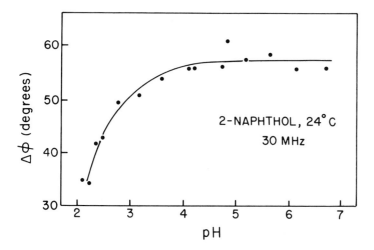

Figure   9.   pH dependence of $\Delta\phi = \phi_R - \phi_F$ for naphthol-
naphtholate [5].

a characteristic phase angle. Then, phase sensitive detection of fluorescence [11] allows the emission from each species to be suppressed. This
is illustrated in Figure 8. By selection of the detector phase angle a
phase sensitive spectrum can be found which overlaps with either the
emission spectrum of naphthol or naphtholate. In this way the phase
angles of each species (ϕ_F and ϕ_R) can be obtained, and the phase
difference ($\Delta\phi = \phi_R - \phi_F$) can be calculated (Figure 9). If the steady
state spectra are known, the phase angles of each state can be
determined, irrespective of the extent of spectral overlap. The decrease
in $\Delta\phi$ seen at low pH is due to the reverse reaction. Specifically,

$$\tan \Delta\phi = \omega/(\Gamma_R + k_2) \qquad (14)$$

At low pH the reverse protonation reaction becomes significant, and
hence $\Delta\phi$ decreases.

PROSPECTUS ON PHASE-MODULATION FLUOROMETRY

To date, most phase shift and modulation measurements have
been performed using a single, or at most a few modulation frequen-

cies. This is because the commerically available instruments operate only at fixed frequencies, which is in turn a result of the use of a crystal operated on its harmonics in the Debye-Sears light modulator. The use of one or a few frequencies severely limits the information content of the experimental data. Basically, with only a few modulation frequencies, the decay law of the sample cannot be accurately determined. Consequently, only a few successful resolutions of complex emissions have been accomplished [12,13].

However, recent advances in lasers and electrooptics have now made it possible to construct phase-modulation fluorometers capable of measurements at continuously variable frequencies [14,15]. With such instruments one may determine the frequency-dependent values of the phase shifts and the demodulation factors. Such measurements can be used to determine the decay law of the sample, irrespective of whether the decay is exponential, multi-exponential or non-exponential. As is the case for pulse fluorometry [16], the experimentally determined decay laws can be used to calculate the time-resolved emission spectra [17]. Such time-resolved spectra have been determined for both proteins and membranes labeled with solvent-sensitive fluorophores [17]. As a result of the almost continuous duty cycle of phase-modulation fluorometry, data adequate to determine the time-resolved spectra can be obtained in relatively short periods of time. The widespread availability of variable frequency phase-modulation fluorometers is likely to result in the increased use of these instruments for the analysis of samples which display complex decay laws.

ACKNOWLEDGEMENTS

This work was supported by Grant PCM 80-41320 from the National Science Foundation. J.R.L. is an Established Investigator of the American Heart Association.

REFERENCES

1. J.R. Lakowicz, Principles of Fluorescence Spectroscopy, Plenum Publishing Corporation, New York, 1983, p. 52 and 383.
2. J.R. Lakowicz and A. Balter, Biophysical Chemistry 16, 99-115 (1982).
3. J.R. Lakowicz and A. Balter, Biophysical Chemistry, 16, 117-132 (1982).
4. R.D. Spencer and G. Weber, Ann. N.Y. Acad. Sci. 158, 361-373 (1969).
5. J.R. Lakowicz and A. Balter, Chem. Phys. Lett. 92, 117-121 (1982).
6. A. Gafni, and L. Brand, Chem. Phys. Lett. 58, 346-350 (1978).
7. T.V. Veselova, L.A. Limareva, A.S. Cherkasov and V.I. Shirokov, Opt. Spectrosc. 19, 39-43 (1965).
8. N.G. Bakhshiev, Yu. T. Mazurenko and I.V. Piterskaya, Opt. Spectrosc. 21, 307-309 (1966).
9. W.R. Laws and L. Brand, J.Phys. Chem. 83, 795-802 (1979).
10. J.R. Lakowicz, H. Cherek, and D.R. Bevan, J. Biol. Chem. 255, 4403-4406 (1980).
11. J.R. Lakowicz and H. Cherek, J. Biochem. Biophys. Methods 5, 19-35 (1981).
12. D.M. Jameson and G. Weber, J. Phys. Chem. 85, 953-958 (1981).
13. J.R. Lakowicz and H. Cherek, J. Biol. Chem. 256, 6348-6353 (1981).
14. E. Gratton and M. Limkemann, Biophysical Journal, submitted (1983).
15. M. Hauser, and G. Heidt, Rev. Sci. Inst. 46, 470-471 (1975).
16. M.G. Badea and L. Brand, Methods in Enzymology 61, 378-425 (1979).
17. E. Gratton and J.R. Lakowicz, in preparation (1983). Also, American Society for Photobiology, June 1983, Madison, Wisconsin.

6

Reversible Electric Field Induced Bistability in Carbon Based Radical-Ion Semiconducting Complexes: A Model System for Molecular Information Processing and Storage

Richard S. Potember, Theodore O. Poehler, Robert C. Hoffman,
Kenneth R. Speck, and Richard C. Benson / Milton S. Eisenhower
Research Center, Applied Physics Laboratory, The Johns Hopkins
University, Laurel, Maryland

I. INTRODUCTION

Organic and polymeric materials are an essential part of the manufacturing of present day electronics and computers. It is estimated that in this year alone over 2.4 billion pounds of organic
and polymeric materials will be used by the U. S. electronics
industry.

The main application of polymers and organics in electronics
is as insulating and structural support materials [1]. Another
important application of molecular materials is for lithographic
resists for IC fabrication. Lithographic processes using organic
polymers are presently the state of the art techniques used to
fabricate high density memories for LSI and VLSI applications
[2]. Polymers also play an important role in computer peripherals,
electrochromic and liquid crystal displays, xerography, and
numerous medical applications [3].

Although organics and polymers play an important role in
electronics, the active element of almost every electronic device
and computer, namely the semiconductor, is presently fabricated
from inorganic materials. This is due to the versatility, speed,
stability, reproducibility, and the familiarity of the electronics
industry in the manufacturing processes and electronic characteristics of inorganic semiconductors.

II. MOLECULAR ELECTRONICS RESEARCH

The rapid development in the semiconductor industry over the past
ten years has been marked by the increasing density and complexity
of semiconductor chip circuitry. In the last few years considerable attention has been given to the possible problems associated
with increasing the density of integrated circuitry. It appears
that many physical/fundamental considerations will, in the near
future, set limits on the size and properties of ICs that can be
manufactured. Some of the limitations arise from processing considerations, materials properties and operational constraints.

Although silicon semiconductor technology has not yet reached
its full potential, future generations of computers may require
revolutionary changes in silicon technology or new technologies
based on totally different types of materials to increase memory
and processing capabilities. One possible class of materials which
may find applications for information processing and storage are
the carbon based semiconductors and conductors.

The replacement of inorganic semiconductors by organic macromolecular, polymeric, or biological materials which perform very
fast switching or gating functions for controlling or modifying
electrical and/or optical signals has recently been termed Molecular Electronics. The science of molecular electronics is a new
interdisciplinary field combining organic chemistry, solid-state
physics and microelectronic engineering. Molecular electronics
offers the potential of revolutionizing electronic and computer

technology by improving computer speed and capacity by making devices from complex molecular structures.

Organic and polymeric materials offer a viable alternative to the traditional inorganics because of their extremely small size, abundance, diversity, ease of fabrication, and potential low cost. Carbon based semiconductors offer an added feature in that it is possible to control the electronic and optical properties of an organic device by altering or tailoring the organic molecular structure before fabricating the device. In molecular electronic devices the electronic and or optical properties are locked into the molecular structure instead of the processing technique. This feature called molecular architecture or molecular engineering may greatly simplify the manufacturing process by reducing the number of fabrication steps and thereby increase density, speed and production yield.

Realizing the enormous potential of molecular electronics, research in this Laboratory has focused on synthesizing and evaluating a wide range of organic and polymeric materials. Materials emphasized in this program are those which exhibit potentially useful electronic, optical or magnetic properties. Many of these more promising high performance materials are then further investigated in a variety of conventional and novel device configurations. It is anticipated that as a part of these efforts, it will be possible to develop a data base of materials, techniques, engineering, and necessary basic scientific principles for future technological development in the area of molecular electronics.

Our current explorations have centered on designing, synthesizing and constructing bistable switching materials which can be used to store and process information. These organic based switching materials are used in devices which rely on very fast reversible electron-transfer reactions to change the solid-state properties of the thin film structures. In these systems the logic or memory operation is performed within the confines of the organic

molecular structure of the device. Three distinct classes of
bistable devices have been demonstrated to date using organic
materials based on charge transfer complexes of the TCNQ type;
these are optical, electrical, and optoelectronic switches [4-
10]. All three classes of devices are activated by electric fields
induced by applied potentials, optical beams, or various combinations of the two.

It has been postulated that the effect of the applied electric
field on the organometallic charge-transfer salt (e.g., copper
TCNQ) is to induce a phase transition to a non-stoichiometric
complex salt contain neutral acceptor molecules ($TCNQ^o$):

$$n[M^+(TCNQ^{\bullet-})] \underset{heat}{\overset{\overset{\text{\large E}}{\longrightarrow}}{\longleftarrow}} x\ M^o + n\text{-}x[M^+(TCNQ^{\bullet-})] + x(TCNQ^o) \tag{1}$$

M = Ag, Cu, Na, K, Li.

The reaction shown in Eq. (1) is reversible and the thermodynamically favored initial phase can be readily reformed by heating the complex phase of the salt. The observed changes in the
electrical and/or optical properties of devices fabricated from
these organic materials is a direct consequence of the electric
field induced reaction shown in Eq. (1).

Various diagnostic techniques including infrared, Auger, X-ray
photoelectron, and Raman spectroscopy have been applied to
specimens from the CuTCNQ family to validate the mechanism shown in
Eq. (1) [6]. For example, the strong Raman bands observed in the
switched and unswitched copper-TCNQ films can be used to differen-
tiate the two species. The TCNQ ν_4 (C=C stretching) Raman bands
are strongly affected by the π-electronic structure of TCNQ. The
fully charged charge-transfer species (b) has a maximum at 1451
cm^{-1}, whereas in the neutral TCNQ species (c), the band is shifted
to 1375 cm^{-1}. This shift is shown in Figure 1. Using this
technique it is easy to distinguish the switched complex from the
unswitched complex [11-16].

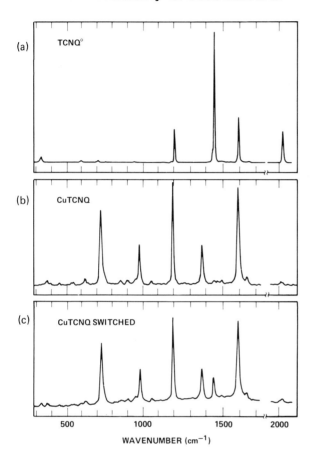

FIGURE 1. Raman bands. (a) Raman spectrum of neutral TCNQ. (b) Raman spectrum of CuTCNQ film. (c) Raman spectrum of CuTCNQ film after exposure to Ar$^+$ laser.

II. OPTICAL PROPERTIES

Spectroscopic techniques reveal that two types of reversible switching effects in these organometallic materials are induced by external electric fields at optical frequencies. The type of effect observed depends upon the irradiance (W/cm^2) of the incident optical beam. At low irradiance levels, an optoelectronic switching from a high to a low electrical impedance state can be induced, while at increased irradiance levels, high optical contrast pat-

terns can be generated directly on the thin film material. In the
first case, small amounts of neutral donor and acceptor molecules
are produced during the irradiation. They remain locked in the
crystal structure of the thin film material and readily recombine
when heated to form the original charge transfer salt. In the
second case, the higher irradiance levels produce macroscopic
amounts of neutral donors and acceptors which are visible by the
unaided eye. Under these high irradiance levels the original
crystal structure is often fractured or damaged by the incident
light, and increased heating is required to reform the original
donor–acceptor complex.

An area of significant interest is the change in the macroscopic optical properties of these organometallic materials in an
electromagnetic field. The potential for large changes in transmission and reflection in CuTCNQ and other members of this family
make these compounds prime candidates as high speed photochromic
filters [17]. The films undergo a change from their respective
blue and violet colors to a rather pale yellow color characteristic
of neutral TCNQ. As shown in Figure 2, a typical CuTCNQ film
formed by solid state diffusion is rather poorly transmitting from

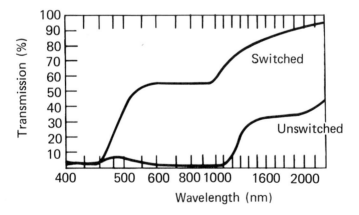

FIGURE 2. Typical CuTCNQ transmission switch – 30 mJ pulses from
frequency doubled Nd:YAG laser (532 nm).

the mid-visible into the near infrared in the vicinity of 1100 nm,
and there is a substantial increase in transmission extending into
the infrared. Upon irradiating the CuTCNQ film with a Nd:YAG laser
at 532 nm, one observes a significant increase in transmission
throughout the spectrum ranging from the mid-visible to the IR.
Extremely dramatic increases are observed in the red end of the
spectrum and in the near IR, those regions which are of particular
interest with respect to many of the principal laser sources.

The transmission properties of an AgTCNQ film are shown in
Figure 3. In contrast to the changes observed in CuTCNQ, the
switched and unswitched films are quite similar in the infrared
portion of the spectrum. Large changes in transmission are noted
in the visible part of the spectrum between 400 and 600 nm. The
switched AgTCNQ becomes strongly transmitting in a band centered at
500 nm, while the unswitched film is poorly transmitting in the
same region.

IV. ERASABLE OPTICAL RECORDING

The interaction of laser radiation with matter for use in optical
storage systems has received considerable attention. A variety of
different media, including photographic films, photoresists, photosensitive polymers, photochromics, thin amorphous films, and

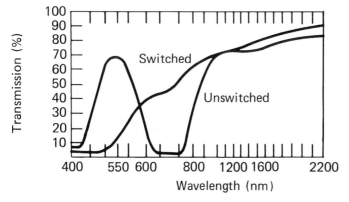

FIGURE 3. Typical AgTCNQ transmission switch – 45 mJ pulses from
frequency doubled Nd:YAG laser (532 nm).

electro-optic materials have been proposed as optical storage systems [18]. The mechanism of optical storage via interaction with
laser light varies considerably from media to media. In all of
these systems, however, the process of optical storage relies on
changes in the optical density, surface morphology, or refractive
index of the photosensitive material. Present optical storage systems have been applied in the fields of document storage, audio and
video reproduction, and direct data collection. Optical storage
systems have not seen wide-spread applications in computer
technology because most optical media are not erasable.

We have recently developed an erasable optical recording media
using the electric field induced switching effect in films of
either copper or silver complexed with the electron acceptors
tetracyanoethylene (TCNE), tetracyanonapthoquinodimethane (TNAP),
tetracyanoquinodimethane (TCNQ), or other derivatives of TCNQ. The
switching in these materials is reversible and fast, with switching
times of less than 10 ns observed in static switching experiments.

Figure 4 shows a cross section of a typical erasable optical
disk media using the TCNQ family of organometallic materials. An
approximately 2000Å thick organic change-transfer salt is deposited

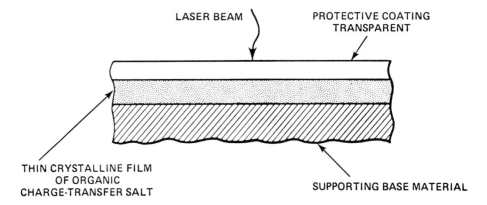

FIGURE 4. Schematic of optical storage system using organic
charge-transfer complexes.

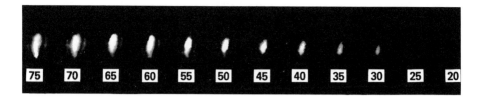

FIGURE 5. Magnified surface of AgTCNQ optical disk showing high contrast between optically switched and unswitched regions for various laser intensities.

We have irradiated a number of other copper and silver TCNQ type change-transfer complexes with the 458, 488 and 514 -nm lines from an argon ion laser, the 532 and 1064 -nm lines from a Nd:YAG laser, the 633 nm line from a He:Ne laser, the 780 nm line from a GaAlAs laser and the 10,600 -nm line from a carbon dioxide laser. In the majority of materials studied, the write threshold power is relatively independent of wavelength throughout the visible and infrared region of the spectrum. The threshold writing power varies between 3 and 150 mW depending upon the specific material choice of metal and acceptor complex. These power levels are well within the capabilities of most moderately powered commercial laser systems.

V. A NOVEL MULTIPLE STATE OPTICAL RECORDING SYSTEM

Existing optical storage media retain information by switching between two states. We have now shown that organic charge-transfer complexes can be used to devise multistate optical recording systems. These media, unlike existing binary media, which store information by switching between two states (commonly called "0" and "1"), can store two or more bits of information in the same area that normally stores only one [19].

To understand how organic molecules can be used to make such a switch, one must first understand the general principles of two stage electron transfer reactions. Examples of two stage electron

on a standard 5 1/4 or 14 inch aluminum, glass or polycarbonate
supporting base using a unique solid-state diffusion process. This
solid-state diffusion deposition guarrantees that the organic film
will be uniform over a wide area and that the size of the individual crystallites (<1μm) will be minimized. The disk is completed
when a thin protective polymer overcoat of polycarbonate is cast
over the organic layer.

An example of optical information storage using this organic
media is shown in Figure 5. The black and white photo shows the
magnified surface of a AgTCNQ optical disk. Note the high contrast
between the optically switched regions (spots) and the unrecorded
background. The spots were made with a 488 nm line from an argon
ion laser; laser power ranged from 75 mW on the left side of the
photo to 20 mW on the right. Spot size was approximately 20 μm.
transfer reactions exist in the reversible radical ion systems.
The formation of quinhydrone, produced from the radical ion semiquinone formed initially from the reduction of quinone is a reversible two stage electron transfer reaction. This reversible redox
system with transfer of two electrons in two separate steps is
shown in Eq. 2.

The chemical structures shown in Eq. 2 have large delocalized
systems in both the reduced and oxidized states. In the semi-
reduced stage, the single electron is strongly delocalized. The
stages differ only by two charges, otherwise they are iso-π-
electronic with one another. An important aspect of this

QUINONE   SEMIQUINONE   QUINHYDRONE

$$\tag{2}$$

reaction is that the redox process occurs only as electron transfer; no rupture of single bonds take place in the reversible
process.

Each radical partner in the system has high thermal stability. The high thermal stability of each electrochemically generated species is important for possible applications in optical
processing and storage because it would allow the stepwise recording and erasure of the "chemically stored" information. It has
been our intent to take this general principle of organic chemistry
and apply it to the metal-TCNQ class of organic semiconducting
materials. By combining the multistage redox aspect found in many
organic amphoteric compounds with the interesting chemical and
optical properties associated with the switching effect observed in
copper and silver TCNQ-type complexes, we can make a totally new
multistate switching material for optical recording applications.

Thin films containing two different radical ion salts mixed in
a nearly 1:1 ratio have demonstrated multiple optical switching
effects. In one experiment, neutral $TCNQ(O-ipr)_2^0$ was added to a
CH_3CN solution containing copper metal foil and was allowed to form
the $CuTCNQ(O-ipr)_2$ complex. The film of $CuTCNQ(O-ipr)_2$ was then
reacted again in a solution of $TCNQ^0$ in CH_3CN. $TCNQ^0$, because of
its higher electronegativity, displaces some of the $TCNQ(O-ipr)_2^0$ in
the film. The result is a mixed complex of the formula
$Cu(TCNQ^{\bar{\cdot}})_x(TCNQ(O-ipr)_2^{\bar{\cdot}})_{1-x}$. The resulting thin polycrystalline
film was then irradiated at with increasing power density
(watts/cm^2) using an argon ion laser source. Changes in the
composition of the film were monitored using Raman spectroscopy.
We have shown in an earlier experiment (Table 1) that the electric
field strength required to cause the switching transition is dependent upon the reduction potential of the acceptor molecule. We
were able to make use of this field strength dependence to
sequentially switch each copper salt from the charge-transferred
complex to a new solid phase containing some neutral acceptor
molecules.

TABLE 1. Relationship between Reduction Potential of the Acceptor and Field Strength at Switching Threshold

Polycrystalline charge-transfer complexes	Reduction potential of the acceptor (E_1 vs. SCE)	Approximate field at switching threshold
Cu-TCNQ(OMe)$_2$	-0.01	2.4×10^3 V/cm
Cu-TCNQ	+0.17	5.7×10^3 V/cm
Cu-TNAP	+0.20	8.2×10^3 V/cm
Cu-TCNQF$_4$	+0.53	1.3×10^4 V/cm

In the multistate optical recording media a single laser source is used to sequentially record and access multiple bits of information at one location on the surface of the film. One method of accessing information is by measuring shifts in the Raman bands between fully charge-transferred species and neutral molecules formed by the stepwise increase in laser writing power.

The results of a typical experiment are outlined in Figure 6. The unrecorded media is first accessed using an Argon ion laser operating at low power level in a Raman spectrometer. Strong ν_4 (C=C stretching) Raman bands are observed for both CuTCNQ and CuTCNQ(O-ipr)$_2$ in the unswitched film shown in Figure 6(a). The two compounds can be easily distinguished from one another because the CuTCNQ(O-ipr)$_2$ stretching frequency at 1390 cm^{-1} is shifted approximately 15 cm^{-1} from the CuTCNQ at 1375 cm^{-1}.

To record a bit of information at a particular location, the power of the Argon ion laser is briefly increased to 20 mW. At this power level the CuTCNQ(O-ipr)$_2$ switches to produce neutral species. The field strength was not great enough to switch the TCNQ. In Figure 6(b) we see that the results of this 20 mW exposure is to produce a new band at 1467 cm^{-1} corresponding to neutral TCNQ(O-ipr)$_2^0$.

To record a second bit of information at this same spot on the media, the power of the Argon laser is now momentarily increased to 40 mW. In Figure 6(c) the result of this increased laser exposure which exceeds the optical switching threshold for both CuTCNQ and CuTCNQ(O-ipr)$_2$ is to produce a new Raman band at 1451 cm^{-1}. This strong band is evidence of the formation of neutral TCNQ species in the film. The two separate reactions producing neutral acceptor molecules can then be reversed using heat to reform their original charge transfer complexes.

This simple laboratory experiment demonstrates the possibility of making a multistate information system using semiconducting organometallic films from the TCNQ family. We believe that multiswitching can be observed in a wide variety of other related radical ion acceptor molecules. The multiswitching effect may be observed in these acceptors by mixing different molecules together or by changing the donor metal. A film composed of copper-TCNQ and

We are presently chemically linking the various acceptor molecules together through σ- and π-bonding systems. We are also synthesizing and studing new molecules which combine the multistate or amphoteric properties of molecules such as hydroquinone in a single strongly delocalized radical ion acceptor molecule.

VI. TWO TERMINAL ORGANOMETALLIC LOGIC ELEMENTS FOR NOVEL COMPUTER ARCHITECTURE APPLICATIONS

Recent excitement in the field of computer hardware design has centered on the possibility of constructing a delocalized content addressable memory using extensive asynchronous parallel processing. An associative memory processor differs from a serial processor in that data is stored and addressed by content rather than specific address location [20-21]. A computer model simulation developed by J. J. Hopfield has shown that content addressable memory is feasable and may offer significant benefits in processing speed, familiarity recognition, error correction, simplified software, and insensitivity to individual logic element failure [22].

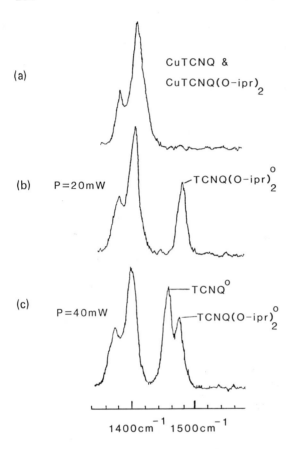

FIGURE 6. Raman spectra of multistate optical switching medium.
silver-TCNQ will switch at different applied electric fields in a
manner similar to the $TCNQ^O-TCNQ(O-ipr)_2^O$ system.

    An associative memory is a device that stores and processes
information in a number of cells. The individual cells are
recorded on and accessed by their content, therefore each logic
element must be able to process as well as store data. These cells
are interconnected with one another in a large matrix or array. In
an associative memory, the storage capacity increases asymptotically in relation to the number of storage elements. This makes high
density arrays essential to the basic architecture. To fabricate
dense circuitry, simply constructed logic elements are advan-

tageous. A basic logic element which might meet these requirements
for associated memory processing is two-terminal bistable electronic switches.

We are currently evaluating the possibility of building an
associative memory using organic molecular materials. We have
previously reported on two-terminal bistable electrical switching
and memory phenomena observed in polycrystalline films of either
copper or silver complexed with the electron acceptors tetracyanoethylene (TCNE), tetracyanonapthoquinodimethane (TNAP), tetracyanoquinodimethane (TCNQ), and various TCNQ derivatives.

The basic configuration of the device consists of a thin
polycrystalline film of a copper or a silver charge-transfer complex sandwiched between two metal electrodes to which an electrical
connection can be made. We have developed an all solid-state
method, compatible with microelectronic techniques, to make these
two-terminal switching devices for possible use in large arrays.
On a cleaned glass or sapphire substrate, an 1 μm thick pad of
copper or silver is vacuum deposited in any geometric pattern.
Registered directly over the metal pad a thin film of a neutral
organic acceptor species is deposited by vacuum sublimation. The
two layered film on the insulating substrate is then heated to
200°C to react the metal with the organic acceptor molecule. The

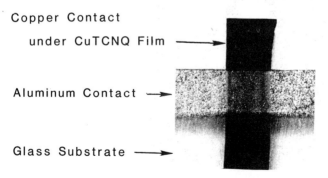

FIGURE 7. Photomicrograph of AgTCNQ bistable switching element.

heat causes a rapid oxidation-reduction reaction to occur in which
the corresponding metal salt of the ion radical acceptor is
formed. As a consequence of the heat treatment, any unreacted
neutral acceptor molecule is sublimed off of the surface of the
charge-transfer complexes. In the final step of the fabrication
process a second metal pad, usually aluminum, is evaporated over
the polycrystalline organic semiconducor. A photograph of a
magnified two-terminal bistable switching element prepared by this
technique is shown in Figure 7.

Using the all solid-state processing technique outlined above
we have been able to prepare a 4 × 4 matix of 1 bit, two-terminal,
bistable storage elements based on organic semiconducting materials. A photograph of the magnified matrix, before electrical
contact was made to the eight metal pads, is shown in Figure 8. In
this photograph the substrate is glass, the first metal deposited
is copper, the acceptor molecule is TCNQ and the top metal pad is
aluminum.

FIGURE 8. Photomicrograph of 4×4 matirx of organic switching
elements.

Threshold and memory behavior is observed in each of these
logic elements by examining current as a function of voltage across
the two terminal structure.   Figure 9 shows a typical dc current-
voltage curve for a 5 μm thick Cu/Cu-TCNQ/Al system.   The trace in
Figure 9 is made with a 10^2-Ω load resistor in series with the
device.   Figure 9 shows that there are two stable non-ohmic resistive states in the material.   Switching occurs when an applied
potential across the sample surpasses a threshold value (V_{th}) of
2.6 V.   This corresponds to a field strength of approximately
8.1×10^3 V/cm.   At this field strength the initial high impedance of
1×10^4 ohms drops to a low impedance value of 200 ohms.   A rise in
current to 4 ma and decrease in the voltage to approximately 1.2 V
along the load line (dotted line) is observed in the Cu-TCNQ system.   The results in Figure 9 are representative of the switching
effects observed in all of the metal charge-transfer salts examined

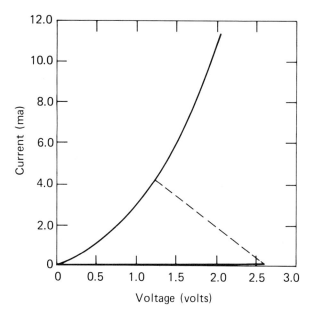

FIGURE 9.   Typical dc current-voltage characteristic showing high-
and low-impedance states for a 5 μm Cu-TCNQ sample.

and is characteristic of two terminal S-shaped or current-controlled negative-resistance switches.

In addition, it has been observed that once the film is in the conducting state it will remain in that state as long as an external field is applied. In every case studied, the film eventually returned to its initial high-impedance state after the applied field was removed. It was also found that the time required to switch back to the initial state appeared to be directly proportional to the film thickness, duration of the applied field, and the amount of power dissipated in the sample while in the low impedance state as well as the choice of donor metal and electron acceptor species.

VII. CONCLUSIONS

We have demonstrated that the unique chemical and physical properties of a particular class of organometallic solids can be used in fast photochromic filters, erasable optical recording media, two terminal bistable threshold and memory logic elements, and optoelectronic switches. These materials may find application in both conventional and novel information processing and storage. Two novel applications of these organics are in optical recording media and high density memories. Organometallic charge-transfer salts provide an avenue to extend the present limits of recording density imposed by conventional laser and optics systems by recording and accessing multiple bits of information at a single spatial location. In this system, the information is stored in and read from specific molecular energy levels in various organic radical-anion species. In the two terminal bistable switches, the rapid resistance changes which accompany the field-induced phase transition may be used in logic elements for associative memories.

In each of the various devices and applications, the concept of molecular architecture is utilized in that chemical structure dictates the specific physical properties observed in each device. This research has established a foundation of materials, and fundamental scientific principles amenable to the design of organometallic molecular electronic devices.

VII. ACKNOWLEDGMENTS

This work was supported in part by the U. S. Naval Air Systems Command and DARPA/AFOSR.

IX. REFERENCES

1. For a general review see J. H. Lai, S. A. Jenekhe, R. J. Jensen, and M. Royer, Solid State Tech., 27, 165 (1984).

2. C. A. Deckert and D. L. Ross, J. Electrochem. Soc., 45C, (1980).

3. G. G. Roberts, IEEE Proc., 130, Pt. I, 197 (1983).

4. R. S. Potember, T. O. Poehler, and D. O. Cowan, Appl. Phys. Lett. 34, 405 (1979).

5. R. S. Potember, T. O. Poehler, A. Rappa, D. O. Cowan, and A. N. Bloch, J. Am. Chem. Soc., 102, 3659 (1980).

6. R. S. Potember, T. O. Poehler, D. O. Cowan, A. N. Bloch, P. Brant, and F. L. Carter, Chem. Scripta 17, 219 (1981).

7. R. S. Potember, T. O. Poehler, D. O. Cowan, and A. N. Bloch, in Proceedings of the NATO Conference on Chemistry and Physics of One-Dimensional Materials, edited by L. Alcacer, Reidel, Boston, 1980, p. 419

8. R. S. Potember, T. O. Poehler, and R. C. Benson, Appl. Phys. Lett. 41, 548 (1982).

9. R. S. Potember, R. C. Hoffman, R. C. Benson, and T. O. Poehler, J. De Physique, 44, C3-1597 (1983).

10. R. C. Benson, R. C. Hoffman, R. S. Potember, E. Bourkoff, and T. O. Poehler, Appl. Phys. Lett. 42, 855 (1983).

11. T. Takenaka, Spectrochim. Acta, 27A, 1735 (1971).

12. A. Girlando and C. Pecile, Spectrochim. Acta, 29A, 1859 (1973).

13. R. Bozio, A. Girlando, and C. Pecile, J. Chem. Soc., Faraday Trans. II 71, 1237 (1975).

14. C. Chi and E. R. Nixon, Spectrochim. Acta 31A, 1739 (1975).

15. D. L. Jeanmaire and R. P. Vanduyne, J. Am. Chem. Soc. 98, 4029 (1976).

16. M. S. Khatkale and J. P. Devlin, J. Chem. Phys., 70, 1851 (1979).

17. T. Hirono, M. Fukuma, and T. Yamada, J. Appl. Phys., 57, 2267 (1985).

18. R. A. Bartolini, Proc. IEEE, 70, 589 (1982); J. J. Wrobel, A. B. Marchant, and D. G. Howe, Appl. Phys. Lett., 40, 928 (1982); K. Nakamura, N. Matzuura, and F. Tanaka, J. Appl. Phys., 51, 5041 (1980); R. A. Bartolini, H. A. Eakliem, and B. F. Williams, Opt. Eng., 15, 99 (1976).

19. W. E. Moerner, IBM Research Report RJ 4648 (49 656), Sna Jose, CA (1985).

20. B. Parhami, Proc. IEEE 61, 722 (1973).

21. G. Palm, Biol. Cybernetics 31, 19 (1980).

22. J. J. Hopfield, Proc. Natl. Acad. Sci., 79, 2554 (1982).

II
Soliton Switching and Molecular Communication

7

Solitons and Polarons in Quasi-One-Dimensional Conducting Polymers and Related Materials

David K. Campbell/Center for Nonlinear Studies and
Theoretical Division, Los Alamos National Laboratory,
Los Alamos, New Mexico

I. INTRODUCTION

In recent years it has become increasingly appreciated that fundamentally nonlinear excitations -- "solitons" -- play an essential
role in an incredible variety of natural systems. These solitons,
which frequently exhibit remarkable stability under interactions
and perturbations, often dominate the transport, response, or
structural properties of the systems in which they occur. In this
article, we will present an introduction to the solitons that occur
in quasi-one-dimensional conducting polymers ("synmetals") and
related systems. The relevance of this subject to molecular electronic devices is twofold. First, many of these materials have
molecular structures similar to possible prototype "molecular
switches". Second, to understand in detail how a molecular electronic device could work, it is essential to have a broad perspective on the nature of possible excitations in a variety of natural
and synthetic molecular materials.

To focus our discussion, we will concentrate on the synthetic polymer "polyacetylene", usually denoted $(CH)_x$. This is arguably the simplest conjugated polymer and certainly the most extensively studied in recent times. Limitations of space prevent us from giving even a sketchy introduction to the present literature, let alone a comprehensive review. Fortunately, a number of recent articles and conference proceedings are available to the reader desirous of further information (1-6).

We begin our discussion, in the next section, with a brief qualitative description of the chemical structure, materials properties, and potential technological applications of polyacetylene. We discuss the existence of two isomers, <u>trans</u>-$(CH)_x$ and <u>cis</u>-$(CH)_x$, and indicate a crucial difference in symmetry which strongly affects the nature of the nonlinear excitations. In Section III we consider <u>trans</u>-$(CH)_x$, show that there are two degenerate configurations (two Kekule structures) for the ground state of the infinite polymer, and then introduce a phenomenological model (7) that indicates how this degeneracy allows "kink" solitons -- that is, nonlinear excitations which over a finite spatial region connect the two different ground state configurations -- to exist in <u>trans</u>-$(CH)_x$. With this motivation, we translate in Section IV from our chemical description to a simple microscopic physical model (8-10), based on a coupled electron-phonon picture, and show that this model admits two kinks of solitons: the "kinks", which were anticipated by our intuitive arguments, and "polarons", which are two-band generalizations of the conventional polaron (11). After a short discussion of the properties of these excitations and their possible experimental consequences, we turn in Section V to a microscopic model for polarons in <u>cis</u>-$(CH)_x$ (12). Here we find that the absence of a degenerate ground state precludes the existence of "kink" solitons; only polaron-type (including multi-polaron) states can exist. Since unlike <u>trans</u>-$(CH)_x$ most quasi-one-dimensional conducting polymers (polydiacetylene, polyparaphenylene, polypyrrole) do <u>not</u>

have degenerate ground states, we conclude that polarons, although more conventional than kinks, are also more generic. In Section VI we relate the polaron found in the coupled electron-phonon model to the (perhaps) more familiar one-band polarons studied in the context of the molecular crystal model (11) and show that the two polarons are always qualitatively the same excitation and, for certain range of the parameters, become quantitatively essentially identical. A brief discussion of the current experimental situation, the possible complications of comparing the simple models with the real materials, and some concluding remarks comprise Section VII.

II. BACKGROUND CHEMISTRY AND MATERIALS PROPERTIES

As the chemical formula $(CH)_x$ suggests, idealized polyacetylene consists of a large grouping of C(carbon)-H(hydrogen) units linked in a linear polymer. The polymer occurs in two isomeric forms: trans-$(CH)_x$ and cis-$(CH)_x$, the structures of which are shown schematically in Figs. 1 and 2.

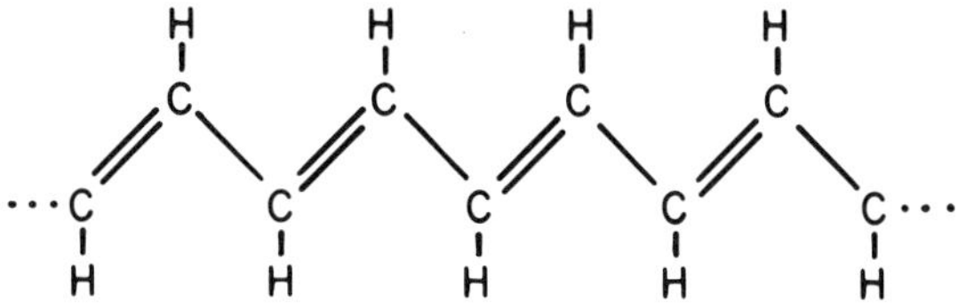

(a)

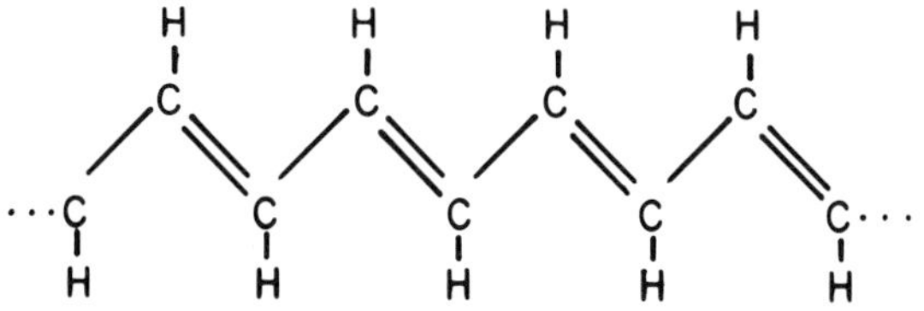

(b)

1. (a) and (b) The two schematic bond structures corresponding to the two degenerate ground states (two Kekule structures) for an infinite chain of trans-$(CH)_x$.

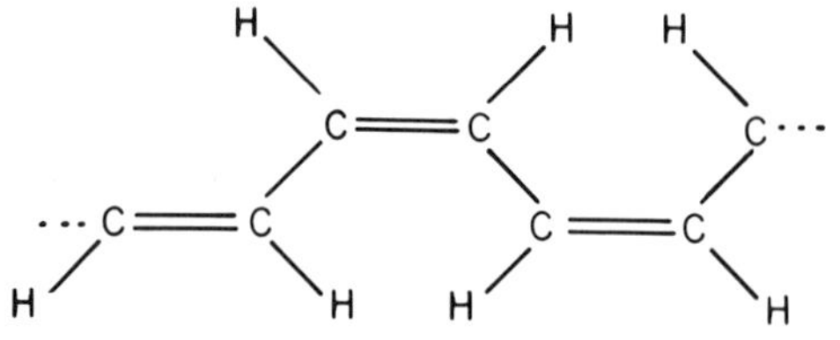

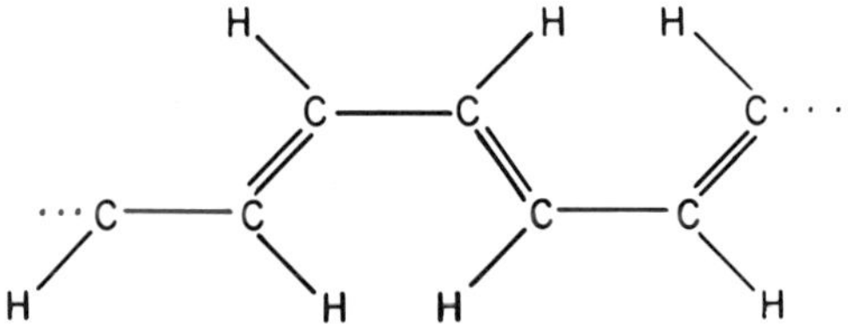

(a)

(b)

2. (a) and (b) The two schematic bond structures corresponding
    to the actual cis-(CH)$_x$ ground state ((a), the cis-cis-
    transoid configuration) and the metastable state ((b), the
    cis-trans-cisoid configuration). The metastable configura-
    tion in (b) has greater energy per (CH) unit and thus in the
    infinite polymer is infinitely higher in energy than ground
    state (a).

    The idealized (infinite) one-dimensional chain chemical

structures shown in these figures summarize conveniently a number

of features contained in the physical models for trans- and cis-

(CH)$_x$ we will introduce later. Consider the case of trans-(CH)$_x$,

Fig. 1. First, the structure shows a carbon backbone consisting

of alternating double and single bonds. Hence the fundamental unit

which repeats contains two carbons and two hydrogens, and the chain

is thus said to be "dimerized". Note that the double bonds are

physically shorter than the single bonds, and thus the schematic

structure indicates that a uniform bond length state is unfavored

relative to bond alternation. Second, taken literally as drawn,

the structures suggest that polyacetylene is indeed a quasi-one-

dimensional material, with the important physical dimension being

along the carbon backbone. Third, the difference between the two

structures shown in Fig. 1 is simply whether, for a given carbon
atom, the double bond lies to the left or to the right, since
all bonds are at the same angle to the backbone axis. For an
infinite chain, it seems intuitively clear that this should make
no difference to the energy of the system; that is, the two Kekule
structures are degenerate energetically. We shall see below that
this is indeed the case and that the ground state of trans-$(CH)_x$,
is two-fold degenerate. This is indicated in Fig. 3(a), which
plots the energy per unit length of the chain as a function of the
value of the bond alternation; notice that the case of uniform bond
length ($\Delta = 0$ in Fig. 3(a)) is unstable.

For cis-$(CH)_x$, the schematic structures shown in Fig. 2 do not
have the obvious symmetry of the trans case and in fact cis-$(CH)_x$
has a unique ground state (the cis-cis-transoid configuration,
shown in Fig. 2(a)). Thus a plot of the energy per unit length
versus bond alternation has the form shown in Fig. 3(b). Although
the energy difference per unit length between the metastable and
ground state is small, for an infinite polymer the total difference
in energy is infinite. We shall later see that the uniqueness of
the cis-$(CH)_x$ ground state configuration has profound consequences
for the types of nonlinear excitations that can exist.

Although all our detailed discussion here will deal with the
idealized material described above, it is important to inject a bit
of reality by describing briefly the nature of polyacetylene as
actually synthesized. In the standard (Ziegler-Natta-Shirakawa)
synthesis,[1] polyacetylene is obtained as a flexible, silver-
colored film. The as-synthesized material is in the cis-isomeric
form. Upon heating, the polymer changes to the trans-isomer, which
is the more stable at room temperature (4, 5). The morphology of
the polymer in the film can be varied substantially by changing the
conditions of synthesis (1-6), but in most cases one finds a matrix
of intertwined fibrils, roughly 200 Å in diameter, with large voids
between the fibrils. Within the fibrils are many individual $(CH)_x$

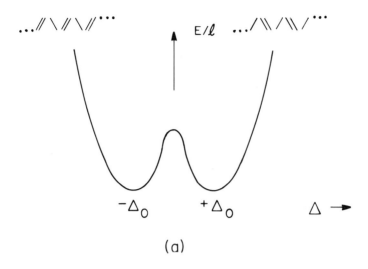

(a)

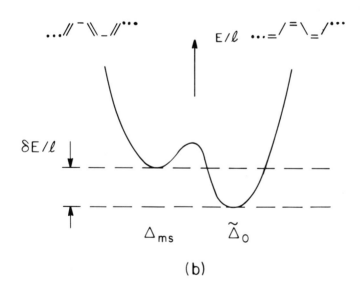

(b)

3. A plot of the energy per unit length versus bond alternation/
   dimerization parameter Δ for spatially constant Δ. (a) For
   <u>trans</u>-$(CH)_x$, note the two degenerate states (corresponding to
   Fig. 2(a) and (b)) at $\Delta = \pm\Delta_o$ and the local maximum (indicating
   instability and corresponding to the hypothetical case of
   uniform bond lengths) at $\Delta = 0$. The height of the barrier
   between $-\Delta_o$ and $+\Delta_o$ is called $\delta V/\ell$ in Fig. 5. (b) For <u>cis</u>-
   $(CH)_x$, note the unique ground state (corresponding to Fig.
   2(a) and the metastable state (corresponding to Fig. 2(b)).

chains; recent experimental results (13) have established that the chain axes are parallel to the fibril axis. There is some cross-linking between individual chains, but the amount is hard to determine quantitatively and seems to be a sensitive function of the details of the synthesis.

Finally, let me remark briefly on one of the principal reasons for technological interest in polyacetylene (4, 5). The conductivity of the as-formed $(CH)_x$ film is on the order of 10^{-8} to 10^{-9} ohm^{-1} cm^{-1}. By chemical doping (with AsF_5, Cl, Li, $\cdots$) at the level of a few percent, the conductivity of the polyacetylene films can be increased to (as much as) $5 \times 10^{+3}$ ohm^{-1} cm^{-1} (5), and thus $(CH)_x$ can be transformed by doping from an insulator to a "synthetic metal". The implications of this for lightweight batteries and electrical and solar cell components may be profound (4, 5).

III. MOTIVATION FOR SOLITONS IN POLYACETYLENE

Before plunging into the details of a microscopic model of polyacetylene, it is useful to discuss some simple phenomenological considerations which motivate the existence of solitons in this polymer. Consider the idealized infinite chain of trans-$(CH)_x$ and recall Fig. 3(a), which shows schematically the energy per unit length versus the bond alternation/dimerization parameter, which for later purposes we have called Δ. Let us construct a simple phenomenological theory (7) -- in the spirit of Landau-Ginzburg theories (14) - which incorporates the features implied by the Fig. 3(a), including the two degenerate minima. Treating Δ as the bond length distortion order parameter,[2] we see that a simple potential energy function that fits the form of Fig. 3(a) is

$$V(\Delta) = \frac{1}{4} (\Delta^2 - \Delta_0^2)^2 . \tag{3.1}$$

If Δ varies in space -- i.e., along the chain -- there will be associated a strain energy -- chemically, this is due to the stretching or compressing of the σ-bonds -- which we can model by a term of the form

$$E_{strain}(\Delta) = \frac{1}{2} \left(\frac{\partial \Delta}{\partial x}\right)^2 . \tag{3.2}$$

Finally, if the bond length distortion varies in time, we would expect a kinetic energy contribution, which in its simplest form would be

$$E_{kinetic}(\Delta) = \frac{1}{2} \left(\frac{\partial \Delta}{\partial t}\right)^2 . \tag{3.3}$$

Combining these three terms leads to an effective energy functional for a general space-time dependent order parameter, $\Delta(x,t)$, of the form

$$E\{\Delta\} = \int dx \left[\frac{1}{2}\left(\frac{\partial \Delta(x,t)}{\partial t}\right)^2 + \frac{1}{2} \left(\frac{\partial \Delta(x,t)}{\partial x}\right)^2 \right.$$

$$\left. + \frac{1}{4} \left(\Delta^2(x,t) - \Delta_o^2\right)^2 \right] . \tag{3.4}$$

Solutions which minimize this energy functional are found by setting the functional derivative $\delta E/\delta \Delta = 0$.

By construction, $E\{\Delta\}$ has two space-time independent solutions -- $\Delta = \pm \Delta_o$ -- corresponding to the two degenerate configurations in Fig. 1; the energy of these configurations is zero. A space-time dependent solution would obey the equation

$$\frac{\partial^2 \Delta}{\partial t^2} - \frac{\partial^2 \Delta}{\partial x^2} - \Delta_o^2 \Delta + \Delta^3 = 0 . \tag{3.5}$$

This equation, which is familiar in soliton circles (15), has "kink"(K) and "anti-kink"($\bar{K}$) solutions of the forms

$$\Delta_K(x,t) = + \Delta_o \tanh\left(\frac{\Delta_o \gamma(x+x_o-vt)}{\sqrt{2}}\right) ,$$

$$\Delta_{\bar{K}}(x,t) = -\Delta_K(x,t)$$

where $\gamma = (1-v^2)^{-\frac{1}{2}}$. For a kink or anti-kink at rest $(v=0, \gamma=1)$, the total energy is[2]

$$E_{K(\bar{K})} = \frac{2\sqrt{2}}{3} \Delta_o^3 , \qquad (3.7)$$

so the K and $\bar{K}$ represent finite energy excitations above the ground state. Notice that the energy <u>density</u> of the kink is localized around $x = -x_o$. Further, since they obey (3.5), the K and $\bar{K}$ are clearly nonlinear excitations. Focusing on the kink solution, we see that for $x \to -\infty$, $\Delta_K \to -\Delta_o$, whereas for $x \to +\infty$, $\Delta_K \to +\Delta_o$. Thus the kink represents a nonlinear excitation which interpolates, over a localized region, between one degenerate ground state and the other; in short, the kink fits <u>precisely</u> the definition of a soliton in <u>trans</u>-(CH)$_x$. Schematically, this excitation is shown in Fig. 4.

Having thus motivated the existence of kink solitons in <u>trans</u>-(CH)$_x$, it is essential to make three points clear before turning to a detailed microscopic model. First, the infinite polymer must be in <u>either</u> the state with $\Delta = +\Delta_o$ <u>or</u> the state with $\Delta = -\Delta_o$; there is no "resonance" between these states. As a consequence, kinks must be made in $K\bar{K}$ pairs from the ground state. To see this explicitly, consider trying to make a single kink localized around $x = -x_o$ on a polyacetylene chain in the $\Delta = -\Delta_o$ ground state. Since the kink satisfies $\Delta_K \to +\Delta_o$ as $x \to +\infty$, whereas the ground state has $\Delta = -\Delta_o$ over all space, to make a <u>single</u> kink would require overcoming the barrier (see Fig. 3(a)) to go from $-\Delta_o$ to $+\Delta_o$ over the whole (semi-infinite) region $-x_o < x < \infty$. Since this barrier has finite energy per unit length, the total barrier to making a single kink is infinite. For making a $K\bar{K}$ pair, however, no such barrier exists, since this excitation can join smoothly with the $\Delta = -\Delta_o$ ground state at both $x \to -\infty$ and $x \to +\infty$. The sketch in Fig. 5 makes this point graphically. As a consequence of the restriction that kinks must be made in $K\bar{K}$ pairs,

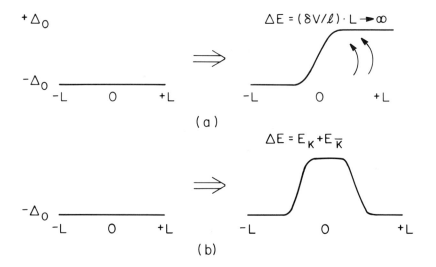

(−Δ_0)   KINK SOLITON   (+Δ_0)

GROUND STATE   GROUND STATE

4. A schematic representation of a kink soliton in <u>trans</u>-(CH)$_x$ showing how it interpolates between the −Δ_0 ground state and +Δ_0 ground state over a finite region.

+Δ_0 $\Delta E = (\delta V/\ell)\cdot L \rightarrow \infty$

−Δ_0 ——————

    −L   0   +L   −L   0   +L

(a)

                                    $\Delta E = E_K + E_{\overline{K}}$

−Δ_0 ——————

    −L   0   +L   −L   0   +L

(b)

5. (a) An illustration of the "topological" barrier against creating a single kink. To reach the state on the right requires overcoming an infinite (as L → ∞) energy barrier. (b) An illustration that creating a kink/anti-kink pair does <u>not</u> require overcoming an infinite barrier.

kinks and anti-kinks are termed "topological" excitations. Second, since K and $\bar{K}$ interpolate between the two degenerate ground states, a necessary condition for the existence of kinks is the existence of this degeneracy. Accordingly, kink solitons should <u>not</u> exist in <u>cis</u>-(CH)$_x$, and we will see in Section V that indeed they do not. Finally, our phenomenological model has motivated the existence of kink-like solitons only. As we shall see in the next sections, more realistic microscopic models predict the existence of another type of soliton: the (non-topological) polaron.

IV. <u>trans</u>-(CH)$_x$: KINKS AND POLARONS

To model microscopically the <u>trans</u>-(CH)$_x$ chain shown in Fig. 1, one must describe the coupled motions of the lattice backbone of C-H units and the electrons that can in principle move along the chain. Of the four valence electrons per carbon, three form relatively deeply bound molecular orbitals in the (CH)$_x$ polymer and thus can, at least in the first instance, be treated as nondynamic. Thus the problem reduces to describing the coupled motion of the (C-H) lattice and the single (π-orbital) electron per carbon that, heuristically speaking, "determines where the double bond goes". This modelling can be done either in a (more realistic) lattice formulation (8, 10, 16) or in a (more analytically tractable) continuum model (9, 17). The continuum theory is derived as the limit for the lattice spacing, a, approaching zero of the lattice model. Fortunately, the results of the two theories are in close agreement, so we can focus here on the more accessible continuum theory, abstracting from the lattice model only the important result of the form of the single (π) electron spectrum, shown in Fig. 6.

The kinematic variables in the continuum electron phonon model (9, 17) of <u>trans</u>-(CH)$_x$ are the bond alternation/band gap order parameter, $\Delta(x)$ (conventionally having dimensions of energy), describing the phonon (i.e., lattice) motions and a two component electron field, $\psi^+ = (\psi^{(1)+}, \psi^{(2)+})$ to describe the π-electrons.

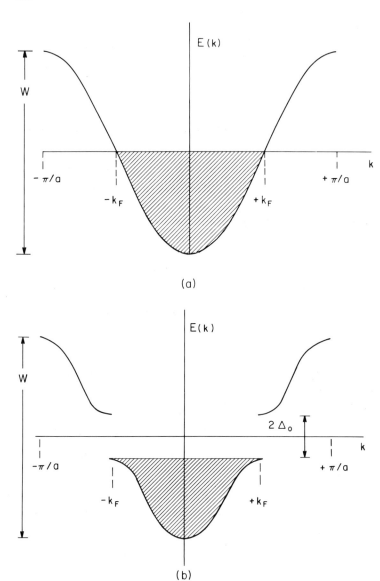

(a)

(b)

6.   The single (π) electron spectrum in <u>trans</u>-(CH)$_x$ in (a) the
     hypothetical uniform (undimerized) case and (b) the actual
     dimerized ground state. The shading indicates filled electron
     states. W is the full band width ($\cong$ 10 eV in <u>trans</u>-(CH)$_x$) and
     $2\Delta_0$ is the full gap ($\cong$ 1.4 eV). Since there is precisely one
     "active" (π) electron per carbon, the band is half filled, so
     $k_F = \pi/2a_0$ where the lattice spacing between (CH)$_x$ units is
     $a \cong 1.22$ Å.

The "adiabatic mean field Hamiltonian" for the microscopic continuum model describing the coupling of these variables has the
form (9, 17)

$$H = \int dx \; \{\frac{\omega_Q^2}{2g^2} \Delta^2(x) + \psi^+(x) [iv_F\sigma_2 \frac{\partial}{\partial x} + \Delta(x)\sigma_3]\psi(x)\} \qquad (4.1)$$

where $\omega_Q^2/2g^2$ is the net effective electron-phonon coupling constant,
σ_i is the ith Pauli matrix, and v_F is the Fermi velocity (in units
with $\not h = 1$, $v_F = Wa/2$). Here and henceforth to simplify the nota-
tion we have suppressed a spin label, s, which should be attached
to every ψ to indicate the two spin states available to an electron in a given energy state. Further, for reasons of later con-
venience, we have not used the standard basis for the electron
wave functions; serious readers will have no difficulty making the
transformation necessary for comparison with the literature (9, 17).
In deriving (4.1), the lattice kinetic energy -- which, as argued
in Section III, would lead to a term proportional to $\dot\Delta^2(x)$ -- has
been explicitly ignored. Obviously, this will have no effect on
the static solutions we discuss below, but it is very significant
for on-going studies involving the dynamics of solitons in $(CH)_x$.

In chemical terms, H represents a simple one-electron Hückel-
type theory. In particular, it contains no direct electron-
electron interactions. Much current theoretical interest (1, 18)
has recently been focused on determining how significant this omis-
sion is. Although, the full story is not yet in, it seems that at
the very least most of the qualitative features of the simple model
in (4.1) survive the inclusion of electron-electron interactions.
Hence (4.1) represents a sensible starting point for studying
trans-$(CH)_x$.

The field equations that follow from the variation of H are
for the single particle electron wave functions

$$\varepsilon_n \psi_n^{(1)} = v_F \frac{\partial \psi_n^{(2)}}{\partial x} + \Delta \psi_n^{(1)} \qquad (4.2a)$$

$$\varepsilon_n \psi_n^{(2)} = -v_F \frac{\partial \psi_n^{(1)}}{\partial x} - \Delta \psi_n^{(2)} \qquad (4.2b)$$

and for the "self-consistent" bond alternation/bond gap parameter

$$\Delta(x) = -g^2 (\omega_Q^2)^{-1} \sum_n' (|\psi_n^{(1)}|^2 - |\psi_n^{(2)}|^2) . \qquad (4.2c)$$

The prime on the summation symbol in (4.2c) indicates that the sum is over all <u>occupied</u> electron states.

How do Eqs. (4.2) reflect the conclusions of our qualitative discussion of the chemical structure of <u>trans</u>-$(CH)_x$? Recall that we anticipated that the configurational energy as a function of (constant) Δ would look like Fig. 3(a), so that the uniform bond length -- equivalently, gapless electronic spectrum as in Fig. 6(a) -- state with $\Delta = 0$ is <u>unstable</u> (the "Peierls instability" (19)) and the ground state is two-fold degenerate. For constant Δ, Eqs. (4.2a) and (4.2b) can readily be solved in terms of plane waves. The eigenenergies are, for $\Delta = \Delta_o$, $\varepsilon(k) = \pm(k^2 v_F^2 + \Delta_o^2)^{\frac{1}{2}} \equiv \pm\omega_k$ so that the electronic spectrum consists of a valence band ($\varepsilon(k) = -\omega_k$) and a conduction band [$\varepsilon(k) = +\omega_k$] separated by a gap of $2\Delta_o$. These subbands are the continuum versions of those shown in Fig. 6(b) for the lattice model. The explicit solutions in the valence band are

$$\psi_k^{(1)}(x) = N_k e^{ikx}(iv_F k) \qquad (4.3a)$$

$$\psi_k^{(2)}(x) = -N_k e^{ikx}\omega_k \qquad (4.3b)$$

with $N_k^{-1} = (2\pi)^{\frac{1}{2}}(2\omega_k(\omega_k + \Delta_o))^{\frac{1}{2}}$.

Using these solutions, one can show that all the results anticipated from the simple chemical structures are indeed correct, that
$E(\Delta)$ varies as $\Delta^2 \ell n \Delta^2$ (9, 10, 20, 21) (which does have the form
shown in Fig. 3(a)) and that the gap equation (4.2c) determines Δ_o
self-consistently via

$$\Delta_o = (g^2/\omega_Q^2) \frac{1}{\pi} \int_{-K}^{K} \frac{\Delta_o \, dk}{(k^2 v_F^2 + \Delta_o^2)^{\frac{1}{2}}} . \tag{4.4}$$

Note the crucial result that the cut-off wave vector, K, is chosen
such that the energy at the cut-off corresponds to the correct π-
electron band width (W $\cong$ 10 eV, as shown in Fig. 6) in trans-$(CH)_x$.
Thus

$$W = 2(K^2 v_F^2 + \Delta_o^2)^{\frac{1}{2}} \cong 2Kv_F, \text{ since } Kv_F \gg \Delta_o .$$

The need for this cut-off comes solely from a linearization of the
spectrum about k_F made in going to the continuum model. As shown
in Fig. 6, the true spectrum of the lattice model automatically
covers a finite range. Solving (4.4) for $Kv_F \gg \Delta_o$ gives

$$\Delta_o = W \exp(-\lambda^{-1}), \text{ with } \lambda = 2g^2 (\pi v_F \omega_Q^2)^{-1} , \tag{4.5}$$

where the parameters have all been introduced previously (see Fig.
6). Since $\Delta_o \cong 0.7$ eV and W $\cong$ 10 eV, Eq. (4.5) implies that
the dimensionless coupling $\lambda \cong 0.38$.

From the considerations of Section III we anticipate that
"kink" and "anti-kink" soliton solutions to Eqs. (4.2) should exist.
They do, and, interestingly, they have the same functional form for
Δ as that found in our phenomenological model:

$$\Delta_K = +\Delta_o \tanh x/\xi_o , \tag{4.6}$$

with $\xi_o \equiv v_F/\Delta_o$, and for the anti-kink $\Delta_{\bar{K}} = -\Delta_K$. The associated
single electron spectrum again consists of extended (modified

plane wave) states in the conduction and valence bands plus an additional, localized "mid-gap" state with $\varepsilon_o = 0$. The explicit form of the mid gap electron wave function for the kink is

$$\psi_o^{(1)} = \psi_o^{(2)} = N_o \text{ sech } x/\xi_o, \tag{4.7}$$

with $N_o = \frac{1}{2} \xi_o^{-\frac{1}{2}}$. The kink excitation and mid-gap electron distribution are shown in Fig. 7(a).

For either K or $\bar{K}$, the "mid-gap" state can be occupied by 0, 1, or 2 electrons, leading to the localized excitations with the bizarre spin/charge assignments indicated in Fig. 7(b). Explicitly, one finds that the neutral kink (K^o) has a single electron in the state ε_o; since all states in the valence band remain spin-paired, the spin of the K^o is 1/2 (8-10, 16, 17)! Similar arguments show that K^+, which has no electrons in ε_o, and K^-, which has two electrons in ε_o, both have spin zero. These results seem to violate conventional solid state spin/charge relations which follow quite generally from the fact that the basic charge-carrying object, the electron, has spin $\frac{1}{2}$ and charge (-e). Nonetheless, recent theoretical studies (22) have confirmed that this effect -- which is related to the concept of "fractional" charge (23) -- is expected, and the search for experimental confirmation of this prediction is on.

The energy of a single kink (or anti-kink) of any charge is $E_K = E_{\bar{K}} = 2\Delta_o/\pi$. However, as noted in Section III, topological constraints require that kinks be produced in $K\bar{K}$ pairs. Thus the minimum energy state involving kinks which can be excited from the ground state is a single $K\bar{K}$ pair with energy $4\Delta_o/\pi$. In a conventional semi-conductor, the lowest-lying excitation would be a particle-hole pair, with minimum energy $E_{ph}^{min} = 2\Delta_o$ (i.e. so that the electron at the top of the valence band is excited to the bottom of the conduction band). Thus one has the exciting theoretical prediction, which has been referred to in several other contributions to this conference, that the elementary excitations from the ground state in trans-(CH)$_x$ are the exotic "kink" solitons.

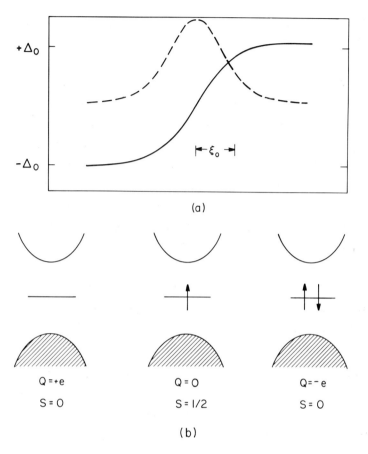

(a)

(b)

7.   (a)   The spatial structure of the gap parameter $\Delta(x)$ for the kink soliton (solid line) and the probability distribution of the localized "mid-gap" electronic state (dashed line).   The characteristic length ξ_0 is indicated.
(b)   The electronic levels and their occupations for the 3 charge states of the kink.   The shading indicates that the valence band is fully occupied.   Q = charge and S = spin.

In addition to considering excitations from the ground state which conserve the total electron number, one is interested in predicting what excitations exist when single electrons are added to, or taken from, the ground state. This is particularly important for understanding the technologically important "doping" process, for the standard dopants either remove or add electrons to the undoped $(CH)_x$ polymer chains. For definiteness, let us consider what happens when a single electron is added to an infinite chain of trans-$(CH)_x$. If one tried to accomodate this electron in the mid-gap state associated with a kink, since kinks must be produced in $K\bar{K}$ pairs, the minimum energy for adding a single electron would be $4\Delta_o/\pi$. Thus it would be energetically more favorable simply to put the electron in the lowest state in the conduction band, since this costs an energy Δ_o. In fact, as we illustrate in detail below, the most energetically favorable state available to a single electron is the "polaron" (12, 16, 21, 24, 25) to which we now turn our attention.

Unlike kinks, polarons represent a localized deviation from one of the degenerate ground states of trans-$(CH)_x$, as illustrated in Fig. 8(a). The explicit form of the polaron solution to Eq. (4.2) can be written in the revealing form (cf. Eq. (4.6))

$$\Delta_p(x) = \Delta_o - \kappa_o v_F \{\tanh \kappa_o v_F(x+x_o) - \tanh \kappa_o(x-x_o)\} \qquad (4.8)$$

where

$$\tanh 2\kappa_o x_o = \kappa_o v_F/\Delta_o . \qquad (4.9)$$

The single electron states for this form of Δ_p again include extended states in the conduction and valence bands, which are plane waves "phase-shifted" by the polaron "potential". The explicit forms are available in the literature (21). In addition, there are two localized electronic states, with energies symmetrically placed at $\varepsilon_\pm = \pm w_o$, where $w_o = (\Delta_o^2 - \kappa_o^2 v_F^2)^{\frac{1}{2}}$. The electronic

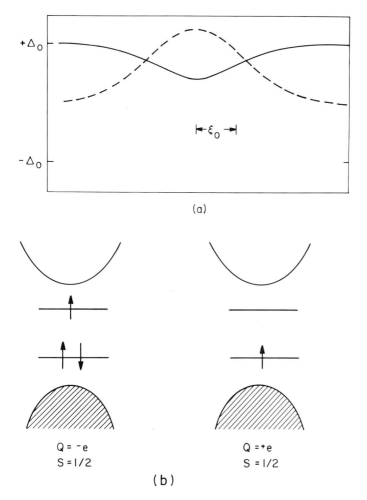

(a)

(b)

Q = -e   Q = +e
S = 1/2   S = 1/2

8. (a) The spatial structure of the bond alternation/band gap
    parameter $\Delta(x)$ for the polaron (solid line) and the probability
    distribution of the localized electronic levels at $\varepsilon = \pm w_0$
    (dashed line). For comparison with Fig. 7, the characteris-
    tic kink length ξ_0 is plotted. (b) The electronic levels and
    their occupations for the "electron polaron" ($Q = -e$, $S = \frac{1}{2}$)
    and the "hole polaron" ($Q = +e$, $S = \frac{1}{2}$).

wave functions for these localized states are, for $\varepsilon_+ = +w_o$

$$\psi_+^{(1)} = N_+[\text{sech } \kappa_o(x+x_o)+\text{sech } \kappa_o(x-x_o)] \qquad (4.10a)$$

and $\psi_+^{(2)} = N_+[-\text{sech } \kappa_o(x+x_o)+\text{sech } \kappa_o(x-x_o)] \qquad (4.10b)$

where $N_+ = (\kappa_o/8)^{\frac{1}{2}}$. For $\varepsilon_- = -w_o$, $\psi_-^{(1)} = \psi_+^{(2)}$ and $\psi_-^{(2)} = \psi_+^{(1)}$.

Interestingly, the polaron configuration for $\Delta(x)$ -- Eq. (4.8) and the associated electron wave functions (21) -- satisfy the electron part -- Eqs. (4.2a) and (4.2b) -- of the continuum equations for <u>any</u> $\kappa_o v_F$ in the allowed range $0 \le \kappa_o v_F \le \Delta_o$. It is the self-consistent gap equation -- (4.2c) -- that determines the specific value of $\kappa_o v_F$ for an actual solution to the coupled equations. Unsurprisingly, the nature of the solution depends on the occupation numbers -- call them n_+ and n_- -- of the gap levels $\varepsilon_\pm = \pm w_o$. Using some aspects of soliton theory (12, 21, 24), one can in effect convert Eq. (4.2c) to an algebraic minimization problem. Apart from simplifying the problem technically, this is very appealing intuitively, for the quantity being minimized is essentially the <u>energy</u> of the full interacting electron-phonon system. One finds that

$$E_P^{trans}(n_+,n_-,\kappa_o) = (n_+ - n_- + 2)w_o + \frac{4}{\pi} \kappa_o v_F$$

$$- \frac{4}{\pi} w_o \tan^{-1}(\kappa_o v_F/w_o) . \qquad (4.11)$$

With $\kappa_o v_F = \Delta_o \sin\theta$ and $w_o = \Delta_o \cos\theta$ -- possible since $w_o^2 + (\kappa_o v_F)^2 = \Delta_o^2$ -- one can show that E_P^{trans} is minimized for $\theta = \theta(n_+,n_-) = (n_+ - n_- + 2)\pi/4$. From this result it follows that for a stable, localized polaron solution the electrons must be distributed in one of two configurations (see Fig. 8(b)):

      (1) $n_+ = 1, n_- = 2$, the "electron polaron" state, with $Q = -e$
               and $S = \frac{1}{2}$;

or (2) $n_+ = 0, n_- = 1$, the "hole polaron" state, with $Q = +e$ and
                     $S = \frac{1}{2}$.

Thus, unlike kinks, polarons have conventional spin/charge relations. For either electron or hole polaron, $\theta = \pi/4$, so $\kappa_o v_F = \omega_o = \Delta_o/\sqrt{2}$ and $E_P^{trans} = 2\sqrt{2}\, \Delta_o/\pi_o$. This is less than Δ_o -- and obviously also less than $4\,\Delta_o/\pi$, the minimum $K\bar{K}$ pair energy -- and hence the polaron is the <u>lowest</u> <u>energy</u> <u>excitation</u> <u>available</u> <u>to a single electron</u> added to a <u>trans</u>-(CH)$_x$ chain.

Using Eqs. (4.8) and (4.11) we can also answer the natural question "What happens if we try to form a "bipolaron" by adding a <u>second</u> electron (or hole) to a polaron state?" For any of the electron-electron ($n_+ = n_- = 2$), hole-hole ($n_+ = n_- = 0$), or electron-hole ($n_+ = n_- = 1$) configurations, E_P^{trans} is minimized for $\theta = \pi/2$, so that $\kappa_o v_F = \Delta_o$, $\omega_o = 0$ and from Eq. (4.9), $x_o \to \infty$. Thus these putative "bipolarons" in <u>trans</u>-(CH)$_x$ actually correspond to an infinitely separated ($x_o \to \infty$) $K\bar{K}$ pair, as Eq. (4.8) would suggest. Hence when a single electron is added to a <u>trans</u>-(CH)$_x$ chain, it should form a polaron, whereas when many excess electrons are present on a single chain, it becomes energetically favorable to form many $K\bar{K}$ pairs. A consequence is that the response and transport properties of lightly-doped -- average of $\leq$ one electron (or hole) per chain -- <u>trans</u>-(CH)$_x$ should be controlled by polarons, whereas more heavily doped samples should reflect the kink dominance.

Our picture of (idealized) <u>trans</u>-(CH)$_x$ thus reflects a fascinating and rich structure of nonlinear, soliton-like elementary excitations. In the next section, we will see how this structure changes when the critical degeneracy of the ground state is not present.

V. <u>cis</u>-(CH)$_x$: MULTIPOLARONS

From the perspective of the nonlinear excitations we are discussing, the most important difference between the <u>trans</u> and <u>cis</u> isomers of (CH)$_x$ is that for <u>cis</u>-(CH)$_x$ there is a unique, non-degenerate ground state, as shown in Fig. 3(b). As we have indicated previously, an immediate consequence is that since there are <u>not</u> two

degenerate ground states for kink solitons to connect, there are
no kink soliton solutions in cis-(CH)$_x$! Similarly, in a wide class
of other quasi-one-dimensional conducting polymers (poly-diace-
tylene, poly-paraphenylene, polypyrrole) with non-degenerate ground
states, there can be no kink solitons. Since this point is so
crucial, let us discuss it from an intuitive viewpoint before pre-
senting the detailed mathematical model.

What would happen if one tried to create a kink/anti-kink pair
from the ground state in cis-(CH)$_x$? Suppose that $\delta E/\ell$ (see Fig.
3(b)) is small -- as it is in cis-(CH)$_x$ -- so that the metastable
and ground state have almost the same energy per unit length. One
could then imagine constructing a configuration -- call it an "anti-
kink" -- that interpolated between $\tilde{\Delta}_o$ (the cis-ground state) and
Δ_{ms} (the metastable state), remained in Δ_{ms} for a spatial distance
d, and then interpolated between Δ_{ms} and $\tilde{\Delta}_o$ (a "kink"). The form
of Δ would look like (4.8), with d = 2x$_o$. For finite d, the configuration would have finite energy, but since $\delta E/\ell > 0$, for d $\rightarrow \infty$,
its energy would become infinite like $(\delta E/\ell)\cdot d$, and hence an independent, "free" K$\bar{K}$ pair can not exist. In this sense, one can
say that the putative kinks in cis-(CH)$_x$ are "confined". Of course,
referring to (4.8), we see that permanently confined K$\bar{K}$ pairs have
the same structure for Δ as polarons. Thus, in cis-(CH)$_x$, only
polaron-type nonlinear excitations will exist.

To make this precise, it is very useful to study an explicit
model of cis-(CH)$_x$ (due to Brazovskii and Kirova (12)) which is a
natural extension of the microscopic electron-phonon Hamiltonian
proposed for trans-(CH)$_x$. One assumes (12, 26) that the gap para-
meter can be written as $\tilde{\Delta}(x) = \Delta_i(x) + \Delta_e$, where the intrinsic gap
$\Delta_i(x)$ is sensitive to the electron feedback (as for trans) but Δ_e
is a constant, extrinsic component (12, 26). This Ansatz can be
motivated in terms of effects arising from molecular orbitals other
than the π-orbital explicitly included in (4.1).

The adiabatic mean field Hamiltonian for cis-(CH)$_x$ then
becomes (12, 27, 28)

$$H_{cis} = \int dx\{\frac{\omega_Q^2}{g^2} \Delta_i^2(x) + \psi^+(x)(iv_F\sigma_2 \frac{\partial}{\partial x} + (\Delta_i(x)+\Delta_e)\sigma_3) \psi(x)\} \ . \quad (5.1)$$

Although in the interest of simplicity we have not indicated this explicitly, it is important to realize that all the physical parameters in the model of cis-$(CH)_x$ -- lattice spacing, effective electron-phonon coupling constant, band width, fermi velocity, band gap -- can have different values from those in trans. Comparing (4.1) and (5.1), we see that with the replacement $\Delta(x) \to \tilde{\Delta}(x) = \Delta_i(x) + \Delta_e$, the electron equations for trans can be converted to those for cis. Thus, the structure of the electron spectrum and the eigenfunctions will be the same. The gap equation is, however, modified to read

$$\Delta_i(x) = \tilde{\Delta}(x) - \Delta_e = -g^2(\omega_Q^2)^{-1} \Sigma'(|\psi_n^{(1)}|^2 - |\psi_n^{(2)}|^2) \ , \quad (5.2)$$

which leads to different constraints on solutions and on bound state eigenvalues. Specifically, the ground state has the unique value $\tilde{\Delta}_o$ determined from

$$\tilde{\Delta}_o - \Delta_e = (g^2/\omega_Q^2) \frac{1}{\pi} \int_{-K}^{K} \frac{dk \ \tilde{\Delta}_o}{(k^2v_F^2+\tilde{\Delta}_o^2)^{\frac{1}{2}}} \quad (5.3)$$

which leads to

$$\Delta_e = \lambda\tilde{\Delta}_o \ \ell n \ \tilde{\Delta}_o/\Delta_o \quad (5.4)$$

where Δ_o is as given by Eq. (4.5).

In confirmation of our earlier arguments, it can be shown directly that there are no kink solutions when the modified gap equation, (5.2), is considered. For polaron configurations, one finds that exactly the same functional form as that in (4.8) does satisfy the cis equations provided κ_o is chosen appropriately.

Again using soliton techniques (12, 21, 27) to convert the gap
equation to an algebraic minimization problem, one finds

$$E_P^{cis}(n_+, n_-, \kappa_o) = (n_+ - n_- + 2)\omega_o + \frac{4}{\pi} \kappa_o v_f - \frac{4}{\pi} \omega_o \tan^{-1}(\kappa_o v_f / \omega_o)$$

$$+ \frac{4}{\pi} \tilde{\Delta}_o \gamma [\tanh^{-1} (\frac{\kappa_o v_f}{\tilde{\Delta}_o}) - \frac{\kappa_o v_f}{\tilde{\Delta}_o}]$$

where $\gamma = \Delta_e / \lambda \tilde{\Delta}_o$ is a measure of the strength of the soliton
"confinement" energy. Introducing θ such that $\kappa_o v_f = \tilde{\Delta}_o \sin\theta$ and
$\omega_o = \tilde{\Delta}_o \cos\theta$, one finds that E_P^{cis} is minimized for solutions to the
equation

$$\theta + \gamma \tan\theta = (n_+ - n_- + 2)\pi/4 ,$$

$$0 \le \theta \le \pi/2 . \qquad (5.6)$$

In the limit of zero extrinsic gap ($\Delta_e = 0$) $\gamma = 0$, and as expected
(5.6) reduces to the result for trans-(CH)$_x$. For $\gamma > 0$, (5.6) --
unlike the corresponding trans equation -- has solutions for all
combinations of n_+ and n_-. As shown in Table 1, these are polaron
and multipolaron states (12, 27). Note that as the effective
polaron occupation number, $N \equiv n_+ - n_- + 2$, increases, both the
polaron width (2 x_o) and depth increase, as one would expect intuitively. Note also that the qualitative nature of these multipolaron
states varies substantially. The (e-e) bipolaron ($n_+ = n_- = 2$), for
example, can only have spin zero and has charge -2 relative to the
ground state. In contrast, the (e-h) bipolaron can exist in either
singlet (s = 0) or triplet (s = 1) forms and, with charge Q = 0
relative to the ground state, could more properly be called an
"exciton". Depending on spin, multipolaron states involving both
e and h are subject to rapid photo-recombination, which renders them
unstable. Nonetheless, in cis-(CH)$_x$ and other similar quasi-one-
dimensional systems, we expect a fascinating variety of polaron
nonlinear excitations.

TABLE

1. Polaron and Multipolaron States in <u>cis</u>-(CH)$_x$

n_+	n_-	<u>N</u>	<u>Q</u>	θ	$E/\widetilde{\Delta}_o$	INTERPRETATION
1	2	1	1	0.38	0.98	polaron (e)
0	1	1	-1	0.38	0.98	polaron (h)
2	2	2	-2	0.71	1.43	bipolaron (e,e)
1	1	2	0	0.71	1.43	bipolaron (e,h)
0	0	2	+2	0.71	1.43	bipolaron (h,h)
2	1	3	-1	0.95	1.96	tripolaron (e,e,h)
1	0	3	+1	0.95	1.96	tripolaron (h,h,e)
2	0	4	0	1.11	2.36	quadripolaron (e,e,h,h)

Caption

The possible polaron and multipolaron states in <u>cis</u>-(CH)$_x$ for the
case of $\gamma = 1$. $N = n_+ - n_- + 2$ is the effective occupation number,
$Q = 2 - n_+ - n_-$ is the charge, θ is the angle defined in the text
(in radians), and $E/\widetilde{\Delta}_o$ gives the full energy of the excitation.
Further interpretation is given in the text.

VI. RELATION TO CONVENTIONAL ONE-BAND POLARON THEORY

Conventional polaron theory, as for example discussed in the context
of the molecular crystal model (11), is based on the intuition that
a distortion of the molecular lattice can create a localized electronic state just below the conduction band and that when this
localized state is occupied by an electron the lattice distortion
is self-consistently stabilized. This situation is shown schematically in Fig. 9. Since this conventional picture focuses only on
the conduction band, we can refer to it as a "one-band" polaron
theory. In contrast, the polaronic excitations we have discussed in
the previous sections, although again involving localized electronic
states and self-consistent lattice distortions, involve both the
(empty) conduction band states and the (occupied) valence band

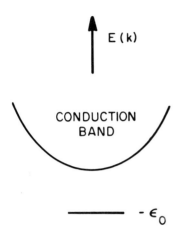

9. A schematic diagram of the spectrum of single electron states
    considered in conventional one-band polaron theory. Below
    the conduction band continuum there is a single localized
    electronic state with energy $E = -\varepsilon_o$ with respect to the
    bottom of the band.

levels, as shown in Fig. 8(b). These are thus "two-band" polarons,
and it is natural to seek to understand the relation of these to
the conventional "one-band" polarons. In this section we show
that these two types of excitations are always qualitatively the
same object, and, for certain values of the parameters, become
quantitatively essentially identical (29). In the ensuing discussion, we shall focus on the electron polaron. With obvious
modifications, the same arguments apply to the hole polaron.

In the continuum limit, the conventional one-band polaron
theory (11) applied to a one-dimensional system (30) leads to an
equation for the localized state electronic wave function $(a_o(x))$
of the form

$$-\frac{d^2 a_o(x)}{dx^2} - y_o(x)a_o(x) = \varepsilon_o a_o(x) \tag{6.1a}$$

where the localized lattice distortion $y_o(x)$ is determined self-consistently in terms of $a_o(x)$ by the equation

$$y_o(x) = \alpha |a_o(x)|^2 , \qquad (6.1b)$$

with α an effective coupling constant. When Eqs. (6.1a) and (6.1b) are combined, they lead to the well-known "nonlinear Schrödinger equation" (30, 31)

$$- \frac{d^2 a_o(x)}{dx^2} - \alpha |a_o(x)|^2 a_o(x) = \varepsilon_o a_o(x) \qquad (6.2)$$

for the localized electronic wave function. The polaron is just the familiar (envelope) soliton solution (31)

$$a_o(x) = (\frac{\alpha}{8})^{\frac{1}{2}} \text{ sech } \frac{\alpha}{4} (x-x_o) , \qquad (6.3)$$

with x_o being the location of the polaron.

To see how this relates to the two-band polaron, we start from the equations of the coupled electron-phonon model -- for definiteness, we choose the trans-(CH)$_x$ case -- and consider a weakly bound polaron. By this we mean an excitation in which $\omega_o \lesssim \Delta_o$, so $\kappa_o v_F/\Delta_o \ll 1$, and for which Δ differs only slightly from its ground state value Δ_o. Thus we write $\Delta \equiv \Delta_o - \tilde{\Delta}$ and study Eqs. (4.2) in powers of $1/\Delta_o$. For electron states near the bottom of the conduction band -- which are those most relevant to our weakly bound polaron -- one has, from (4.2), $\varepsilon_n = \Delta_o + 0(k^2)$. Thus for these states $\varepsilon_n + \Delta \cong 2\Delta_o$ and (4.2b) implies

$$\psi_n^{(2)} \cong \frac{1}{2\Delta_o} [-v_F \frac{\partial \psi_n^{(1)}}{\partial x}] , \qquad (6.4)$$

so that $\psi_n^{(2)}$ is of order $(1/\Delta_o)$ relative to $\psi_n^{(1)}$ and further, this leading term in $\psi_n^{(2)}$ can be calculated directly from $\psi_n^{(1)}$. Hence

to leading order in $1/\Delta_0$, only the Eq. for $\psi_n^{(1)}$ (4.2a) remains.[3]
Focusing on a weakly bound state with

$$\varepsilon_0 \equiv (\Delta_0^2 - \kappa_0^2 v_F^2)^{\frac{1}{2}} \cong \Delta_0 - \kappa_0^2 v_F / 2\Delta_0 + \cdots , \qquad (6.5)$$

and substituting for $\partial \psi_0^{(2)} / \partial x$ by differentiating (6.4) -- which is
valid to leading order in $\kappa_0 v_F / \Delta_0$ -- we obtain, using (6.5),

$$\left(\frac{-\kappa_0^2 v_F^2}{2\Delta_0}\right) (\psi_0^{(1)}) = \left(\frac{-v_F^2}{2\Delta_0}\right)\left(\frac{\partial^2 \psi_0^{(1)}}{\partial x^2}\right) - \tilde{\Delta} \psi_0^{(1)} , \qquad (6.6)$$

which clearly has the form of a Schrödinger equation with potential
$\tilde{\Delta} \equiv \Delta_0 - \Delta$.

Turning to the self-consistency equation for Δ, Eq. (4.2c), we
rewrite it slightly in the following suggestive way:

$$\Delta(x) = \Delta_0 - \tilde{\Delta}(x) = \frac{-g^2}{\omega_Q^2} (|\psi_0^{(1)}|^2 - |\psi_0^{(2)}|^2)$$

$$\frac{-g^2}{\omega_Q^2} \left(\sum_{\varepsilon_n < 0, s} |\psi_n^{(1)}|^2 - |\psi_n^{(2)}|^2 \right) . \qquad (6.7)$$

We have already seen that $\psi_0^{(2)}$ is $0(1/\Delta_0)$ relative to $\psi_0^{(1)}$, and so
to the order we are working we can drop it in (6.7). But what
about the (infinite!) sum over the states with negative energy?
From Fig. 8(b) one sees that for small $\kappa_0 v_F / \Delta_0$, these states are
all separated by a "large" energy ($\cong 2\Delta_0$) from the state at $\varepsilon_0 = +\omega_0$. Further, these states are all fully occupied in the electron
polaron configuration, just as they are in the ground state. This
motivates the "frozen valence band" (FVB) approximation, which is
indicated graphically in Fig. 10. Here one argues that since
$\Delta(x)$ differs only slightly from its ground state value, the
shifts in states in and near the valence band are small, and one

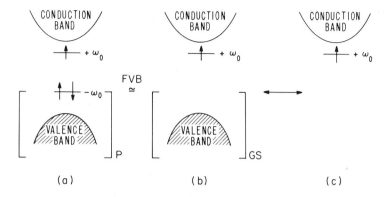

(a)   (b)   (c)

10. A graphic illustration of the frozen valence band (FVB)
     approximation used in reducing the two-band polaron theory to
     a one band model. The actual full one-electron spectrum for
     the electron polaron configuration (a) is approximated (FVB)
     by replacing the states below the fermi energy by the values
     in the ground state (b) so that the spectrum corresponds
     directly to that considered in the conventional one-band
     theory (c).

can approximate the sum over all states with energies less than

zero by its value in the ground state. But, from (4.2c) and (4.4),

this value is just Δ_o! Thus, within the FVB approximation, the sum

on the right of (6.7) cancels the Δ_o on the left and we are left

with

$$\tilde{\Delta}(x) = \frac{g^2}{\omega_Q^2} \, |\psi_o^{(1)}|^2 \, , \tag{6.8}$$

which when inserted into (6.6), yields precisely the same nonlinear

Schrödinger equation found in the one-band polaron theory. The

proof that the FVB approximation is actually valid to leading order

in $(\kappa_o v_F / \Delta_o)$ is available in the literature (29). Interested

readers can verify directly, by expanding the forms of $\psi_o^{(1)}$ and

$\Delta_p(x)$ for $\kappa_o v_F / \Delta_o \ll 1$, that (6.8) holds for our polaron solutions,

provided that $g^2 / \omega_Q^2 = (2\kappa_o v_F / \Delta_o)$.

We have established the precise equivalence of the two po-
larons in the limit $\kappa_o v_F / \Delta_o \ll 1$. When is this limit valid in the
class of quasi-one-dimensional conducting polymers we have been
considering? To answer this question, we need to recall from
Sections IV and V that the full coupled electron-phonon equations
uniquely determine $\kappa_o v_F / \Delta_o$. For trans-$(CH)_x$, for example, $\kappa_o v_F /
\Delta_o = 1/\sqrt{2}$, and the weakly bound polaron limit is clearly not valid.
On the other hand, for cis-$(CH)_x$ (and similar systems), in the limit
that $\gamma = \Delta_e / (\lambda \bar{\Delta}_o)$ is large -- so there exists a strong soliton confinement energy -- the single (and even bipolaron) solutions can
have $\kappa_o v_F \ll \Delta_o$, and hence the two-band polaron will look virtually
identical to its one-band counterpart.

Finally, it is appropriate to mention that the nonlinear
Schrödinger equation structure that arises as a limiting case of
the coupled electron-phonon theory occurs in a wide variety of
quasi-one dimensional systems other than conducting polymers.
An example particularly relevant to the present audience is the
Davydov soliton in α-helix proteins (32), which is discussed in
some detail by Scott Layne in these Proceedings. Here the two
coupled degrees of freedom are the "vibron" describing the vibrational excitation of an amide-I bond and the longitudinal sound mode
describing the deviation from equilibrium of the hydrogen-bonded
backbone of the protein. The resulting soliton has the same form
as the one-band polaron discussed here. This example provides a
clear illustration of the genericity of the concept of solitons and
of the importance of understanding these nonlinear excitations.

VII. IMPLICATIONS AND DISCUSSION

The rich structure of theoretically predicted nonlinear excitations
and the striking constrasts between the cis- and trans-isomers
suggest that polyacetylene should be an ideal experimental testing
ground for a variety of concepts which apply to many quasi one-
dimensional materials. As one might expect, however, the reality
of the situation is more complicated than our simple models have

suggested, and although large numbers of experiments (1-6) have indeed been carried out, the interpretation of the results in terms of the theoretical concepts (e.g., kink solitons and polarons) is controversial and still not firmly established.

In part, this situation arises from limitations of the theoretical models. The simple one-dimensional adiabatic mean field models we introduced in Sections IV and V ignored quantum fluctuation effects of the lattice, electron-electron interactions, coupling between the chains, and the effects of disorder caused by intrinsic or extrinsic defects. All of these omissions are currently being corrected (1-6), but the final word on their combined effect on the theoretical picture generated from the simple model has not been said.

In larger part, the gap between experimental results and theoretical interpretation arises from properties of the actual material not reflected in the simple models. Thus, for example, the as-synthesized material contains an inprecisely known number of cross-links between chains; these cross-links destroy the π-orbital conjugation assumed in the simple models. A second example of "real world" complication arises in the study of chemically "doped" polyacetylene. Because of the strength of the intrachain chain coupling (the σ-bonds) and the fibrillar nature of the material, "doping" the chains (by adding small amounts of, for example, AsF_5 or Na) does not destroy the material but can be viewed as simply the addition or removal of electrons from the chain. Since this should produce precisely the nonlinear excitation we have discussed previously, it might seem that comparative doping studies of cis- and trans-$(CH)_x$ should resolve all questions about the nature of the excitations in these materials. Unfortunately, upon doping cis-$(CH)_x$ undergoes (at least partial) isomerization to trans-$(CH)_x$, and thus the interpretation of the comparative doping experiments is difficult.

All these complications notwithstanding, there is now a large body of experimental data that can be understood in terms of the

theoretical concepts related to the kink and polaron solitons. In
many cases the "soliton interpretation" is either not unique or not
definitive, and hence considerable controversy still rages. As a
result, it is impossible to summarize the full situation both
briefly and accurately, and thus we shall simply describe the types
of experimental tests that have been made and refer the serious
reader to the appropriate literature (1-6) for details. We shall
discuss four points: (1) the creation of mobile spins upon thermal
isomerization of cis- to trans-$(CH)_x$ (33-35); (2) the existence of
spinless charge carriers in doped trans-$(CH)_x$ (36); (3) photo-
luminescence and photoconductivity (37); (4) optical absorption
(30-40) and infrared activity (41, 42) both in photo-induced
experiments (43-44) and in doped materials (45-47).

First, when as-synthesized cis-$(CH)_x$ films are carefully
isomerized to trans-$(CH)_x$ by controlled heating, there is a substantial increase in the magnetic susceptibility with little charge
in the conductivity (33-35); further, the spins giving rise to the
susceptibility are mobile (33-35). In the soliton framework this
can be understood as the creation, during the isomerization process (5), of neutral kink solitons in the trans-$(CH)_x$ which, as we
discussed in Section IV, carry spin but no charge.

Second, when trans-$(CH)_x$ is doped in the region of 1% to 6%,
the magnetic susceptibility goes down as the conductivity goes up
(36). This is consistent with first changing all the neutral kink
solitons to charged solitons (which carry no spin) and then continuing to form charged solitons upon further doping.

Third, a comparison of the photo-luminescence and photoconductivity of cis- and trans-$(CH)_x$ reveals striking contrasts of
a kind qualitatively expected from the soliton picture (5). In
trans-$(CH)_x$, where the electron-hole pair formed by the initial
photon should quickly relax into a soliton/anti-soliton pair (16, 48)
which can then escape, the photoconductivity is high and no photo-
luminescence is observed (37). In contrast, in cis-$(CH)_x$ where the

electron-hole pair should relax to a bipolaron (exciton), which remains localized and threfore subject to recombination, one observes photo-luminescence but little photoconductivity (37).

Fourth, optical absorption in the presence of kink solitons (38-39) or polarons (40) should be modified by the existence of the localized electronic states in the gap. For doped materials, there is evidence for both polaron (46) (at very low dopant concentrations, as anticipated in Section IV) and kink soliton (45) absorption. Similarly, photo-induced experiments (43-44) reveal structure within the gap. The infrared activity associated with excitations causing this in-gap absorption is consistent with the soliton picture (41), although this interpretation is not unique (42).

In summary, although there is a substantial body of experimental data which the soliton framework can explain, there remains disagreement (49) over the uniqueness and/or definitiveness of the soliton interpretation.

Having thus (very briefly) sketched the experimental status of solitons in polyacetylene, let me turn to one final point which is particularly relevant for the present symposium: namely, the role of solitons in transport phenomena in conducting polymers. There are two important aspects to this problem: (1) the dynamics of solitons in the models of $(CH)_x$; and (2) the manner in which solitons, assuming they exist, appear in the actual conducting polymers.

We have thus far said nothing about the dynamics of solitons in $(CH)_x$. In large part, this is because the neat analytic results available for static solutions simply cannot, for technical reasons, be found for dynamic solitons. Thus, for example, in the microscopic models of $(CH)_x$ (unlike the phenomenological model of Section III) there is no known analytic solution for a moving kink (or polaron). Numerical simulations reveal (16, 30) that this dynamics can be quite complex, with many phonons accompanying a moving soliton and soliton-soliton interactions being important (30).

Thus the simple picture of a soliton moving ballistically along the polymer is too naive, and even within the models one needs much further study of soliton dynamics.

More important than these (essentially technical) problems is the fact that in the real conducting polymer material the charged solitons are actually _bound_ ("pinned") by Coulomb attraction to the dopants that formed them. Hence they cannot contribute to the DC conductivity (at least for electric fields below a large depinning field) by physically moving. Instead, it has been argued (51) that conductivity arises by variable-range electron hopping from (pinned) charged solitons to neutral ones. Whatever the actual mechanism, it is clear that the naive picture of kinks and polarons running up and down the polymeric chains is simply not applicable.

In conclusion let me stress that, despite theoretical uncertainties and experimental complications in real materials, the subject of quasi-one-dimensional conducting polymers remains one of the most exciting areas of condensed matter physics. And perhaps one day these materials _will_ play roles in molecular electronic devices.

NOTES

1. For a more thorough description of synthetic techniques, catalysts, and the resultant morphologies, see the reviews in Refs. (1-6).

2. Since our discussion is motivational, we shall not complicate it by inserting the necessary factors to make all the terms in the energy expression have consistent dimensions. In the sections on the microscopic models, we will be precise regarding dimensions.

3. Note that this particular reduction holds only for states with $\varepsilon_n \cong \Delta_o + 0(k^2)$. For states in the valence band, $\varepsilon_n \cong -\Delta_o + 0(k^2)$, and thus $\psi_n^{(2)}$ is the large component and $\psi_n^{(1)}$ is determined in terms of it. Furthermore, for large k (i.e., states far away from the band edges) one clearly cannot make this reduction.

REFERENCES

1. Proceedings of the International Conference on Synthetic Conductors and Superconductors in Low Dimensions, Les Arcs, France, December 1982, J. Phys. Colloq. (to be published, 1983).

2. Proceedings of the International Conference on Low-Dimensional Conductors, Boulder, Colorado, August 1981, Mol. Cryst. Liq. Cryst. 77, 1981.

3. Physics in One Dimension, eds. J. Bernasconi and T. Schneider (Springer Verlag, 1981).

4. A. J. Heeger and A. G. MacDiarmid, in The Physics and Chemistry of Low Dimensional Solids, ed. L. Alcácer (Reidel, 1980), pp. 353-391.

5. S. Etemad, A. J. Heeger, and A. G. MacDiarmid, Ann. Rev. Phys. Chem. 33, 443-469 (1982).

6. D. Baeriswyl, G. Harbeke, H. Kiess, and W. Meyer, Chapter 7 in Electronic Properties of Polymers, eds. J. Mort and G. Pfister (Wiley, 1982).

7. M. J. Rice, Phys. Lett. 71A, 152 (1979); M. J. Rice and J. Timonen, Phys. Lett. 73A, 368 (1979); E. J. Mele and M. J. Rice, Chemica Scripta 17, 21 (1981).

8. A. Kotani, J. Phys. Soc. Japan 42 408 and 416 (1977).

9. S. A. Brazovskii, JETP Lett. 28, 606 (1978) (trans. of Pisma ZhETF 28, 656 (1978)); Sov. Phys. JETP 51, 342 (1980) (trans. of ZhETF 78, 677 (1980)).

10. W. P. Su, J. R. Schrieffer, and A. J. Heeger, Phys. Rev. Lett. 42, 1698 (1979); Phys. Rev. B22, 2099 (1980).

11. T. Holstein, Ann. Phys. 8, 325 and 343 (1959); L. Friedman and T. Holstein, Ann. Phys. 21, 494 (1963); D. Emin and T. Holstein, Ann. Phys. 53, 439 (1969).

12. S. A. Brazovskii and N. N. Kirova, JETP Lett. 33, 4 (1981) (trans. of Pisma ZhETF 33, 6 (1981)).

13. See the contribution of D. Bott et al. in Ref. 1.

14. V. N. Ginzburg and L. D. Landau, ZhETF 20, 1064 (1950) (in Russian); discussed at length in A. L. Fetter and J. D. Walecka Quantum Theory of Many-Particle Systems (McGraw-Hill, 1971).

15. For a discussion of the various contexts in which this equation has arisen, see, for example, D. K. Campbell, J. F. Schonfeld, and C. A. Wingate, Physica D (to be published, 1983).

16. W. P. Su and J. R. Schrieffer, Proc. Nat. Acad. Sci. 77, 5526 (Physics) (1980).

17. H. Takayama, Y. R. Lin-liu, and K. Maki, Phys. Rev. B21, 2388
     (1980); J. A. Krumhansl, B. Horovitz, and A. J. Heeger, Solid
     State Commun. 34, 945 (1980); B. Horovitz, Solid State Commun.
     34, 61 (1980) and Phys. Rev. Lett. 46, 742 (1981).

18. S. Mazumdar and S. Dixit, Phys. Rev. Lett. (to be published,
     1983); J. Hirsch, Phys. Rev. Lett. (to be published, 1983).

19. R. E. Peierls, Quantum Theory of Solids (Clarendon Press,
     Oxford, 1955); D. Allender, J. W. Bray, and J. Boudreau,
     Phys. Rev. B9, 119 (1974).

20. See contribution of W. P. Su, S. Kivelson, and J. R. Schrieffer
     in Ref. 3.

21. D. K. Campbell and A. R. Bishop, Phys. Rev. B24, 4859 (1981)
     and Nuc. Phys. B200[FS4], 297 (1982).

22. S. Kivelson and J. R. Schrieffer, Phys. Rev. B25, 6447 (1982).

23. R. Jackiw and C. Rebbi, Phys. Rev. D13, 3398 (1976); R. Jackiw
     and J. R. Schrieffer, Nuc. Phys. B190[F53], 253 (1981); W. P.
     Su and J. R. Schrieffer, Phys. Rev. Lett. 46, 738 (1981); M.
     J. Rice and E. J. Mele, Phys. Rev. B25, 1339 (1982).

24. I. V. Krive and A. S. Rozhavskii, JETP Lett. 31, 610 (1981)
     (trans. of Pisma ZhETF 31, 647 (1980)).

25. J. L. Bredas, R. R. Chance, and R. Silbey, Mol. Cryst. Liq.
     Cryst. 77, 319 (1981).

26. M. J. Rice, Phys. Rev. Lett. 37, 36 (1976).

27. A. R. Bishop and D. K. Campbell, in Nonlinear Problems:
     Present and Future, eds. A. R. Bishop, D. K. Campbell, and
     B. Nicolaenko (North Holland, 1982) p. 195.

28. I. V. Krive and A. S. Rozhavskii, Sov. J. Low Temp. Phys. 7,
     449 (1981) (trans. of Fiz. Nizk. Temp. 7, 921 (1981)).

29. D. K. Campbell, A. R. Bishop, and K. Fesser, Phys. Rev. B 26,
     6862 (1982).

30. G. Whitfield and P. Shaw, Phys. Rev. B14, 3346 (1976).

31. For an early but extensive review see A. C. Scott, F. Y. F.
     Chu, and D. W. McLaughlin, Proc. IEEE 61, 1443 (1973).

32. For a pedagogical introduction and survey, see A. C. Scott,
     Phys. Rev. A 26, 578 (1982).

33. I. B. Goldberg, H. R. Crowe, P. R. Newman, A. J. Heeger, and
     A. G. MacDiarmid, J. Chem. Phys. 70, 1132 (1979).

34. B. R. Weinberger, E. Ehrenfreund, A. Pron, A. J. Heeger, and
     A. G. MacDiarmid, J. Chem. Phys. 72, 4749 (1980).

35. B. Francois, M. Bernard, and J. J. André, J. Chem. Phys. 75,
     4142 (1981).

36. S. Ikehata, J. Kaufer, T. Woerner, A. Pron, M. A. Druy, A. Sivak, A. J. Heeger, and A. G. MacDiarmid, Phys. Rev. Lett. 45, 1123 (1980).

37. L. Lauchlan, S. Etemad, T.-C. Chung, A. J. Heeger, and A. G. MacDiarmid, Phys. Rev. B24, 3701 (1981).

38. N. Suzuki, M. Ozaki, S. Etemad, A. J. Heeger, and A. G. MacDiarmid, Phys. Rev. Lett. 45, 1209 (1980).

39. S. Kivelson, T.-K. Lee, Y. R. Lin-liu, I. Peschel, and Lu Yu, Phys. Rev. B25, 4173 (1982).

40. K. Fesser, A. R. Bishop, and D. K. Campbell, Phys. Rev. B 27, 4804 (1983).

41. E. J. Mele and M. J. Rice, Phys. Rev. Lett 45, 926 (1980).

42. B. Horovitz, Solid St. Commun. 41, 729 (1982).

43. J. Orenstein and G. L. Baker, Phys. Rev. Lett. 49, 1043 (1982).

44. See the contribution of Z. Vardeny, J. Orenstein, and G. L. Baker to Ref. (1).

45. S. Feldblum, J. H. Kaufman, S. Etemad, A. J. Heeger, T.-C. Chung and A. G. MacDiarmid, Phys. Rev. B 26, 815 (1982).

46. See the contribution of S. Etemad to Ref. (1).

47. S. Etemad, A. Pron, A. J. Heeger, A. G. MacDiarmid, E. J. Mele, and M. J. Rice, Phys. Rev. B 23, 5137 (1981).

48. See the contribution of J. R. Schrieffer to Ref. (1).

49. See, for example, Y. Tomkiewiez, T. D. Schultz, H. B. Brom, A. R. Taranko, T. C. Clarke, and G. B. Street, Phys. Rev. B 24, 4348 (1981).

50. P. Lomdahl, A. R. Bishop, and D. K. Campbell (in preparation).

51. S. Kivelson, Phys. Rev. B 25, 3798 (1982).

8

Soliton Switching and Its Implications for Molecular Electronics

F.L. Carter, A. Schultz and D. Duckworth / Naval Research
Laboratory, Washington, DC

The maximum velocity of a soliton down a conjugated chain is
unfortunately limited to the velocity of sound, however, other
soliton characteristics, such as i: low energy dispersion and ii:
conformation changes, (bond distance) gives solitons far reaching
implications for molecular electronics. To orient the reader,
the soliton for conjugated systems, along with the polaron, will
be discussed in simple valence bond terms. The possible motion of
solitons pass saturated carbon atom defects and past other
solitons will then be considered.

A general scheme for testing soliton generation by
capacitance measurements is then offered. A microlaser output
device, soliton driven, will then be considered along with a
related soliton amplifier. After noting the possibility of
synthesizing a soliton driven mechanical actuator, the remainder
of the paper is devoted to parallel arrays of soliton driven

finite-state machines prepared from soliton valves. In parallel
arrays one might want to consider the simultaneous action of two
or more solitons, however, here we will restrict ourselves to the
completed and successive actions of single solitons.

THE SOLITON PSUEDOPARTICLE

The term "soliton" is derived from the phrase "soliton wave"
first described by the hydrodynamicist J. Scott-Russell almost
150 years ago and has come to imply a non-linear phenomenon
involving the non-dispersive transport of energy in a dispersive
medium. An excellent review of the general concept has been
provided by A.C. Scott, F.Y.F. Chu and D.W. McLaughlin (1). In
general, and in molecular systems in particular, soliton
propagation has a maximum velocity of the speed of sound but its
velocity is often only a fraction of its maximum. At least two
modes of soliton phenomena have been postulated for molecular
systems. A.S. Davydov and N.I. Kislukha (2) suggested that
soliton propagation down the α-helix through the amide-hydrogen
bond system was a method for sending signals in biological
systems originating from ATP bond breaking. This possibility has
been supported by the numerical studies of A. Scott and coworkers
(3). The second mode of soliton propagation is in conjugated
systems like trans-poly-acetylene (4) and this soliton will form
the basis for this paper.

Before proceeding, however, the reader should note: 1) the
examples offered here are primarily conceptual in nature and
currently are untested; 2) nevertheless, they have intuitive
bases that are firmly and broadly supported by the tetravalency
of carbon and experimental and theoretical reaction kinetics; and
3) if in their simplicity, these examples and concepts were to be
found lacking, the basis for conformational changes in MEDs would
not be weakened.

As will be indicated in the succeeding sections on soliton switches and valves the conformation changes induced by the passage of solitons are a powerful source of mechanisms for an MED technology. Phenomenologically a soliton in $(CH)_x$ can be readily handled as a psuedoparticle of constant energy and momentum. Here we note that the important result of the propagation of the soliton through a conjugated system is the exchange single and double bonds. (A conjugated system is one in which single and double bonds alternate throughout.) In conjugated systems solitons may be positively charged, radical, or negatively charged according to the number of unbonded electrons in a p-π carbon orbital of zero, one, or two, respectively, with the corresponding spins of 0, 1/2, and 0. In a motionless state charged solitons are well known to chemists as carbonium (+) and carbanium (-) ions.

UP AND DOWN SOLITONS

The exchange of single and double bonds effected by a moving soliton is indicated in Fig. 1a where the + charged soliton is moving to the right towards a section of trans-$(CH)_x$ whose double bonds have a positive slope. However, it leaves behind (to the left) trans-$(CH)_x$ whose double bonds have a negative slope.

In this behavior a propagating soliton is like a moving domain wall in a crystal. However, it is of interest that the center of the moving soliton (marked by a plus sign in Fig. 1a) moves to the right two carbons at a time and is always associated with the upper carbons in $(CH)_x$ and not with the lower carbons (5,6). This soliton we call an "UP" soliton.

There is another important point that is evident from the chemistry but not from the physics (6). That is, an UP soliton can propagate only to the right if the slope of the double bonds is positive as in Fig. 1a (and only to the left if the double

a.    + soliton

b.    average
            solitary wave   energy

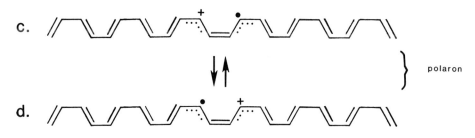   } polaron

Fig.1. The soliton and polaron in a conjugated $(CH)_x$ chain are
        like an undispersed solitary wave but only soliton propagation
        results in a permanent conformation change. Parts c and d are
        but two of several valence bond representatives of a charged
        radical polaron.

bond slope is negative). Similarly, a DOWN soliton can propagate

only to the right if the slope of the double bonds is negative,

etc. This set of restraints is the direct result of the

tetravalency of the carbon atom. (Conceptionally, a high energy

soliton could propagate via a cyclopropyl radical and obviate

these restrictions, however, such a soliton is less likely.)

Carbon can have a valency of four or less but never higher as it

is not energetically feasible.

"END" EFFECTS

Recent x-ray data for trans-$(CH)_x$ suggest that the bond length

difference Δ between a single bond and a double bond may be as

small as $\Delta = 0.07$ (7) or as large as $\Delta = 0.10$ (8). The smaller

difference may be taken as a minimum value since twinning (now
well established in $(CH)_x$) and the statistical averaging
characteristics of the x-ray technique can act together to give a
diminished observed Δ value. Nevertheless, it has been known for
a long time that as the conjugation in polyenes increases the
value of Δ diminishes at the chain center. On the other hand,
one can say that an end effect exists in polyenes such that Δ
increases to $\Delta = 1.54 - 1.23 = 0.21$ A at the termination of the
conjugation. Such "end" effects are important because they imply
a change in the behavior of the soliton as it approaches a chain
end or a defect (6). The results of W.P. Su, J.R. Schrieffer and
A.J. Heeger (4) suggest the effective soliton psuedoparticle mass
is proportional to $m_e \propto (\mu)^2/(a\ell)^2$, where $\mu \sim \sqrt{3\Delta/4}$, a is the
chain repeat distance, and ℓ is the number of carbons in the
soliton. The soliton length is then $a\ell$. Given $\mu = 0.03$ A, the
soliton length is appreciable involving about 14 carbon atoms.
However, as a soliton approaches an sp^3 defect or chain end its
effective mass increases due to the "end" effect and its
effective length diminishes as a pseudoparticle whose momentum is
constant. Therefore upon approaching a defect a soliton not only
becomes a more appropriate size to interact with a saturated
carbon defect but also slows down appreciably. This change in
velocity at the point of impact with the defect may permit the
hydrogen atom transfer as is indicated in Fig. 2. Note that the
hydrogen atom moves in the opposite direction as the soliton and
that the soliton changes from an UP to a DOWN soliton. The
hydrogens on the saturated carbon defect are activated by the two
adjacent double bonds. In a manner similar to this 1,2 hydrogen
shift a 1.3 hydrogen shift could occur for a hydrogen on an
adjacent pair of saturated carbons and enable a soliton to pass
such a defect (6). In such a case a UP soliton stays an UP
soliton, etc. If solitons are able to pass single and double
defects then one speculates on the nature of the soliton that

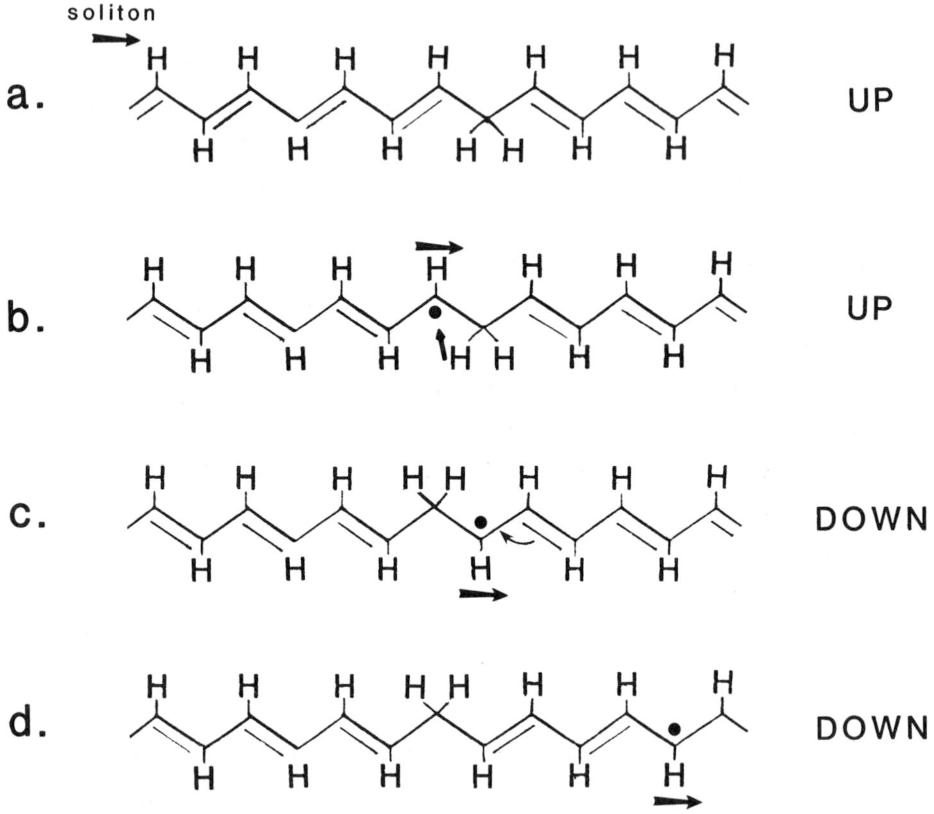

Fig. 2. A soliton might propagate past a sp^3 carbon defect via a
hydrogen radical 1,2 shift. In the process the UP soliton would
be converted into a DOWN soliton. In Fig. 2 all hydrogens are
explicitly shown while in Figs. 1 and 3 their presence is
understood.

might transport itself in a polymer of periodic defects where a
hydrogen would move one or two atoms per soliton passage.

Fig. 3 shows that two different kinds of solitons might pass
one another on the same molecular wire, in this case, changing
between an UP radical soliton and a charged DOWN one.

TESTING SOLITON GENERATION

While solitons might be generated by photoabsorption, such a
technique does not have the spatial resolution generally desired

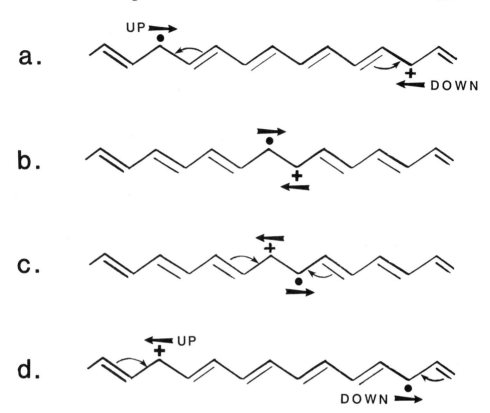

Fig.3. The chemical structures offered in this figure suggests that
      a radical soliton can pass one another by exchanging positions at
      adjacent carbon atoms.

for control of molecular electronic devices. An alternative
method for the generation of solitons that is more amenable to
MED technology has been suggested (9) and involves proton
tunneling in a electric field. This suggestion might be tested
as indicated in Figs. 4 and 5.

      As before, in Fig. 4, when the electric field E is
sufficiently strong the phenol hydrogen will tunnel as a proton
to the pyridyl nitrogen generating a charged pair of incipient
solitons (Fig. 4b) which will propagate through the polyacetylene

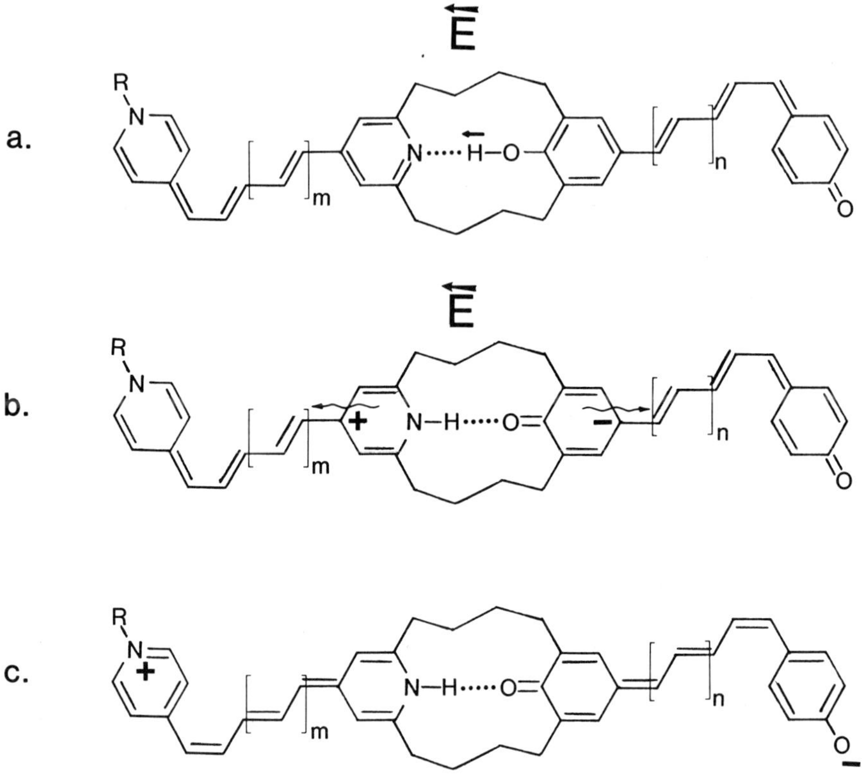

Fig.4.   Proton tunneling in a hydrogen bond in the presence of an
    electric field might be employed as a generator of charged
    soliton pairs.   Such a molecule as above with m ≈ n ≈ 12 in a
    Langmuir-Blodgett film arrangement and in the presence of a
    photon field could be used to test soliton generation.

chains to the chromophores terminating each chain (Fig. 4c).   If
a Langmuir-Blodgett or Sagiv film can be made of long chain
molecules as indicated in Fig. 5, then proton tunneling could be
readily studied as a function of electric field and incident
photon energy by observing either a change in capacitance of the
film or a change in the light absorption of the terminating
chromophores.   Clearly, in the presence of a varying electric
field such a film could be used as a color sensitive optical
detector by noting the change in capacitance as a function of

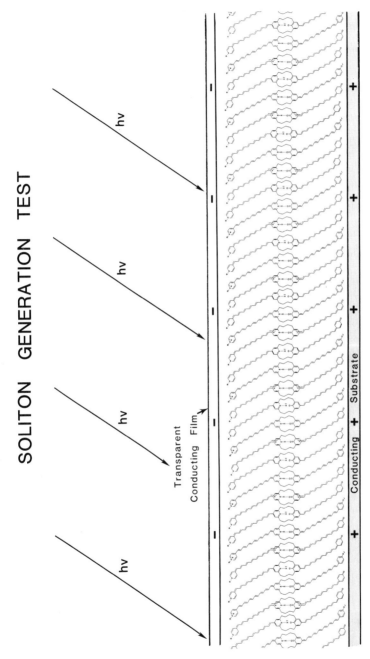

Fig. 5. Proton generation of charged solitons can be tested by observing capacitance changes as a function of field and photon energy in the sandwich film arrangement indicated here.

field. If the film were to be reticulated and the capacitance of each cell were monitored one would have the equivalent of a monolayer camera film.

In part, soliton generation can be seen as the inverse process of a soliton meeting an sp^3 defect as discussed earlier. As the hydrogen bonded proton starts tunneling to the pyridyl nitrogen the effective soliton masses are high and their lengths in the aromatic rings are short (Fig. 4b), however, as they enter the trans-$(CH)_x$ chains they become more typical solitons until they arrive at the chromophores.

The testing scheme of Fig. 5 can be made position-sensitive by either reticulating the conducting capacitor films into small individual elements or by not terminating all the polyacetylene chains with chromophores but letting some of them penetrate the conducting films to computing units or automata as will be discussed later. If the film of Fig. 5 is reticulated and the voltage ramped across each subdivision, then the capacitance will undergo very large changes when the electric field energy plus the energy of the incident photons is equal to the proton tunneling energy (more properly to the soliton generation energy). That is, we will have a monolayer position-sensitive camera film. In a latter section we will think of this device as input for a 3-d array of cellular automata.

MOLECULAR MICROLASER

In addition to soliton linked input devices, soliton linked output devices are of interest.

A highly directional optical output was suggested in 1979 (10a, p. 147) in the form of a linear array of oriented excited molecular chromophores. However, a technique for producing the

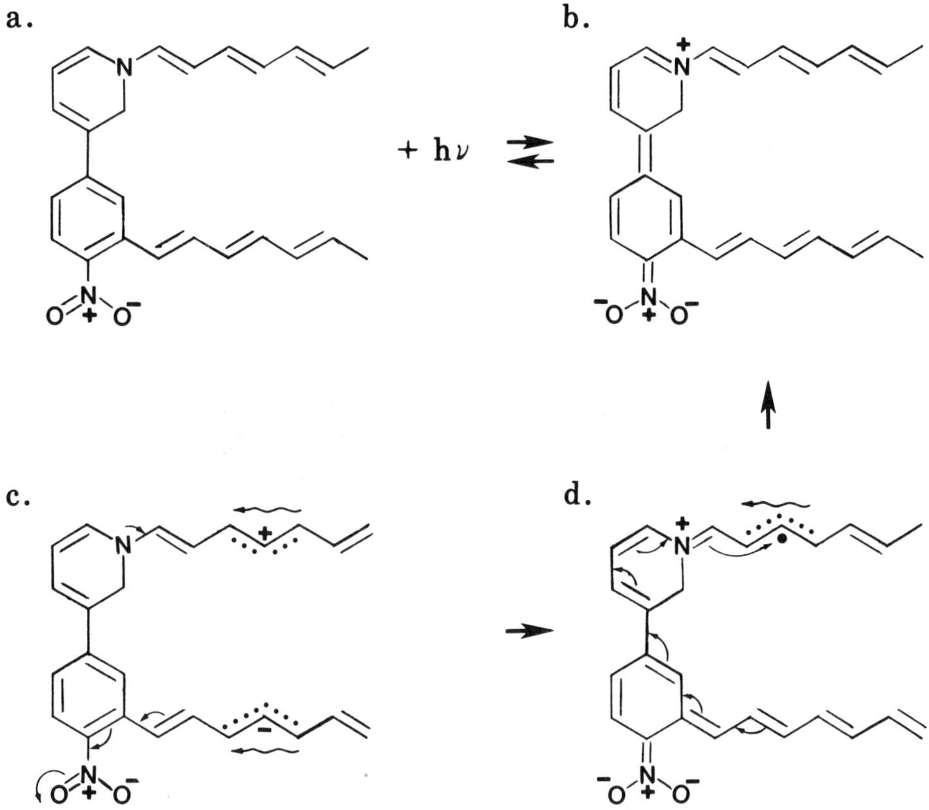

Fig.6. Two charged solitons are employed to activate an embedded chromophore. The photon release is triggered here by a radical soliton.

excited state and for triggering the photon release was not discussed. This situation was remedied (11) as indicated in Fig. 6. In part a to part b of the figure we illustrate the photoactivation of a chromophore to an excited zwitterionic state. In the series c to d to b we see: c) the excitation of the chromphores by two approaching charged solitons; d) the rearrangement of the internal bonding conformation of the excited chromophore by the approaching radical soliton; and in b) the return of the excited chromophore to the ground state with the

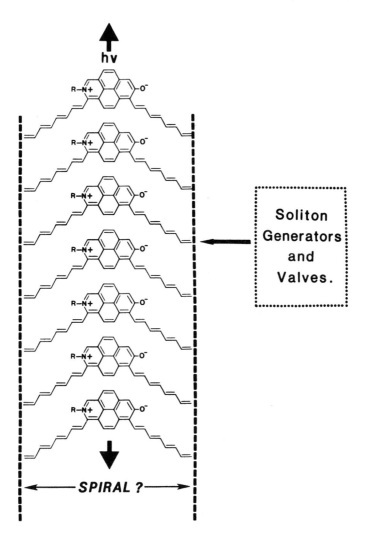

Fig. 7. An array of soliton activated chromophores might behave as a highly directional microlaser.

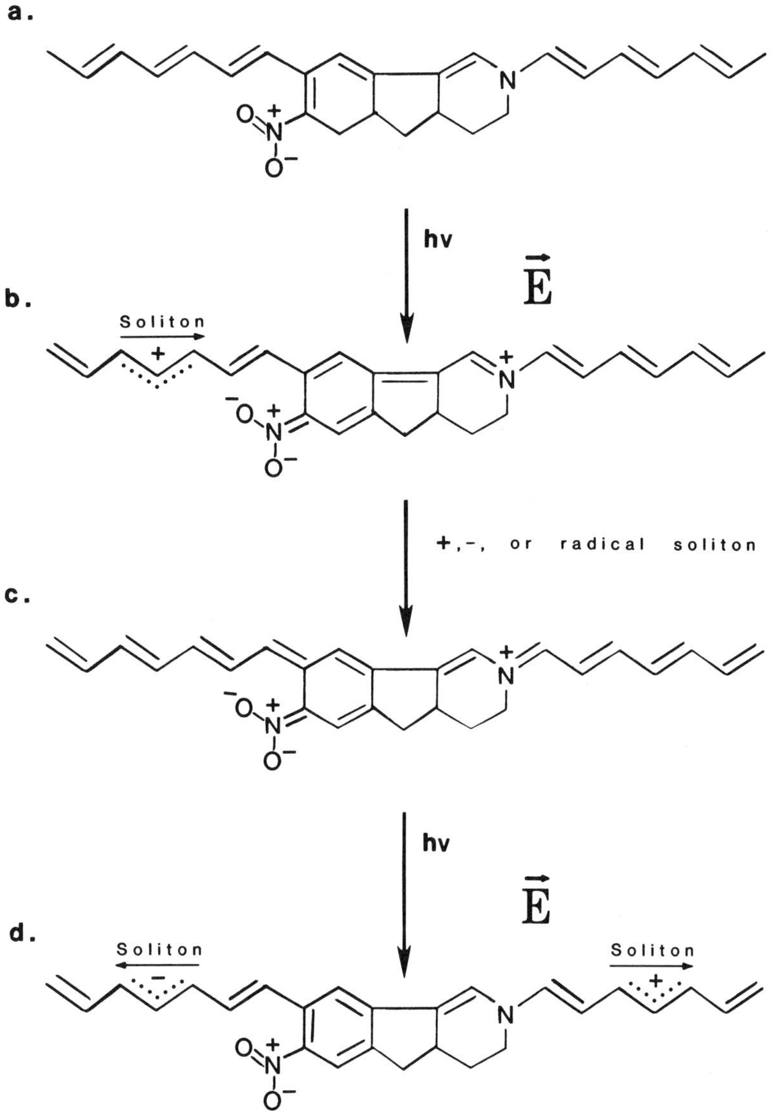

a.

b.

c.

d.

Fig. 8. Soliton amplification is obtained by the passage of a
soliton (positive in Fig. 8b) through a photoactivated
chromophore embedded in a conjugated system. In the presence of
an electric field the rearranged photoactivated chromophore can
emit two soliton as in part d.

release of the photon. Thus, in Fig. 6 the chromophores are
excited by two charged solitons and photon release triggered by a
radical soliton.

Now by combining the directional output concepts suggested
in Ref. 10a with the soliton excitation and triggering
considerations offered above we have the components of a
soliton powered laser. By spiraling the chromophores about the
vertical axis one may be able to achieve an enhanced
directionality or circular polarization.

SOLITON AMPLIFICATION

In order to argue that soliton phenomenon might play an important
part in molecular electronics it is necessary to formulate a
method of soliton amplification. This might be found in a
reorganization of the soliton-triggered light emitter just
described. In Fig. 8 we have another chromophore imbedded in a
trans-polyacetylene chain. After putting this chromophore in an
excited state with the light of the correct wavelength (Fig. 8a),
a charged soliton, the one to be amplified, is sent through it
(Fig. 8b and c). This soliton passage rearranges the bonding of
the chromophore so that the electric field can extract two more
charged solitons: another positive one departing to the right,
and a negative one departing to the left (Fig. 8d). The original
soliton to be amplified (Fig. 8b, left) can either be positive,
negative or radical. The only feature required is the
reorganization of the chromophore bonding after photoactivation.

The amplification provided in Fig. 8 amounts to 200%. Via
the use of an array of such devices any desired amplification
level can be obtained. The field of solitons will not be uniform
however. As indicated above this type amplifier can be used to
change any soliton (radical or charged) into one of a particular
charge.

9. This figure of a mechanical actuator illustrates the use of soliton propagation to produce a motion in groups A and B of about 2.0 Å. Such mechanical motion in the actuator group will also produce a delay in the soliton propagation.

THE SOLITON MECHANICAL ACTUATOR

K.E. Drexler, in both this conference (12) and earlier (13), has drawn our attention to the mechanical aspects of molecular machinery. In this section we note how soliton transport might be used to actuate such motion. In the earlier article (13) on "molecular" enginering, Drexler reminds us of R. Feynman's delightful paper entitled, "There's Plenty of Room at the Bottom" (14). Feynman proposes a series of miniaturizations in which small machines are used to build even smaller machines, eventually down to the molecular level. By making use of the principles of self-organization and self-synthesis as discussed herein (15), the chemist may be able to leapfrog the purely mechanical approach suggested by Feynman and go directly to the molecular level. Therefore, in the spirit of a mechanical device, we suggest below a mechanical actuator controlled by the passage of a soliton.

The actuator illustrated in Fig. 9 is a combination of two ideas: The first we have seen earlier in the discussion of "smart" molecules by Haddon and Stillinger (16), where the information state depends on the location of the labile groups L_i, which are not necessarily hydrogen. The second concept arose earlier in the consideration of what happens when a soliton in trans-polyacetylene meets one or two sp^3 (or $-CH_2-$) defects. If a hydrogen undergoes a 1,2 shift (or 1,3 shift for two successive saturated atoms), then the soliton can pass the defect, although it may be temporarily delayed.

The mechanical device of Fig. 9a is illustrated as being actuated by a positive soliton approaching from the left. When the plus charge reaches the pyridyl nitrogen on the left, it attracts the R group, possibly ($>CH-$), from the opposing nitrogen and the soliton continues to the right, as in Fig. 9b. The motion of the R group by at least 2.0 Å to the left drags with it the groups A and B. Presumably, the movement or position of either A or B controls or actuates some other unspecified event at a later date. Thus, we have seen how the passage of a soliton in a conjugated system can give rise to a definite group relocation by the breaking of one bond and the establishment of a second bond.

SOLITON VALVES AND FINITE-STATE MACHINES
Soliton switching can occur in a variety of direct and indirect ways. As originally proposed in the 1st Molecular Electronic Device (MED) Workshop in 1981 (17a) it involved the changes

induced on one conjugated chain by a soliton transport in a cross-conjugated chain. The packing density of switches in such systems could be as high as 10^{18} gates/cm^3. The soliton valve is also a direct soliton switch and was introduced at the same time (17a). This switch, the valve, will form the basis or point of initiation for the remainder of this paper. However, before proceeding along that route, we would like to mention two less direct soliton switches. One was introduced at this proceedings by a graduate student in EE, i.e. P.M. Groves (18), and this switch avoids some of the objections to soliton switching in cross-conjugated systems. The forth switch proposed involves moving the charge in a cyanine dye group by the motion of a nearly charged soliton (19).

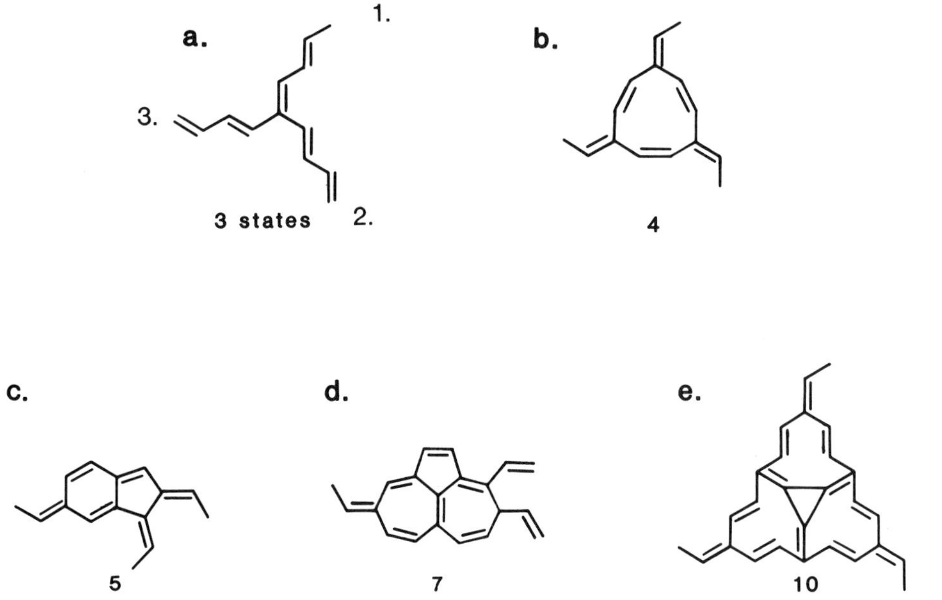

Fig. 10. The valve of Fig. 10a is used in various cyclic configurations, Fig. 10b-e, to obtain different number of states in finite
state cellular automata via soliton propagation down their three
$(CH)_x$ chains. Strain can be relieved in these configurations by
ring expansion, however, the number of bonds between valves
should be kept even or odd as in Fig. 10c to Fig. 10b.

Let us now return to soliton valves and indicate how a variety of finite-state machines can be assembled from configurations of valves. As illustrated in Fig. 10a a valve consists of a single carbon atom connected to three semi-infinite chains of trans-$(CH)_x$. For the simple valve each soliton propagation corresponds to group operation (i.e. a 120° rotation) in that the double bond at the central carbon atom is rotated to a different chain. However, soliton propagation between one pair of the chains is precluded by the presence of two adjacent single bonds (between the left and bottom chains in Fig. 10a), hence the soliton valve is related to, but not isomorphous with, a common three-way valve. A cyclic configuration of three valves separated by an odd number of conjugate bonds (Figs. 10b and 10c) is also of special interest in that every soliton propagation is a group operation with the corresponding point group being D_2. This group has the symmetry of an object with three perpendicular two-fold axes. The appropriate group table is in the upper left corner of Table 1, outlined by dotted lines.

Table 1
Group Table for Three Valves, Bicyclic Configuration

	E	ab	ac	bc	a/b	a/c	b/c	R
E	E	ab	ac	bc	a/b	a/c	b/c	R
ab	ab	E	bc	ac	R	b/c	a/c	a/b
ac	ac	bc	E	ab	b/c	R	a/b	a/c
bc	bc	ac	ab	E	a/c	a/b	R	b/c
a/b	a/b	R	b/c	a/c	E	bc	ac	ab
a/c	a/c	b/c	R	a/b	bc	E	ab	ac
b/c	b/c	a/c	a/b	R	ac	ab	E	bc
R	R	a/b	a/c	b/c	ab	ac	bc	E

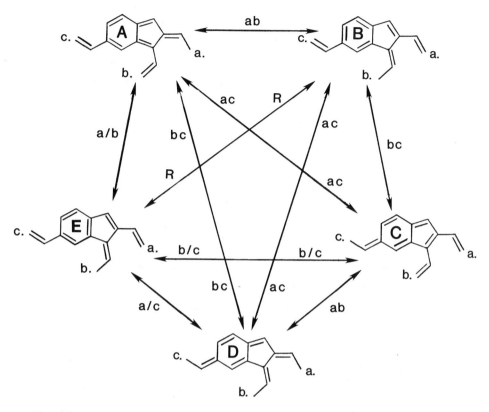

Fig. 11. All five of the different states available to this bicyclic
configuration of valves are readily accessible by soliton
propagation.

While the simple valve has three different states and the
cyclic configurations of three valves (Figs. 10b and 10c) has
four states, the bicyclic configuration of five valves (Fig. 10d)
has five states as illustrated in Fig. 11. Four of these states
are isomorphous with those of Fig. 10b, but the fifth state (E),
illustrated in Fig. 11 involves changing the bonding conjugation
in the common edge of the two rings, (i.e. compare states, B and
E). Soliton paths involving this common edge are indicated by the
slash, as in a/b. The corresponding group table is shown in
Table 1 and contains as a subgroup the group for Fig. 10b. All
the operations except R correspond to soliton propagations.

There is a complication at this time and that is indicated
by the ambiguity existing when a soliton in the simple valve of
Fig. 10a approaches the branching carbon from the upper chain.
This ambiguity might be decided on the basis of pseudoparticle
momentum, UPness or DOWNness, or charge considerations, but the
preferred path is not clear at this time. The passage of a
soliton through a valve, however, does reverse its UPness or
DOWNness.

The results summarized in Table 1 suggest that configurations of valves could not only serve as elements (states) in
non-binary finite-state machines but that the results of soliton
propagations can correspond to complex group operations. In
other words, configurations of valves might represent the
smallest computing element or cellular automata. This will be
discussed further below.

Finally, we note that the tetracyclic configuration of nine
valves (Fig. 10e) has ten different states as indicated by Fig.
12. The path of soliton propagation is indicated by the dotted
lines. For a carbon-based system some of the indicated bond
angles are highly strained for the configuration of Fig. 10e
indicating that the number of bonds, N, between valves must be
increased while retaining the oddness or evenness of N, as in the
relationship between Figs. 10c and 10b. This configuration (Fig.
12) might be of interest in a decimal-based system. As noted
earlier rather different circuit diagrams for soliton switching
have been worked out by Groves (18) even with the above
ambiguity; these include Boolean gates, inverters, and a fan
memory.

Identification of all possible finite state configurations
composed of different numbers and arrangements of valves can be

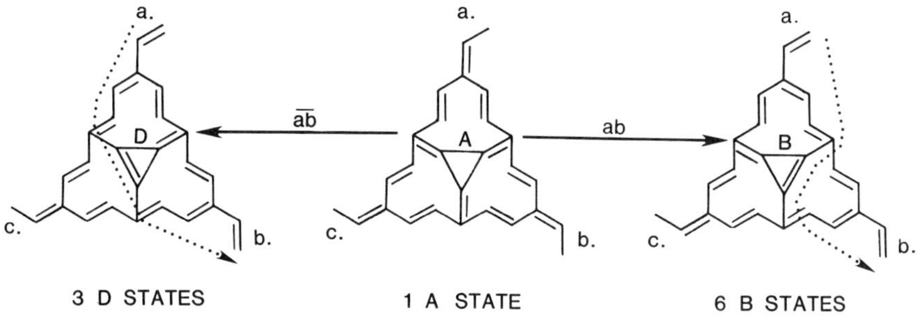

3 D STATES   1 A STATE   6 B STATES

Fig. 12. The three types of states available by different soliton
paths (dotted lines) are illustrated. A total of 10 states are
all accessible by soliton propagation. However, preselecting the
soliton path constitutes an unsolved problem.

systemized of course. An 8-state configuration identified by a
PASCAL program is illustrated in Fig. 13. The number sequences
joining different configurations is the soliton path through the
numbered valves. As in Fig. 10 and Fig. 12 the structure shown
is not necessarily a realistic one as it may be too strained.
However, strain may be relieved by increasing the number of
conjugated carbons between the numbered valves. The isomorphism
is not changed provided this is done in a way that maintains the
evenness or oddness of the conjugation between valves.

Below it is interesting to consider the implications of
large scale arrays of finite state machines, cellular automata or
other processors, however we will note that not only do
configurations of valves lead to new computing elements but that
solitons might play special roles intercommunicating between
molecular switching elements.

A VALVE SORTER AND MEMORY ELEMENT
Slavish imitation of semiconductor logic is not necessary. This
is indicated by the discovery by one of us (D.D.) that valves can
be configured into a sorter and memory device. A cyclic
version of this is indicated in Fig. 14 where the numbers 1

8-STATE CONFIGURATION

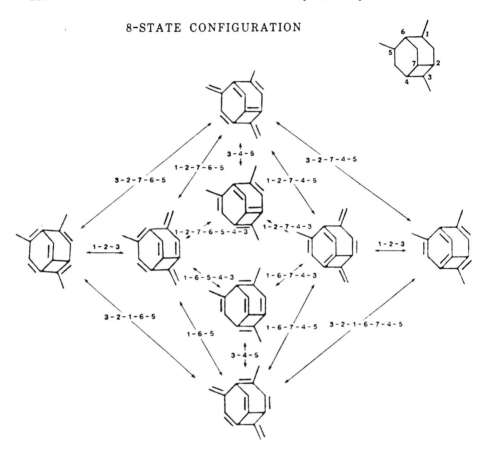

Fig. 13. Carbon atoms functioning as valves are numbered in the upper right hand corner. In contrast to the other valve configurations this set was identified by a computer program written by one of us (DD).

through 10 (and X) symbolizes long chains of polyacetylenes. This device operates as follows: assume a soliton is sent to the device down one of the chains, for example number 3, as in Fig. 14a. When the soliton reaches the sorter it cannot turn right because of two adjacent single bonds: it can only go left. Similarly, it cannot escape via the lower numbered chains because

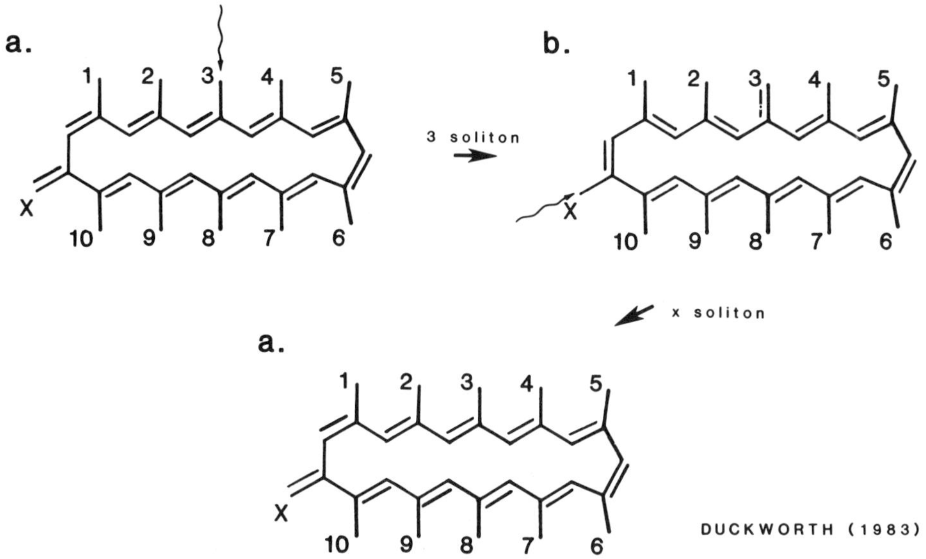

Fig.14. This soliton valve sorter and memory element remembers a
soliton's "arrival point" (number 3 in Fig. 14a) and returns a
soliton to that point from x (Fig. 14b) thereby restoring the
sorter to its original condition.

of two adjacent single bonds. The soliton can only escape out
the X chain leaving in its track the configuration modified as in
Fig. 14b.

On the other hand, a soliton sent up the chain X can only
exit out chain 3 (see Fig. 14b) because of the need to follow
conjugated pathways. This restores the memory element or sorter
to its original configuration, compare Fig. 14a and 14c.

In short, this device is the molecular analogue of the
SWITCH statement in ALGOL or the CASE statement in PASCAL.
Furthermore, it is a rather elegant but economical method of
achieving the high level functions.

CELLULAR AUTOMATA

The prospects of a future coupling between applied cellular
automata, finite state machines and/or molecular electronic
devices is very exciting in terms of both massive parallel
processing and new physical and operational computer
architecture. A few words concerning the concept of cellular
automata is in order. The interested reader is referred to the
recent review of S. Wolfram (20,21) and the paper of Preston (22)
for a computer application of cellular automata in pattern
recognition problems.

The simplest example of cellular automata is a linear row of
two-state cells that communicate only with their nearest
neighbors. What we are interested in is the progress of the cell
contents at each time step. In this case, each cell has only two
states (e.g. 0 or 1, or dead or alive) and what happens at the
next time step depends only on its current state and the state of
its closest neighbors. For example, if we represent a live cell
and its two neighbors on each side as (011) then under the modulo
2 rule (neighbors only) it survives as (1) in the next time
sequence. That is, each automata is a very small commputer, in
this case a finite-state machine of only two states. Given all
possible variations of the states of a cell and its two
neighbors, we have 2^8 = 256 possible rules. One popular and
interesting rule is indicated below:

111	110	101	100	011	010	001	000
0	1	0	1	1	0	1	0

where all possible conditions of the center cell and its two
neighbors are arranged numerically above the bars and the
resultant state of the center cell in the next time sequence is
below the bars. The numerical interpretation of the states below
the bars is 132 to base 8 or 90 to the base 10. This then is

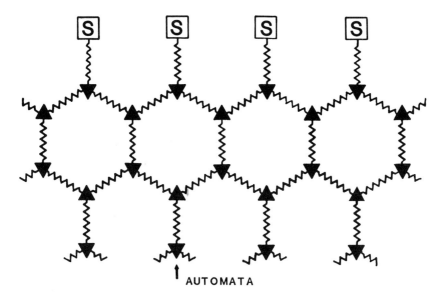

Fig.15. A two-dimensional network array could be formed of the
triangular soliton automata like those in Fig. 10 and Fig. 13.
The S squares indicate a controlled input source of solitons,
possibly an optically stimulated proton-tunnel soliton generator
as in Fig. 25.

Rule 90 according to Wolfram (20) or the modulo 2 rule. If the
restriction that (000) gives only the 0 state (that is, death
begets death) and that left and right symmetry applies, then the
number of cases is reduced by the restriction $a_1a_2a_3a_4a_2a_5a_40$ to
2^5 rules. Even these very simple rules give rise to some very
interesting cases of self-organization which are exhibited as
fractals, topics of concern to much of modern technology. As we
shall see below, however, two- and three-dimensional arrays of
cellular automata are of potentially even greater interest.

MOLECULAR TECHNOLOGY

If one assumes that a molecular technology will develop that
permits one to synthesize in place a single automata, such as
those based on the soliton valves in Fig. 10, then it is clear
that at the same time, one would be able to prepare a
two-dimensional array of cellular automata using the same

reaction cell and reagents. Such an array could involve hundreds
or thousands or tens of thousands of cellular automata on a side.
The concepts of modular chemistry permits one to speak of three
dimensional arrays as well.

The simplest MED automation must be the soliton valve of
Fig. 10a with its three possible states and three input/output
lines. This automaton and the others of Fig. 10 can be used to
form a triangular automata network like that of Fig. 15. If the
interconnecting $(CH)_x$ chains are neglected then these automata
are very small indeed, the valve is a single carbon atom, less
than 10 $Å^3$, the cyclic configurations of three valves, less than
400 $Å^3$. If the volume of an electric tunnel switch is taken as
an upper limit then a practical guide for the volume of a MED
active element might range from $5x10^4$ to $4x10^6$ $Å^3$. If a MED
automaton was 4000 Å on an edge then the automaton could contain
between $16x10^3$ and $13x10^5$ active elements, as noted above. When
it becomes technically feasible to synthetically prepare one such
microprocessor or automaton then it is clear that two- and
three-dimensional arrays of such processors with associated
memory elements can be fabricated. It is significant that modern
x-ray technology is approaching the capability of being able to
determine the structure of a periodic array of such MED automata
on a submillimeter sample (19).

INTERAUTOMATA COMMUNICATION

Two of the problems with current semiconductor computer
technology are (1) that the two-dimensional fabrication
techniques exhaust chip area with the interdevice wires, and (2)
that the parasitic capacitance between these wires severely
limits the effective switching time. However, with cellular
automata arrays interactions are primarily between neighbors
rather than global, hence the number of interconnections are
small and the distances are short, leading to close-packed
components and high computing speeds.

Communication between MED automata could be achieved by electrical conduction along $(SN)_x$ wires with associated potential differences or via optical micro-lasers as discussed above as in Figs. 6 and 7. However, for short distances soliton propagation has definite advantages in spite of its low velocity. As two examples of this, in Figs. 16 and 17, a single chain soliton reverser is employed to modify control groups for electron tunnel switches. (For a discussion of control groups see Ref. 10b or Ref. 24.) In Fig. 16 an approaching soliton drives the positive charge from the control group of one tunnel switch to that of a second switch, then goes around the single chain reverser and returns both tunnel switches to their original states. By controlling the lengths L_1 and L_2 the timing of these events can be controlled within limits.

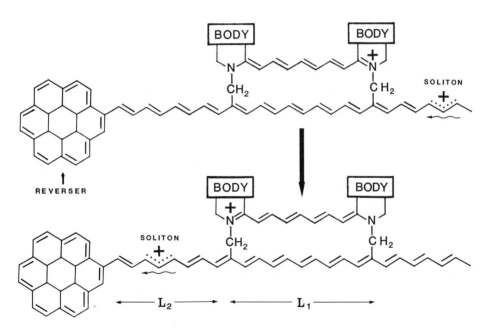

Fig.16. A charged soliton can be used (top) to turn one tunnel switch off and another on if they share a cyanine-dye like control group. After the soliton (bottom) is returned to the right by the soliton reverser, both switches will be restored to their original states. The timing of these events is controlled by the travel lengths L_1 and L_2.

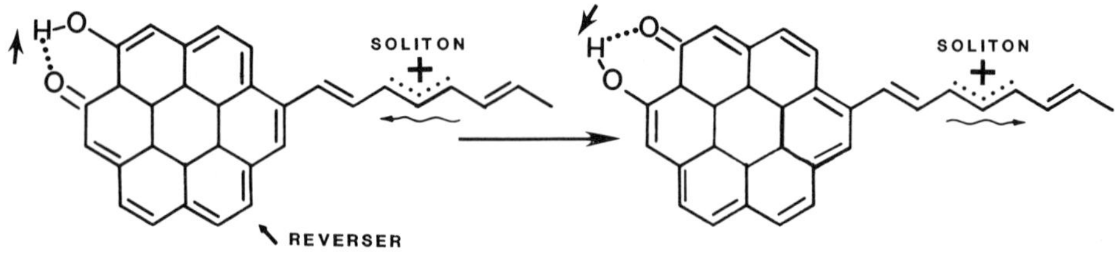

Fig. 17. In this control group, the soliton reverser is part of the
enol-keto tautomer. A second soliton will traverse the reverser
in the opposite direction and restore the original dipole moment
direction.

In the second case (Fig. 17) the control group employs the
dipole field of a hydrogen bond in an enol-keto tautomer.
However, in this case the control group is built as part of the
single chain reverser. The first soliton slows down and flips
the dipole down and the second soliton flips it back to the
original position. These two examples serve to point out that a
single chain of $(CH)_x$ can be used to communicate with
neighboring automata or components at near negligible energy loss
(even the soliton returns). At the high MED densities
considered, such energy loss considerations are very important.

PARALLEL PROCESSING

In Fig. 15 we indicate a triangular array of cellular automata or
finite state computing machines, possibly based on soliton valves
as in Fig. 10. The input to this array is a source of solitons.
A two-dimensional source of such solitons might be the
reticulated monolayer camera film of Fig. 5 based on proton
tunneling. This would permit the possibility of the optical data
being processed in parallel by a three dimensional array of
automata as in Fig. 18. Such a three-dimensional array processor
could reduce data in a manner comparable to the optic nerve, that
is, an increasingly sophisticated analysis as the data and its
processing becomes increasingly remote from the optical source.
Note that each of the parallel processors in Fig. 18 have an

MULTIPLE PROCESSING OF VISUAL DATA

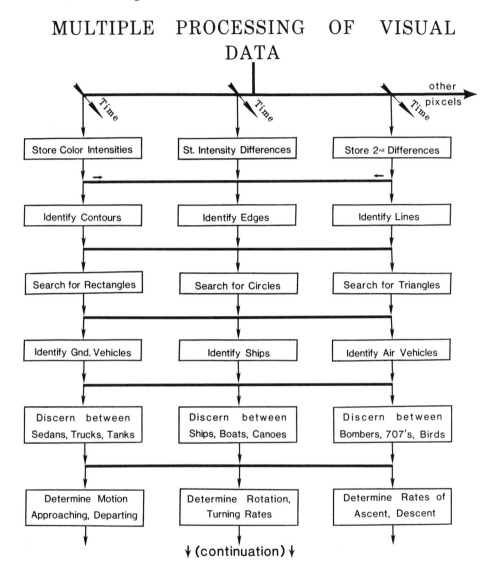

Fig.18. The optical data of a cathode ray tube or other data source like that of Fig. 5 might be rapidly processed and interpreted by a three-dimensional array of automata with increasing processing and memory capabilities.

associated memory capacity as well, and that they may become
larger and more sophisticated processors as the data is refined.

The effectiveness of parallel processing is very clearly
indicated by the abilities of a basketball player under the net.
He can jump, receive, and dunk a basketball all before he returns
to the floor, thanks to the miracle of parallel processing. All
this with a nervous system with a subsonic soliton-like signal
transport system.

Another point to consider is that while conformation
switching times are inherently slower than purely electronic
devices, even with soliton switches (200 Å apart) the switching
times are very easily in the subnanosecond time scale.

REMARKS ON ENERGY

First we might note that unlike some current logics (e.g., TTL)
information stored on devices like the bistable chemical memory
element of Ref. 10b requires no additional energy to maintain the
information once the data is stored. This is because the
potential valleys associated with a particular conformational
state can be designed to be electron volts deep.

An important result of the above is that an accumulation of
the energy tails associated with the switching of nearby
components is unlikely to cause loss of information if the MED is
properly designed. As an illustration, compare the difference in
conjugation associated with the usual enol-keto control group
with perhaps 3 carbons (see Fig. 6 of Ref. 23) with that of Fig.
17 with greater than 17 carbons. Obviously, a small control
group is much more easily inverted by thermal effects than that
of Fig. 17. It is difficult to imagine a switching mechanism for
Fig. 17 that does not involve a soliton. Thus, the passes

associated with the potential valley for the enol-keto system of Fig. 17 is at least 1.3 eV high, roughly sixty times room temperature thermal energy.

This brings us to a consideration of switching energies. While R. Keyes (24) indicates that current computers have a switching energy of 10^8 to 10^{10} kT, C.H. Bennett estimates that RNA prepared DNA at an energy cost of only 20-100 kT per nucleotide (25). Although conformation changes due to soliton propagation are formally dissipationless it is reasonably anticipated that acoustic phonons are generated when a soliton travels through reversers, control groups, valves, and, because of confinement trapping effects, even along lengths of trans-polyacetylene chains (26). Thus, we expect that MED will operate more in the Brownian regime of Bennett than in that of modern computers or the reversible regime of E. Fredkin and T. Toffoli (27).

LOOKING TO THE FUTURE

In summary, we note that switching systems built on soliton phenomenon can achieve very high gate densities like 10^{18} gates/cc and can operate with very low dissipation of energy. Furthermore, it appears that soliton input and output schemes can be devised in addition to the soliton driven molecular analogues of sophisticated mathematical operations. The relatively slow speed of soliton switches can be overcome by parallel processing as demonstrated by the marvelous eye-hand-foot coordination in our athletic youths. As noted above if such molecular switching systems are fabricated in three dimensional periodic arrays then the molecular structure of such main frames can be studied by an extension of conventional x-ray techniques (19) where the entire central processing unit might be the x-ray sample. Another possible new direction in computer architecture is suggested by

the cellular automata work of Barker in this proceedings (28).
The local coupling required by cellular automata is in excellent
accord with the architecture (see Fig. 15) that arises naturally
from soliton finite-state machines or elements as in Fig. 10.
These concepts together suggest future computations might become
almost hardware free by having the computation float in a 'sea of
cellular automata'.

REFERENCE

1. A.C. Scott, F.Y.F. Chu, and D.W. McLaughlin, Proc. IEEE,
    61, 1443 (1973).

2. A.L. Davydov and N.I. Kislukha, Phys. Stat. Sol. (b) 59,
    465 (1972); Sov. Phys. JETP, 44, 571 (1976).

3. A.C. Scott, Phys. Rev. A. 31,3518 (1985); ibid. 26, 578
    (1982); J.M. Hyman, D.W. McLaughlin, and A.C.
    Scott, Physica 3D, 23 (1981).

4. W.P. Su, J.R. Schrieffer and A.J. Heeger, Phys. Rev. Lett.
    42, 1698 (1979); Phys. Rev. B22, 2099 (1980).

5. C.T. White, "Effects of Strained Carbon-Carbon Bonds on the
    Electronic Structure of Polyacetylene", Bulletin of APS,
    Abstract JJ7, Spring Meeting, Washington, DC, 1980; "Some
    Effects of Internal Coordinates on the Properties of
    Non-Simple Metals and Semiconductors", in NRL Program on
    Electroactive Polymers, Second Annual Report, Ed. R.B. Fox,
    NRL Memorandum Report 4335, p. 24.

6. F.L. Carter, "Solitons and SP^3 Defects on Trans-Polyace-
    tylene", in Polymers Electroactifs, Eds. P. Bernier and B.
    Payet, CNRS, at Ecole d'Hiver, Font-Romeu, Jan. 1982, Vol. 1,
    p. 146.

7. C.R. Fincher, Jr., C.E. Chen, A.J. Heeger, A.G. MacDiarmid
    and J. Hastings, Phys. Rev. Lett. 48, 100 (1982).

8. J.B. Lando and M.K. Thakur, "Synthesis and Characterization
    of the Dimer (1,11-Dodecadiyne): Structure of the 8-Poly-
    diacetylene Dimer", in this proceedings.

9. F.L. Carter, "Conformational Switching at the Molecular
    Level", in Molecular Electronic Devices, Ed. F.C. Carter,
    Marcel Dekker, Inc., New York, N.Y. 1982, p.51.

10. (a) F.L. Carter, "Problems and Prospects of Future
    Electroactive Polymers and 'Molecular' Electronic Devices",
    in the NRL Program on Electroactive Polymers, First Annual
    Report, Ed. L.D. Lockhart, Jr., NRL Memo Rpt. 3960, p. 121
    (1979); (b) "Further Considerations on 'Molecular' Electronic
    Devices", Second Annual Report, Ed. R.B. Fox, NRL Memo Rpt.
    4335, p.35 (1980).

11. F.L. Carter, "Toward Computing at the Molecular Level" in
    Microelectronics--Structure and Complexity, Ed. Raymond
    Dingle, Plenum Press, NY, 1983, in press.

12. K.E. Drexler, "Molecular Machinery and Molecular Electronics
    Devices", in this proceedings.

13. K.E. Drexler, Proc. Natl. Acad. Sci., USA 78 (1981) 5275.

14. R. Feynman, "There's Plenty of Room at the Bottom", in
    Miniaturization, Ed. H.D. Gilbert, Reinhold, New York, (1961)
    p. 282.

15. F.L. Carter, "New Directions for Chemists Toward Modular
    Chemistry and Molecular Lithography", in this proceedings.

16. R.C. Haddon and F.H. Stillinger, "Molecular Memory and
    Hydrogen Bonding", in Molecular Electronic Devices, Ed., F.L.
    Carter, Marcel Dekker, Inc., N.Y., N.Y., 1982, p.19.

17. (a) F.L. Carter, Ed., Molecular Electronic Devices, Marcel
    Dekker, NY, NY, Dec. 1982; (b) 2nd International Workshop on
    Molecular Electronic Devices, 13-15 April 1983, NRL,
    Washington, DC, this proceedings.

18. M.P. Groves, "Dynamic Circuit Diagrams for Some Soliton
    Switching Devices", in these proceedings.

19. F.L. Carter, Physica 10D (1984) 175-194.

20. S. Wolfram, Rev. Mod. Phys. 55, 601 (1983).

21. N.H. Packard and S. Wolfram, J. Stat. Phys., (USA) 38,
    901-946 (1985).

22. K. Preston, (IEEE) Computer, January, (1986) p. 36.

23. F.L. Carter, "From Electroactive Polymers to the Molecular
    Electronic Device Computer", in VLSI--Through the 80's and
    Beyond, Ed. Denis McGreivy, IEEE Computer Society,
    Washington, DC, (1982).

24. R.W. Keyes, Proc. IEEE, 69, 267 (1981); Internal J. of
    Theor. Phys., 21, 263 (1982).

25. C.H. Bennett, Intern. J. of Theor. Phys., 21, 905 (1982).

26. W.P. Su and J.R. Schrieffer, Proc. Natl. Acad. Sci. USA,
    77, 5626 (1980).

27. E. Fredkin and T. Toffoli, "Conservative Logic", Mass. Inst.
    of Tech., Report MIT/LCS/TM-197 (1981).

28. J.R. Barker, "Complex Networks in Molecular Electronics and
    Semiconductor Systems", in this proceedings.

9

Dynamic Circuit Diagrams for Some Soliton Switching Devices

Michael P. Groves / Department of Computing Science,
University of Adelaide, Adelaide, South Australia

ABSTRACT

Great interest in Molecular Electronic Devices was shown at the last
WMED. Several possible devices and useful chemical reactions were
mentioned and the problems of manufacture and connections to these
devices were discussed. This paper deals with soliton switching in
polyacetylene. To do this some of the devices mentioned by Forrest
L. Carter in his paper for the first WMED will be discussed and some
new devices will be introduced. These devices or components are then
combined to form Boolean gates and the address selectors and storage
cells of a molecular memory. This paper describes a system where
each bit of information transfer or each switching operation is
accomplished by one or two solitons. Thus in addition to the
extremely high component packing densities found with all molecular
electronic devices this system offers minute power consumption. The
paper describes dynamic circuit diagrams which can show the changing
states of the components. These were found to be a great help in
designing and checking new circuits.

INTRODUCTION

If electronic components are to continue getting smaller a day will
come when they are the size of small molecules (10-100 atoms each).
Because enormous chemical structures are manufactured daily in
biological systems it is not unreasonable to assume man will be able
to construct and connect these Molecular Electronic Devices (MEDs).

The first Med, a rectifier, was proposed by Aviram and
Ratner(1). The last workshop on molecular electronic devices(2)
forms a general background for this paper. In particular, Ulmer's
paper (3) discussed the construction of Meds by genetic means,
McAlear and Wehrung's paper (4) considered the positioning of Meds
on a chip, Carter's paper (5) looked at soliton generation and
switching and Guenzer's paper (6) discussed the reliability problems
of very small devices. Part of a later paper (11) by Carter dealt
with electrical connections between metals and conductive polymers.

SOLITONS IN POLYACETYLENE

Polyacetylene (Fig. 1) consists of a chain of carbon atoms held
together by alternating double and single bonds. Each carbon atom
is also bonded to a hydrogen atom. Polyacetylene has two stable
states, which differ in the position of the alternating double and
single bonds with respect to the atoms. Polyacetylene is highly
suitable for the "wires" to interconnect Meds because it is an
electrical conductor; because of its two state nature, and because
of the tremendous range of conjugated chemical structures.

Solitons where introduced by Rice (7). The mathematical nature
and models for them are discussed in (8,9). Solitons, doping and

FIGURE 1. Trans-Polyacetylene: The small arrows represent the
movement of a pair of electrons the large arrows repesent the
passage of a soliton.

electrical conduction in polyacetylene are described by MacDiamid and Heeger (10). A soliton is a moving wave which interconverts the two states of polyacetylene. It effectively picks up one arrangement of bonds and lays down the other. The effect of solitons on the state of a polyacetylene chain is shown by the arrows in figure 1.

It should be noted that two successive solitons (in the same or opposite directions) will leave the state of a polyacetylene chain unchanged. There are three charge states for solitons, they are positive, negative and neutral.

The soliton is highly suitable as a charge carrier, firstly because it may be transmitted over chemically large distances without loss and secondly as solitons behave as quantised particles they are suited to transmitting digital information.

JUNCTIONS

Two simple components whose properties are provided by the nature of polyacetylene are described.

The first is the junction (Fig. 2), where three polyacetylene chains are joined to a central carbon atom. The junction allows the branching and interconnection of polyacetylene chains. The junction under the name of soliton valve was introduced by Carter (5).

The junction has three states which are each characterized by the position of the double bond on the central carbon atom. Solitons passing through the junction cause interconversions between these three states(Fig. 2).

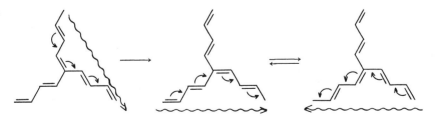

FIGURE 2. The three states of a junction and some of the solitons which cause interconversions between these states.

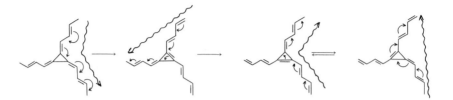

FIGURE 3.   A ring junction can transmit solitons in any of its four states.

Only alternating double and single bonded paths can transmit solitons. So no solitons can pass between the two chains conected to the central carbon by single bonds as the path between these chains includes two consecutive single bonds. Alternatively we can say that all solitons must pass through the central carbon's double bond.

RING JUNCTIONS

It is often useful to have a form of junction called a ring junction (Fig. 3 left) which will pass solitons in any direction. That it can pass any soliton in any direction in the first state shown is clear as the shortest path between any two arms always consists of alternating double and single bonds. Once such a soliton is passed we get one of the other three possible states for a ring junction (Fig. 3). In these states it is still possible to pass solitons between any two arms, because when the shortest path is not suitable the path around the ring (which is one bond longer) will be suitable.

So a ring junction can pass solitons in any direction whatever state it is in. Chemists should note that it is not necessary to form a three membered ring as shown, as a five or seven membered ring is logicaly equivalent.

DYNAMIC CIRCUIT DIAGRAMS

Because polyacetylene has a two state nature components designed to make use of this will also have a range of possible states and their properties will depend on their state. It is useful to introduce

FIGURE 4. The symbols for a polyacetylene chain (left), a junction
(middle) and a ring junction (right).

dynamic circuit diagrams for this system. I call them dynamic
because they can show the changing states of a circuit. These
diagrams have each component represented by a symbol which shows the
state of the component. These symbols are designed so they can
readily be altered to correspond to changes in state of the actual
chemical device.

The symbol for a polyacetylene chain (Fig. 4 left) is just a
line. A junction (Fig. 4 middle) is represented by three lines
meeting at a dot. The extra short line is called a dash and
represents the position of the double bond to the central carbon
atom. Because the properties of a ring junction are the same in all
its states, there is no need for state information in the ring
junction symbol and it is simply represented by a small circle at
the intersection of three lines (Fig. 4 right).

SOME DEVICES

The next two sections discuss two types of devices. Firstly the sid
which is used to generate solitons. Then the switch which is used to
control the flow of solitons. In both cases the devices will be
introduced as logical entities independent of the chemical
structures that could be chosen to make them and then some possible
chemical structures will be given.

Sids

The first component (Fig. 5) is the SID (soliton storage device)
which is similar to Carter's(11) soliton generator. When a potential
difference is applied to a clear sid a pair of positive and negative

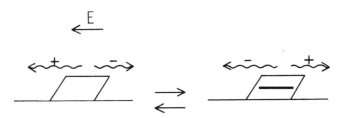

FIGURE 5. The logical symbol for a sid is shown in its two possible states with the solitons the sid generates when it changes state.

solitons are generated and travel in opposite directions away from the sid (Fig. 5). This gives the higher energy dashed state (Fig. 5).

When the ends of a sid are electrically connected, that is the potential is removed and both ends are connected to a common earth, then the sid will generate another pair of oppositely charged solitons moving in the reverse directions to the first pair. This will return the sid to the more stable clear state.

There is a close analogy between a sid and a capacitor that will only hold one electronic charge. The sid is "charged", that is goes from its ground or clear state to its dashed state by the application of a potential difference. Had the potential difference been applied in the opposite direction it would have no effect on the sid. When the potential difference is removed the sid remains in its dashed state that is it remains "charged". However when the two ends of a charged capacitor or sid are conected the sid will discharge and return to its clear state.

Sid Structures

A way of making a sid which closely resembles a capacitor is to have a group that stablizes a positive charge near one that stablizes a negative charge, such a sid is shown in figure 6. In this sid the seven and five membered rings will accept positive and negative charges respectively to form stable aromatic six PI electron systems.

A potential difference applied to the sid will create a pair of oppositely charged solitons which will move in the directions shown (Fig. 6) to give the higher energy charged (logically dashed) state. When the ends of the sid are electrically connected the sid will spontaneously revert to the neutral state by generating a further pair of solitons as shown in figure 6.

Carter's (11) soliton generator becomes a sid when the following two changes are made. Firstly the hydride ion (H⁻) hopping is stopped either by chemical and structural means or by never applying sufficent potential to induce it. Secondly the soliton generator must be of the second type described by Carter in that there is to be a chemically induced bias between the two states and this bias has to be large enough to spontaneously produce solitons so the sid can return to the more stable state.

A further sid structure is shown in figure 7. This structure could have a slight difference in the overall mechanism of reaction. Because of the intermediate anion both solitons would not neccassarily be produced simultaneously. The benzene ring induces the bias neccasary for the sid to return to ground state.

I have choosen the overall descripion of the mechanisum of a sid's operation to be consistent with Carter's (5,11). However there is a possibility that this sid could operate by passing a negative soliton via the intermediate anion right through the sid as indicated by the little arrows (Fig. 7). Also it is by no means

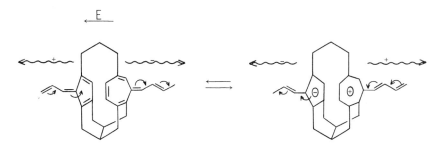

FIGURE 6. The charging and discharging of a sid that resembles a capacitor. The rings were chosen to stablize specific charges.

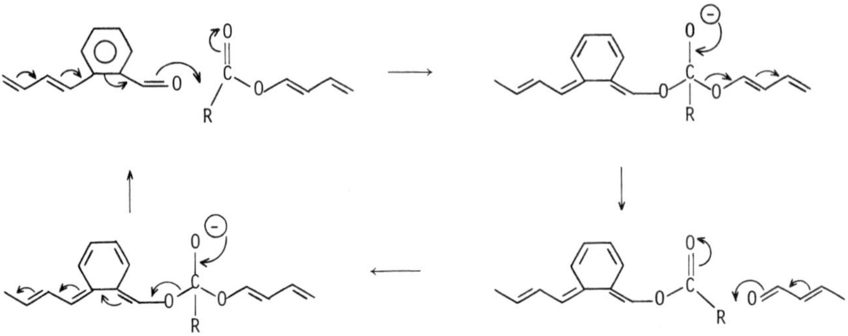

FIGURE 7. A further sid and its charge discharge cycle. Because of uncertainties in the mechanism the solitons are not shown.

certain what mechanism would be used when the sid discharges. However both mechanisms are for this paper logically equivalent.

Switches

The second component is called a switch (Fig. 8) because it can control the conduction in the two chains which run through it. A clear switch (Fig. 8 middle) can pass a soliton down either chain to give the dashed state for that chain (Fig. 8 left and right). A second soliton down the same chain will return the switch to clear again.

The switching rule is that both chains cannot have a dash at the same time. So when one chain is dashed the other chain is blocked and cannot pass solitons. The switch can only be cleared by a soliton down the dashed chain.

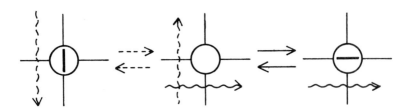

FIGURE 8. The logical symbol for a switch is shown in its three allowed states. A fourth state which would resemble a plus sign is not allowed.

There is no electrical connection between the two chains through a switch, but there is a physical connection which is that the conduction in each chain is controlled by the state of the other chain.

Switch Structures

In designing a switch it is necessary to have the two chains interact so that they can't both be in the same state (of their two possible states) at the same time.

The switch is shown in figure 9 in its clear position where the top part can pivot in the plane of the paper around the central "hinge". The nitrogen atoms can form a single bond in place of a carbon double bond, but only when they are within some distance of each other.

The point of the switch is that when a soliton is sent along either of the chains and the nitrogen – nitrogen bond formed, the other chain cannot pass a soliton as the nitrogen atoms are held firmly outside their bonding distance. Another soliton sent on the first chain will clear the switch.

As a switch must effectively distinguish between single and double bonds on a chain, any difference in properties between single and double bonds could be the basis for designing a switch. In addition to the switch discussed here there are also possibilities

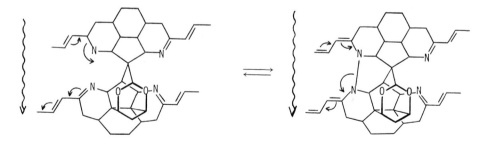

FIGURE 9. A hinge type switch structure. The oxygen atoms (marked by O) play no role in the switching and are only present to avoid the possibility of hydrogen atoms which could interfere with the switching sites.

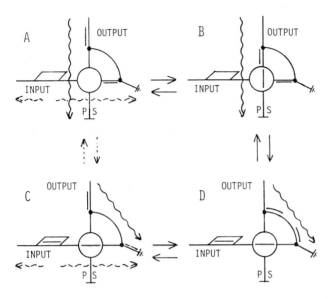

FIGURE 10. The four states of an inverter along with all the solitons that cause changes between these states. The output is considered ON in the top two states A and B and OFF in the bottom two states C and D.

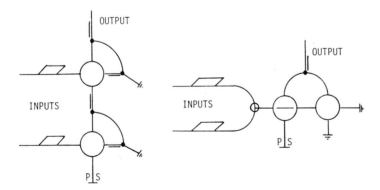

FIGURE 11. (left) The NOR gate shown here in its rest state (both inputs OFF and the output ON) was formed by placing two inverters in series. (right) The Exclusive OR gate. If either inputs is turned ON the output will turn ON. If both input are switched ON the output will turn OFF.

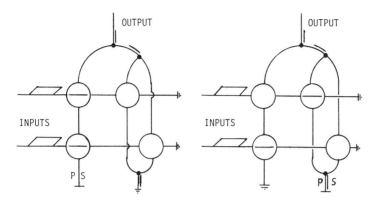

FIGURE 12. (left) An AND gate. A few simple changes in the design of the AND gate gives the OR gate with top input inverted shown on the right.

based on the stereochemical requirements of double bonds and the electron rich nature of double bonds.

COMBINING THE COMPONENTS

Now that the basic components (chains, junctions, switches and sids) have been introduced, we will consider ways they can be combined into structures that perform Boolean operations. In the following diagrams (Figs. 10-12) there are chains marked PS and earth (the standard symbol is used for earth). The PS chains are conected to a power supply which is at a potential (positive or negative with respect to the earths) suitable for causing soliton generation at the sids.

For the logic system to be described the inputs of all devices are connected through a sid to earth. These sids will produce solitons when a potential difference is applied to or removed from the inputs. These solitons are used to control the other components in the device.

INVERTER

The inverter (Fig. 10) is first shown in state A, in its rest state with no potential applied to the input. The output is at a raised potential and can therefore cause the sids in the driven devices to generate solitons. One such soliton can pass in from the output through the top junction, the switch and on to the power supply to give state B. A second soliton will return the inverter to state A. After an even number of solitons have been passed, the inverter will be in state A.

When in state A, a potential on the input will cause solitons to be generated and flow as indicated by the dashed arrows (Fig. 10) to give state C. Now the chain from the power supply to the output is blocked by the switch, so the output has been turned OFF. However the chain from the earth to the output through the two junctions is now clear and will pass solitons. So the sids in the driven devices can discharge to earth through this chain. Solitons along this chain will cause the state to alternate between C and D as shown in figure 10. After an even number of solitons have been passed the inverter will have returned to state C.

When the potential is removed from the input and the input is earthed, the sid will discharge by generating solitons which travel in the directions of the dashed arrows (Fig. 10). This returns the inverter to the rest state A where the output is again ON.

To summarise when the input to the inverter is OFF the output is ON and allows an even number of solitons in through the output and on to the power supply. When the input is ON the output is OFF and the inverter allows an even number of solitons in through the output and on to earth. So an inverter could charge - discharge an even number of sids and hence drive the inputs to a number of other inverters or gates.

Boolean Gates

A NOR gate (Fig. 11 left) is easily formed by connecting the output of one inverter to the power supply of another. Given a NOR gate most electronic digital components can be made. Another gate that also requires a minimum of components is the exclusive OR gate (Fig. 11 right).

Should the reader wish to check these gates it is recommended that matches are broken up into small pieces which can act as moveable tokens for the dashes in the diagrams. In working through these diagrams particular attention should be paid to the following general rule for all components.

When a soliton travels along a dash the dash is removed. When a soliton travels through an empty positon which may contain a dash then a dash is inserted.

The output of any given gate is normally used to drive the inputs of other gates. When the output of the first (driving) gate is switched OFF it must also be earthed so the sids on the driven gates can discharge.

An AND gate is shown in figure 12 left with both inputs OFF. This gate is in effect composed of two subgates. The two switches on the left form an AND type subgate which will connect the output to the power supply if and only if both inputs are ON. The two switches on the right form an OR type subgate which will connect the output to earth if either input is OFF. The output is always connected to either the power supply or the earth but never to both.

If the bottom earth connection and the power supply connection on the AND gate are interchanged an OR gate is formed. Figure 12 right shows an OR gate with the lower input inverted, that is, the Boolean expression for this gate is Upper OR (NOT Lower). The inversion is accomplished by constructing the gate so the chain between the sid and the first switch on the input to be inverted is one bond longer or shorter than the chain for the normal input. This inversion is not explicitly shown in the diagram (Fig. 12 right) but is easily inferred from the position of the dashes. This brings to light an important point that the dashes not only represent the state of a structure but also give information on the nature of the structure.

MEMORIES

First we will consider address selectors then memory cells and finally these will be combined into an example memory.

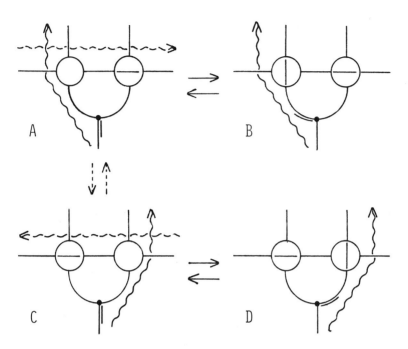

FIGURE 13. The selector in its four possible states. IN each diagram the input chain is the bottom vertical line, the outputs are the top vertical lines and the select chain is the horizontal line. In states A and B input solitons will pass out the left output while in states C and D they would pass out the right output.

Selectors

A memory has an array of memory cells each containing a bit (1 or 0) of information. Before reading or writing on an cell it is necessary to select the appropriate cell. In the simplest possible case we have two memory cells and depending on the state of a chain in the selector, wish to send solitons to a specified one of these cells, so it may be read.

Such a selector is shown in its four possible states in figure 13. The input chain is the bottom line, the two alternate outputs are the top vertical chains and the select chain (which determines which output the input solitons use) is the horizontal line.

A soliton on the input chain of the selector in state A is directed out the left output because the right output is blocked by the dashed switch. This will give state B. In state B solitons on the select chain are blocked by the left switch. Further solitons on the input chain will also be directed out the left output and will cause the selector to alternate between states A and B. After an even number of input solitons the selector will return to state A.

When in state A a soliton on the select chain will give state C. All input solitons are now directed out the right output and cause the selector to alternate between states C and D. An even number of input solitons will return the selector to state C, where a soliton on the select chain will return the selector to state A.

To summarise: solitons on the input chain of a selector will be directed out the left or right outputs depending on the state of the select chain. The state of the select chain can only be changed by a soliton on that chain and then only when an even number (or none) of input solitons have been passed.

Memory Cell

If we look again at the selector we find all that is necessary for a memory cell. Define the left outputs of the selectors in (Fig. 13) as logical "0" and connect them to earth. (i.e. throw these outputs away). Define the right output as logical "1"and connect it to the memory output.

We then associate the state of the write (select) chain which blocks the right output (Fig. 13 A and B) with the memory cell holding a zero bit and the state of the select chain which blocks the left (earthed) output, as holding a "one" bit.

A pair of solitons applied to the input will perform a non destructive read on the contents of the cell. Single solitons applied to the write chain will write on the cell, by changing its state from logical "0" to "1" or vice versa.

Writing (Fig. 15) is complicated by the need to erase the information held in the store. To clear the cell the output from the cell is temporarily connected to the write chain via a path through switches and junctions external to the cell and not shown here. When

READING 0   READING 1

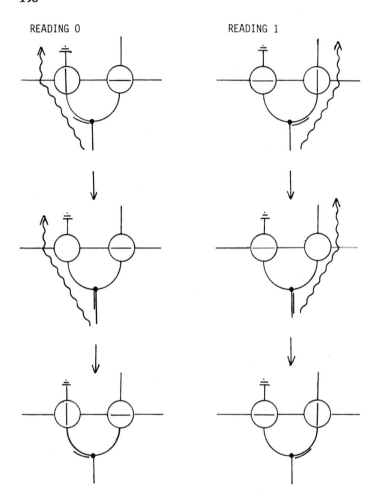

FIGURE 14. The operations required to read a memory cell. For the
cells shown the input chain is the bottom vetical line, the output
chain is the top vertical line and the write chain is the horizontal
line. The three cells on the left show the results of reading a
cell containing a "0" while the cells on the right demonstrate
reading a "1".

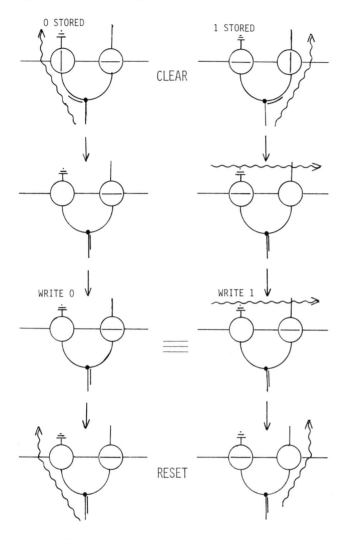

FIGURE 15. The operations required to write on a memory cell shown
in two streams running down the page. First the cell is located via
a read soliton then it is cleared and the bit to be stored is
written on it and then a further read soliton resets the cell. Note
that after the clear operations are complete the left and right
cells (third from top) are in the same state.

a single read soliton is applied to the input of a cell containing a logical "1" the soliton goes out the output and is sent back onto the write chain to change it from a logical "1"to a logical "0". If the cell contained a logical "0" the soliton is sent to earth and there is no change. In both cases the write chain is now in the logical "0" position.

Writing on the cell in this state, is simply accomplished by applying 0 or 1 solitons to the write chain to store a logical "0" or "1" in the cell.

The write (and clear) operations are completed by applying another read soliton to the cell's input. This soliton is not required for anything, so the output, from the memory is temporarily earthed, to throw this soliton away.

The Fan Memory

The fan memory (Fig. 16) is a simple way to construct a memory from selectors and memory cells. The memory shown holds four separately addressable bits. For a computer's registers, there would, for example, be 32 of these fan memories side by side, addressed in parallel and each giving one bit in each word, to hold in total four 32 bit words.

The diagram (Fig. 11) has the dashes lightly drawn so the reader may insert broken matches in their place, to personally illustrate the following discussion with examples.

The lower half of figure 11 shows the three selectors on two chains needed to address the four memory cells in the top half. These cells from left to right contain 0110. The selectors are all in their default or rest position which would select the left most (zero address) memory cell.

To select a memory cell for reading or writing, the number (in binary) from 0 - 3 (00 - 11) is applied as potentials (power supply for binary 1 and earth potential for binary 0) to the sids on the select chains. This causes soliton generation on the chains at power supply potential. These solitons will change the state of the

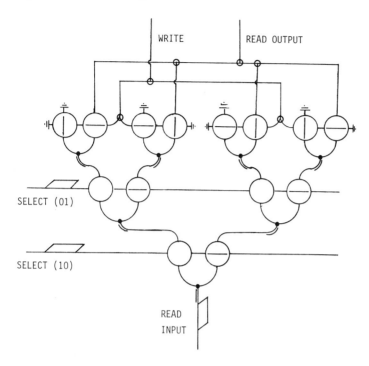

FIGURE 16. A fan memory which holds four separately addressable bits. The top part of the diagram is the four memory cells that hold the information. The bottom part contains the three selectors neccasary to address these cells. The cells from left to right or from address "0" to "3" (binary 11) contain 0 1 1 0 respectivly.

select chains and so connect the read chain (at the bottom of Fig. 16) to the appropriate memory cell.

For example, to read the third memory cell from the left, a soliton is applied to the lower select chain, making 10 or binary two on the two select chains. When a potential is applied to and removed from the read chain sid it generates a pair of solitons which are directed to the third memory cell, which then directs them to the output, as the memory cell contains a stored "1". To complete the read solitons are generated to return the selectors to their default position, by removing the potentials on the select chains (that is by earthing these chains).

To write a one on the second memory cell from the left, a potential is applied to the top select chain, forming the binary one 01. When a potential is is applied to the read input the soliton generated is directed via the selectors to the second memory cell, which passes the soliton on to the output. For this part of the write cycle in which we wish to clear the memory cell, the output is connected to the write chain.

While the write chain is connected to all the memory cells the only one that can pass a soliton is the one that was just freed by the read. So the soliton on the write chain is passed through that cell to earth and in so doing changes its state to a logical "0" in the process.

Thus the cell has been cleared. Now to write a one on the cell a soliton is applied to the write chain and can once again only be passed by the freed memory cell. This soliton changes the cell's state to logical "1" and passes on to earth.

A logical "1" has been written on the desired sixth memory cell and all that remains is to return the whole system to its rest state. This is accomplished by removing the potential on the read chain. This generates a soliton which is directed by the selectors to the memory cell. The memory cell passes it onto ground via output, which is earthed for this part of the cycle. Then finally the selectors are returned to their rest position by removing the potentials on the select chains.

The fan memory was chosen because it has the simplest and clearest structure. There are other possibilites including matrix systems using separately addressed rows and columns. The fan memory requires a little more than four switches per bit stored.

The information in the memory is simply addressed by potentials on parallel chains to represent a binary number, with each chain carrying a digit. The output from the memory is in the convenient form of pairs of solitons on parallel chains.

It is worth noting the following points about this type of memory. The retention of stored information requires no current to the memory cells. Reading a memory cell does not erase the information in it. The nature of the memory cell only allows two values to be stored. There are no permanent intermediate values possible, so there can be no confusion as to which value is stored.

CONCLUSIONS

The components which have been described, switches sids, are intended as logical types. Many other chemical structures for them are possible, in particular I am currently considering simpler switch structures.

In a system where every component has a range of stable states circuit diagrams that can show the state and changes in state are virtually essential. I found their use a great help in designing and checking new circuits. Indeed, deciding on good symbols for the components was closely followed by advances in the structures I could form.

The system described, like other molecular electronic devices, offers opportunities for extremely high component packing densities. Because there are no power hungry resistive elements and because a single soliton can cause a chain to change state from logical '0' to '1' or vice versa, devices made from these components would have a minute power consumption. Also, as well as being able to form Boolean gates and other devices, the system described is naturally suited to forming non volatile memories and to addressing those memories.

ACKNOWLEDGEMENTS

I thank Dr. John G. Sanderson for advice and encouragement, Dr. R. H. Prager for chemical discussions and Dr. Forrest L. Carter for a preprint.

REFERENCES

1. A. Aviram and M. A. Ratner, "Molecular Rectifiers", Chem. Phys. Letters, $\underline{29}$, 277 (1974).

2. F. L. Carter, Ed., Proceedings of WMED (1981), NRL Memorandum Report 4662 (1981); also available from Marcel Dekker, New York, NY (1982).

3.   K. M. Ulmer, "Biological Assembly of Molecular Ultracircuits", in Ref.2, p.167.

4.   J. H. McAlear and J. M. Wehrung, "Biotechnical Electronic Devices", in Ref.2, p.127.

5.   F.L. Carter, NRL Memorandum Report 3960, p.121 (1979); NRL Memo. 4335, p.35 (1980); "Conformational Switching at the Molecular Level", in Ref.2, p.53.

6.   C. S. Guzener, "Reliability Problems Caused by Nuclear Particles in Microelectronics", in Ref.2, p.223.

7.   M. J. Rice, Phys. Lett. 71A, 152 (1979).

8.   W. P. Su, J. R. Schrieffer and A. J. Heeger, Phys. Rev. B 22, 2099 (1980).

9.   H. Takayama, Y.R. Lin-Liu and K. Maki, Phys. Rev. B 21, 2388 (1980).

10.   A. G. MacDiamid and A. J. Heeger, "Recent advances in the chemistry and physics of Polyacetylene: Solitons as a means of stablizing carbonium ions and carbanions in doped (CH)x", in Ref.2, p.208.

11.   F. L. Carter, "Further Considerations on 'Molecular' Electronic Devices", The NRL Program on Electroactive Polymers, NRL Memorandum Report 4335, R. Fox, September 15, 1980.

10

The Modification of Davydov Solitons by the H-N-C=O Group

Scott P. Layne/Center for Nonlinear Studies
Los Alamos National Laboratory, MS-B258
Los Alamos, New Mexico

PREFACE

The concept that ATP hydrolysis energy could be transported along a
one-dimensional molecular chain in an alpha-helical protein was
first developed by Davydov & Kislukha (1). Davydov proposed his
solitary wave (or soliton) model of energy transport after a 1973
conference of the New York Academy of Sciences. A central issue of
this meeting dealt with a "crisis" in bioenergetics and centered on
the question of "how can energy be transported in biological
systems?" Since this conference, substantial progress has been made
with the analytic and numerical analysis of Davydov solitons.
Studies have verified the particle-like and stable quality of soliton behavior. Therefore, the time is right for the development of
further theoretical biological concepts based on soliton propagation. Already, Davydov (2) has proposed a model of muscle contraction based on soliton propagation in myosin. However, in this paper
I wish to propose a new model of general anesthesia and a new model
of glycoprotein dynamics based on soliton propagation in the
cytoskeleton.

If for a moment it is assumed that solitons form with a normal amount of biological energy, then two questions seem highly relevant: 1. Should one look exclusively for solitons on the same alpha-helix or for a combination of both solitons and altered solitons (below threshold pulses, reflected pulses and perturbed pulses) on the same helix? 2. Once a soliton forms on an alpha-helix, where does it go and what does it do in a "living" macromolecule?

This paper proposes that one should look for a variety of energy profiles on the same alpha-helix. It makes a central assumption that H-N-C=O groups which are in the vicinity of an alpha-helix, but which are not necessarily members of the one-dimensional chains within the helix, can alter the local propagation dynamics on the helix. Therefore, once a soliton forms on an alpha-helix, an extrinsic H-N-C=O group can modify its behavior. This alteration of dynamics might be a rather useful device for living macromolecules, especially if one portion of an alpha-helix were free to propagate solitons while an adjacent portion was critically hindered from propagation. Under these circumstances, the particle-like qualities of a soliton would allow for a variety of behaviors. Conceivably, a soliton is capable of moving through other solitons without decay and it is capable of reflecting without a loss of integrity. Therefore, in total, this paper considers four interrelated behaviors of solitons on dynamically nonhomogeneous alpha-helical proteins based on their special capabilities:

1. Soliton formation critically hindered via alteration of amide-I anharmonicity.
2. Soliton reflection and/or destruction by altered amide-I anharmonicity.
3. Solitons as "energy wells" for long-range nonlinear behavior within glycoproteins.
4. Soliton energy "modulation" via acetylamide groups that are inherent to glycoprotein cores.

Before going into the details of the proposed model, I would like to outline the format of this paper: Section I introduces some

basic facts concerning cellular architecture and behavior with relation to the model. Section II discusses the current evidence for
solitons in living organisms. Section III discusses solitons in
relation to general anesthesia. Section IV discusses solitons in
relation to glycoprotein structure. Section V introduces the concept of self-organizing behavior in glycoproteins that are powered
by solitons at their cores. Section VI proposes experiments to
verify the model. Section VII poses open questions that are developed in this paper.

I. INTRODUCTION

Neurons are filled with numerous filamentous proteins that weave
throughout the entire intracellular volume. These filamentous
proteins also span the lipid bilayer of the neuron and serve to
couple the extracellular space with the interior of the cell. The
complex three-dimensional association of these elements constitutes
the so-called cytoskeleton of the nervous system. Recent structural
analysis of a number of these cytoskeletal proteins has demonstrated
that they exhibit considerable alpha-helical character, which bears
a similarity to the keratin molecule (3). Therefore, it seems
reasonable to postulate that many of these filamentous macromolecules are alpha-helical proteins which are capable of conducting
solitons over much of their length.

In conjunction with the above findings, it appears that many of
these skeletal proteins are cross-linked among one another into
bundles that extend over hundreds to perhaps thousands of angstroms
($\overset{o}{A}$). These same filamentous elements also appear to associate dynamically among themselves, changing their shape in accordance with
localized variations of the Ca^{2+} to Mg^{2+} concentration ratio within
the cytosol and in accordance with localized variations of ATP concentration. In the absence of Ca^{2+}, the trabecular meshwork is
slender and elongated, but after pretreatment with 10mM Ca^{2+}, the
trabecular connections appear broken and the axoplasm condensed.
In addition, electron micrographs have demonstrated that "bridging
structures" occur between bundles of cytoskeletal elements and mitochondria. In other words, as the filaments continue to weave their

way through the cytosol they appear to touch mitochondria as they
pass by them. These observations of structure again lead to the
hypothesis that the cytoskeleton may play a functional role in the
transport of ATP hydrolysis energy via its alpha-helical components.
It also suggests that this transport of energy could be regulated
considerably by divalent cations and, in particular, by Ca^{2+}.

Expanding this concept of energy regulation to molecules other
than Ca^{2+} furnishes an entirely new framework for understanding the
molecular mechanism of action of anesthetic agents that are capable
of forming hydrogen bonds. Soliton propagation unifies the mechanism of action of such diverse agents as barbiturates, hydantoins,
succinimides and urethanes. It also furnishes a molecular mechanism
of action for gamma-aminobutyric acid (GABA) and supports its role
as a major inhibitory neurotransmitter in the central nervous
system. The forenamed substances are capable of interfering sterically with either the formation or propagation of Davydov solitons
via two-point hydrogen bonding across the hydrogen bonds that
stabilize the rotational geometry of an alpha-helix. The common
denominator of activity is the H-N-C=O group for the anesthetic
agents and the separate O=C and N-H groups for GABA. By binding
across the hydrogen bonds of an alpha-helix, these agents alter
amide-I spine anharmonicity, increase dynamic soliton mass, "leak"
longitudinal sound energy and behave as external amide-I dipoles in
relation to the alpha-helix.

Soliton propagation invites an entirely new model for explain-
ing the dynamic behavior of glycoproteins that are associated with
alpha-helices. The polysaccharide domain of glycoproteins is capable of tapping the soliton's longitudinal sound energy through
ubiquitous acetylamide groups that are typically located near the
hydrogen bonds of the alpha-helix. In the polysaccharide (polyelectrolyte) domain of glycoproteins, the interplay of electrostatic
repulsion, thermal coiling and/or intramolecular attraction makes
it possible to convert the longitudinal sound energy from a soliton
into mechanical work by changing the shape of the macromolecule.
In these movements, divalent cations play an essential role in non-

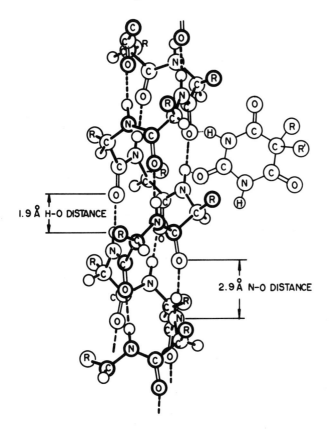

Fig.1. Barbiturate's two-point interaction with the hydrogen bonds
of the alpha-helix.

linear and long-range associations within the polysaccharide domain.
The ability of the carbohydrate portion of glycoproteins to expand
with $H^+ = K^+ \geq Na^+$ or to shrink under the influence of $Ca^{2+} > Mg^{2+}$
offers a possibility of transforming chemical energy into mechanical
work by an isothermal process (4).

Glycoprotein behavior also has important implications for
intracellular signaling via solitons and the neuronal skeleton.
Approximately 65-70% of glycoproteins are found on the cellular
surface, which comprises about 5% of the total cell volume, whereas
the remaining 30-35% is associated mostly with the mitochondria and

their immediate microenvironment (5). Although it is difficult to
quantify the glycoprotein content of the intracellular space, because of problems with purification and separation, the above data
clearly demonstrates that glycoproteins are associated primarily
with the neuronal membrane and secondarily with mitochondria.
Therefore, the cytoskeleton may serve either signaling or "integrative" functions between the cell surface and the mitochondrial
microenvironment via soliton propagation initiated by glycoprotein
modulation at the membrane surface. A more precise explanation of
this mechanism follows in the text.

In the lipid bilayer of the neuronal membrane, intramembranous
particles are inserted which vary markedly in size and in chemical
structure. These intramembranous particles are associated with the
larger pool of glycoproteins and provide the basis for three important membrane functions (6). First, their presence within the lipid
bilayer causes adjoining portions of the lipid bilayer to become
more rigid. This occurs through long-range and nonlinear interactions between charged sites on the intruding particle and charged
sites on the tails of the lipid molecules. Second, the outer ends
of the intramembranous particles protrude from the surface of the
lipid bilayer as strands of glycoprotein. These strands have
terminal carbohydrate (sialic acid) groups which are associated with
multiple fixed negative charges. The surface of the membrane thus
behaves as a polyanionic-polyelectrolytic sheet with strong affinity
for cations. Third, the intramembranous particles can be caused to
move laterally within the lipid bilayer. These movements of the
particles can be triggered by binding of cationic molecular sites
to the protruding strands of glycoprotein and requires the expendi-
ture of metabolic energy. Actin-binding proteins subjacent to the
membrane bilayer are responsible for this movement (7). The
soliton-glycoprotein model presented in this paper describes a
mechanism that is consonant with membrane behavior. It outlines a
means of transductive coupling between protruding membrane particles
and the cytoskeleton that involves both cationic flux and energy
transduction.

Finally, the above mentioned nonlinear interactions have been implicated by a number of experiments that involve extremely low frequency (ELF) and modulated microwave fields which induce their effects at or near thermal equilibrium (8). Like hydrogen bonding anesthetic agents, these weak nonionizing fields can have a profound influence on glycoprotein dynamics. They can interfere with either the soliton power supply or the condensed phase states within the glycoprotein (9). Moreover, a "stable" soliton power supply for polysaccharide displacements invites the possibility of a variety of self-organizing and limit-cycle behaviors for these macro-molecules as outlined below.

II. SOLITONS IN LIVING ORGANISMS

Central to the soliton concept is the fact that the amide-I reson-ance (N-C=O) is intrinsic to every peptide group of every protein and, therefore, might act as a potential well for storage and trans-port of biological energy. However, the amide-I resonance has not been seriously considered as a potential "well" of biological energy because the line widths of a typical amide-I absorption peak implies a lifetime (due to linear coupling between amide-I bonds) on the order of 10^{-12} seconds. This is much too short for normal bio-logical mechanisms. So Davydov's (10) concept is that the lifetime of these vibrations can be markedly increased on an alpha-helical protein by: 1. introduction of localized amide-I bond energies by ATP hydrolysis which induces longitudinal sound waves on the alpha-helix, and 2. longitudinal sound waves act as potential wells to trap the bond energy and prevent its dispersion.

Hyman, Mclaughlin and Scott (11) point out that according to linear analysis, energy transported by amide-I resonances should spread out from the effects of dispersion and rapidly become dis-organized and lost as a source of biological energy. However, in the nonlinear analysis by Davydov, the coupled excitation propagates as a localized and dynamically self-sufficient entity, constituting a soliton. For such a coupled excitation to be viable, certain threshold conditions must be satisfied. The nonlinear coupling between the amide-I spines must be sufficiently strong and the

amide-I vibrations must be energetic enough for the retroactive
coupling to take hold. Below the threshold, a soliton cannot form
and the dynamic behavior will be essentially linear -- that of an
exciton. Above this threshold, soliton formation is possible with
lossless energy transduction.

In 1979, numerical studies were carried out at the Los Alamos
National Laboratory which confirmed a theoretical prediction that
certain "threshold" conditions on the nonlinearity must be satisfied for a Davydov soliton to be viable. This suggestive finding
led Scott (12) to improve and expand Davydov's original numerical
model. The modified version of Davydov's equations included ten
additional dipole-dipole coupling terms between adjacent amide-I
groups in the alpha-helix. The expanded model was a better representation of true helical symmetry. The primary aim was to decide
whether it would be reasonable to find solitons on typical alpha-
helical proteins. Since ATP hydrolysis produces approximately 0.48
ev of useable energy, it was assumed that two (0.205 ev) amide-I
quanta launch the energy pulse on two aligning spines of an alpha-
helix. For the hydrolysis of one ATP molecule, the numerical
computations on the modified Davydov model demonstrated soliton
organization at a critical level of anharmonicity which is quite
similar to the self-consistent field calculations of the formamide
dimer. But more importantly, the calculations showed that the internal dynamics of the soliton is governed primarily by two frequencies. This observation predicted a Raman spectrum which is in close
agreement with measured laser-Raman spectra of metabolically active
Escherichia coli (13). Organisms that are not metabolically active
do not demonstrate such Raman active modes, so induction of amide-I
resonances by laser light is not responsible for the observations.
Only cells metabolizing glucose demonstrated the wave numbers shown
below. Based on a soliton "interspine oscillation" with a period
of 2×10^{-12} seconds (spectral energy, E_1 = 17 cm^{-1}) and a second
"longitudinal speed" component of the soliton with a period of
$8/3 \times 10^{-13}$ seconds (spectral energy, E_2 = 125 cm^{-1}), Scott (14) has
constructed a spectrum for the internal dynamics of a soliton which

is very similar to Webb's measurement of metabolically active
Escherichia coli. In addition, a recent Fourier analysis of the
soliton's internal dynamics by Lomdahl (15) yields similar agreement.

There should be little doubt that an artificially constructed
and stimulated alpha-helix can support and conduct solitons in an
experimental procedure. Given enough energy a helix will maintain
such a pulse. Nevertheless, this is not the central issue. The
primary question is whether a normal amount of biological energy
is sufficient to form a soliton.

III. SOLITONS IN RELATION TO ANESTHESIA

The alpha-helix consists of two spirals. The major spiral, which
forms the body of the helix, has the repeating atomic units: NCC -
NCC. The minor spiral(s), which form the amide-I bonds that hold
the helix together, have the repeating atomic units: H-N-C=O $\cdots$
H-N-C=O. Pauling (16) has described the alpha-helix in two stable
configurations depending on the rotational angle of the molecule.
Proteins with either 3.61 or 5.1 amino acids per turn create hydrogen bonds with an average nitrogen to oxygen bond length of 2.9 $\overset{\circ}{A}$.
The hydrogen bond energies are about 8 Kcal/mole, where it is
assumed that each nitrogen atom forms a hydrogen bond with another
oxygen of another residue on the protein. The stability of the
alpha-helical structure in the noncrystalline phase depends solely
on the interactions between adjacent residues. The number of amino
acids per turn of the molecule (tightness of the helix) would be
affected somewhat by a change in hydrogen bond distance and by
interactions of the helix with neighboring small molecules. These
interactions, in this instance, by the two-point hydrogen bonding
of a barbiturate (or other forenamed) molecule(s) might cause small
torques in the helix, deforming it into a configuration with a
different number of amino acids per turn. Since Davydov solitons
may travel down the three amide-I spines of a helix, barbiturates
might alter profoundly either the propagation of formation of
solitons at the location where the anesthetic molecule binds to the
protein. Even if the amide-I position shift were only minor, barbi-

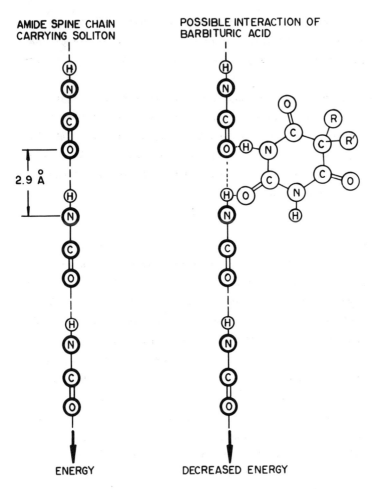

AMIDE SPINE CHAIN
CARRYING SOLITON

POSSIBLE INTERACTION OF
BARBITURIC ACID

2.9 Å

ENERGY   DECREASED ENERGY

Fig.2.   Barbiturate's interaction with the one-dimensional amide-I
         spine system that supports solitons in the alpha-helix.

turates would still alter amide-I bond anharmonicity and the one-
dimensional relationships that are critical for holding a soliton
together.

   Classical molecular potential energy conformational studies
have confirmed that the one-point hydrogen bonding ability of barbiturates, hydantoins and succinimides is unrelated to either their
hypontic or anti-epileptic activity. Further, these studies suggest
that hydrogen bonding hypnotic and antiepileptic agents may require

the participation of more than one carbonyl or amide hydrogen for
the production of pharmacologic activity (17). This suggestion of
a two-point hydrogen bonding interaction supports the soliton model
of anesthetic activity.

Based on a 200 micromole concentration for anesthetic action
and on a uniform volume distribution, with an approximation of the
alpha-helix as a 5 $\overset{o}{A}$ cylinder, it is calculated that one barbiturate
molecule interacts with the helix at an interval of about 10^5 turns
of the major spiral in a 5.41 $\overset{o}{A}$ per turn molecule (18). Therefore,
barbiturate interactions with the alpha-helix will make little
difference to its overall ability to collect energy from its cytoplasmic environment because an alpha-helix surrounded by ATP could
introduce longitudinal sound energy from a myriad amide-I bonds.
Clearly then, the importance in the overall dynamic must lie in
what happens to the soliton at the few points where barbiturates
bind to the helix. With this in mind, it might be concluded that
soliton behavior (blockade, dissipation, altered propagation and/or
reflection) at these disrupted areas on the molecule holds the key
to understanding.

When a soliton encounters a barbiturate (or other forenamed)
molecule bound to the helix, it can be held in place while the
altered spine "leaks" energy from the nonlinear pulse. Similarly,
when a soliton enters an altered length of helix, it can degenerate
into a "linear" pulse or emerge with a lesser amount of energy.
This action would serve to stop soliton propagation through the
helix or to shorten the lifetime of a pulse that emerged from the
altered spine. It is also possible that a polymer soliton would
reflect off the altered bonds within a conducting molecule. Scott
(12) has shown that solitons which reflect off the raw end of an
alpha-helix degenerate into "critically damped" oscillators. This
is probably due to interactions within the tail of the reflected
energy. Presumably, in all of the above mechanisms, the energy that
is robbed from the soliton by a pharmacologic agent would disperse
and become disorganized.

In addition, the soliton model of general anesthesia presents
an attractive mechanism of action for GABA. Much as with barbi-

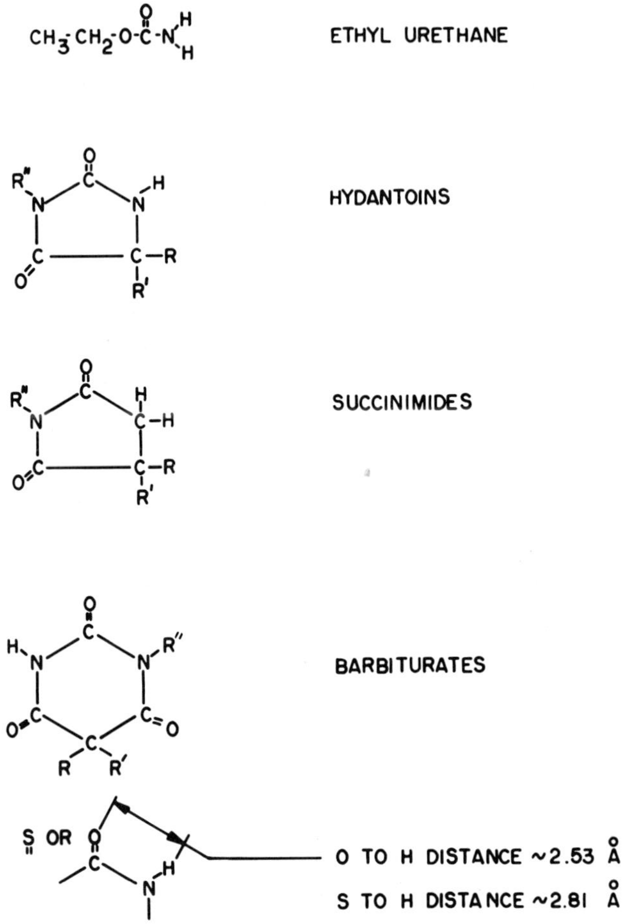

ETHYL URETHANE

HYDANTOINS

SUCCINIMIDES

BARBITURATES

O TO H DISTANCE ~2.53 Å

S TO H DISTANCE ~2.81 Å

Fig.3.   Anesthetic and anti-epileptic agents that are capable of
         forming two-point hydrogen bonds. The presence of either an
         alkyl or aryl group at R and R' confer increasing lipoid
         solubility. The presence of a hydrogen atom at R'' promotes
         two-point bond attachments whereas the presence of an alkyl
         (usually methyl) group at R'' sterically chokes such attach-
         ments. When R'' equals a hydrogen atom then succinimides form
         two, hydantoins form three and barbiturates form four possible
         two-point hydrogen bond attachment configurations. Generally
         sulfur forms weaker hydrogen bonds than oxygen so thiobarbi-
         turates have two relatively weaker bonding configurations (46).
         For all figures the bond distances are based on values from
         Dickerson and Geis (37).

turates, GABA is capable of assuming a two-point interaction with the amide-I spine of an alpha-helix. However, unlike barbiturates GABA is not a stiff planar ring. Nor is it a strict H-N-C=O dipole. Hence, its four carbon backbone would be capable of mutual vibrations with a Davydov soliton and can be viewed as a resistor which acts in parallel with the amide-I spine. A number of publications have pointed out that barbiturates potentiate the inhibitory action of GABA in neurons (19,20). These publications have also gone on to suggest that GABA and barbiturates act on the same receptor site but they have not specified its molecular configuration. Since GABA is not an optically active molecule, it is reasonable to propose that one of its sites of action is across the one-dimensional spine systems of an alpha-helix. Therefore, this model implies that GABA operates as an energy modulator within the cytoskeleton of neurons. In the upcoming explanation of glycoprotein behavior it will be seen that GABA acts as an inherent "barbiturate" of the brain.

In the proposed model, barbiturates induce anesthesia by disrupting energy transport in the alpha-helical proteins of the cytoskeleton. At the neuronal membrane, these cytoskeletal proteins protrude through the lipid bilayer as patches of glycoprotein. The next section of this paper will explain how glycoproteins utilize solitons in their cores to power polysaccharide displacements. However, at this point it is necessary only to point out that this disruption of energy would result in an overall decrease in polysaccharide (polyelectrolyte) motion in a glycoprotein. Since the intercellular space of the brain is filled extensively by a glycoprotein matrix (21), a relative loss of polysaccharide (polyelectrolyte) motility should result in an increased intercellular impedance. Adey has measured a marked increase in CNS intercellular impedance during barbiturate anesthesia (22,23,24). Therefore, this experimental result is supportive of the proposed model. In line with these measurements, Darbinjan (25) has found that the reticular formation appeared to be one of the most sensitive regions of the brain to barbiturate anesthesia. The cortex, which is thought to be relatively oligosynaptic when compared to the reticular activ-

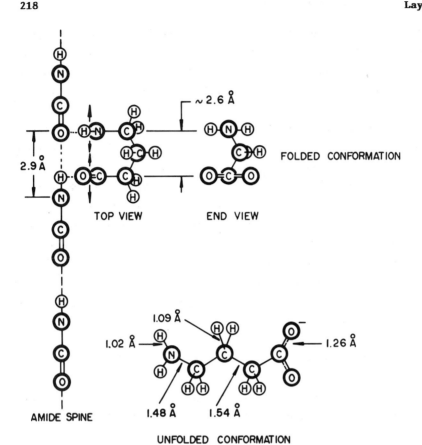

Fig.4. Energetic or nonequilibrium conformation of gamma-amino-
        butyric acid that is capable of interacting with the amide-I
        spine system. The GABA molecule is flexible so it forms a
        two-point attachment that rocks open and closed while it is
        in resonance with the spine. Unlike barbiturates, GABA is
        not a stiff planar ring.

ating system, was not as sensitive to barbiturate anesthesia. This
implies that it was mainly the pronounced surface area of the reticular activating system which accounted for the difference in
sensitivity. Since the neuronal membrane is known to be covered by
a glycoprotein sheet, this result is consistent with a loss of
polysaccharide motility. These results are also in agreement with
Pauling's (26) statement that a loss of consciousness which occurs

during sleep or general anesthesia might be a result of either a
decrease in the activity of the "exciting mechanism" or an increased
impedance of the "suporting mechanism". But unlike Pauling's model
of general anesthesia, which concerns clathrate formation due to
non-hydrogen bonding anesthetic agents, the soliton model of
anesthesia deals with agents that are capable of hydrogen bonding
to the cytoskeleton.

In summary, up to the present there has not been a satisfactory
model to account for barbiturate's diverse activities. Theories
have implicated barbiturate's hydrogen bonding ability as a mode of
action but they have failed to provide a specific context for
activity. The soliton theory offers a unifying mechanism of action
for a number of diverse agents, such as, barbiturates, hydantoins,
succinimides and urethanes. It accounts for the observation that
barbiturates (and antiepileptic agents) appear to enhance GABA's
inhibitory activity. It accounts for the observation that barbiturates display no stereospecific activity at anesthetic concentrations (20), since the one-dimensional spine system of an alpha-
helix cannot be stereospecific. It accounts for the observation
that N-methylated derivatives of barbituric acid are "ultrashort"
hypnotics (27), since N-methylation will "choke" the number of two-
point bonding configurations in the anesthetic ring (see Fig. 3).
During general anesthesia, barbiturates also appear to depress ATP
utilization over generation. In order words, tissue concentrations
of ATP, ADP, AMP and PCr remain unchanged or increase slightly with
narcosis. This has led Siesjo (28) to speculate that barbiturates
disrupt a tightly-coupled and high-gain feedback circuit within the
neuron. This description of a high-gain system is consistent with
the soliton-glycoprotein model. Both the soliton and the polysaccharide (polyelectrolyte) portion of a glycoprotein operate as
long-range and highly nonlinear systems. Since these two systems
are joined at the core of the glycoprotein (an explanation will
follow), it is reasonable to expect that their mutual interactions
will display both high-gain and tight-coupling. This effect may
be conveyed through the filamentous proteins of the cytoskeleton.

IV. SOLITONS IN RELATION TO GLYCOPROTEINS

In this section it will be postulated that some glycoprotein macromolecules contain an alpha-helical protein core which is capable of
supporting solitons. Based on this assumption, a mechanism of
energy transfer from the protein core to the polysaccharide domain
will be developed. As in previous discussions, it will be argued
that the H-N-C=O group is central to this energy transfer process.
Once the energy of a soliton is transferred from the protein core
to the polysaccharide domain, the polyelectrolyte model of
Katchalsky (4) will be applied to explain glycoprotein dynamics.
As Katchalsky has pointed out, polysaccharides perform work by
displacing individual sugar molecules from their equilibrium
positions. In polysaccharides with many fixed negative charges,
this work occurs through an isothermal process. However, before
going into an explanation of this dynamic process, it will be
necessary to review four important points concerning glycoprotein
structure:

1. Glycoproteins consist of both straight and branching
    chain polysaccharide molecules that are bound covalently
    to a polypeptide backbone. The polysaccharide portion of
    the macromolecule forms a "cloud" of polyelectrolytes
    around the core protein which carries a fixed negative
    charge that binds cations, particularly divalent cations,
    in a long-range cooperative manner with the following
    affinities: $H^+ = K^+ \geq Na^+$ and $Ca^{2+} > Mg^{2+}$. Polyelectro-
    lytes expand and contract under the influence of cations
    and in general, the condition $Ca^{2+} > Mg^{2+}$ causes contrac-
    tion, while $H^+ = K^+ \geq Na^+$ all promote expansion (29,30,
    31,32).

2. Three predominant carbohydrate-protein linkages are asso-
    ciated with glycoproteins (33). It has been demonstrated
    that 90% of carbohydrate-protein linkages found in rabbit
    brain are associated with the asparagine-N-acetyl-
    glucosamine complex while the remaining 10% of carbo-

hydrate-protein linkages are associated with either
serine or threonine residues (34,35).

3. Finne (36) has demonstrated that molecules of N-acetyl-
galactosamine are particularly concentrated nearer the
core of glycoproteins than on the periphery of the macromolecule. This configuration is also apparent upon in-
spection of the variety of glycoprotein structures, which
are illustrated by various reviews, and leads to the
impression that glycoproteins are constructed with
multiple acetylamide molecules near the core of the
molecule.

4. Asparagine is classified as a neutral amino acid residue
and hence, its side chain can be found both inside and
outside a protein molecule. Usually, neutral polar residues are found outside the molecule although they can
reside inside the protein if their polar groups are
"neutralized" by hydrogen bonding to other like residues
or to the carboxyl group on the main chain (37).

Asparagine, most likely, would not often enter the interior of the
alpha-helix, nor perhaps into the vicinity of its corresponding
hydrogen bond, without some form of energetic stimulation. When it
is serving as an attachment residue for a polysaccharide chain, it
may enter the vicinity of the hydrogen bonds only when the outer
"cloud" of polysaccharides (polyelectrolytes) forces it into a
configuration of higher internal rotational energy. This, in turn,
might happen only when Ca^{2+} causes the polyanionic sugar "cloud" to
contract down into a tighter volume around the core protein. Under
these circumstances, asparagine would be capable of neutralizing
the amide-I spine in its region, and thereby be capable of altering
both hydrogen bond anharmonicity and the geometry of the spine
system which carries solitons.

Since asparagine attachment residues are always associated with
one or more N-acetylglucosamine(s), these adjacent sugar mole-
cule(s), with their associated acetylamide group(s), would also
move closer to the center of the glycoprotein under the influence

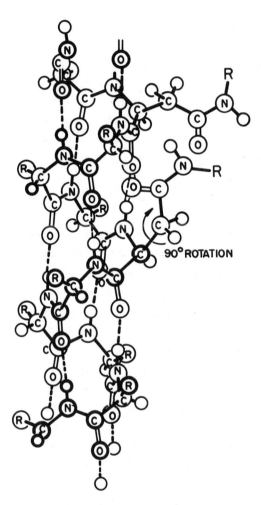

Fig.5.   Asparagine side group rotating about its alpha- to beta-
         carbon single bond to enter the vicinity of the amide-I spine
         system that supports solitons.   The side group's equilibrium
         position is shown above and the higher internal energy
         conformation is shown below.   R equals N-acetylglucosamine
         and its associated polysaccharide chain.

of Ca^{2+}. This movement of sugar molecules could result in a greater association between amide-I spines and acetylamide moities. With this, acetylamide groups can be seen as the active media of energy transfer from alpha-helix to polysaccharide cloud which is under the overall direction of Ca^{2+} in relationship to polysaccharide cooperative behavior. The amide group of asparagine can be seen as a facilitator of energy transfer.

If the effects of asparagine side residues were not offset by some other factor, Ca^{2+} could bind to the polyanionic sugar "cloud" and shut down soliton propagation completely through the glyco-protein core or make it impossible for an unaltered soliton to traverse the modified segment of alpha-helix. Fortunately, the properties of acetylamide overcome the opposing forces of asparagine and Ca^{2+} by creating a dynamic balance of energies. As Ca^{2+} in-duces polysaccharide contraction and compresses (rotates) aspara-gine's amide group nearer to the amide-I spine, it also causes the myriad of associated acetylamide groups to crowd closer to the amide-I spines. Much like barbiturates, the acetylamide groups are capable of forming dipole-dipole interactions with the amide-I spine which absorbs soliton energy. However, unlike barbiturates, this energy would not be randomized to the surrounding medium. Instead, the "leaked" alpha-helical energy would go for use in the poly-saccharide "cloud" of sugar molecules with the possibility of transforming chemical energy into mechanical work by an isothermal process. Since there are so many acetylamide groups around the core of a glycoprotein, they are excellent candidates for "leaking" energy from the soliton in the alpha-helical core into the outer polyelectrolyte matrix by, at least, several mechanisms:

1.   Changing the boat-chair conformation of nearby sugar groups which are on the order of 6 Kcal/mole, based on the C1 to 1C conformation of β-D-glucopyranose (38).

2.   Rotating sugar molecules around their somewhat sterically hindered glycosidic linkages. The extent of this rota-tion varies in different polysaccharides. The energy of

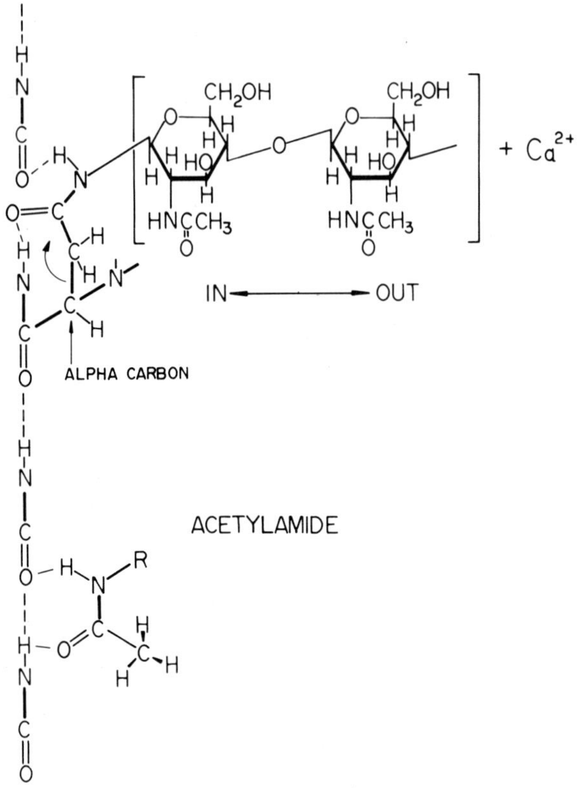

Fig. 6.   Amide-I spine system in relation to asparagine-N-acetyl-
glucosamine - β - 1 - 4 - N - acetylglucosamine which forms
linkages with other negatively charged sugar molecules and
initiates a large percentage of carbohydrate attachment
points.   Interaction of calcium with the negatively charged
sugar molecules causes torsion and compression of the
asparagine side group into the interior of the alpha-helix
with subsequent alteration of amide-I spine anharmonicity
and/or tapping of soliton energy via acetylamide groups.   R
equals glucose or galactose.

these rotations is difficult to quantitate and many polysaccharides, such as cellobiose, are restricted to narrow
values of θ and ψ glycosidic rotational angles. Nevertheless, these energies are around 0.5 Kcal/mole, based on
the cellobiose transition from $\theta = -40°$ and $\psi = -40°$ to
$\theta = -40°$ and $\psi = 0°$ (39).

3. Shifting the hydrogen bonding relationships between adjacent sugar molecules via mechanisms (a) and (b) and also
via direct injection of energy into hydrogen bonds. Bond
energies for single hydrogen bonds are approximately 5-8
Kcal/mole. Based on quantum chemical studies of di-N-
acetylglucosamine it has been demonstrated that hydrogen
bonds are important in stabilizing the mutual orientations
of the two associated pyranose rings. However, the bulky
N-acetyl substituents of the sugar molecule have been
found to be relatively less important than the adjacent
hydroxyl groups in forming hydrogen bonds (40).

4. Shifting of Ca^{2+} binding within the polysaccharide "cloud"
via direct injection of energy into Ca^{2+} associated linkages and also by changing the overall conformation of the
polysaccharide molecules. Reid (31) has proposed that
alginates coordinate Ca^{2+} primarily when the sugar molecules arrange themselves into the "egg-box" configuration
which requires a 1C ring conformation by participating
sugar molecules. It then appears that a change in the
long-range ordering of polysaccharide molecules results
either in release or acceptance of Ca^{2+} within the polysaccharide matrix.

It should be understood that not all sugar molecules harbor
negatively charged groups at physiologic pH, so the affinity for
Ca^{2+} will vary from sugar to sugar. Generally, the preferred coordinating number for Ca^{2+} is eight. However, it has been found that
this number will vary with the coordinating species. For unidentate
ligands the preferred, although not exclusive coordinating number,
is six. For multidentate ligands (chelates), the preferred coor-

dination is seven or eight (41). In addition, the Ca^{2+} - 0 bridge distance is shorter for unidentate ligands than for multidentate chelates. For proteins, Ca^{2+} coordinates with bridge distances ranging from 3.8 to 5.3 Å. This distance appears to depend on the degree of hydration and on the affinity of the coordinating groups for Ca^{2+}. It has been shown that carbonyl groups coordinate Ca^{2+} predominately and that nitrogen groups do not bond with divalent cations. In addition, and contrary to common assumption, there is no clear correlation between the number of carbonyl groups and Ca^{2+} affinity (42). However, there does appear to be a correlation between acidic strength and Ca^{2+} affinity and, in general, weak anions demonstrate the formation constant sequence of $H^+ = Na^+ \geq K^+$ and $Mg^{2+} > Ca^{2+}$. For stronger acid anions the order is reversed, that is, $H^+ = K^+ \geq Na^+$ and $Ca^{2+} > Mg^{2+}$ (29). Polysaccharides chelate Ca^{2+} as a stronger acid anion with complex coordination numbers ranging from six to eight. In such complexes, the usual number of water ligands range from zero to three.

V.   GLYCOPROTEINS AND SELF-ORGANIZING BEHAVIOR

This glycoprotein model proposes that asparagine residues and acetylamide groups act in dynamic concert in the process of tapping energy from the strictly one-dimensional domain of the soliton to the nonlinear and long-range phase space domain of the polysaccharides. It may be possible that the polysaccharide domain displays various forms of behavior depending upon the following parameters:

1. Polysaccharide mass.
2. Density of soliton energy supply.
3. Pattern of energy supply: sporadic vs. constant.
4. Ion concentrations: Ca^{2+}, Mg^{2+}, H^+, Na^+, K^+.
5. Polyelectrolyte acidity (affinity).
6. Polyelectrolyte ionization.
7. Energy trap characteristics of asparagine-acetylamide system.
8. Presence of applied electric fields with isothermic interactions.

Polyelectrolytes which are not coupled to an energy supply are evaluated by (a) The interplay of electric forces, (b) Brownian motion of molecular chains, and (c) Intermolecular van der Waals' forces (4). When glycoproteins are coupled with soliton energy supplies, the system can be approximated phenomenologically by equations that demonstrate both driving and damping. The driving force is provided by the soliton, whereas the damping force is provided by the polyelectrolyte. For such an approximation, the mathematical development of self-organizing systems outlined by Haken (43) assists in understanding the complex behaviors. Based on the fundamental relation

$$\dot{q} = -\gamma q + F(t) \ .$$

Where

      q = A general order parameter describing the state of the system. This order parameter responds with respect to the applied force $F(t)$.

      γ = Phenomenological damping constant.

      $F(t)$ = Force which is introduced into the system by the soliton.

      t = time

Which yields the integral

$$q(t) = \int_o^t e^{-\gamma(t-\tau)} F(\tau)d\tau + C_o e^{-\gamma t} \ .$$

This relation models a number of long-lived systems and "slave" short-lived systems that are expressive of self-organizing behavior. When dealing with a complex system, a number of mode or order parameters are incorporated into the above relation. This establishes a hierarchy of one damped subsystem over the next system. The importance of the above basic relation lies in the following: all the damped modes follow their order parameters adiabatically and the behavior of the whole system is determined by the behavior of a few order parameters. Thus, even very complex systems are subject to mathematical modeling and may show well regulated behaviors. Inclusion of various order parameters into the above relationship

also allows for a bifurcation of behaviors. Therefore, complex
systems like the polysaccharide matrix can operate in different or
evolving modes, which are defined by the behaviors of the order
parameters. The above equations are also important from the point
of view of memory storage because the size of the order parameter
fluctuations are crucial for the overall performance of the system.
Acting in two opposite ways, order parameter fluctuations can affect
both adaptability and reliability. Adaptability, or ease of
switching, requires large order parameter fluctuations and flat
potential curves, whereas reliability, or difficulty of switching,
requires small order parameter fluctuations and deep potentional
curves.

The above phenomenological model is included in order to introduce an intuitive sense of importance to self-organizing behavior
in glycoproteins. It is not intended to be a strict representation
of glycoprotein dynamics. However, it does point out that self-
organizing behavior in glycoproteins may be quite important to the
mechanism of memory storage on the cellular membrane. The concept
that memory storage may be contained in the glycoprotein pool on
the cellular surface has been addressed by Adey (6). Also an
insightful model of soliton waves on the lipid bilayer of a cell,
which satisfies the Sine-Gordon equation, has been developed
Lawrence and Adey (9). This splay wave model is particularly
significant because it develops a mathematical approach whereby
large sections of membrane (and glycoprotein) may be coordinated
by one-dimensional waves.

VI. EXPERIMENTS TO VERIFY THE PROPOSED MODEL

Work is now under way to verify the existence of Davydov solitons
in isolated alpha-helical proteins. The basic strategy involves the
launching of an energy pulse (in the form of vibrational or chemical
bond energy) from one end of an alpha-helix and measuring the movement of that pulse. This is done by following the course of the
induced molecular perturbation with an extremely rapid sequence of
laser light flashes and noticing the concomitant Raman scattering.
If solitons are found in homogeneous polypeptides, then more com-

plex molecules will be examined. For example, the role of the
asparagine side chain can be examined by placing one or more asparagine residues in the center of a polyalanine helix and observing the
soliton as it moves through this region. Similarly, it is possible
to attach an N-acetylglucosamine moiety to the asparagine side
chain and perform an identical experiment. Biological alpha-helices
of known structure can also be examined. The goal is to work on
increasingly complex systems.

The activity of barbiturates, hydantoins, succinimides, urethanes and GABA can also be examined by using a similar method.
Again, a soliton can be launched from one end of an alpha-helix and
can be examined with various concentrations of anesthetic present
in the experimental medium. By gradually increasing the concentration of anesthetic in the experimental medium, it sould be
possible to "short circuit" soliton propagation. In addition,
Fig. 3 shows that barbiturates have four, hydantoins have three,
succinimides have two and urethanes have one potential H-N-C=O
bonding configuration across an alpha-helix. Therefore, equimolar
concentrations of these agents should demonstrate a "titration"
curve of soliton perturbation. Also, selective N-methylation of
these agents should decrease their activities in a predictable
manner. The best experimental results should come from procedures
that contrasted N-methylated derivatives against non-methylated
derivatives, since the problems with controls would be largely
avoided.

It is also conceivable that symmetric Davydov solitons could
be transformed into asymmetric solitons at the point of perturbation. Since the symmetric soliton has a higher energy than the
asymmetric soliton, its conversion to the lower energy propagation
could be hastened by an energy robbing interaction. An asparagine
side chain, an acetylamide moiety or an anesthetic interaction with
the alpha-helix could perform this conversion function. Symmetric
solitons can be likened to a three phase conductor, whereas an
asymmetric soliton can be likened to a two phase conductor. Therefore, a different Raman spectra should be obtained for each type
of propagation.

Most assuredly, it will be difficult to develop a completely synthetic soliton-glycoprotein experiment. Starting with a known glycoprotein of biological origin would be the easiest first step. However, the soliton-glycoprotein system can be verified in part by using molecular orbital theory to calculate the equilibrium conformation of the dynamic species. If solitons are found to propagate with physiologic energy supplies, then the soliton-glycoprotein model could point the direction to new mechanisms of energy utilization and regulation.

VII.   OPEN QUESTIONS DEVELOPED BY THE MODEL

Soliton-Glycoprotein Behavior at Points of Altered Amide-I Spine Anharmonicity?

a.   "Energy trap" behavior, meaning, soliton resonance $\geq$ acetylamide resonance $\gg$ long-range polysaccharide resonance. In this particular instance a soliton is caught by the acetylamide and asparagine side groups that are connected to an alpha-helix. The energy of the soliton is then directed to the polysaccharide matrix in order to power glycoprotein displacements. This allows a glycoprotein to perform as an isothermal macromolecular machine.

b.   "Pseudo-termination" behavior, meaning, the combination of asparagine side chains and acetylamide interactions (plus GABA and barbiturate interactions) might alter an alpha-helix to the extent that it cannot propagate or even "energy trap" a soliton. In this particular instance, the soliton could reflect from the altered area of helix. Preliminary computer analysis by Scott (12) indicates that solitons interact with their tail as they reflect from the raᵥ end of a helical molecule. Therefore, a reflected pulse is not as robust as a nonreflected soliton. In general, a reflected soliton has a lifetime = $(0.54 \times 10^{-12})n_{max}$ seconds, where n_{max} = maximum number of unit cells on a helix. A simple calculation of the reflected distance is illuminating. Since a mitochondrion is about 10^4 Å in diameter, or about 1800 turns on a 5.41 Å per turn alpha-helix, signaling as a result of a secondary chemical event might be feasible over a useful subcellular distance. By the simple relation shown below, the maximum signaling distance might be:

maximum distance = maximum time × velocity

$$D_{max} = T_{max} \times V_{reflection}$$

from Hyman, <u>et. al.</u> (1981): $V_{reflection} = 1.26 \times 10^3 \dfrac{meters}{second}$

Assume, $n_{max} = 1800 = 10^4 \overset{o}{A} \cong 1$ mitochondrial diameter

Therefore:

$$D_{max} = (0.54 \times 10^{-12})(1800) \ sec \times 1.26 \times 10^3 \dfrac{meters}{second} \times 10^{10} \dfrac{\overset{o}{A}}{meter}$$

$$D_{max} = 1.2 \times 10^4 \overset{o}{A} \cong 1 \ \text{mitochondrial diameter}$$

It appears that intracellular signaling, as a result of a reflected energy pulse, might be feasible over a distance of one or more mito-chondrial diameters, depending on the maximum length of the alpha-helical protein. As an example of such a secondary chemical event, McClare (44) has proposed a reversible (or isoenergetic) exchange of energy between a pulse on an alpha-helical protein and the activated $ADP/Mg^{2+}/Pi$ complex. If this exchange of energy were sufficiently dense, it is conceivable that a thermodynamic reversal of ATP hydrolysis could affect localized concentrations of ATP/ADP within the cytosol. Therefore, further numerical investigation of Davydov soliton behavior, particularly at points where reflection is likely, seems justified in order to clarify the "energy-trap" versus "pseudo-termination" question.

c. "Energy regulation" behavior, meaning, the dynamic displacements of the asparagine side chains and acetylamide groups within glyco-proteins may act to modulate soliton propagation at the glyco-protein. This behavior would depend on the self-organizing proper-ties of the glycoprotein and can be visualized as a dynamic mechan-ism which is intermediate to energy absorbtion and reflection. Hence, it is conceivable that a glycoprotein may convert a soliton into a variety of energy profiles.

Cytoskeletal Movements in Relation to Solitons?

Glycoproteins that are coupled to soliton power supplies may affect cellular architecture by several mechanisms:

a. Glycoproteins are polyelectrolytes that are capable of chelating large stores of Ca^{2+}. Therefore, they may influence cytoskeletal architecture by altering microenvironmental Ca^{2+} flux. Modulated by solitons, glycoproteins may release or gather Ca^{2+} as they function as isothermal macromolecular machines. This Ca^{2+} flux may influence actin-binding proteins that are sensitive to divalent cations. These proteins are believed to regulate cell architecture and motility.

b. Glycoproteins are known participants in the antigen-antibody complex. They have also been implicated in the adhesion of one cell surface to another. Therefore, it seems reasonable to speculate that intracellular glycoproteins may adhere to one another which may serve some useful function in cytoskeletal migration. Asymmetric Davydov solitons may play a functional role in such associative movements by displacing only a portion of a polyelectrolyte at any one time. Asymmetric polysaccharide displacement may result in a change of "adhesiveness" of one portion of polyelectrolyte in relation to another. This change in adhesiveness may influence glycoprotein movements within the cytosol.

Biological Alpha-Helices in Relation to Solitons?

Alpha-helical proteins are a common "generic" structure in living organisms. However, biological alpha-helical proteins are not homogeneous and idealized helical structures. Their shapes are distorted by interactions with multiple side groups and by interactions with other neighboring molecules and ions. The dynamic stability of Davydov solitons in these approximate alpha-helices is not known. Hence, a valid criticism of the soliton model would argue along the line presented in the last few sentences.

Nevertheless, the special particle-like and stable qualities of solitons makes them excellent candidates for surviving the distortions of symmetry that are inherent to biological alpha-helices. These distortions may be visualized either as static or dynamic. A static distortion is an unchanging or permanent alteration of true alpha-helical symmetry. This type of distorsion may function as an "energy guide" for a pulse on an alpha-helix. It could reproduce

a similar energy loading and unloading function with each propagation on the molecule. It may direct a soliton (or a below threshold
pulse) to a specific site on the molecule where it is utilized for
a specialized function. In contrast to the mechanism of energy
modulation via a fixed structure, is the mechanism of energy modulation via a dynamic structure. This is exemplified by the anesthesia and glycoprotein models presented in previous sections. The
dynamic alpha-helix is created by altering the positions of H-N-C=O
groups that are extrinsic to the amide-I spines of the alpha-helix.
In this particular instance, energy loading and unloading is flexible and subject to "useful" change.

In closing, the distance that a soliton may propagate down a
real alpha-helix is open to question. In globular proteins the
longest alpha-helical components are 35-40 $\overset{\text{o}}{\text{A}}$ (44). Therefore, there
is a question as to whether or not a soliton will form in a globular
protein, since a soliton's length is approximately 10 H-N-C=O groups
($\sim$50 $\overset{\text{o}}{\text{A}}$) in an alpha-helix. In the interwoven and alpha-helical
rich cytoskeleton of the neuron, the maximum range of propagation
could be significantly longer.

ACKNOWLEDGMENT

It is a pleasure to thank W. Ross Adey, Alwyn Scott, Peter
Lomdahl and Robert Birge for stimulating conversations.

REFERENCES

1. A. S. Davydov & N. I. Kislukha, Phys. Stat. Sol. (b) <u>59</u>, 465
   (1973).

2. A. S. Davydov, Phys. Scripta <u>20</u>, 387 (1979).

3. R. J. Lasek & M. L. Shelanski, <u>Neuroscience Res. Prog. Bull.</u>
   <u>Vol. 19 No. 1</u>, MIT Press, Cambridge, 1981, pp. 32-82.

4. A. Katchalsky, <u>Connective Tissue: Intercellular Macromolecules</u>
   <u>(Symposium New York Heart Assoc.)</u>, Little, Brown & Co., Boston,
   1964, p. 9.

5. L. Warren, <u>In Biological Roles of Sialic Acid</u>, (A. Rosenberg
   & C. Schengrund, eds.), Plenum Press, New York, 1976, p. 103.

6. W. R. Adey, BioSystems <u>8</u>, 163 (1977).

7. A. Weeds, Nature 296, 811 (1982).

8. W. R. Adey, Physiological Rev. 61, 435 (1981).

9. A. F. Lawrence & W. R. Adey, Neurological Research 4(1&2), 115 (1982).

10. A. S. Davydov, Biology and Quantum Mechanics, Pergamon, Oxford, 1982, p. 185.

11. J. M. Hyman, D. W. McLaughlin & A. C. Scott, Physica 3D 1&2, 23 (1981).

12. A. C. Scott, Physical Review A 26(1), 578 (1982).

13. S. J. Webb, Physics Reports 60, 201 (1980).

14. A. C. Scott, Physica Scripta, 25, 651 (1982).

15. P. S. Lomdahl, L. MacNeil, A. C. Scott, M. E. Stoneham & S. J. Webb, Physics Letters 92A(4), 207 (1982).

16. L. Pauling, Science 134, 15 (1961).

17. H. J. R. Weintraub, International J. Quantum Chem. Quantum Biol. Symposium 4, 111 (1977).

18. S. H. Roth, In Molecular Mechanisms of Anesthesia, Progress in Anesthesiology Vol. II, (B. R. Fink, ed.), Raven Press, New York, 1980, p. 119.

19. R. A. Nicoll, Proc. Natl. Acad. Sci. USA 72, 1460 (1975).

20. J. L. Barker, L. M. Huang, J. F. MacDonald & R. N. McBurney, In Molecular Mechanism of Anesthesia, Progress in Anesthesiology Vol. II, (BR Fink, ed.) Raven Press, New York, 1980, p. 79.

21. E. G. Brunngraber, Neurochemistry of Aminosugars: Neurochemistry and Neuropathology of the Complex Carbohydrates. Charles C. Thomas, Springfield, 1978, Chapters 7 & 13.

22. W. R. Adey, R. T. Kado & J. Didio, Experimental Neurol. 5, 47 (1962).

23. W. R. Adey, B. G. Bystrom, A. Costin, R. T. Kado & T. J. Tarby, Experimental Neurol. 23, 29, (1969).

24. W. R. Adey, Progress in Physiol. Psychol. 3, 181 (1970).

25. T. M. Darbinjan, V. B. Golochinsky & S. I. Plehotkina, Anesthesiology 34, 219 (1971).

26. L. Pauling & R. B. Corey, Proc. Natl. Acad. Sci. USA 37, 205 (1951).

27. J. Andriani, The Chemistry and Physics of Anesthesia, 2nd Ed. Charles C. Thomas, Springfield, 1962, Chapter 19.

28. B. K. Siesjo, Brain Energy Metabolism, John Wiley & Son, Chichester, 1978, p. 233.

29. M. N. Hughes, The Inorganic Chemistry of Biological Processes, John Wiley & Son, London, 1972, Chapters 3 & 8.

30. D. A. Rees, In Functional Linkage in Biomolecular Systems, (F. O. Schmitt, D. M. Schneider, D. M. Crothers, eds.), Raven Press, New York, 1975, p. 35.

31. D. S. Reid, In Ions In Macromolecular and Biological Systems, (D. Everett & B. Vincent, eds.), University Park Press, Baltimore, 1978, p. 82.

32. T. W. Barrett & W. L. Peticolas, J. Raman Spectroscopy 8, 35 (1979).

33. R. Kornfeld & S. Kornfeld, Annual Rev. Biochem. 45, 217 (1976).

34. R. U. Margolis, R. K. Margolis & D. M. Atherton, J. Neurochemistry 19, 2317 (1972).

35. R. K. Margolis & R. U. Margolis, Biochimica et Biophysica Acta 304, 421 (1973).

36. J. Finne, Biochimica et Biophysica Acta 412, 317 (1975).

37. R. E. Dickerson & I. Geis, The Structure and Action of Proteins, Harper & Row, New York, 1969, pp. 13-14.

38. P. J. Winterburn, In Companion to Biochemistry, Selected Topics for Further Study (A. T. Bull, J. R. Lagnado, J. O. Thomas & K. F. Tipton, eds.), Longman Group, London, 1974, p. 307.

39. R. E. Christoffersen, Quantum Mechanics of Molecular Conformations, (B. Pullman, ed.), John Wiley & Son, London, 1976, p. 274.

40. J. S. Yadav, G. Barnickel & H. Bradaczek, J. Theor. Biol. 95, 285 (1982).

41. D. W. Urry, Ann. N.Y. Acad. Sci. 307, 3 (1978).

42. R. H. Kretsinger, Neuroscience Res. Prog. Bull. Vol. 19 No. 3, MIT Press, Cambridge, 1981, pp. 236-237, 298-302.

43. H. Haken, Synergetics: An Introduction 2nd Ed, Springer-
     Verlag, Berlin, 1978, Chapter 7.

44. C. W. F. McClare, Ann. N.Y. Acad. Sci, 227, 74 (1974).

45. J. S. Richardson, In Advances In Protein Chemistry Vol. 34,
     (C. B. Anfinsen, J. T. Edsall & F. M. Richards, eds.),
     Academic Press, New York, 1981, p. 167.

46. P. C. Jocelyn, Biochemistry of the SH Group, Academic Press,
     London, 1972, p. 98.

11

Synthesis and Characterization of the Dimer (1,11-Dodecadiyne): Structure of the 8-Polydiacetylene Dimer

Jerome B. Lando and Mrinal K. Thakur/Department of Macromolecular Science, Case Institute of Technology, Case Western Reserve University, Cleveland, Ohio

INTRODUCTION

Extensive work has been performed on diacetylene polymerization in the past few years.[1] The unique characteristic of these poly-merizations is that they can result in a large, nearly defect-free single crystal consisting of fully extended polymer chains.[2] This polymerization is achieved by crystallizing the monomer and then ex-posing crystals to radiation (γ-rays, x-rays, UV or visible) or thermal annealing to initiate polymerization. Chain propagation occurs only by a 1:4 addition of the monomer in the crystalline phase. The general reaction scheme is given in Figure 1. Two mesomeric structures of the polymer backbone are possible. The acetylenic form is usally energetically more favorable and is observed more often.[3] The cumulenic form, however, does exist in some cases where it may be favored by the reaction kinetics[4] or the packing of side groups.[5] In addition to the solid state reaction, true single crystals can be formed by simultaneous polymerization and crystallization either from solution or from the solid state if one controls the nucleation step

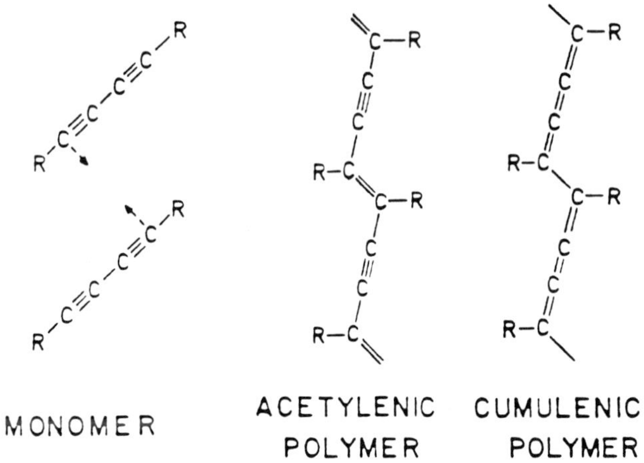

MONOMER   ACETYLENIC   CUMULENIC
                    POLYMER   POLYMER

Figure 1   Diacetylene Reaction Scheme.

sufficiently.[6]

The diacetylene polymerization of the macromonomers
$((CH_2)_n-C\equiv C-C\equiv C)_x$ (n = 5,6,8) has been recently reported.[7,8,9] The
macromonomers were synthesized by an oxidative coupling of (1,11-
dodecadiyne) through a step growth reaction. In the present work a
low molecular weight material (dimer) was synthesized.

$$\underset{R}{\underline{HC \equiv C - (CH_2)_8 - C\equiv C-C\equiv C-}}\ \underset{R}{\underline{(CH_2)_8 - C \equiv CH}}$$

This dimer acts as a monomer for a subsequent diacetylene polymerization initiated by Co^{60} γ-irradiation. Besides the central diacetylene group, the hope was that the terminal acetylene groups
could very well participate in a polymerization reaction. Therefore
a major objective of this work was to study the structure and reaction of these triple bonds along with that of the diacetylene groups
by a detailed x-ray structure analysis of a single crystal.

CRYSTALLIZATION AND PHASE TRANSITIONS

Microscopic single crystals of the dimer have been obtained
from a hexane solution of the mixture at 4°C. Electron diffraction
patterns from these crystals have been recorded for different dosages
of γ-irradiation and consequent diacetylene polymerization. This

diffraction analysis was very useful for studying transitions in the course of the diacetylene polymeiization. A distinct intermediate between the monomer and the polymer has been observed for a range of conversion. The diffraction patterns at three different compositions are given in Fig. 2a, b and c respectively. Before polymerization the b*c* lattice net is orthogonal. In the intermediate the intensity distribution changes considerably, as does the monoclinic angle (α). In the final or completely polymerized phase, however, the b*c* lattice net is orthogonal. The observed intermediate may be a solid solution of the monomer in polymer or the polymer in monomer. The structure in the intermediate appears to be considerably disordered, resulting in diffuse peaks in the diffraction pattern. In the initial (monomer) and the final (polymer) phases, however, the lattice distortions are negligible and consequently the diffraction maxima are sharp. The unit cells corresponding to the different phases are given in Table 1. A more accurate unit cell for the polymerized dimer as determined from x-ray measurement is given in Table 2, along with other crystal data.

X-RAY EXPERIMENTAL

A well-formed plate-like polymerized crystal of dimensions 0.5 x 0.05 x 1.5 mm was chosen for x-ray data collection. Preliminary Weissenberg and oscillation photographs indicated that it crystallizes in a monoclinic system with the chain axis (unique axis) lying

TABLE 1. Unit Cell Dimensions of the Different Crystalline Phases of Dimer

Crystal phase	a (Å)	b (Å)	c (Å)	α
Dimer monomer (before irradiation)	28.25	7.15	5.24	101.5°
Intermediate	Unknown	7.92	5.12	96.2°
Polymerized (to completion)	Unknown	8.21	4.93	90.0°

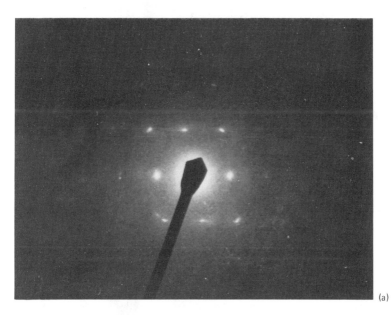

(a)

Figure 2 (a) electron diffraction pattern of the dimer (unirradiated);
(b) electron diffraction pattern of the intermediate phase; (c) electron
diffraction pattern of the polymerized dimer.

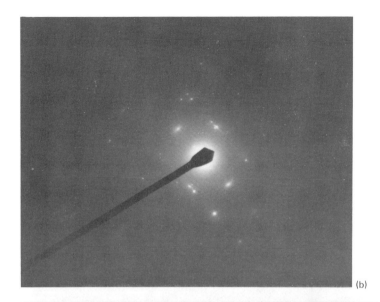

(b)

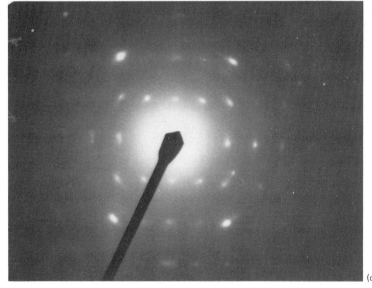

(c)

TABLE 2.   Crystal Data of the Polymer

Crystal system:	Monoclinic
Space group:	$P_{2/n}$; z = 2
Lattice parameters:	a = 26.779(8)Å
	b = 8.254(10)Å; b = 8.21Å $\Big\}$ Electron
	c = 4.906(6)Å ; c = 4.93Å $\Big\}$ Diffraction
	γ = 119.558(12)°
Unit cell volume:	v = 943.26Å^3
Mol. wt. of a repeat:	M = 344 g/mole
Calculated density:	d_c = 1.133 g/cm^3
Measured density:	d_m = 1.12 g/cm^3
Mass absorption coefficient:	μ = 1.33 cm^{-1}

along the long edge of the crystal plate. The intensity data were
collected using an automated Syntex $P2_1$ four circle diffractometer
operating with a graphite monochromator and MoK$_\alpha$ radiation (λ =
0.71073Å). Accurate unit cell dimensions and the orientation
matrix were calculated from the least squares refinement of the
angular positions of twenty higher order reflections. The intensity
of three standard reflections (117,3-31 and 66-2) was periodically
monitored after each 50 reflections to check the crystal and the
diffractometer stability. No significant deviation in intensity was
observed. The data were collected by an ω-scan tecnhique at rates
varying from 1.5 to 4.0°min^{-1}, such that more time was devoted to
less intense reflections. A total of 5278 data points ranging from
h = -14 to 0, k = -10 to 19, and ℓ = 0 to 13 were collected. No
absorption correction was applied because the mass absorption coefficient was 1.33 cm^{-1} and a side scan (ϕ scan) over a few reflections showed no significant deviation in intensities. Lorentz and
polarization corrections were applied. Reflections with intensities
greater than 2.5 times the corresponding standard deviation were

counted as observed. The total number of observed reflections was 926. The refinement of the structure was accomplished with these data points. The agreement between the observed and the calculated structure factors was characterized by the reliability index or residual:

$$R = \frac{\Sigma \left| F_0 - F_c \right|}{\Sigma F_0}$$

Refinement and Results:

The structure was solved essentially by a trial method. Additional information from various other sources was particularly useful in this attempt. Structures of similar materials have been solved by the present authors[7,8,9] as well as by others.[5] A comparison with all these known structures and crystal data facilitated the initial steps towards the solution. The unit cell dimensions provided the preliminary information regarding the orientation of the chains and the side groups. The c-axis perfectly matches the diacetylene chain repeat so the diacetylene backbone should be oriented along the c-axis. The a-repeat is approximately twice the length of the side group $(CH_2)_8-C\equiv CH$, for an all-_trans_ conformation. Therefore it was a reasonable assumption that the side groups should be oriented along the a-axis. There are two chains per unit cell and a glide plane perpendicular to the c-axis. Thus one chain should run through the center of the ab plane. The unit cell being monoclinic (c-axis unique), the side group orientation, if assumed along the a-axis, is perpendicular to the main polymer backbone. The perpendicular distance between the two neighboring side groups (planar zigazg) is a little more than 4Å, in close agreement with the packing of polyethylene chains. Based on all these internally consistent logical justifications, a preliminary model was constructed where the polymer backbone was placed along the c-axis and the side chains along the a-axis. Several other models with minor and major modifications of this preliminary one were also investigated. Calculations were made on a VAX 780 computer.

Initially the linked atom least squares (LALS) program originally developed by S. Arnott and co-workers[11] was used with a hundred

ac Projection

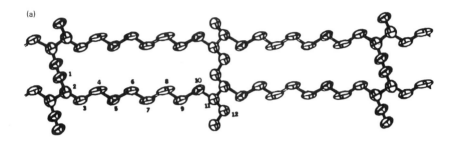

bc Projection

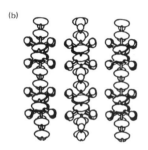

Figure 3 (a) ac projection of polymerized dimer (only a section has
been shown for clarity); (b) bc projection of polymerized dimer.

randomly selected reflections for a preliminary refinement of the
bond angles, dihedral angles and the overall isotropic temperature
factor. The minimum residual obtained from this refinement was 21%.
The existing bad contacts were minimized by varying the orientation
of the terminal acetylene groups, which subsequently led to a residual
of 19%. The calculated coordinates from the LALS output were
subsequently used for a full matrix least squares refinement over
all 926 observed reflections. The starting residual was found to
be 35%. A free variation of the coordinates of all the carbon
atoms decreased it to 19% in a few cycles. The individual isotropic temperature factors were subsequently varied and the hydrogen
atoms added to the carbons. The resulting residual was 12%. At
this point a problem was apparent. The terminal acetylene hydrogens of the two consecutive molecules in the unit cell were found
to be too close together. The hydrogen of one seemed to be almost
overlapping with the carbon of the other. This difficulty was
eliminated by considering the possibility of a chemical (addition)
reaction among the acetylene groups through the high energy photo-
initiation (γ-ray) and the consequent formation of a polyacetylene
backbone parallel to the c-axis. The calculated coordinates of
the acetylene carbons from the structure refinement were in close
agreement with that corresponding to a _trans_ polyacetylene chain.
The hydrogen atom position was now shifted sideways by 120°. The
ideal coordinates for this hydrogen atom, appropriate to a polyacetylene chain, were calculated separately and incorporated for
refinement. The final residual after refining the anisotropic
temperature factors was found to be 0.08%. The polyacetylene carbon atoms were restricted to isotropic (spherical) temperature
factors because their refined anisotropic (ellipsoidal) temperature
factors were too elongated along the chain direction without any
significant improvement of the residual. The ac and the bc
projections are given in Fig. 3a and 3b. Half of the unit cell with
no hydrogen atoms is shown in the ac projection for clarity. The
deviation of the carbon atom positions from the ac plane ($\vec{b} = 0$)
are given in Table 3, which indicates that the deviations are within

Table 3. Deviation of the Atom Positions From the ac Plane

Atom	Deviation (Å) from the ac Plane (y=0)	r.m.s. Deviation (Å)
C1	-.0030	
C2	-.0318	
C3	-.0331	
C4	0.0107	
C5	-.0179	
C6	0.0263	0.021
C7	-.0000	
C8	0.0332	
C9	0.0092	
C10	0.0268	
C11	0.0025	
C12	-.0229	

experimental error. All the non-hydrogen atoms are essentially coplanar in the ac plane.

Discussion:

The final structure is composed of sheets at y = o & ½ b of regular two-dimensional networks of polydiacetylene and polyacetylene chains interconnected by eight methylene units (the planer zigzag). The interplanar distance is approximately 4Å. The bond lengths along the diacetylene direction are consistent with the acetylenic form.

One of the most important results of this structure analysis is the measurement of bond lengths among the polyacetylene carbons. This is the first time that polyacetylene bond lengths have been experimentally determined. (See Table 4) There are additional important physical properties deducible from the structure. As men-

Table 4. Bond Lengths Along the Diacetylene and the Polyacetylene
Directions

Bond Type	Bond Length (Å) Along	
	Polydiacetylene Chain	Polyacetylene Chain
C − C	1.451(10)	1.463(9)
C = C	1.359(9)	1.357(13)
C ≡ C	1.248(6)	

tioned earlier and shown in the ac projection, the structure is
composed of sheets of polydiacetylene and polyacetylene chains,
lying parallel to the ac plane. The second layer of atoms or the
sheet in the middle of the unit cell is lying $\frac{1}{2}(\vec{a} + \vec{b})$ away from
the original sheet at the ac plane. Consequently, the polydiacety-
lene and the polyacetylene chains are only 4Å apart along the
b-direction. A considerable overlap between these chains may be
expected. In addition to the π overlap among the polyacetylene
carbons, this interchain overlap may lead to a high electrical conductivity. It should not be surprising if the conductivity is even
higher than that of the undoped polyacetylene by itself. The results of the preliminary investigations of its conducting properties
are presented in the following section.

A two-point measurement of the conductivity of polymerized dimer
of 1,11-dodecadiyne was made on three single crystals, approximately
1.6mm long along the chain direction (c-axis). The single crystals
were too small for a four-point measurement. In order to make a
four-point measurement, the material was crystallized in the form of
large spherulites (diameter ∿ 1cm) as shown in Fig. 5. Initially a
two-point measurement was made on these spherulites by placing the
contact points diametrically across. An optical birefringence test
indicated that the polymer chains were approximately radially
oriented in the spherulites. The spherulitic samples were subsequently collected and compressed in the form of a film, suitable
for a four-point measurement. A digital electrometer with range
from 10^{-11} to 10^{-1} amp was used for the current measurement and the

Figure 4   Spherulitic growth of polymerized dimer.

voltage was recorded by a digital Keithly multimeter. The experiment was repeated from time to time for 2-3 months and no significant decay of conductivity was noticed. A preliminary study of the temperature dependence of the conductivity showed that the conductivity falls with decrease of temperature. At liquid nitrogen temperature (-196°C) the conductivity was too low to be measured ($< 10^{-7}$ Ω^{-1} cm^{-1}) on our apparatus.

The conductivity from the two-point measurements on the single crystals was approximately 10^{-2} Ω^{-1} cm^{-1} along the chain direction (long edge). The actual conductivity should, however, be higher than this observed value because in a two-point measurement the contact resistance is not eliminated. Thus the observed resistance was higher than the actual. The spherulitic samples also exhibited a similar value of conductivity for two-point measurement. The I-V characteristics from the four-point measurement were linear or ohmic at room temperature and the calculated conductivity was approximately 2×10^{-2} Ω^{-1} cm^{-1}. Since the four-point measurement was not made on a single crystal, the measured value is a resultant over somewhat different chain orientations. Comparison of the conductivities of the oriented ($>10^{-2}$ Ω^{-1} cm^{-1}) and pressed spherulite ($\sim 2 \times 10^{-2}$ Ω^{-1} cm^{-1}) samples shows that the degree of anisotropy in conductivity is probably not very large. The remarkable stability of the conductivity indicates that the material is not significantly susceptible to oxidation in air or the attack of water. The decrease of conductivity with a decrease in temperature is a characteristic of a semiconducting material. Therefore the polymerized dimer may be classified as the first good (low band gap) organic semiconductor undoped.

The electrical properties of a crystalline material depend upon its electronic structure, and ultimately on the chemical constitution of its repeat unit. Therefore to elucidate the electrical properties of the polymerized dimer our study of its crystal structure was coupled with an investigation into its structure-conductivity relationship. As described previously, the polymerized dimer is composed of sheets of alternating polydiacetylene

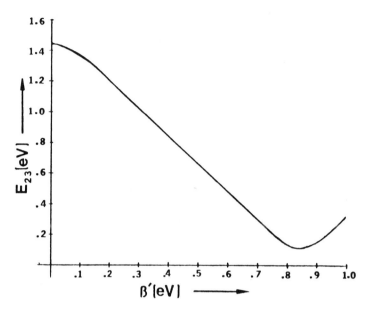

Figure 5 Dependence of the band gap (E_{23}) on the
interchain interaction (β').

and polyacetylene backbones. The nearest neighbor distance between
a polydiacetylene and polyacetylene chain is 4Å along the b-direction. All chains have a planar conformation with the planes
parallel to the ac surface. Geometrically the p_z orbitals of the
backbone carbons are directed approximately along the b-axis. Thus
the chains in neighboring sheets (4Å apart) along the b-axis should
have a considerable π-electron overlap and this overlap is probably
responsible for the high conductivity. Originally polyacetylene
has a conductivity of less than 10^{-5} Ω^{-1} cm^{-1} and polydiacetylene
has less ($\sim 10^{-7}$ Ω^{-1} cm^{-1}). But in the specific system of the
polymerized dimer, the systematic coupling of the two linear chains
in a regular lattice enhances the conductivity by more than 3 orders
of magnitude. The coupling acts like a dopant intercalated between the chains. The interchain distance in the a-direction being
much larger (~ 13.5Å), the conductivity along the a-axis should be
lower than that in the b- and c-directions. Yet the overall anisotropy in conductivity should not be too high.

The π-electron band structure of this material was analyized using Huckel formulation and the effective band gap was calculated as a function of the interchain coupling (Fig. 5). The band gap, as estimated is approximately 0.7 ev.

ACKNOWLEDGEMENT

The partial support of this work by the Office of Naval Research under contract N00014-77-C-1234 is gratefully acknowledged.

REFERENCES

1. Wegner G. Makromoleculare Chemie 134, 219 (1970).

2. Wegner, G., Z. Naturforsch. 24b, 824 (1969).

3. Boudreaux, D. S., and Chance, R. R., Chem. Phys. Letts. 51, 273 (1977).

4. Boughman, R. H., personal communication, 1982.

5. Enkelman, V., and Lando, J. B., Acta Cryst., B34, 2342 (1978).

6. Wegner, G., E. W. Fischer and Munoz-Escalona, Makromol. Chem. Suppl. 1, 521 (1975).

7. Thakur, M. K., and Lando, J. B., Structure Property Relationship of Polymer Solids, P. Anne Hiltner, Editor, Plenum.

8. Day, D. R. and Lando, J. B., J. of Polym. Sci., Polym. Letts., 19, 227 (1981).

9. Thakur, M. K. and Lando, J. B., Macromolecules, 16 143 (1983).

10.Day, D. R., and Lando, J. B., Macromolecules, 13, 1483 (1980).

11.Arnott, S., Campbell Smith, P. J., Acta Cryst. A34, 3 (1978).

12

How Do We Talk to Molecular Level Circuitry ?

Albert F. Lawrence/Advanced Technolgy,
Design Development Lab, Hughes Aircraft
Company, Long Beach, California

I. INTRODUCTION

A major problem with any hypothetical molecular-level machine is communication. A computer built from an assembly of nanometer circuits
would lose some of its intrinsic appeal if input and output required
a roomful of high precision lasers and high sensitivity detectors.

Solving the input-output (I/O) problem forces us either to consider
means of communication between structures at the micrometer scale and
structures at the nanometer scale or to devise ways to bypass problems of scale altogether. We will outline two approaches from the
many possible. The first approach involves a method which may be
used to relax requirements for structuring circuit elements by using
the time domain in the form of spectral information to communicate
with molecular structures. I/O by optical means is highly compatible
with a content addressable memory (CAM) architecture.

   The second approach involves the use of highly sensitive
Josephson junction detectors built to sub-micron dimensions. These

253

detectors may respond to magneto-optical or electro-optical events occuring in small populations or even single molecules. Because successful application of the second method requires an advance in our understanding of quantized nonlinear systems coupled through a radiation field, we will limit our discussion to some of the theoretical issues.

II.   CONTENT ADDRESSABLE MEMORY

Content addressable memory is the most highly parallel computer architecture presently proposed in the literature. In this design the central processor broadcasts read and write requests to every memory cell simultaneously. Access to a given memory cell is determined by the content of that cell. Content addressable memory is a highly efficient scheme for a large class of computations. We may also argue that this memory design is more compatible with molecular electronics than is conventional gate technology. In the former approach, information is coded in the physical states of individual molecules.

Before we address the molecular physics, we will review the operations which are required for content addressable memory to perform parallel computations efficiently. Our discussion follows that of C. Foster (5). The structure of a simple CAM is pictured in Figure 1. The memory is divided into a central control unit and a number of memory cells. The central control unit may broadcast a mask and a comparand to all memory cells simultaneously. Memory cells may also be accessed according to a fixed priority scheme. The following operations are available in this prototype CAM.

SET: Turn on all tag bits.

COMPARE: Where the mask contains ones compare bits of comparand to bits of every cell simultaneously. If comparand and cell disagree, reset tag bit where mask has zeros, do not do any comparisons.

REPORT: Tell control unit if any of the cells are responders.

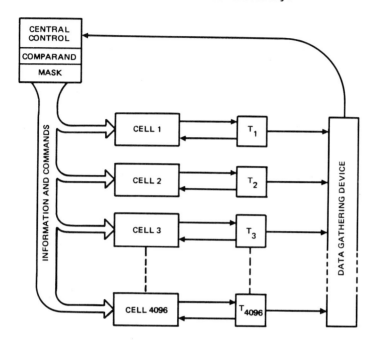

Figure 1 Overview of CAM (after Foster, 1976).

> FIRST: Keep the first responder. Reset all others.
>
> WRITE: In those portions of the word where the mask cont-
>   ains ones, copy the contents of the comparand into
>   each cell that is a responder; do not change state
>   of tag bits. In those portions of the word where
>   the mask contains zeros, do not change the original
>   values of any of the cells.

A number of algorithms may be made bit serial, rather than word
serial by using these five operations. The following arithmetical
operations may be performed by a sequence of instructions which
depend only on word length.

> Adding a constant to every word in memory.
>
> Multiplying every word in memory by a constant.
>
> Squaring each word in memory.
>
> Performing table-driven transforms on every word in memory.

Each of these operations may be performed in fixed time, independent
of the data in memory. In addition, a list may be sorted very rapidly in a CAM.

Foster has proposed additional operations which may be added to
this rudimentary CAM. One may extend the REPORT operation to an
operation COUNT which tells the control unit how many responders
there are. A CAM with this operation may be used to perform vector
operations such as sums over an array. Another possibility is to
design the memory as an array with nearest neighbor interactions.
Communication may be established by MOVE instruction, that moves
activity (sets tag bits) north, east, south or west one position.
Such a CAM would perform relaxation algorithms in a highly parallel
manner. This would permit extremely rapid solution of large classes
of partial differential equations.

II.A. Advantages of Content Addressable Memory

In general, when N data elements must be processed, use of a CAM
increases processing speed by a factor of N/long (N). Because a
variety of vector and sorting algorithms may be performed in a highly
parallel fashion with CAM, it is well adapted to tasks which must be
performed in real time. CAM may also be used to parallelize computations in plasma physics, hydrodynamics and molecular physics, all
of which challenge the capabilities of computers presently used for
numerical modeling.

Content addressable memory has significant advantages in addition to speed. Programming for a CAM is generally easier than for
a Von Neuman type processor, since the processing is straight-for-
ward, and search procedures may be eliminated. A reasonable estimate is that average coding costs can be reduced by 20-25%. Because
coding costs for a given computer installation outweigh hardware
costs by an order of magnitude or more, CAM will pay for itself even
if it is several times more expensive than conventional coordinate
addressable memory.

II.B. CAM and Molecular Electronics

Building a CAM from today's semiconductor technology presents several
problems. First, more gates are required for CAM than for coordinate

addressable memory. This increases hardware costs and decreases
reliability. Design of a CAM may present problems to the engineer
because the control circuitry is unfamiliar and complex in comparison
to conventional memory devices. Access to a particular word is
slower in CAM than in coordinate addressable memory and the problem
gets worse as memory capacity increases. All of these problems may
be overcome through clever design, but such techniques add to complexity, thus affecting cost and reliability unfavorably.

Materials which admit faster switching times and greater circuit
density must be sought as an alternative to present semiconductor
technology. The most desirable materials would perform one or more
of the basic CAM operations through fast transformations in molecular
states.

Our basic paradigm in seeking an ideal substance is a material
composed of molecules which have a series of sharply defined stable
absorption lines. Absorption at one wavelength of the series should
be independent of absorption at another, with each event producing a
second molecular transition. These transitions may be detected in a
second series of lines representing the scattering spectrum of the
molecules. A memory composed of this material will function as a
CAM, for which the read and write operations are on different wavelengths.

Because building memory elements on the molecular level is
beyond current technology, the CAM approach affords a means to make
the most efficient use of the properties of our hypothetical substance. Storage of words of information corresponds to existence of
sub-populations of molecules in corresponding conformational or
energy states. Because memory operations involve well-defined sub-
populations, memory state transitions should appear deterministic.
Therefore, storage of information on discrete molecular units need
not require the organization of the material at the molecular level.
Independent storage avoids most of the problems which arise in building a coordinate addressable memory on the molecular level.

In order to develop a memory with the complete set of CAM capabilities we must be able to write to a small sub-population of the
molecules of our ideal substance. One may organize the memory in

such a way that writing is directed to particular spatial locations.
This approach would be more compatible with coordinate addressing
than with content addressing. Spatial addressing creates problems
with addressing all active memory locations simultaneously. An
alternative is to make the writing dependent on a local field level.
Biassing by an external field may be used as a method of selecting
the "active" sub-population.

One possible mechanism for field-dependent selection is through
soliton-mediated energy transfer. Davydov has shown in his continuous model that solitons in the protein α-helix will form for a
restricted range of the nonlinear coupling between amide-I excitations and phonons along the α-helix (5). Outside this range solitons
will not form. This observation has been verified by A. Scott (9) in
his numerical work with the discrete Davydov equations.

The coupling between phonons and amide-I excitations may be
modulated by an external electric field. We have begun numerical
calculations to determine whether solitons may be gated through
polarization of the carboxyl dipoles in the protein α-helix.

Another of the operations in the basic CAM we have described is
selecting the FIRST responder. First responders may be selected by
means of a field gradient. Standing wave fields created by crossed
laser beams may be used in this manner. For example, chemical activation by an interference pattern is the basis of recording holograms.
Microcircuit patterns have also been etched with the aid of this
technique.

II.C. Some Specific Memory Systems

A wide variety of molecules undergo reversible photochemical reactions
when exposed to light. Because laser technology currently exists
which provides for optical discrimination in cross sections which far
exceed area discrimination of magnetic read/write devices, photochemical memory storage may well represent the most faborable avenue
for investigation.

At very low temperatures (4-10K) most molecules will yield sharp-
line electronic spectra. Some molecules, however, maintain broad,
structureless electronic spectra because of internal degrees of free-

dom (4). These systems will often exhibit the unusual property of laser induced sharp-line hole burning. In particular, a high intensity laser beam can be used to modify the absorption spectrum of the molecular ensemble such that an optical window is produced at the laser excitation frequency. A multi-wavelength laser source can then be used to burn holes at discrete wavelengths and write information. A less intense beam of light can then be used to nondestructively read the information. This technique is currently being explored by IBM San Jose to produce optical disk memory devices. The major problem with this technique, however, is associated with the difficulty of accurately "removing" the hole without producing a new hole at a different wavelength.

There are a wide variety of molecules which will translocate protons across a barrier or through a membrane environment when stimulated by light (2). The former are characterized by intramolecular photostabilizers of the type used by NASA and Jet Propulsion Laboratory, Pasadena to prevent photochemical damage to spacecraft. The latter are biological systems.

A simple and compact memory storage device can be imagined which contains a proton pump in enclosed phospholipid vesicles. These vesicles would contain a pH indicator dye inside. When a vesicle is exposed to light, protons will be pumped out of the vesicle into the external medium changing the pH and the color of the indicator. A laser beam would write information by activating the pump at the absorption maximum. The information would be nondestructively read by monitoring a second wavelength corresponding to the absorption maximum of the indicator dye. A wide variety of indicator dyes are available which provides for any wavelength region from 380nm to 750nm to be used for the interogating wavelength. An interesting feature of this proposed memory storage device involves the mechanism of clearing the "bit" from state "1" to "0". The use of a laser wavelength corresponding to an absorption maximum of the phospholipid membrane will induce a phase transition which will open a transmembrane channel to the external medium. Protons will then flow through the membrane and increase the pH inside the vesicle. Hence, there

are three laser wavelengths associated with the three separate processes of read, write and clear.

The above mentioned systems all suffer from the potential problem that the memory storage devices must be prepared in sufficient
purity to guarantee reliable operation. This criterion may be viewed
as both a synthethis and a purity problem, because these two requirements are nominally linked regardless of preparation techniuqe.

One of the potential advantages of utlilizing certain biomolecules is that many such compounds can be synthesized in vivo and then
isolated in large quantities. The large scale preparation of some
proteins at desired levels of purity is well within the reach of
current technology. Nonetheless, most of the biomolecules discussed
above appropriate to the devices outlined above can not be prepared
in the required quantity nor purity using present technology.

III.   TECHNICAL PROBLEMS

Building high density, high speed molecular memory devices requires
the optimization of several diverse technical areas. These include

   1. Molecular environment.
   2. Optical (laser) technology.
   3. Electronic amplification discrimination technology.

III.A. Stability

A major problem with proteins or other bio-molecules is that they
are extremely sensitive to their environment. Many active proteins
need to be at a lipid - water interface or in a lipid membrane to
maintain their activity. Water-soluble portions of proteins are
sensitive to the concentrations of a number of cations and to the
Ph level of the solution. Materials which might be attractive for
their spectral or chemical properties may be unsuitable because they
would denature or loose their activity in the solid state.

One way around this difficulty is to build a device from a succession of monolayers. Using techniques such as deposition of monolayers in a Langmuir-Blodgett trough, a succession of mono-molecular
films might be deposited upon a substrate. Layers of active proteins
may, in this way, be sandwiched between layers of other substances

which serve to protect and stablize the proper environment for conformational or spectral activity. Monolayer techniques are presently
under development at a number of sites.

III.B. Cryogenic Problems

A number of the above molecular mechanisms are only operative for
data storage at low temperatures. Accordingly, the problem of maintaining a low temperature is a significant problem in terms of both
technological constraints and cost effectiveness. It should also
be noted that the above mentioned memory storage techniques will lose
much, if not all, of the stored data, should the temperature rise
50 degrees above nominal. In the case of the low temperature (80°K)
systems we have described this imposes restrictions on the cryogenic
subsystems which must maintain low temperatures over extended time
periods. The above requirement therefore presents both a technological problem and an economic disadvantage which need to be overcome.

III.C. Laser Technology and Optical Problems

Although research grade lasers and optics are currently available
which provide the necessary intensity and mode stability required for
utilizing some of the above mentioned memory systems, these systems
are not, in general, commercially available. Exceptions include CW
gas lasers, which are relatively inexpensive and can be designed to
provide TEM00 output, and diode lasers which are also inexpensive,
but suffer from limited wavelength range and poor TEM control. The
importance of maintaining a Gaussian beam profile must be emphasized.
Poor focusing properties and beam divergence problems prevent the use
of lasers that operate in higher than TEM01 or TEM10 modes.

Another approach would be to incorporate diode lasers into the
micro-circuitry. Acousto optical interactions may be used to steer
optical probes in thin-film wave guide devices. Optical focussing
and pulse shaping may also be performed by ultrasonic waves.

III.D. Electronic Limitations

All of the above mentioned memory systems have the potential of
extremely high densities. These densities should exceed by two to

four orders of magnitude those densities currently available using
magnetic (disk) storage devices. Accordingly, the bandwidths required of the interrogating electronics must be extremely high, on
the order 10-100 GHz. Modulation of optical beams at these rates has
been accomplished by a number of methods. The simplest technique
compatible with microcircuit technology is based on Mach-Zender inter-
ferometers. Interferometric techniques have been used in demonstra-
tions of a wide variety of high speed devices. Analog to digital
converters, running up to 10 GHz or above have been reported at
Lincoln Laboratories. Thus research grade optoelectronic devices
currently exist which approach these speed requirements, but these
systems are not cost effective. Accordingly, it is essential that
one take the electronic limitations into account in designing any
molecular memory system. One technique that should prove helpful in
optimizing the effective bandwidth of the interrogating electronics
is to use separate read, write and clear data paths. This will divide the required bandwidth by a factor of two (not three as might
be anticipated).

IV.   JOSEPHSON JUNCTION DEVICES

An alternative approach to the problem of communication with nanometer structures utlizes the sensitivity of Josephson junction devices. Detection devices based on superconductive circuits with two
Josephson junctions have proven more sensitive to magnetic fields
than any other class of devices. This device is termed the direct
current super conducting quantum interference detector (DC-SQUID).
Similarly superconducting loops with a single junction have been
used extensively as high sensitivity electromagnetic field detectors.
The latter are termed radio frequency (RF) SQUIDS.

The basic principle behind the use of super conducting loops for
magnetic field detection is expressed by the fluxoid quantization
equation (1).

$$\iint_S B \cdot d\sigma + \frac{m}{2C^2 p} \oint_\Gamma J_s \cdot d\ell = n \Phi_0$$

$$B \qquad J_s \qquad \Phi_0 = \frac{h}{2e}$$

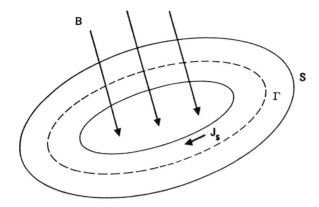

In this equation

 is the magnetic field

 is the current in the loop and

 is the flux quantum.

The fluxoid threading a superconducting loop is quantized. If we
assume that the superconducting material is thicker than the penetration depth of the magnetic field in the superconductor, the fluxoid
quantization equation may be written

$$n \, \Phi_0 = \int\limits_S \int B \cdot d\sigma$$

Thus the magnetic flux threading a superconducting loop is quantized
if the loop is larger than a few microns.

SQUID devices have been successfully employed to detect quantum
--level changes in magnetic flux. In particular these devices have
been useful for measuring magnetic transitions of various types in
solid samples. Such a scheme might be the basis of an output device,
in which a squid is used to detect changes in magnetization of a
small ensemble of molecules.

The major questions to be answered are

1. What types of molecular magnetic transitions may be detect-
    ed by a SQUID?

2. How might such transitions be conveniently induced?

3. What are the intrinsic limitations in sensitivity?

4. What are the thermal stability requirements?

Compounds such as the dipthallocyanines are convenient for investigations into magnetic transitions.

This scheme is subject to the same technical problems as Josephson junction work in general, i.e., the necessity of keeping the working circuitry at liquid helium temperatures. In addition to cryogenics, materials, and cycling ($4^{o}K$ to room temperature) problems, working with organic materials gives rise to additional difficulties. These include problems of fabrication and chemical compatibility even if the organics are used in bulk.

A second application of Josephson junction devices is related to the RF SQUID. The basic approach is to build a tunnel junction or microbridge which would resonate with a small population of molecules. The Josephson junction device must be scaled to nanometer dimensions in order that quantum--level flux changes might be detected. Effects of flux changes of the correct order of magnitude have been reported for niobium bridge structures.

In order to determine whether it is possible to put small ensembles of molecules into resonance with Josephson junctions, several theoretical questions arise.

These include:

1. Are there long-lived molecular excitations of the correct energy? The energy levels must be comparible with the band gap in the superconductor.

2. What are the dynamics of a system of coupled molecular oscillators and a Josephson junction oscillator? Both components might be highly nonlinear.

3. Are there any reasonable approximations which do not depend on many-body theory. Precise results from many-body theory for molecular oscillators are practically impossible to obtain, given the present state of computer simulation.

Oscillations in the band from 100 GHz to 1 THz may be detected by Josephson junctions. These frequencies are high enough that times required for detection of a wave packet would not affect device speeds.

Numerical simulations at our laboratory indicate that short segments of protein -helix (21-30 peptide units) may exhibit sustained oscillations at about 100 GHz. These observations arise from the Davydov-Scott model of solitons in the protein α-helix. In this model amide I excitations are coupled nonlinearly to the phonon mode of the protein. Under proper conditions excitation of the amide I mode in the protein chain will induce formation of a solitary wave having soliton-like properties. This solitary wave propagates as a bound system of amide I excitations and phonons. A solitary wave induced in the α-helix traverses the segment repeatedly because it is reflected at the ends. During the course of reflection, the solitary wave disperses somewhat, and energy is pumped into a standing wave mode (phonon) of the entire molecule. The phenomena is recurrent, in that the dispersion of the amide I excitation goes through a maximum at the same time the energy of the standing wave is maximum. Subsequent to the phase of maximum energy of the standing wave mode the amide I excitation becomes more localized and the amplitude of the standing wave decreases. Given the right choice of parameters in the model, this recurrence has been observed to persist for several hundred cycles.

Assuming that nonlinear modes of proteins couple into a radiation field, the second problem is to describe interactions between Josephson junctions and small ansembles of molecules. A convenient conceptual model for a Josephson junction interacting with a radiation field is given by the paired equations (8).

$$\left(i\hbar \frac{\partial}{\partial t} - e V_0 \right) \psi_L = -1/2 \, \mathscr{P} \, \mathscr{E}_J^+ \, \psi_R$$

$$\left(i\hbar \frac{\partial}{\partial t} + e V_0 \right) \psi_R = -1/2 \, \mathscr{P} \, \mathscr{E}_J^- \, \psi_L$$

The term $\mathcal{P}$ arises from the tunneling interaction in the junction and may be regarded as a dipole operator, $\mathcal{E}$ arises from the positive and negative frequency components of the vector potential, V_o is the bias voltage and ψ_R, ψ_L are the wave functions of right and left superconductors respectively. These equations are generally used to describe two-level oscillators in a radiation field. For this reason they are a convenient departure point for a treatment of the interactions between molecular oscillators and Josephson junction oscillations.

Another approach to the problem of describing the Josephson-junction-molecular oscillator system is to describe the system classically and then quantize. This approach runs into the problem that we are dealing with extended systems. As an example, the continuum approximation of the Davydov soliton is described by the nonlinear Schrodiglner equation. This system is quantizable, but unfortunately no satisfactory general theory for treating extended nonlinear systems is available. Assuming that the classical equation of motion for the molecules is one of the soliton equations, we may be able to find an infinite number of conserved quantities. If this is so a quantum description via the Feynman path integral may be available. This has been shown for the so-called sine-Gordon equation (7). The basic conjecture is that path integrals for soliton systems are solvable by WKB techniques precisely because each of these equations exhibit an infinite number of conserved quantities. Since code conserved quantity implies a symmetry the manifold of solutions of the soliton equation should factor into an infinite dimensional analogue of a system of harmounious oscillators. Such a result would put a powerful technique in the hands of those who desire to gain a conceptual understanding of molecular electronic systems.

ACKNOWLEDGMENTS

The author wishes to acknowledge the assistance and advice of Robert Birge, Department of Chemistry, University of California, Riverside, and Terence Barrett, Department of Biophysics, University of Tennessee, and Naval Research Laboratory in preparation of this manuscript.

REFERENCES

1. A. Barone and G. Paterno, Physics & Applications of the Josephson Effect, Wiley-Interscience, New York, 1982.

2. T. Barrett, Physics Letters 91A:139, 1982; Physics Letters 92A:309, 1982; Private Communication, 1982.

3. R.R. Birge, Private Communication, 1983.

4. R. R. Birge, D. F. Bocian and L. M. Hubbard, J. Am Chem Soc, 104:1196, 1982.

5. A. S. Davydov, Physica Scripta, 20:387, 1979.

6. C. Foster, Content Addressable Parallel Processors, Van Nostrand-Reinhold, New York, 1977.

7. B. Hasslacher and A. Neveau, Rocky Mountain Journal of Mathematics, 8:341, 1978.

8. D. Rogovin and M. Scully, Physics Reports, 25:175, 1976.

9. A Scott, Structure & Dynamics: Nucleic Acids and Proteins, Adenine Press Guilderland, New York, 1983, p. 389.

III
Bridging Technologies: From Silicon Down

13

On Predicting Feasible Molecular IC Structures and Concepts

Don L. Kendall / Instituto Nacional de Astrofisica, Optica y
Electronica, P.O. Box 51, 72000 Puebla, Pue., MEXICO

I.   INTRODUCTION AND ABSTRACT

The questions that will be treated in this paper are believed to
be important and perhaps even essential for the successful
fabrication of molecular integrated circuits. The emphasis will
first be placed on a special kind of Multi-Use Substrate (MU-
STRATE) and second on a scheme called Signal Induced Feedback
Treatment (SIFT) to test and then locally modify molecular
electronic functions by an electron-beam feedback heat-treatment.
Both the special substrate and the SIFT scheme are feasible using
existing technology, although neither has yet been fully accomplished in the laboratory. Of course, the implementation of
these two precepts to make useful molecular or semiconductor IC's
will require intensive innovative effort and concept development
in many fields, most especially in the field of chemical
synthesis. An attempt will also be made to map out in broad-
brush form some of the corollary technologies and methodologies

and even new types of logic devices that will be necessary to accelerate progress in this embryonic field. The extremely difficult problems that will be encountered due to heat removal, electric field breakdown, fringing effects, interconnect capacitance, and other physical limitations will also be mentioned. However, the primary emphasis will be on a structure that is technologically feasible and useful. From this one can perhaps work backwards and make the required inventions that will allow the work to advance. One possible example involving tunnel diode logic will be given. It should be noted that when and if the MU-STRATE approach becomes a reality, the fabrication methods will be radically different from those of today. Most or all of the lithography and normal fabrication steps will be shifted to making the substrate. This will be followed by e-beam and other fabrication and testing processes that will form the final circuits. Nevertheless, when one tries to forecast the direction of such a new and complex field, it is well to acknowledge the truth of Paul's well known statement: "For now we see through a glass, darkly" (1).

The next part of the paper will be dedicated to a look at the author's specialty of producing very deep and very narrow grooves in single crystal silicon. In particular, several ideas from a review article (2) that might somehow be related to the fabrication of MED's will be discussed. The most important possibility is a structure that may have potential as a high brightness x-ray source.

In the next section, several experiments are discussed that are presently underway or that have been proposed to be undertaken in the author's laboratory in Mexico. These are supported by a grant from the Volkswagen Foundation[1].

a) A high flow-through environmental particle filter.

b) A voltage controlled semipermeable membrane.

c) A n-i-p-i superlattice (3) based on Si.

d) Arrays of precise angle grain boundaries in Si.

e) Application of the 19th century Russell Effect to Si (4).

f) Adsorption on steps of almost atomically flat surfaces.

g) Special apertures in Si to modify the thermal and length
   distribution of molecules.

These experiments are discussed in advance of their completion in
the spirit of encouraging more rapid progress in their
accomplishment. Furthermore, it would appear that several could
contribute in some way to the field of molecular electronics.
This is especially true of the Russell Effect (Experiment e
above). Such experiments should serve to remind us that an
emerging field such as molecular electronics is vulnerable to
many changes of directions.

Finally, a few observations are made regarding the impact
that the techniques and concepts discussed in this paper might
have on Si and other semiconductor technologies. In fact, it is
possible that IC's based on these inorganic technologies will
derive significant benefits from the successful development of
the special substrates and the feedback treatment methods that
are discussed in this work. These inorganic technologies and the
tremendous inertia they possess will no doubt provide the
greatest challenge to the emergence of molecular electronics in
the foreseeable future. In spite of this, the potential benefits
make the quest imminently worthwhile. These benefits will accrue
not only in electronics, but also in chemistry, biology, physics,
and other fields.

A MULTI-USE SUBSTRATE (MU-STRATE)

Several assumptions are made here regarding the nature of a
Multi-Use subSTRATE for very high density MED circuits. This
will be referred to as a MU-STRATE. The ideal MU-STRATE might
be:

1. Monocrystalline.

2. Oriented in a close-packed direction.

3. Beam-accessible from top and bottom.

4. Very thin, yet self-supporting.

5. Matable with substrates and packages from above and below.

6. Interlaced with closely spaced highly conductive isolated paths from top to bottom on a regular orthogonal array.

7. Interlaced with isolated resistor paths from top to bottom.

8. Capable of discretionary interconnections from either side.

9. Readily coolable.

10. Equipped with special pads to provide different voltage levels and clocking signals to different subsystems.

Not all of the above characteristics of the so-called ideal substrate are necessary, although each has certain advantages depending on the methods chosen to form and operate the active devices. No attempt will be made to discuss the specific advantages of the above features. Instead, a particular structure will be proposed that meets most of the criteria. This structure can be fabricated using existing technology, and is thereby amenable to direct testing, either for MED or standard Si or perhaps GaAs IC fabrication. First, the structure will be specified in general terms, then a possible fabrication sequence is presented in enough detail to give the reader some idea of what is feasible using a wide range of Si based technologies. The particular structure, the proposed fabrication sequence, and the SIFT testing and heat treatment scheme are the results of the author's long term interest in three dimensional structures and in semiconductor diffusion phenomena. Other workers will devise different fabrication methods and approaches, but there will likely be a great deal of commonality in the final structures and concepts.

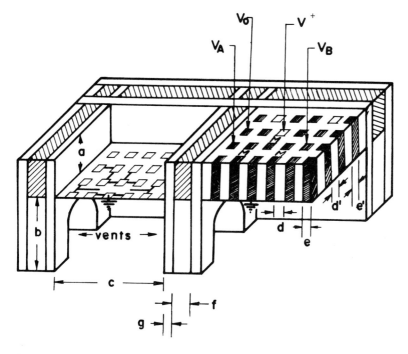

1. A sectional view of the Multi-Use Substrate (MU-STRATE).

    A somewhat schematic view of the proposed MU-STRATE is shown
in Fig. 1. Before discussing the fabrication process, it will
be noted that the structure meets all ten requirements of the
so-called "ideal" substrate except criteria 2 and 6. The desire
to have the surface oriented in a close-packed direction
(criterion 2), is based on the possible use of the
Langmuir-Blodgett (LB) Technique for depositing specific groups
of atoms on the substrate (5). This method allows one to deposit
onto a substrate a close-packed monolayer or multilayer of
aliphatic organic molecules (6). The ideal substrate might
therefore be monocrystalline with its surface lying in a close
packed plane and with its atomic separation having the same
spacing (or some sub-multiple thereof) as the diameter of the
polar head of the aliphatic molecule. This would allow a sort of

epitaxy between the LB layers and the underlying substrate and perhaps would result in extremely uniform LB layers having interesting interface or bulk properties. Nevertheless, such a precise restriction on the diameter of the aliphatic head is simply not realistic nor necessary for the present applications. Furthermore, for metal-oxide-semiconductor (MOS) technologies, the close-packed (111) plane of Si is the worst choice of the major crystal planes since it has the highest surface state density.

As an interesting aside to the above issue, it has occurred to the author (and probably to others) that an aliphatic organic compound having a head that meets the above restriction could perhaps make a very useful contribution as a nucleating agent for the growth of quasi-single crystal Si (or other crystal) on an inexpensive amorphous or polycrystalline substrate. In the case of Si growth in the (111) direction, the appropriate diameter would be some integer multiple of the distances between the coplanar atoms in the (111) plane, namely 3.84 A. Thus, one might design a compound with a head of diameter 3.84, 7.68, 11.52 or 15.36 A, etc., then carefully adsorb a monolayer (or several ordered monolayers) onto an amorphous substrate, then heat the substrate to the desired deposition temperature. The "residue" of the dissociated aliphatic heads would remain on the proper spacing (on average) to serve as an ordered nucleating agent for the subsequent Si vapor deposition. It might be desirable to incorporate Si or even certain refractory elements into the compound to ensure that a stable residue remained during the high temperature nucleation process. It is known that certain growth conditions favor the textured growth of Si crystallities in the [111] direction on amorphous substrates (6). The addition of a properly ordered and spaced nucleating agent should encourage the crystallites to all face the same direction on the parade ground, thereby forming a sort of quasi-single crystal material.

However, an excellent extended single crystal material is not
likely, even with the best nucleation, unless an atomically flat
amorphous substrate is used.

As regards criterion 6 mentioned above, the MU-STRATE shown
in Fig. 1 meets this criterion very nicely except it may not be
"orthogonal". This is because, for the particular fabrication
method suggested below, the substrate is aligned with its surface
in the (110) plane so that the perpendicular {111} planes form
the walls of the very narrow grooves that must be formed by
chemical etching. These two {111} planes form an angle of 70.5°
with each other rather than 90°. However, the devices are as
readily formed by the subsequent e-beam treatment on this "tilted
array" as they are on a completely orthogonal 90° arrangement.
Furthermore, there are other feasible fabrication methods that
will result in a 90° arrangement, as will be shown.

The important parts of the MU-STRATE could perhaps be formed
in arbitrary directions of almost any material by using plasma
etching and related technologies. The use of plasma technologies
would allow 90° patterns and also the use of (100) Si would
result in a lower surface state density at Si/SiO_2 interfaces
if one chose to use MOS technology. However, the very small
conductive feedthroughs shown in Fig. 1 have depth-to-width
ratios of 10 to 40, which is presently difficult to obtain in Si
using plasma-type technology. Nevertheless, a new method called
Ion-Beam-Assisted-Etching (IBAE) has achieved depth-to-width
ratios of 50 in GaAs (7), so there is indeed great hope that the
MU-STRATE structure will eventually be produced using some dry
etching technology. It should be mentioned that GaAs would also
be a good material for a MU-STRATE.

The wet chemical etching with simple solutions of KOH/H_2O
at 44 wt.% has already been shown to etch downward into very
narrow grooves in (110) Si at a rate that is 400 times faster
than the lateral dissolution rate (8). This allows a maximum

depth-to-width ratio of 200, which is more than adequate to allow fabrication of the proposed MU-STRATE. In addition, the wet-chemical approach is more amenable to large batch processing and can be undertaken with a minimum of capital investment. For these reasons, the emphasis in this work will be placed on wet-chemical processes. As will be seen, several of the processes simply cannot be done using the powerful dry technologies. Having said this, it is likely that the "dry" etching and deposition technologies will eventually play an important role in the MU-STRATE fabrication. The most important point does not lie in the fabrication method, but in the inherent features of the MU-STRATE itself.

Now turning to the MU-STRATE shown in Fig. 1, it should be noted that the figure is not drawn to scale. Realistic dimensions for the various regions are the following:

a = 0.5 to 4 μm

b = 50 to 200 μm

c = $\underline{c}'$ = 50 to 200 μm

d = d' = e = e' = 0.05 to 0.40 μm

f = 10 μm to 50 μm

g = 5 μm

The number of the large cells in 1.0 cm^2 varies from about 2000 to 20,000 when $\underline{c}$ is 200 μm and 50 μm, respectively. These estimates take into account the area occupied by the (relatively) large support and supply buses. Using the $\underline{smallest}$ dimensions of 0.05 μm for $\underline{d}$ and $\underline{e}$, the area density of conducting vias is about $\underline{8 \times 10^9 / cm^2}$ for the 200 x 200 μm cell size (again taking into account the supply buses). On the other hand, using the $\underline{largest}$ dimensions of 0.4 μm for $\underline{d}$ and $\underline{e}$, the corresponding area density of $\underline{vias}$ is $\underline{8 \times 10^7 / cm^2}$ for the 50 x 50 μm cell size.

The two underlined numbers above give the upper and lower extremes for the area densities of the conducting vias themselves

for the two extremes in the lateral dimensions of the vias. If
we further assume that 8 vias are required to make a single NAND
or NOR logic element, we find that the maximum memory element
density is about $10^9/cm^2$. This represents an improvement of
about 10^3 in area density over the megabit/cm^2 that is
presently attainable (in the laboratory) using the best of
today's optical lithographic techniques (9). The lower figure
underlined above (assuming 0.4 µm geometry) represents a 10 times
improvement over the presently attainable area density of memory
elements. This relatively modest improvement in density over
today's most dense circuits makes one appreciate the almost
incredible strides that the industry has taken since the
invention of the integrated circuit in 1958 (10).

However, the area density is not the whole story. The
MU-STATE approach has other significant advantages over the
present way of making IC's. The largest advantages are probably
[1] thinness, [2] coolability, [3] beam-accessibility from both
sides and [4] vertical supply buses. The first two make it
capable of larger and more usable volume density, and [3] and [4]
make it more underlined above manufacturable and testable than today's IC's.
This will be made clearer in a later section.

Let's dwell a moment on the volume density of logic
elements. One can imagine that it might be possible to stack 10
of today's megabit memories into one 1 cm^3 of space, although a
hefty amount of forced air or liquid cooling would be required
since the chips are not well designed for heat removal. Thus,
today's volume density of logic elements is limited to about
$10^7/cm^3$. On the other hand, the MU-STRATE appears at first
glance to be capable of a volume density of up to $10^{13}/cm^3$!
However, this would require eliminating the support structure
completely, that is, making b = 0 in Fig. 1. It would also
require making the multi-via membrane thickness 1.0 µm thick and
then carefully putting a stack of 10^4 of these very thin chips

into one essentially solid block. This stacking and
interconnection challenge might be physically possible, but to
operate this might little block of $8x10^{13}$ vias and 10^{13}
logic elements with its 10^4 supply buses is another story.
There is no way for the heat in the center of this block to be
removed except by heat conduction through almost solid Si.

To put the heat removal problem in perspective, a high
performance Complementary Metal Oxide Semiconductor (CMOS) array
of area 1 cm^2 requires at least 50 μA/gate at say 1.0 V (11).
Also, present packaging techniques can only dissipate about 1.0 W
(11,12). Assuming that only 10% of the gates are toggeling at a
given time, one can arrive at a maximum gate density per cm^2
of about $2x10^5$ (11). By making very optimistic assumptions
regarding future improvements, one might extend this gate density
by a factor of 1000 to $2x10^8$ gates/cm^2. For example, the
package might be equipped with liquid freon cooling that allows
it to dissipate 20W (12) thereby gaining a factor of 20 in power
dissipation. In order to actually accomplish this factor of 20,
one would probably need to incorporate heat pipes into the
structure (2), and/or use vertical etching to produce radiating
vanes on the back side of the chips (13) (similar to the
MU-STRATE). One should keep in mind the fact that 15 W light
bulbs are still marketed in many countries! Another factor of 50
might conceivably be gained by reducing the toggeling to a much
lower fraction of the devices and at the same time reducing the
gate current significantly. This factor of 50 decrease in
average gate current may never occur for this family of
circuits, but it is used here to obtain a sort of upper limit on
the potential improvements in gate density/cm^2 for MOS
technology.

The estimate of $2x10^8$/cm^2 for the maximum tolerable
area density of CMOS gates is larger than a published estimate of
$3x10^7$ gates/cm^2, but the latter (14) was based on chip

dissipation of only 1 W/cm^2. Increasing this to 20 W/cm^2 brings the two estimates into general agreement. A similar calculation (14) for a Read Only Memory (ROM) gave a density of 10^8 gates/cm^2, which would become $2x10^9$ gates/cm^2 if the larger dissipation value were used. The latter is about 10 times greater than the number of letters in the Encyclopedia Brittanica, which suggests that the textual content of this 24 volume set might someday be encodeable on a single ROM chip. Similar estimates have been made for the maximum area density of bipolar logic in a ROM configuration (14).

There are other effects that become dominant at these very high packing densities such as fringing fields between vias, oxide-breakdown, metal migration, drain-substrate breakdown voltage, and substrate doping fluctuations (14,15). These all become important at the very high gate densities discussed above and even at considerably lower gate densities. For these reasons, one can say almost unequivocally that CMOS and other MOS technologies will be limited to gate densities of less than about $10^8/cm^2$. The volume density of gates will be similarly limited to less than $10^8/cm^3$. However, one might be able to indeed fabricate the chips at a density of $10^6/cm^2$ and then stack 100 into a functional block of $10^8/cm^3$ using the MU-STRATE or some similar approach.

One might imagine that reducing the operating temperature by a factor of 100 to 3°K might reduce the power consumption by an equivalent factor. In a fundamental sense this must be true for some unspecified logic system since the thermal energy that must be overcome by the triggering voltage (or other) source is much smaller. However, n-channel MOSFET's actually consume slightly more power at 77°K than they do at 300°K, although there are significant gains in speed and other parameters (16). Thus, each technology and logic system must be evaluated on its own merits at low temperature.

Fabrication Method

One possible sequence of fabrication steps for making the
MU-STRATE is the following:

1. Orientation Dependent Etch (ODE) narrow grooves in p+ Si
    to depth a along y direction (2,8).

2. Fill narrow grooves with epitaxial n or p Si along y
    (17-19), i.e., Chemical Vapor Deposition (CVD).

3. Repeat 1 along x direction.

4. Repeat 2 along x direction.

5. Remove p+ where desired using concentration dependent
    etch (CDE) (20).

6. Oxidize walls of CDE etched holes.

7. Fill holes with refractory metal using CVD.

8. ODE backside of slice to depth b along y direction.

9. Fill backside of slice with n-Si.

10. Repeat 8 along x direction.

11. Repeat 9.

12. Repeat 5 on backside.

13. Repeat 7 on backside.

14. Remove n-Si from backside with etchant that doesn't
    attack refractory metal.

15. Etch vents in support structure to allow easy air or
    vapor movement.

A number of lithography, oxidation, and silicon nitride steps
are omitted for simplicity. The only processes above that are
not presently used in semiconductor laboratories are steps 7 and
14. Neither of these should be difficult to develop. The
following comments on each of the critical steps will help
clarify the fabrication process.

Step 1 uses the capability of KOH/H_2O to etch at least 400
times faster in the vertical direction as compared to the lateral
direction (8). It requires the alignment of the narrow etch
windows along the (110)/(111) intersection to a precision better

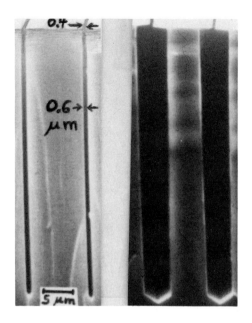

2. SEM views of anisotropically etched samples showing the
difference between a very well aligned mask and one misaligned
by 2.1°.

than 0.1°. This is readily accomplished by using a well cleaved

reference flat on one edge of the slice. Figure 2 shows the

difference in groove width between a well aligned groove and one

having a misalignment of 2.1°.

Step 2 involves the epitaxial refill of single crystal Si in

very deep grooves (17-19). It is accomplished by adding an

etching gas like HCl to the depositing mixtures (SiCl$_4$ and

H$_2$, for example) until the top surface of the slice is in a

sort of quasi-equilibrium that is closely balanced between the

tendency toward deposition and the tendency toward vapor etching.

This creates a near-zero deposition rate at the top surface, but

still allows deposition in the bottom of a deep groove (17).

Steps 3 and 4 (and later steps 10 and 11) are necessary

because of the nature of the ODE process. This allows the

etching of a vertical groove along a very long window opening,

but the etching of a short window opening is very quickly blocked
against further etching by a set of very slow etching $\{111\}$
planes (2).

Step 5 is based on the ability of certain mixtures of the
acids HF, HNO_3, and acetic to etch heavily doped Si of either n^+
or p^+ at rates at least 1000 times faster than lightly doped Si
(20). In the process steps listed above, p^+ and n^+ refer to
concentrations of about $2x10^{19}/cm^3$. Higher concentrations are
avoided since such high carrier concentrations are known to block
the ODE process mentioned earlier (21, 22) (for either p^{++} or
n^{++}).

Step 7 will require the development of a CVD process for
metal deposition that is based on a similar principle to that
discussed for step 2 above. Alternatively, it will require an
electrochemical plating process that heavily favors deposition in
holes. The latter is made more difficult by the dielectric layer
on the groove walls (Step 6).

The total fabrication process for the MU-STRATE involves 4
to 6 lithography steps, 4 epitaxial Si processes and 1 CVD metal
process. A savings of at least one lithography and one epitaxial
step can be made by using the dry etching processes mentioned
earlier. This complexity should be compared to the 12
lithography steps presently necessary for a modern CMOS
fabrication process. However, the Vertical Etching and Epitaxial
Refill (VEER) technologies above are not presently used in
production processes. The epi-refill of Si (as well as metal)
needs to be made less sensitive to gas phase conditions (17), but
this should not be an overly difficult task.

A perspective view of two of the cells in Fig. 1
illustrates most of the important features of the MU-STRATE. The
vias shown with hatched lines are those in which the Si has been
removed and replaced by metal (steps 5 to 7). Those without
hatching represents Si p^+ vias that were protected during steps

5 to 7. The particular location of metal and p^+ vias has no particular significance other than to show that either or both are possible options. The resistance of the metal vias will be of the order of 20 ohms. The p^+ vias will be of the order of 5000 ohms (for $d = e = 0.1$ μm, and $a = 1.0$ μm). The lateral connections between vias can be made on either or both sides by patterned metal deposition, by scanned ion implantation and/or localized e-beam heating and/or proton bombardment to cause out-diffusion. The latter out-diffusion process can be very selective in terms of which impurities are removed and can provide an excellent "epitaxial" layer.

As an example of the aforementioned epitaxial layer process for the case shown in Fig. 1, the p^+ vias might be surrounded by closely compensated n-type Si doped with $10^{19}/cm^3$ phosphorous and $9 \times 10^{18}/cm^3$ indium. During the local bombardment of protons, the slice temperature would be raised to 800° to 1000°C and the vacancy concentration would be greatly increased in the region of proton penetration (about 0.5 μm at 50 keV). Thus, the much more rapidly diffusing P would diffuse out of the surface leaving the slower In atoms to form a p^+ bridge between the neighboring p^+ vias (23). For the complementary situation of forming an n^+ bridge in compensated p-Si, one might out-diffuse B from regions doped with fast diffusing B and slow diffusing Sb (23). This interesting process has been demonstrated for the proton enhanced out-diffusion of B in Si (24). In principle, the out-diffused B could be replaced by localized ion implantation of B if the proton bombardment process went too far. This alternation of localized out-diffusion and implantation processes could form the basis of a very useful technology in IC manufacture, namely Discretionary Erasure and Writing of connecting Links (DEWLINKS).

Another way to perform the so-called DEWLINKS process is to first ion-implant B at relatively high dosage at room temperature

into the n-type regions of the MU-STRATE. This will cause the
formation of an amorphous very high resistivity layer across the
top surface of the MU-STRATE. Then the substrate would be heated
locally with a low energy e-beam at high beam current (25). This
highly localized annealing process (25,26) will form the desired
p^+ bridge between the p^+ vias. To "reverse" the annealing
process, one could again ion implant locally and later anneal
with the e-beam if desired. It will be noted that ion
implantation in precise locations is one of several ways that one
can IMPRINT specific atoms for subsequent use.

The DEWLINKS function is an important one and ways to
perform it efficiently must be found before significant progress
can be made using the SIFT technology to be described in the next
section. One would think that either inorganic or organic
conducting layers might be deposited locally (into fusible links,
etc.) by various means and then manipulated by e-beam and other
beam techniques. It would also be very useful to have a
"Scanning Molecular Beam Deposition" (SMBD) system. This SMBD
beam (another IMPRINTATION scheme) is shown schematically in
Fig. 3, along with ion, photon, and electron beams, all of
which can impinge on either or both sides of the MU-STRATE. This
beam-accessibility aspect of the MU-STRATE was a central feature
of a proposal made earlier, as was the support structure for the
very thin remaining Si (27). In that work, a number of devices,
as well as isolated regions and conducting regions, were proposed
using ion-implantation from both sides of the very thin regions.
In a related patent from the same period, ion implantation of N
and/or O was proposed as a method of obtaining a buried
dielectric under monocrystalline Si as well as electrically
isolated pockets of very thin Si in a thick substrate (28). This
patent, together with those mentioned earlier (18, 19, 23, 26,
27) anticipated many of the features of the MU-STRATE, including
some of the SIFT concepts to be discussed in the following

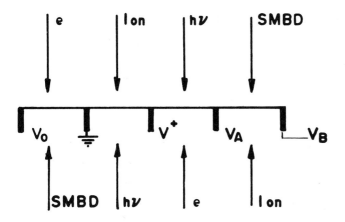

3. Section of the MU-STRATE showing different types of beams impinging on both sides.

section. However, the dense array of metal and silicon vias, the incorporation of isolated vertical bus bars into the support structure, the cooling vents in the support structure, and the DEWLINKS processes are proposed here for the first time.

SIGNAL INDUCED FEEDBACK TREATMENT (SIFT)

First, a bit of background may help to understand the nature of the challenge. A number of difficult problems were encountered during the early fabrication of the Si vidicon. The Si target consists of an array of 10^6 diodes in an area of 2.5x2.5 cm. One of the most vexing problems was a severe non-uniformity in the dark current in the form of a "swirl" pattern. This was due to a large degree to the extremely variable oxygen concentration in the Si due to turbulence in the melt during the growth of the crystal by the Czochralski method. This oxygen variation caused a local variation of resistivity of about 20 to 30%. It was well known that the resistivity of n-type Si could be lowered by heat treatments in the range of 400 to 550°C (29). Alternatively, the resistivity could be raised by heating to above 600°C. On this basis, the author suggested that the dark current associated with each diode be monitored by the e-beam, and then the current in

the beam increased to raise the local temperature to 550°C until
the resistivity was lowered to some predetermined standard value
for each diode in the array. The suggestion was not met with
great enthusiasm since the adjustment of the 10^6 diodes would
have required a total time of 10^7 seconds or about 116 days.

Sometime later it became clear that there was a much more
efficient way to do the localized heat treatment. This involved
storing the dark current pattern, then feeding it (or its
negative) directly back to the slice itself but at a much higher
current. This pattern (or its negative) caused the slice to be
heated only where it was needed, namely in those regions where
the dark current was largest (or smallest). This alternating
testing and heating process was done iteratively until the target
gave an adequately uniform dark current. This analog type
process required only a few minutes instead of 116 days for the
digital type process mentioned earlier. This procedure became
the basis of two broad patents involving testing and heat
treatment using e-beams, laser beams, and other testing and
heating means in both analog and digital forms (26). It is a
modification of this Signal Induced Feedback Treatment (SIFT)
approach that is proposed here to help fabricate Molecular
Electronic Devices (MED's) and to interconnect them to form
molecular IC's. The concept has not been attempted for molecular
devices, but it appears to provide a point of departure for a
wide range of fruitful experiments.

The necessary elements for a useful SIFT process are [1] a
MU-STRATE or some similar substrate, [2] molecules that can be
switched from a conductive to a non-conductive state by a
voltage, a current, or a pulse of light, [3] information on the
analog signals obtained with e-beam testing of Si-based IC's of
similar dimensions, [4] an e-beam testing machine with at least
one e-beam and preferably two e-beams impinging from opposite
directions. The other three beams shown in Fig. 3 will be

essential for certain applications, but the e-beam is probably
adequate to show feasibility of the main ideas.

With regard to the above elements, the need for the
MU-STRATE in such tests should be clear by now, although for
demonstrating feasibility, the dimensions could be much larger
than those discussed earlier. A simpler patterned substrate that
is beam-accessible from only one side might also be of some
value. Relative to [2] through [4] above, a radically different
approach is suggested here. This approach attempts to model some
of the analog processes of nature while adding some reasonable
constraints based on today's technological limitations. As a
good starting point, one might consider measuring the electrical
"signatures" of a standard Si^4 Random Access Memory (RAM) or
Read Only Memory (ROM) array with a Scanning Electron Microscope
(SEM) with low beam energy. The low beam energy is necessary to
avoid damaging the insulating gates (30). The RAM or ROM might
be provided with <u>blanket contacts</u> in one of several forms. The
simplest would be a thin uniform conducting layer (organic or
inorganic) that would provide a small amount of leakage across
the whole surface. Another useful contact arrangement would be a
"patterned blanket" in which the dimensions between the various
source, gate, and drain contact windows were taken into account.
In one example of such a pattern, the SEM might induce a
particular temporary conductive pattern by impressing a prestored
pattern of horizontal and/or vertical stripes of scanned
electrons onto the memory or logic array. While this charged
(weakly conducting) pattern is decaying, the SEM can be probing
and storing and perhaps impressing other patterns at other
voltages. Finally, subtraction, addition, multiplication, and
other data manipulation techniques would be applied to the
various stored signals.

The purpose of the above rather unusual exercise is to
unambiguously identify the signal associated with a potential

NAND or NOR or other logic element. The rotational orientation of these elements on the slice should also be a part of these electrical signatures, since the subsequent wiring will be greatly simplified if it proceeds only in specific directions across the slice. This wiring procedure might be learned on a Si slice and then later applied to an MED structure. This procedure would focus on identifying the good and bad elements and interconnecting the good ones progressively from one side of the chip to the other. The interconnections might be performed by one of the DEWLINKS schemes mentioned earlier or by metal deposition followed by e-beam "burn and test" methods that progress from one edge of the slice to the other.

After the SIFT procedures are worked out on a Si or other known array, one would instigate similar methods on specifically placed molecules on a MU-STRATE or some simpler pre-patterned substrate. The first step would be designed to determine an optimum heat treatment for the MED's. Then the use of the special test schemes would identify those few molecular aggregations (say 2% of the total) that happen to form NOR gates (or NAND, or whatever) facing in a particular direction on the slice. To give a specific example, this 2% might show up as bright patches in a pattern of the leakage current to ground after scanning at 700 volts and then subtracting the equivalent leakage current at 600 volts. The negative of these bright patches would then be fed back in brief pulses to the sample at a higher current level so that the remaining 98% of the surface of the slice is heated to the optimum heat treatment temperature (say 300°C) for a fraction of a second. The original good 2% would not be affected since they would be automatically blanked out by the signal reversal scheme. Then the same special test would be repeated. After this first brief heat treatment, or Signal Induced Feedback Treatment (SIFT), an additional 1.5% might be favorably oriented. After a few hundred iterations one

might have 90% in the proper state and additional iterations
would add very few good ones. Then the DEWLINKS scheme worked
out on the Si array would be applied to the MED array, perhaps
interconnecting the regions on the bottom side of the MU-STRATE
to avoid damaging the somewhat delicate molecular aggregates on
the top side. Some of these bottom-side linkages are indicated
on the left side of Figure 1. The proper molecular linkages are
denoted by M on the right side of the same figure. The
triggering (or input) voltages V_A and V_B, the supply voltage
V^+, the output voltage V_O, and the ground connections are
shown in the positions they might occupy for the molecular NOR
gate suggested by Carter (31). The connections to the large
supply buses are not shown.

The most important aspect of this whole procedure is the use
of analog methods along with beam feedback modification. The
use of digital techniques during testing and "sifting" is
certainly not excluded, but the digital methods require a large
amount of storage and they are often much slower to implement
(32) than a well designed analog-feedback process. A good
example of this analog-digital difference is the "few minutes
versus 116 days" for the SIFT process used on the Si vidicon
target discussed earlier. An example from nature might be the
relatively short time of a few decades to adapt a butterfly wing
of a given species to its local environment. One can postulate,
although not necessarily prove, that nature's digital techniques
would have taken much longer. Another illustrative example of a
very efficient analog process is the superposition of a negative
color transparency on top of its complementary positive
transparency (using two slide projectors). Even though the
original transparency may have many thousands of different shades
and colors (of Indian corn, for example) it can be instantly
"uniformized" by superposing its complementary negative. In
fact, this example is almost a replica of the feedback process
discussed earlier for the Si vidicon.

Redundancy

Redundancy is one solution to the resulting problem implied
earlier of having only 90% good logic elements after the SIFT
process is terminated. For example, if a given one or zero is
stored five times instead of just once, the chance of an
erroneous reading of the stored reading may plummet from 0.1 to
10^{-5}. Since the volume density of memory elements attainable
using the MU-STRATE exceeds the normally attainable volume
density by a factor of perhaps 10^6, this factor of 5 sacrifice
for redundancy is not too critical. A similar comment could be
made for the extra interconnect area that these dense memories
require.

Continuous Adjustment of Parameters in Operation (CAPO)[2]

Another problem that one anticipates in MED circuits is device
parameter instability. This is best illustrated by the offset
voltage shift one often obtains in a differential amplifier (DA)
as the DA ages or undergoes high energy particle irradiation.
Since the offset voltage often depends on how well the threshold
voltages of two MOSFET transistors (or the gain of two bipolar
transistors) are matched, methods are being developed (33) to
regularly monitor the offset voltage and to intermittently
readjust one of the threshold voltages (or one of the bipolar
transistor gains) so that the offset voltage of the DA stays
acceptably low (34). Such Continuous Adjustment of Parameters in
Operation (CAPO)[2] is costly in terms of extra circuitry, but
may be especially useful for MED circuits.

Power Consumption and New Memory Systems

As was discussed earlier, the power consumption of CMOS and other
MOS< as well as bipolar, circuitry puts a severe upper limit on
the memory or logic element density. New memory or logic or
signal propagation schemes must be devised before the

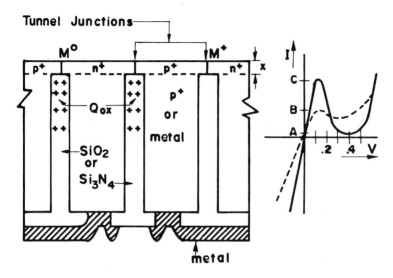

4.   Possible tunnel diode logic element based on the MU-STRATE.

technologically attainable element densities can be fully
utilized.   One possible useful type of memory cell is shown in
Fig. 4.   This is based on the extremely small junction area
that could be made by locally melting a very thin layer across
the top of the MU-STRATE of Fig. 1.   When a thin top layer (say
0.01 μm) of the structure is melted with a pulsed low energy high
current e-beam (25), or an ultraviolet laser, the liquid layer
can spread across the whole structure.   The ensuing rapid cooling
and recrystallization should form good tunneling junctions at all
the metal-semiconductor or p^+n^+ boundaries (depending on
whether the conducting vias are metal or semiconductor,
respectively).   In the metal-semiconductor case, the tunnel diode
characteristics will be better formed if the metal contains a
p-type impurity such as B.   In the case where the conducting via
is p^+, tunnel diodes may already exist at very low operating
temperatures (say 4.2°K).   However, due to interdiffusion during
the epitaxial deposition of n^+ Si, the characteristics may be
improved by melting and recrystallizing the surface layer.

In either the metal-semiconductor or the p^+n^+ junction
case, the I-V characteristics can be modified by injecting
charges or charged particles into the junction region or the
regions near the junction. For example, the solid line shown on
the graph of Fig. 4 might be modified to the dashed line by
implanting positive ions, Q_{ox}, into the thin dielectric between
the metal and the semiconductor. Alternatively, molecules on the
surface in the different charged states, M^0 or M^+, may induce
a similar change in the device characteristics. The logic state
would be detected by measuring the current (C versus B of Fig. 4)
while impressing 0.1 V across the junction.

The most important aspect of such a logic storage and
detection scheme is the extremely low power consumption per
junction. For example, if the p^+ and n^+ concentrations were
both 3×10^{18} cm^{-3} in Ge, and the operating temperature were
4.2°K, the peak current density at V = 0.1 V would be 0.1 A/cm^2
(35). (The current density for Si would be very similar.) For
the dimensions e = e´= 0.1 μm (Fig. 1) and X = 0.01 μm (Fig. 4),
the power for each four-sided diode is only 4×10^{-13}W. Such low
power consumption would allow a volume density of $5\times10^{13}/cm^3$
even if all devices were operated simultaneously (assuming 20 W
power dissipation). If these junctions were each provided with a
metal-oxide gate where the molecules M are shown, more power
would be required. However, this too would be very small if the
thickness of the bridge X were of the same order of thickness as
the tunnel barrier, and also of the induced inversion and
depletion layers in each half of the bridge. This Gated Tunnel
Diode (GTD) is a particularly interesting device in that it
combines aspects of the Junction Field Effect Transistor (JFET),
MOSFET, and the tunnel diode. Other useful devices can no doubt
also be devised around the MU-STRATE structure. The GTD is meant
to be a primer for such concepts. In addition, one can readily
envision high value dielectric capacitors by connecting many

metal vias in parallel (36), as well as high value resistors by
connecting the semiconductor vias in series (19), or high value
integrated inductors by connecting the metal vias into
rectangular coils (37). The use of the latter three devices in
conjunction with MOSFET (or GTD) switches opens up a large number
of possibilities in analog as well as digital circuits. In
summary then, the MU-STRATE forms the basis for a broad range of
circuit concepts, as well as for testing MED ideas.

Before closing the discussion of the MU-STRATE, it should be
pointed out that other workers are pursuing a somewhat similar
approach to three dimensional circuits, albeit on a geometrical
scale several hundred times larger. This is the work at IBM that
is designed primarily (at present) to detect open and shorted
devices (38). They use a single scanning e-beam system on one
side of a thick three dimensional array of chips as well as
"flood beams" of electrons that impinge from either or both sides
of the array. This is an important concept and many of the
detection schemes and structural arrangements are directly
applicable to this work, and vice versa.

OTHER DEEP GROOVE APPLICATIONS
Reference 2 discusses a wide range of applications for deep and
narrow grooves in Si and other materials. These include the
vertical multijunction solar cell, high value capacitor,
diffraction gratings, particle sieves, gas chromatography system,
ink-jet printer, high voltage rectifier stack, solid state
inductor, three dimensional IC structures, vertical channel
thyristors, infrared polarizers, high finesse infrared
interference filters, bacteria and virus filters and membranes
and a possible x-ray laser structure. The only ones that will
be treated here are the two underlined above. Also, several
relevant experiments will be discussed that are presently
underway with the sponsorship of the Volkswagen Foundation[1].

X-ray or Gamma-Ray Laser (2)

The so-called x-ray laser structure deserves special mention
because of the possibility that a high brightness x-ray source
might be based on such an approach. The basic structure is shown
in Fig. 5. The angle θ is meant to indicate that the slice may
be tilted off the (110) plane toward the (111) plane by several
degrees. The advantages of such a configuration are the
following: [1] The interior of the structure is opened up to the
pumping radiation (electrons, protons, neutrons, photons) while
maintaining its crystal perfection. [2] The x-ray (or gamma-ray)
emitter can be chosen to have a wavelength suitable for anomalous
transmission of x-rays (very low absorption losses) in the Si
lattice. [3] Almost any element(s) can be deposited with good
epitaxy onto the walls of the grooves by using chemical means or
ion and molecular beam epitaxy. [4] The effective spacing of the
Si lattice can be adjusted to the emitted wavelengths by tilting
the crystal before vertical etching. [5] The "effective
solubility" of deposited impurities can be much higher than the
bulk solubility. For example, if the grooves are separated by
0.2 μm, a monolayer of Fe on each wall gives an average
concentration of about $10^{20}/cm^3$, whereas its maximum solid
solubility in Si is only $3\times10^{16}/cm^3$ (39). Almost any element

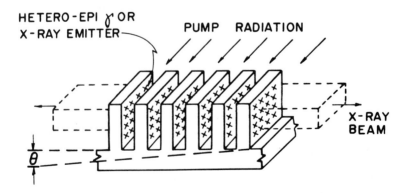

5. Proposed high-brightness x-ray source.

can be considered as a potential candidate for a deposited x-ray
or gamma-ray (Mossbauer) emitter in this structure. One of the
interesting possibilities is Si itself, which can be transmuted
to P by neutron or proton bombardment. The $K\alpha$ x-ray of P of
6.155 A is close to twice the (111) spacing of 3.135 A of Si. If
θ of Fig. 5 is chosen to be 11°, the match of the half
wavelength of the x-ray with the effective lattice spacing is
nearly perfect and the absorption losses for the x-rays will be
negligible. Thus, if such a structure is pumped with high energy
protons and later with electrons, the emitted x-ray beam along
the axis will be extremely parallel, thereby insuring very high
beam brightness. Note that this process does not require
population inversion or stimulated emission of x-rays or gamma
rays, although any help or this sort would certainly be welcomed,
as would a high fraction of recoiless transitions for a Mossbauer
gamma emitter. The relative ease with which such x-rays (or
gammas) could be modulated should also not be overlooked, for
example, by very slight warping of the structure by an optical
input or by capacitive or resistive inputs on the structure
itself.

Particle Filter
A very high flow-through particle filter for removing smoke,
pollen, asbestos, etc., from air and liquids can be made from Si
by extending the grooves in Fig. 2 all the way through the slice.
However, Si is too fragile for many applications so we are
presently fabricating a mold from Si and then making the filter
from a tough plastic such as polyimide. The Si mold is then
sacrifically dissolved. One of the goals of this project is to
produce a submicron filter for use inside the nostrils that will
have at least 70% direct open area for air flow.

A n-i-p-i Superlattice
By using the VEER process (Vertical Etch and Epi-Refill), it is
possible to produce thousands of pn junctions in series (17-19).

If these p and n portions both consist of relatively high
resistivity Si, the space charged (or "intrinsic") zones will
overlap considerably. When this overlap is significant, a
n-i-p-i superlattice is formed that has some unique properties
(40). For example, the effective band gap becomes larger when
high intensity light is absorbed in such material, which means
that the material becomes more transparent at high intensities of
certain wavelengths. Such material is being produced for the
Volkswagen program.

Precise Angle Grain Boundaries

Imagine that a deep groove structure is produced and then
physically bent at high temperatures. It is then filled with
epitaxial Si by the VEER process. The resulting structure can be
designed to have a family of small angle grain boundaries with
many mismatch dislocations of either edge or screw type, or a
mixture thereof. These grain boundaries will have precisely
controlled mismatch angles and they will be characterized in
terms of their conductance and carrier recombination properties.
Such knowledge is important for understanding polycrystalline
solar cells, but may also contribute to the MED field by
providing a new kind of conductive linkage in a solid.

Russell Effect Applied to Silicon

In 1897 Russell (41) discovered that a scratched or abraded Al
slab in contact with a photographic plate would darken the film
after several hours. Some 59 years later this effect was shown
to be 400 times stronger on an abraded Si slice than on an Al
plate (42). It was also explained at that time to be a sort of
catalytic effect in which atmospheric O_2 and H_2O were
catalyzed at the freshly abraded surface of Si or Al to form
hydrogen peroxide H_2O_2. The effect was also extremely
damaging to the film in the first Skylab mission due to the
exposure of the Al film support (43).

We have recently studied the Russell Effect on freshly
etched (111), (110), and (100) Si surfaces[3]. We have also
studied it as a function of separation between slice and film.
The film darkening is negligible for distances greater than about
1 cm. We have also observed the effect on Si slices that are
partially covered with SiO_2. The SiO_2 blocks the process.
This "emission" of H_2O_2 from specific unprotected regions of
a Si slice may find application in initiating local molecular
reactions on a closely neighboring substrate. The smallest
lateral dimensions over which this process has been "resolved" on
a piece of film is about 10 μm, but this can probably be made
smaller if access ways for the reactants and products are
provided. Note that this is another type of IMPRINTATION.

Voltage Controlled Semipermeable Si Membrane

Very thin semipermeable ion and molecular species membranes can
be made by etching deep vertical grooves part of the way through
a Si slice (as in Figure 2), then etching a companion parallel
set of grooves into the opposite side of the slice, being careful
to avoid the grooves that entered from the front side (2). We
are presently producing similar membranes by using V-shaped
grooves that are obtained on (100) surfaces when the etching
windows are aligned along $\langle 110 \rangle$ directions. These grooves are
easier to align than the vertical grooves and also easier to
inspect for defects. Such a structure is shown in Figure 6. We
will study the flow of ions such as Na^+, K^+, and Cu^+
through these very thin solid Si membranes. The solubility of
these donor-type ions can be enhanced by a factor of about 10^8
at 100°C by doping the membrane with the p-type impurity B to
about $10^{20}/cm^3$, under which conditions the fast diffusing
Cu^+ ions have a solubility of almost $10^{20}/cm^3$ themselves.
This allows the passage of a large ionic current of about 1
mA/cm^2 at 100°C across the membrane in a simple
concentration-cell arrangement (2).

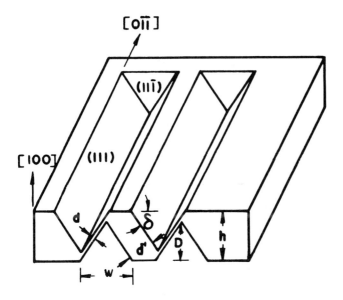

6.   Voltage controllable semipermeable membrane structure.

We will also introduce very small pores into some of these
membranes by anodization in HF/H_2O solutions (44) so that the
passage of molecules and viruses can be studied. Finally, we
expect to introduce p^+ and/or n^+ layers on both sides of the
membrane so that the flow of ions, charged aggregates, and
polarizable species of various types can be studied under the
conditions of applied voltage. These pores incidentally have
diameters ranging from a few tens to a few hundred nm, but their
entrance diameters are only about <u>one</u> nm (44). Control of
these entrance diameters and pore sizes in such membrane
structures would appear to open up a large range of potential
applications, extending to a sort of voltage controlled
"filtration spectroscopy". The possibility of simulating some of
the properties of the axon membrane is also of considerable
interest.

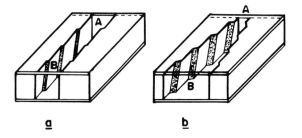

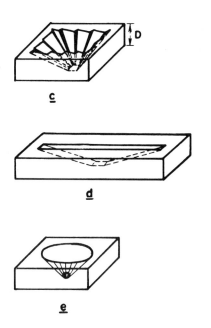

7. Technologically feasible apertures in silicon.

Special Apertures in Si and Molecular Manipulation

The ability to make certain types of grooves and apertures in a
single crystal material like Si raises a number of thought
provoking questions. For example, a slightly misaligned window
opening on a (110) slice produces a deep narrow groove having
small misalignment steps separated by expanses of almost
atomically flat {111} surfaces. These steps face "north" on one
wall and "south" on the opposite wall of the grooves, as in
Fig. 7a (2,8). *Is it possible that long chain molecules in
such an enclosure would tend to coil up in a counterclockwise
spiral and thereby decrease their average diameter?* If this
occurs, it will be due to the molecular collisions on opposite
walls being directed in opposing directions. This assumes that
the reflections from the intervening flat regions on the two
walls will compensate each other and not contribute to the effect
(except to weaken it somewhat).

The second structure shown in Fig. 7b of two diverging
walls raises a similar question: *Would long chain molecules
that enter the groove at A be forced further into the groove by
the wall collisions (all toward B), and would this tend to
elongate the molecules?* This question, as well as the one asked
earlier about molecule "tightening", appears amenable to a two
dimensional Monte Carlo type calculation of the effect of
molecular collisions on an elongated "pool table" having
appropriately directed steps on the walls. If either tightening
or elongation occurs, energy might be accumulated and stored in
such structures. Furthermore, when the molecules are pushed or
diffused out of their containers, they could perform useful
mechanical work as they return to their normal diameters.

The three apertures shown in Fig. 7c, 7d and 7e are also
capable of fabrication in Si slices. Those in 7c and 7d can be
made on (100) and (110) slices respectively, using anisotropic
etchants, and 7e can be made on any orientation slice using an

isotropic etchant and by imposing a temperature gradient across the slice during etching. The first question that might be asked about such aperatures is the following: *Is it possible to change the velocity distribution of N_2 and O_2 and other molecules in the vicinity of such asymmetric apertures?* Sunlight would be allowed to fall on either side of the structures so as to develop a temperature difference and one might also modify the thermal emissive properties of either or both sides. There is no intent to <u>gain</u> any energy here (as in the famous Maxwell Demon proposal), only to <u>use</u> the incoming solar energy to perhaps separate cold from hot molecules. In principle, such an effect might produce a sort of "solar cooling" or perhaps by inversion of the structures a new type of solar heating. Such a question bears some relation to the simple Hilsch Tube that successfully separates hot and cold molecules by injecting air along the circumference of a flat cylindrical container (45). The faster molecules are extracted from a hole near the circumference of the cylinder and the slower molecules diffuse to the central axis of the cylinder where they are extracted at another opening.

Still more provoking questions can be asked about the particular structure shown in Fig. 7c. This is shown with walls whose top edges are slightly misaligned from the ⟨110⟩ directions. If these apertures are properly produced (and this is not a trivial exercise because of the strong tendency to form {111} facets), each wall will have misalignment steps facing in a clockwise (CW) directon (for example). *What will happen to long chain molecules that are allowed to diffuse slowly from top to bottom through a group of such apertures?* The collisions that cause the formation of the so-called boundary layer and which lead to near-zero molecular velocity near the boundary might introduce a measurable CW torque to the diffusing molecules (as well as to the slice itself!). The shape of the molecules might also be affected.

However, it is the second question about the structure 7c
that most stimulates this author. *Is it possible to produce
naturally or artificially, some sort of clockwise vortex (say of
the molecular or boundary layer type) in such apertures?* The
answer depends, of course, on the size and shape of the aperture,
as well as the height and density of the misalignment steps
relative to the thickness of boundary layer. The size of the
small square hole in the bottom of the funnel shaped aperture
can be as small as a few nm, depending on the slice thickness D
and the size of the opening in the top masking oxide, so long as
it is remembered that the 4 walls of the funnel subtend 54.74°
with the top (100) surface. The upper opening will be larger
than the small one by 2.828 D (see Fig. 7c) or about 283 μm
larger for a 100 μm thick slice. The misalignment steps are
typically 50 nm to 500 nm in height depending on the misalignment
angle θ and the etching conditions (2,8). However, in principle,
the step heights can be reduced to the Si bi-layer thickness of
about 0.3 nm by using chemical additives or polish-etch
conditions. The spacing between the steps of height h will be
approximately h cot θ. Incidentally, classical calculations will
almost certainly fail when the step heights become as small as
the Si bi-layer. However, it is also this range of step heights
that may give the biggest effects, as such steps may also be
involved in the Giant Raman Effect.

Within the above constraints it should be possible to
calculate the expected molecular kinetics in the vicinity of such
an aperture. Since the boundary layer thickness in air at
atmospheric pressure is much larger than the height of the
misalignment steps, and first order answer will certainly be that
no significant CW vortex will be formed. However, the situation
should be much different if the 4 walls are sequentially heated
in pulses corresponding to the average time required for a
molecule to move from one wall to the next. The enhancement of

the effect will be still larger if only the vertical "risers" on the steps are heated. This should be technologically feasible using various schemes (e.g., directional evaporation of conductors, etc.). In any case, it may be possible under some appropriate set of conditions to initiate and sustain a sort of "driven micro-vortex" in the CW direction in each aperture. If such a situation can be induced in single crystal apertures (or even in non-single crystal apertures), the "micro-tornadoes" in a structure with many apertures would be expected to produce a pressure difference between the top and bottom of the structure. This would serve to thrust the structure of 7c either upward or downward, depending on whether the funnel shaped apertures face downward or upward, respectively.

Whether or not the unusual proposal above can ever be used as a means of lift or locomotion, the effect of such apertures in single crystal materials appears to merit some serious calculations and experiments. For example, the effect of such apertures (either artificially driven or not) in an airfoil would be of interest to the question of air safety in the presence of wind-sheer forces. In such a case, the single crystal structure would probably be used as a mold for a metal or organic airfoil.

The questions raised in this section are highly speculative. It behooves one to ask if there are any realistic conditions of pressure, viscosity, aperture form, step height, and/or other factors that would give a useful result. The possible applications discussed here are [1] long chain molecule shape modification for energy storage, [2] hot and cold molecule separation for energy conversion, [3] torque effects on a solid body, and [4] the possible "lift" of a driven vortex in a funnel shaped aperture. There must be many more kinds of applications and structural arrangements for which the use of such unique apertures should be seriously considered.

Adsorption on Misalignment Steps

The simple anisotropic etching model presented for etching near
the {111} planes (2) suggests the possibility that particular
metals, enzymes, or other molecules might be adsorbed or
otherwise deposited along the misalignment steps (see Fig. 7a).
For example, Si catalysis experiments could be performed using a
series of etched Si groove (filter type) structures at different
misorientation angles and having particular step heights produced
with different chemical treatments. More useful catalysis
experiments could be performed if Pt or other metals were
electrolessly plated or ion-implanted onto the step-risers of
these samples. If the step heights were of the order of the Si
bi-layer (about 0.3 nm), adsorption of certain metals or
molecules along these steps might result in dense arrays of one
dimensional strips having unusual superconducting, catalytic, or
other properties. A similar conjecture could have been made
regarding conduction along the impurity precipitates in the
parallel dislocation arrays proposed earlier. However, it is not
yet clear whether dislocations (decorated or not) have any
conductance in the normal sense (46), although there certainly
remains some hope that they do (47).

Deposition along misalignment steps has certain advantages
compared to precipitation along dislocations. For example, the
depositions are more easily controlled and manipulated. Physical
access for subsequent chemical (and molecular electronic)
applications is also greatly facilitated.

CONCLUSIONS

Archimedes is purported to have said, "Give me a place to stand
and I will move the earth". In the first part of this work, the
Multi-Use Substrate was an attempt to provide "a place to stand"
for the envisioned molecular electronic devices and circuits. In
the next section a new method of making and testing integrated

circuits was proposed (SIFT)that will hopefully provide enough leverage to "move the earth" at least a few meters in this field. This entails the use of feedback principles at a much higher level of complexity than is common today. If such methods are successfully developed, IC's will be produced without "designing" them in the normal sense, although general functional goals will be pursued during their fabrication. Other important aspects of such circuits that are discussed briefly are the large amounts of redundancy required to allow for significant internal storage errors, as well as the probable need for circuit branches to continuously monitor and adjust the unstable device elements. Such methods are believed to be especially appropriate for MED's, but they may also be necessary for Si and GaAs before they can progress into the extremely high circuit densities made possible by modern technologies.

The severe limitation provided by the power dissipation of modern IC's was also discussed. Very significant gains in power dissipation (by a factor of 100 or more) have recently been attained (13). Similar gains are probably attainable using integral heat pipes in the back side of similarly etched structures (2) or in the Multi-Use Substrate proposed here. Nevertheless, devices and circuits having much lower dissipation levels will eventually have to be developed if circuit densities are to take advantage of the technologically producible structures like the Multi-Use Substrate. These potential volume densities are in the range of 10^{12} memory elements/cm^3 for the MU-STRATE proposed here, which should be compared with today's maximum attainable volume density of about 10^7 elements/cm^3. By working backward from the MU-STRATE structure, a new tpe of logic element was proposed that meets the greatly reduced power requirements. This is based on extremely small area tunnel diodes (either gated or ungated) which dissipate about 10^{-13} W per gate when operated at low temperature.

Another badly needed technology that was highlighted is the Discretionary Erasure and Writing of connecting Links (DEWLINKS). Two ways to perform this important function were presented. These involve either alternating localized ion implantation and localized annealing steps, or alternating localized proton enhanced outdiffusion and ion-implantation steps.

Several experiments being performed in the author's laboratory under the sponsorship of the Volkswagen Foundation[1] were also discussed. These included a high flow-through environmental particle filter for smoke, pollen, and asbestos. A voltage controlled semipermeable Si membrane was also reported which may allow the axon membrane to be simulated, as well as the controlled filtration of virus and molecules. In another experiment, the epitaxial refill of deep grooves allows the formation of doping (n-i-p-i) superlattices as well as parallel arrays of either edge or screw dislocations in controlled-angle grain boundaries. Fabrication and characterization of such structures should have important repercussions in understanding a wide range of new devices.

The darkening of photographic plates in localized regions by the emission of hydrogen peroxide from freshly etched or abraded Si (the Russell Effect) was also characterized in terms of resolution capability, separation between film and plate, and efficacy of various film sensitization schemes.

Finally, several speculative proposals were presented, the most important of these relative to MED's may be the high brightness modulatable x-ray source. This would allow the direct writing of submicron circuit patterns without the necessity for high vacuum.

Another group of speculative concepts were discussed which involved the unique properties of small apertures produced in single crystal materials, especially Si. Possible applications identified were the separation of hot and cold molecules using solar energy, the shrinkage or expansion of the average radius of long chain molecules for possible energy storage, a miniature diffusion pump, and a possible new type of airfoil structure having very unusual properties. The adsorption of metals, enzymes and other species on the misalignment steps was also discussed, especially as these might serve as one dimensional conductors and/or catalytic sites.

Of the several concepts presented in this work, the Multi-Use Substrate (MU-STRATE), and the Signal Induced Feedback Treatment (SIFT) have the most relevance to the fabrication of molecular electronic devices. The experiments in progress that were discussed are expected to make measurable and useful scientific and technological contributions over the long term, though not primarily in the MED field. The most important of this latter group are probably the two Vertical Etch and Epitaxial Refill (VEER) experiments, one dedicated to making n-i-p-i superlattices and the other to generating dense arrays of dislocations in precisely controlled-angle grain boundaries. Of the speculative proposals, the most promising may be the high brightness x-ray source. The most interesting is almost certainly the new kind of airfoil using small apertures in single crystal material. We are fortunate to live in an age when most of these ideas can and probably will be tested, both theoretically and experimentally.

ACKNOWLEDGEMENTS
The author wishes to express his sincere appreciation for the assistance of Graciela Rosas de Guel and other colleagues in the Microelectronic Laboratory at I.N.A.O.E. in the studies undertaken for the Volkswagen Foundation.

FOOTNOTES

1. This work was partially supported by the Volkswagen
    Foundation.

2. A capo is a physical clamp placed on the bridge of a guitar
    to adjust the musical key upwards in frequency. The CAPO is
    a type of electrical parameter clamp.

3. The film is dipped for 15 seconds in a 1.5% solution of
    NH_4OH, rinsed for 30 seconds in methanol, then the film is
    only partially dried.

4. The Si IC layout would be restructured on the basis of the
    types of molecules on the MU-STRATE.

REFERENCES

1. Paul, 1 Corinthians 13:12, KJV, ca. 100 A.D. Translated from
    Greek.

2. D.L. Kendall, Ann. Rev. Mater. Sci., 9, 373 (1979). Ed.
    R.A. Huggins, Palo Alto, CA.

3. G.H. Dohler, Physica Scripta., 24, 430 (1981).

4. G. Rosas de Guel, D.L. Kendall, Y.M. Mirza and M. Reynoso,
    Proceedings of the 2nd Latin American Symposium on Surface
    Physics, Puebla, Mexico; October 1982.

5. M. Pomerantz, in Molecular Electronic Devices, Ed. F.L.
    Carter, Marcel Dekker, Inc., New York, 1982, p. 279.

6. T.I. Kamins and T.R. Cass, Thin Solid Films, 16, 147
    (1973); T.I. Kamins, J. Electrochem. Soc., 127, 686 (1980).

7. M.W. Geis, G.A. Lincoln, N. Efremow, and W.J. Piacentini, J.
    Vac. Sci. Technol., 19, 1390 (1981).

8. D.L. Kendall, Appl. Phys. Lett., 26, 195 (1975).

9. Y. Tarui, IEEE Trans., ED-27, 1321 (1980).

10. J.S. Kilby, U.S. Patent 3,138,743 (1964).

11. F.B. Micheletti, IEEE Trans., ED-25, 857 (1978).

12. R.W. Keyes, IEEE Trans., ED-26, 271 (1979).

13. D.B. Tuckerman and R.F.W. Pease, IEEE Electron Dev. Lett., 2, 126 (1981).

14. B. Hoeneisen and C.A. Mead, Sol. State Elect., 15, 819 (1972); also 15, 891 (1972).

15. P.K. Chatterjee, G.W. Taylor, R.L. Easley, H.S. Fu, and A.F. Tasch, Jr., IEEE Trans., ED-26, 827 (1979).

16. R.H. Dennard, F.H. Gaensslen, E.J. Walker, and P.W. Cook, IEEE Trans., ED-26, 325 (1979).

17. R.K. Smeltzer, D.L. Kendall, and G.L. Varnell, Conf. Rec. 10th IEEE Photovoltaics Spec. Conf., Palo Alto, CA, 1973; R.K. Smeltzer, J. Electrochem. Soc., 122, 166 (1975).

18. D.L. Kendall, F.A. Padovani, K.E. Bean, and W.T. Matzen, U.S. Patent 3,969,746 (1976).

19. D.L. Kendall and M.M. Judy, U.S. Patent 4,065,742 (1977).

20. H. Muraoka, T. Ohhashi, and U. Sumitomo, in Semiconductor Silicon 1973, Eds. H.R. Huff and R.R. Burgess, ECS Softbound Series, Princeton, N.J. (1973).

21. J.B. Price, ibid., p. 339; A. Bohg, J. Electrochem. Soc., 118, 401 (1971).

22. E.D. Palik, J.W. Faust, Jr., H.F. Gray, and R.F. Greene, J. Electrochem. Soc., 129, 2051 (1982).

23. D.L. Kendall, U.S. Patent 3,810,791 (1974).

24. P. Baruch, J. Monnier, B. Blanchard, C. Castaing, Appl. Phys. Lett. 26, 77 (1975).

25. R.F.W. Pease, D.J. Bartlelink, N.M. Johnson, and J.D. Meindl, Appl. Phys. Lett., 35, 463 (1979).

26. D.L. Kendall, U.S. Patents 3,674,995 (1972) and 3,725,148 (1973).

27. D.L. Kendall and J.C. Knowles, Jr., U.S. Patent 3,936,329 (1976).

28. R.A. Stehlin, R.J. Dexter, D.L. Kendall, J.M. Pankratz, U.S. Patent 3,897,274 (1975).

29. C.S. Fuller and R.A. Logan, J. Appl. Phys., 28, 1427 (1957).

30. D.C. Shaver, J. Vac. Sci. Technol., 19, 1010 (1981); F. Mizuno, S. Mori, and K. Satoh, ibid., 19, 1019 (1981).

31. F.L. Carter, reference 5, p. 121.

32. H. Wohltjen, ibid, p. 231.

33. E. Anaya and D.L. Kendall, Research Project, INAOE, Puebla, Mexico.

34. L. Valdivia and D.L. Kendall, Proceedings of IEEE MEXICON, Mexico City, 1977, p. V-43.

35. S.M. Sze, Physics of Semiconductor Devices, J. Wiley, New York, 1969, p. 177.

36. D.L. Kendall and W.T. Matzen, U.S. Patents, 3,962,713 (1976); 4,017,885 (1977).

37. D.L. Kendall, U.S. Patent 3,881,244 (1975).

38. H.C. Pfeiffer, G.O. Langner, W. Stickel, and R.A. Simpson, J. Vac. Sci. Technol. 19, 1015 (1981).

39. F.A. Trumbore, Bell System Tech. J., 39, 205 (1960).

40. G.H. Dohler, Physica Scripta, 24, 430 (1981).

41. W.J. Russell, Proc. Roy. Soc. (GB), 61, 424 (1897).

42. A.J. Ahearn and J.T. Law, J. Chem. Phys. 24, 533 (1956).

43. G.R. de Guel, M. Reynoso, D.L. Kendall, and M.Y. Mirza, Extended Abstracts, Electrochem. Soc., Wash. DC, 1983; M.E. VanHoosier, J.D.F. Bartoe, G.E. Brueckner, N.P. Patterson, and R. Tousey, Appl. Opt. 16, 887 (1977).

44. Y. Watanabe, Y. Arita, T. Yokoyama, and Y. Igarashi, J. Electrochem. Soc. 122, 1351 (1975); Y. Arita and Y. Sunohara, ibid., 124, 285 (1977); T. Unagami and M. Seki, ibid., 125, 1339 (1978); T. Unagami, ibid., 127, 476 (1980).

45. Scientific American, November 1958, "The Hilsch Tube".

46. H.J. Queisser, in Defects in Semiconductors, Eds. J.W. Corbett and S. Mahajan, North Holland Materials Research Conference, Boston, 1982.

47. W. Shockley, Sol. State Technol. 26, 75 (1983).

14

Elemental and Molecular Microanalysis in the Sub-Micrometer Spatial Domain

Dale E. Newbury/Center for Analytical Chemistry, National Bureau of Standards, Washington, DC

I. INTRODUCTION

The understanding and development of microscopic structures has been traditionally aided by the development of microanalysis techniques. Microanalysis or microbeam analysis is the analysis of selected regions of a sample, where the lateral dimensions of the region analyzed are on the order of micrometers. In general, microanalysis can be achieved by either of two routes: (1) The primary radiation can be focused into a micrometer-sized probe so that local excitation is obtained for analysis. (2) The secondary radiation can be focused to produce a true optical image of the excited region of the sample. In this case, the sample may be irradiated over a large area. Local information is obtained by aperturing in the image plane to isolate a particular area of interest. "Microanalysis" refers only to the lateral specificity of an analysis technique and does not imply any additional characteristic such as a particular level of sensitivity (e.g., "trace" analysis which indicates a capability for the measurement of a constituent which is present as a minute fraction) or

spatial sensitivity in depth (e.g., surface analysis, where information is derived from the outer atomic layers of a sample).

Currently, a wide variety of techniques is available for comprehensive analysis, elemental, isotopic, and molecular, at the micrometer scale. These analysis techniques include electron probe microanalysis (EPMA), secondary ion mass spectrometry (SIMS), scanning
Auger microanalysis (SAM), laser Raman microanalysis, and laser
microprobe mass analysis (LAMMA). The capabilities and limitations
of these techniques have been described in detail previously (1). As
the dimensions of electronic devices are reduced below one micrometer
and eventually to molecular dimensions, microanalytical techniques
must undergo further development to be able to obtain information on
such a scale. In this paper, consideration will be given to current
capabilities and future prospects for development of analysis techniques in the sub-micrometer spatial domain.

II. ELECTRON BEAM TECHNIQUES

A Monte Carlo electron trajectory simulation of the interaction of an
electron beam with a copper target at a beam energy typically used in
electron probe microanalysis (e.g., 20 keV) is shown in Figure 1(a)
(2). Due to elastic scattering, the electron beam is degraded spatially to produce an interaction volume with dimensions of approximately 1 micrometer. For materials of lower atomic number at the
same beam energy, the interaction volume is larger due to the greater
range of the primary electrons, as illustrated for carbon in Figure 2,
where the interaction volume has a diameter of approximately 5 micrometers. In the EPMA, analysis is accomplished by means of spectrometry
of characteristic x-rays produced during inelastic scattering of the
beam electrons which results in the ionization of inner atomic
electron shells. As long as the electron energy exceeds the critical
excitation energy of the shell of interest, ionization can occur. The
distribution of K-shell ionization events within the interaction volume is shown in Figure 1(b). The sampling volume, which is the region
from which the secondary radiation originates, is nearly as large as
the interaction volume of the primary radiation, despite the relatively high energy needed for excitation of the K-shell in this case
(8.98 keV).

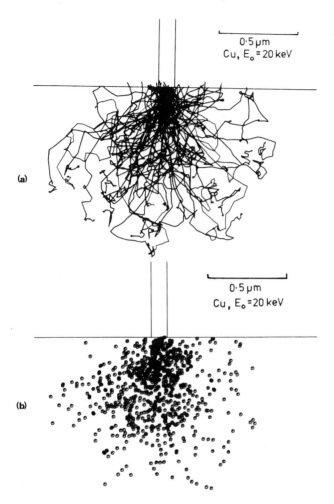

Figure 1 Monte Carlo electron trajectory simulation of beam interaction in a copper target at a beam energy of 20 keV. (a) Electron
paths. (b) Sites of inelastic scattering events resulting in K-shell
ionization events.

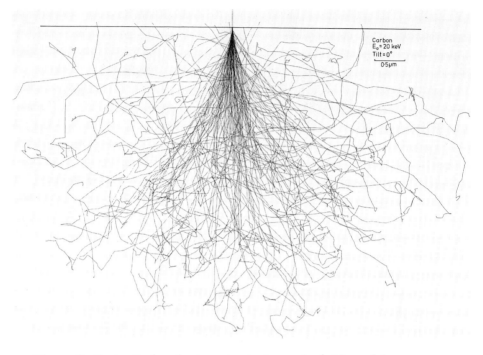

Figure 2 Monte Carlo electron trajectory simulation of beam
interaction in carbon.

In scanning Auger microscopy, analysis is carried out by means
of spectrometry of electrons of characteristic energy. Auger electrons are emitted from an atom as a result of electron transitions in
the de-excitation process following an inner shell ionization. The
Monte Carlo calculation of the distribution of the inner shell ionization events (e.g., Figure 1(b)) is thus a distribution of sites of
possible characteristic x-ray and Auger electron generation. While
x-rays can escape from ionization sites at any depth and retain their
characteristic energy, Auger electrons are subject to inelastic
scattering processes. With an inelastic mean free path on the order
of 1 - 3 nm, only Auger electrons generated near the surface can
escape with their characteristic energy, which greatly reduces the
sampling depth compared to x-ray analysis under the same conditions.
However, the lateral sampling size is subject to the same limitations

as x-ray analysis since backscattered beam electrons which result from
multiple elastic scattering can generate Auger electrons as they exit
the surface.

The physical limitations on spatial resolution imposed by the
interaction volume which results from scattering effects restrict
conventional EPMA analysis to a sampling volume on the scale of 1
micrometer. In order to improve the spatial resolution to allow
microanalysis on a sub-micrometer scale, two approaches are possible,
depending on the form of the sample. (1) For samples in the form of
unsupported thin films, operation at high beam energies (100 keV and
higher) is useful. (2) For samples in the form of bulk solids,
operation at low beam energies (3 keV and lower) can result in sub-
micrometer resolution.

A. High Beam Energy Analysis

The advantage of operating with high beam energies and unsupported
thin films is illustrated in Figure 3, which shows Monte Carlo electron
simulations for two materials, silicon and gold in the form of films
100 nm thick, at beam energies of 100 and 400 keV. The spatial resolution of the primary interaction volume is seen to be of the order
of 30 nm or less, even for a strongly scattering material such as
gold. For low atomic number targets such as silicon at high beam
energies, scattering of the primary beam is reduced to the point that
the interaction volume is defined almost exclusively by the diameter
of the focused probe. Unfortunately, at these very high spatial
resolutions secondary scattering effects can be the limiting factor
on the minimum size of the interaction volume. Fast secondary
electrons, that is, conduction band electrons ejected with energies
up to one-half of the beam energy, are scattered at right angles to
the beam and propagate laterally into the foil, degrading spatial
resolution, as illustrated in Figure 4 (3). Characteristic and
bremsstrahlung x-rays generated within the primary interaction volume
can also propagate laterally to cause secondary fluorescence of the
foil constituents. Despite these secondary effects, analytical
resolutions finer than 50 nm are readily achievable.

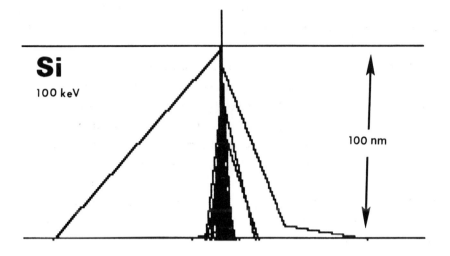

(a)

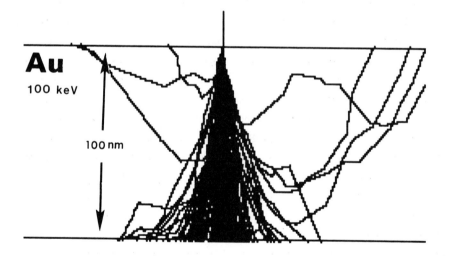

(b)

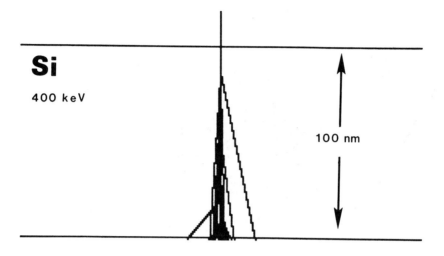

(c)

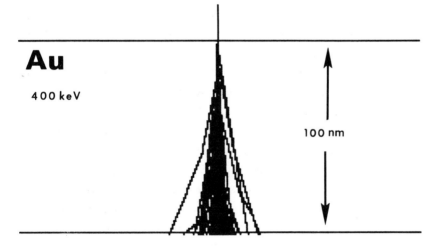

(d)

Figure 3   Monte Carlo electron trajectory simulation of beam interaction in thin foils: (a) silicon at 100 keV; (b) gold at 100 keV;
(c) silicon at 400 keV; (d) gold at 400 keV.

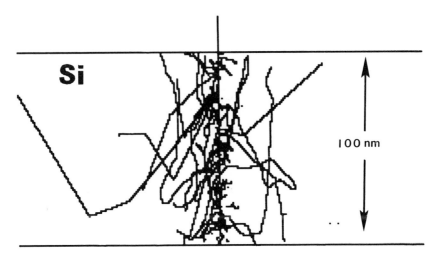

Figure 4 Monte Carlo electron trajectory simulation of beam interaction with the generation of fast secondary electrons. Sample:
silicon at 100 keV.

    The technology for microanalysis at high beam energies is now
available in the so-called analytical electron microscope (AEM), which
combines conventional high resolution transmission electron microscopy (TEM), scanning transmission electron microscopy (STEM), scanning
electron microscopy (SEM), and electron diffraction techniques in a
single instrument (4). For analytical work, the AEM generally incorporates an energy dispersive x-ray spectrometer and an electron energy
loss spectrometer for elemental analysis. Because of the inherent
peak-to-background limitations as well as spectrometer resolution
characteristics, these analytical techniques are generally limited to
sensitivities of 0.001 mass fraction and are thus not considered a
form of trace microanalysis. However, due to the small mass of the
sample which is excited, typically 1-100 fg, the minimum mass which
can be detected is on the order of 10 ag - 1 fg, depending on operating conditions and matrix effects. With a suite of multi-element
standards, quantitation by empirical and theoretical methods is
possible with relative errors of 20% or less. While the x-ray
spectrum contains only atomic information as a result of the high
energy shells used, the electron loss spectrum contains potentially a

complete characterization of electron energy levels from inner shell to outer shells, including valence and conduction band electrons, as well as molecular bonding levels.

B. Low Beam Energy Microscopy and Analysis

If the specimen of interest is in the form of a bulk solid, then the use of high beam energies to achieve sub-micrometer spatial resolution is not possible due to their large range. For solid specimens, sub-micrometer spatial resolution is possible through the use of electron beams of very low energy in the range 1 - 5 keV. The electron range R in a solid target is described by an equation of the form (2):

$$R = 0.0276 \ A \ E^{1.67}/Z^{0.89}\rho \quad \text{(micrometers)} \qquad [1]$$

where A is the atomic weight, E is the incident beam energy, Z is the atomic number, and ρ is the density. Because of the strong dependence of the range on the beam energy, lowering the beam energy from 20 keV to 3 keV reduces the range in a silicon target from 4.7 μm to 200 nm. Lateral resolution has a similar scale to that of the range, and thus operation at low beam energy substantially improves the spatial resolution.

This improvement in spatial resolution is not without cost. The brightness of an electron source is proportional to the beam energy, and hence for a given probe size, the current is reduced proportionally as the beam energy is reduced. Moreover, a completely different strategy must be used for x-ray analysis. In order to ionize a particular electron shell, the beam energy must exceed the critical excitation energy, E_c. Operation at the conventional beam energy of 20 keV provides sufficient energy to excite K, L, and/or M x-rays in the energy range 1 - 10 keV which are suitable for the analysis of all elements with Z $\geq$ 11 (sodium). When the beam energy is reduced to a value of 3 keV, the analyst is required to measure L and M x-ray lines with energies below 1 keV in order to detect elements with Z > 16. Such low energy x-rays are strongly absorbed in the specimen, even when formed at shallow depths, as well as in windows of x-ray

spectrometers. The advent of high efficiency windowless or ulta-thin
window silicon (lithium) energy dispersive x-ray spectrometers has
increased the accessibility of the energy range from 200 eV to 1 keV
(5). The basis for quantitation in this energy range is under study
with the development of analytic expressions for the x-ray generation
depth distribution function, which is vital for proper correction of
matrix effects (6). Besides the possibility of x-ray spectrometry,
the surface layers of the sample can be examined simultaneously by
electron spectroscopy, including Auger and secondary electron spectroscopy and scattered primary electron energy loss spectroscopy (7).
Because of the low energy of the electron transitions which are
measured, elemental characterization is augmented by the possibility
of molecular microanalysis. In addition to microanalysis at individual locations, scanning images can be prepared with the characteristic
emission or loss signals (7). Finally, by means of the electron
channeling effect, crystal orientations can be measured by generating
selected area electron channeling patterns (8).

Although high beam energy microanalysis techniques have undergone
intensive development in recent years, interest in low beam energy
microscopy and microanalysis is just beginning. This field should see
extensive development in the near future.

III. ION BEAM TECHNIQUES

Several ion beam technologies currently exist which can provide microbeam analysis at the 1 - 10 micrometer spatial scale. These techniques
include secondary ion mass spectrometry (SIMS) in the form of the ion
microprobe and the ion microscope, low energy ion scattering spectrometry (ISS), high energy ion Rutherford backscattering (RBS), and
high energy ion-induced x-ray spectrometry (9). In order to reduce
the spatial resolution below 1 micrometer the only ion beam technique
which shows promise is SIMS. At the intermediate ion beam energies
utilized in SIMS, typically 5 - 20 keV, the range of the primary ion
is only of the order of 10 - 50 nm. Since the primary ion undergoes
mainly inelastic scattering with very little elastic scattering, the
ions do not spread laterally. Thus, if a sub-micrometer beam can be

focused, its lateral spatial resolution will not be degraded by
scattering. Moreover, the secondary ions detected in SIMS are low in
energy, which restricts their maximum escape depth to a few nm from
the surface, which again preserves the high spatial resolution.

Recently, high brightness liquid metal ion sources have been
developed which can provide sub-micrometer beams with the proper
lenses (10). These sources have been incorporated into scanning ion
microscopes which utilize the high yield of secondary electrons
generated under ion bombardment to form scanning ion/electron images
(11). By collecting secondary ions in the manner of an ion microprobe,
an analytical spectrometry can be obtained at high lateral spatial
resolution. In considering the utility of such a beam to do useful
analysis on a sub-micrometer scale, attention must be given to the
generation and collection of secondary ions. Since sputtering is a
destructive process, a certain volume of material must be removed in
order to detect a particular level of a constituent. The volume to
be removed can be estimated from the equation:

$$\pi r^2 \, d \, \rho^* \, C_i \, s^+ \, \epsilon = n \qquad\qquad [2]$$

where r is the beam diameter, d is the depth to be sputtered, ρ^* is
the atomic density (atoms/cm^3), C_i is the concentration of the constituent of interest, s^+ is the yield of sputtered ions per number of
sputtered atoms, ϵ is the combined transmission and detection efficiency of the spectrometer, and n is the number of ions which must be
counted. If we consider that 100 secondary ions must be detected to
give a reproducibility of $\sigma = 10\%$, and that the factor $\epsilon = 0.01$, then
for a beam radius of 100 nm equation [2] can be used to predict the
depth of material which must be sputtered to detect a given concentration in a silicon target. These values are listed in Table 1 for
two different secondary ion sputter yields. Examination of Table 1
reveals that depth resolutions better than 100 nm can only be obtained
for high ion yield species when the constituent is present at high
concentrations. Thus, high spatial resolution SIMS can be expected
to be limited in its applicability to sub-micrometer characterization.

TABLE 1. Sputtering Depth to Detect 100 Ions of a Constituent with a 100 nm Beam.

Concentration	Ion Yield = 0.1 Depth	Ion Yield = 0.001 Depth
1	0.06 nm	6 nm
0.1	0.6	60
0.01	6	600
0.001	60	6000

IV. PHOTON BEAMS

Spectroscopies based on scattered, absorbed, or emitted photons provide a powerful means of characterizing the molecular constituents of a sample (12). As such, these photon spectroscopies complement the predominantly atomic (elemental) analysis obtained with electron or ion beam excitation. The photon techniques which can be applied at the micrometer spatial level include laser Raman microanalysis, microspectrofluorimetry, and micro-infrared microscopy. Due to the relatively long wavelengths of the photons utilized in these techniques, it is difficult to characterize sub-micrometer regions, except for the use of ultraviolet wavelengths. However, the mass sensitivity of ultraviolet-excited fluorescence is so great that it is possible to detect mono-molecular layers doped with an appropriate fluorescent dye. Moreover, compared to the electron and ion beam techniques, photon excitation is generally much less destructive. Thus, the non-destructive optical microanalysis techniques have considerable potential for characterizing molecular films for molecular electronic devices.

An alternative microanalytical technique based on photon beam excitation is laser microprobe mass analysis (LAMMA) (13). In LAMMA, a pulsed beam of ultraviolet photons evaporates the sample and ionizes some of the sample atoms, which are subsequently analyzed in a time-of-flight mass spectrometer. For a solid, bulk sample the region evaporated has a volume of several cubic micrometers. When the specimen is in the form of a thin film, the lateral extent of the evaporated region can be made as small as 0.5 μm. Of particular interest in the LAMMA technique is the capability for analysis of molecular as well as elemental constituents. The LAMMA has been recognized as one of the only forms of organic mass spectrometry with spatial resolution on the micrometer scale. Currently, experiments are being conducted to demonstrate the applicability of LAMMA to detecting monomolecular layers and to determine the limits of lateral spatial resolution (14).

REFERENCES

1. D. E. Newbury, "Microanalysis in the Scanning Electron Microscope: Progress and Prospects", SEM/1979/II, 1.

2. J. I. Goldstein, D. E. Newbury, P. Echlin, D. C. Joy, C. Fiori, and E. Lifshin, Scanning Electron Microscopy and X-ray Microanalysis, Plenum, New York, 1981.

3. D. C. Joy, D. E. Newbury, and R. L. Myklebust, J. Micros., 128, RP1 (1982).

4. J. J. Hren, J. I. Goldstein, and D. C. Joy, Introduction to Analytical Electron Microscopy, Plenum, New York, 1979.

5. P. J. Statham, Energy Dispersive X-ray Spectrometry, National Bureau of Standards Special Publication 604, Washington, 1981, p. 141.

6. J. D. Brown, A. P. von Rosenstiel, and T. Krisch, Microbeam Analysis/1979, San Francisco Press, San Francisco, 1979, p. 241.

7. C. Le Gressus, H. Okuzumi, and D. Massignon, "Changes of Secondary Electron Image Brightness under Electron Irradiation as Studied by Electron Spectroscopy", SEM/1981/I, 251.

8. D. C. Joy, D. E. Newbury, and D. L. Davidson, J. Appl. Phys., 53, R81 (1982).

9. A. W. Czanderna, ed., Methods of Surface Analysis, Elsevier, Amsterdam, 1975.

10. V. E. Krohn and G. R. Ringo, Secondary Ion Mass Spectrometry-Fundamentals and Applications, Japan Society for Promotion of Science, 1978.

11. R. Levi-Setti, "Secondary Electron and Ion Imaging in SIM", SEM/1983, to be published.

12. T. Hirschfeld, Microbeam Analysis-1982, San Francisco Press, San Francisco, 1982, pp. 247-252.

13. R. Wechsung, F. Hillenkamp, R. Kaufmann, R. Nitsche, and
    H. Vogt, "LAMMA - A New Laser Microprobe Mass Analyzer",
    SEM/1978/I, 611-620.

14. R. A. Fletcher, I. Chabay, D. Weitz, and J. Chung, J. Chem.
    Phys., 1983, to be published.

15

Interfacing Silicon Microelectronic Substrates with Biological Systems

B. L. Giammara and G. A. Rozgonyi /
Microelectronics Center of North Carolina
Research Triangle Park, North Carolina

R. H. Propst / Biomedical Engineering and Mathematics
Curriculum, University of North Carolina, Chapel Hill,
North Carolina

J. S. Hanker / Dental Research Center, University
of North Carolina, Chapel Hill, North Carolina

INTRODUCTION

In order to interface microelectronic substrates with living cells or tissues, the micromorphology of the substrate surface, as well as other factors regulating the adhesion of cells and tissues, must be considered (1). Cell adhesiveness is a fundamental cell property that is important in cell migration. The adhesion of cells to extracellular substrates is not uncommon. Although some cells can grow in suspension, others require a substrate to which they can adhere and on which they can subsequently spread in order to grow. The electrical and molecular effects controlling cell adhesion to substrates are complex; adhesion is known to be a reversible, energy-dependent stepwise process in which attachment of the negatively charged cell surface is influenced by various factors. The presence of divalent cations, serum factors, cellular glycoproteins such as fibronectin, and the chemical nature of the substrate surface can be important parameters for attachment (2,3). When fabricating ordered arrays of 2 or 3 dimensional molecular electronic devices, the nature of the substrate surface becomes an

325

especially important factor. Recent progress in the fabrication of biotechnical electron devices with protein lines has been described by McAlear and Wehrung (4) and Hanker and Giammara (5). Of especial interest is the report by Cooper et al. (6) that thin films of silicon monoxide applied to a plastic Petri dish in the form of an oriented array of fine lines provided a suitable substrate on which mouse neuroblastoma cells could be grown as ordered, well-defined connected chains of cells in a pattern controlled by the boundaries of the films.

The present study describes an approach to biochip fabrication by the use of photoengraved semiconductor microelectronic substrate chips for the purpose of interfacing with living cells. Current technology is capable of producing fine line trenches in thermally grown SiO_2 on silicon substrates, as well as making grooves in the silicon itself. The depth and periodicity of the trenches can be varied on a single substrate chip from one half micron to ten microns or larger. Graphoepitaxy of electrodeposited tin onto SiO_2/Si substrates has recently been reported by Darken and Lowndes (7). The expression "graphoepitaxy" has been used by Smith and Flanders (8) to describe a mode of crystal growth where the periodicity and amplitude of the artificially engraved substrate influence the orientation of the growing layer. Chips used in our approach can be used in a passive substrate mode without electronic bias. Alternatively, they can be used in an electronically active mode where surface charge is generated by an applied biasing voltage in order to stimulate "nucleation" and growth of deposited biological materials.

The effects of applied steady electric fields on the orientation and growth of cells have long been known. Ingzar in 1920 (9) found that an electric field could orient nerve growth. Weiss (10) suggested that this effect was due to mechanical guidance provided by the oriented micelles in the culture medium resulting from application of the field. It is now known that the effects of electric fields on the orientation of neurite growth, in a simple culture of dissociated cells, are principally due to the direct action of the field on the neurons (11), although chemical differences in the medium and the physical nature of the substrate are also of importance (12). In order to study the effects of electric fields on the attachment and movement of cultured cells, two devices were designed. The first consisted of an ordinary glass microscope slide upon which a conducting pattern was silk-screened to allow the application of a step-wise decrease in

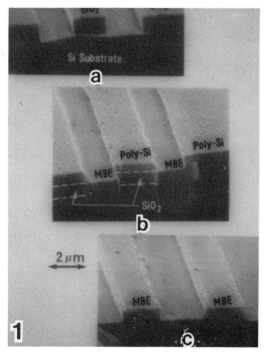

FIGURE 1. Scanning electron micrographs of cleaved samples following different processing steps: a) Patterned SiO_2 layer on Si substrate produced by photolithography and reactive ion etching. Cultured cells were subsequently applied to this substrate. b) Sample after molecular beam epitaxy (MBE) deposition. Epitaxial growth in oxide channels. Polycrystalline growth on top of oxide. c) Sample after stripping off oxide and poly-Si leaving the epitaxial silicon pattern. (From Bean and Rozgonyi, App. Phys. Letters, 1982). Bar = 2 μm.

voltage across the slide. The second employed essentially the same design configuration, but instead of silk-screening a pattern on the microscope slide, it utilized metallic adhesive tape to attach the silicon chips and to provide the electric field across the chips. This was used to determine the effects of an applied electric field on cultured connective tissue cells.

The purposes of this study were to ascertain: 1) whether cells would grow on silicon microelectronics semiconductor substrates; 2) the effects of cellular exudates or the requirement for interfacing proteins for adhesion of cells to these substrates; 3) the importance of cell type and size and device line geometry for epitaxial-like directed growth of cells on these semiconductors; and 4) the effects of an applied electric field.

MATERIALS AND METHODS

Preparation of Substrates

Fabrication of SiO$_2$ Chips

The photoengraved semiconductor microelectronic silicon substrates were used from the fabrication sequence reproduced in Figure 1. The pattern has a repeat distance of 2 μm lines, i.e., both the SiO$_2$ lines and Si substrate channels were 2 μm in width. Silicon substrates compatible with these concepts, see Figure 1a, were provided by Bean and Rozgonyi (Bell Laboratories) (13) using molecular beam epitaxy (MBE) through a photoengraved oxide masking layer. Deposition of silicon with simultaneous boron ion implantation produced p-type patterned epitaxial silicon films (see Figure 1c) on n-type silicon substrates. The resolution provided by the mask and conventional photolithographic processes produced engraved substrate chips with line widths down to 500 nm. (For our purposes of integrating readily available semiconductor micro-electronics technology with new concepts in biomolecular technology, the substrate chip in Figure 1a was used in the passive deposition growth mode, or in combination with original devices described below that allowed the application of an electric field.) It was planned that future experiments would use the p/n junction structure of Figure 1c biased to electronically perturb or activate the silicon substrate itself for influencing cell deposition processes.

Applied Field Patterning

By Voltage Slides

By photolithographic techniques, a silver pattern, using a standard thick film silver-palladium ink, was silk-screened onto ordinary glass microscope slides which were allowed to dry. They were then baked into the glass by processing through a muffle furnace at 750°C. The silver pattern (Fig. 2a,b) provided a 1.5 cm distance between the anode and cathode and the attachment of electrodes that resulted in a stepwise decrease in voltage. Included in the pattern was a control voltage band or zero electrical field.

By Taped Voltage Devices

These devices consisting of metallic adhesive tape on glass microscope slides have the same capabilities as the Voltage Slides. They allow silicon chips (arrow) to be secured to the slides and result in the same pattern of step-wise decrease in voltage to zero. Resistors and alligator chips were attached to the tape (Fig. 3).

Procedure for Coating Glass or Silicon Dioxide (Chip) Substrates for Bio-interfacing

Substrates were treated with four different bio-interfacing coatings to promote adhesion of cells. These were prepared in the following manner:

1. Fibronectin (human, Collaborative Research). 1 mg was reconstituted to 1 ml with distilled water. A one to three dilution was made with Minimum Essential Medium (MEM, Eagle), Gibco #410-1100 (with or without 10% Fetal Bovine Serum, FBS).

2. Poly-L-lysine. A 10% solution of poly-L-lysine hydrobromide (Sigma P 1399) was diluted one to three with MEM (with or without 10% Fetal Bovine Serum).

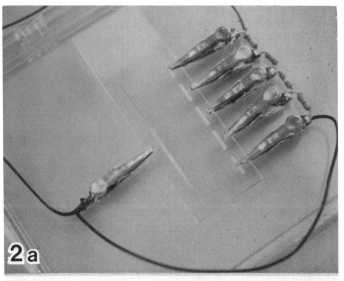

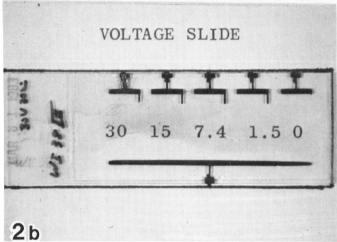

FIGURE 2.   a) Applied field pattern showing the attachment of electrodes to the Voltage Slide.   b) The silk-screened pattern provided a stepwise decrease in voltages over a 1.5 cm distance between anode and cathode.

FIGURE 3.   An electric field was applied in stepwise decreasing voltages across silicon chips (arrow) as on the Voltage Slides.

  3.   Bovine serum albumin (BSA). A one to three dilution of
       22% BSA in MEM was made with or without 10% FBS.

  4.   Human placental collagen (Sigma type VI), 3.3 mg in 1 ml
       0.2N HCl, stock solution, was diluted one to three with
       MEM (with or without 10% FBS).

Preparation of Cells

     NCTC Clone 929 (Strain L) mouse connective tissue fibroblast-
like cells or WI-38 human diploid lung fibroblasts were used. The
medium was rinsed off with Hank's buffered salt solution (HBSS) and
the flasks were trypsinized to remove cells. Twice, these cells
were spun down, rinsed and recovered by centrifugation with 1½ ml
MEM containing 0.1 ml Pen/Strep and 10% Fetal Bovine Serum.

Application of Cells to Substrates

     Glass microscope slides, or SiO_2 chips, either coated with
the various proteins, or uncoated, were thoroughly moistened with
MEM, and drained. The cells were applied immediately after
draining. They were then incubated for various periods of time

(from 1 hr to 4 days) in a cell culture incubator at 37°C (constant humidity and air/CO_2 mixture).

Application of Electric Field

During the incubation period, an electric field was applied to some of the previously described slides and SiO_2 chips. Either 30 or 60 volts was applied with a Heath Model PS-4 regulated power supply. It was important to keep the slides moist with MEM throughout the incubation period and to prevent media bubbling at the higher voltages.

Light Microscopy and Cell Counts

After application of electric fields, the Voltage Slide or Taped Voltage Devices were rinsed in phosphate-buffered saline (PBS) and immersed in phosphate buffered 3% glutaraldehyde until processed for staining. Gill's No. 3 Hematoxylin (Sigma) was used for cell visualization for light microscopy. Nomarski interference microscopy was performed with a Zeiss light microscope and Polaroid photographs were taken at 2mm intervals across the slide from anode to cathode and under the four areas of voltages applied plus the control. Cell counts were performed on these light micrographs.

Scanning Electron Microscopy

For scanning electron microscopic examination, the Voltage Slides, or Taped Voltage Devices (after subjection to electric fields) were rinsed in phosphate buffered saline (PBS) and fixed in phosphate buffered 3% glutaraldehyde. After rinsing in PBS they were dehydrated by the use of graded alcohols and Freon 113 and critical point dried in a Bomar critical point dryer followed by sputter coating in a Hummer V with 100 nm of gold-palladium. A JEOL 35C was used for the scanning electron microscopy at 18KV.

RESULTS

Adherence of Cultured Connective Tissue Cells to Device Patterns

A.   In the Absence of an Applied Electric Field (Passive Mode)

The bio-coatings used to interface the silicon chips, i.e., fibronectin, poly-L-lysine, bovine serum albumin or human placental collagen, appeared to have little, if any, appreciable effect on the adherence of L 929 mouse connective tissue or WI-38 human diploid lung fibroblasts to silk-screened Voltage Slide patterns or to silicon semiconductor substrates. The cell type itself, however, appeared to have a crucial effect on the conformation of the attached cells.

In Figure 4 (a-d) the WI-38 fibroblasts were grown on the silicon substrates of Figure 1a and a coating of fibronectin, collagen, bovine serum albumin or poly-L-lysine was used as the interface between the cells and the silicon. In Fig. 4d, where poly-L-lysine was used, there was a slight tendency for spreading. Although by Nomarski imaging the tendency for cell extensions did not appear to be dramatic, the same cell line by darkfield microscopy (Fig. 5) revealed cellular extensions conforming to the channels of the silicon chip substrate. For the most part, however, the WI-38 cells retained a rounded shape rather than flattening and elongating. By scanning electron microscopy (Figs. 6a,b), it was noted that even though the cells did not assume the elongated shape, their cellular processes or attaching tendrils could be actively anchoring the cells. The coating material, shown in Fig. 6c, can be seen filling in and rounding out the corners of the 2 micron wide channels. A cellular tendril can be seen at this higher magnification as well as the approximately 1000Å thickness of the coating (arrow).

Although applied to the silk-screened patterns or to silicon chips in MEM of identical composition, the L 929 mouse connective tissue fibroblasts did not retain the spherical shape as did the WI-38 human diploid lung fibroblasts (Figs. 4a-d) but stretched and elongated (Fig. 7) as fibroblasts are known to do upon tight attachment to a substrate. Indeed, both light and scanning electron micrographs (Figs. 8a,b; 9a,b) showed a distinctly larger number of fibroblasts attached and growing in the areas of pattern lines than in adjacent areas of the silicon chips. The processes of the attached fibroblasts were apparently adhering to the device lines and the major axes of the elongated cells were principally

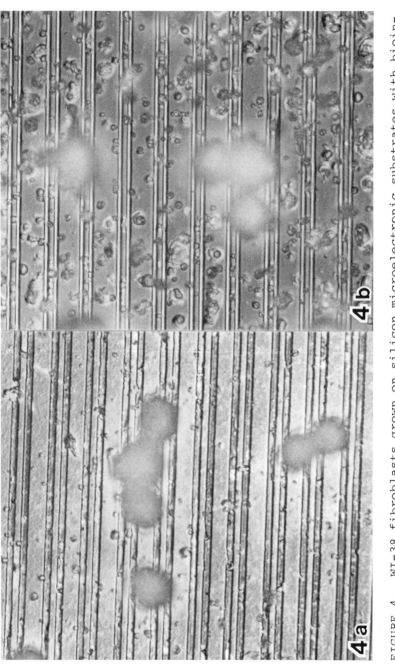

FIGURE 4. WI-38 fibroblasts grown on silicon microelectronic substrates with biointerfacing coatings: a) fibronectin, b) human collagen, c) bovine serum albumin, d) poly-L-lysine (note cell spreading). 600X.

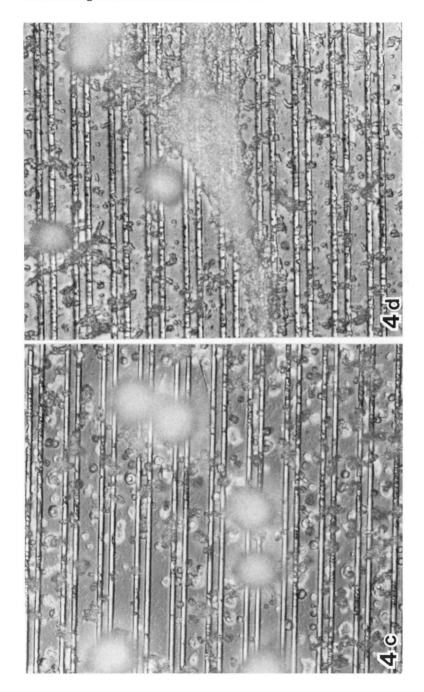

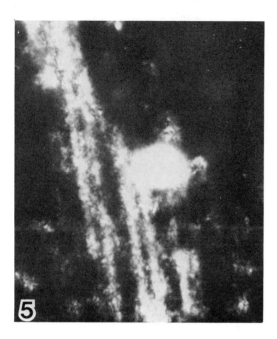

FIGURE 5.   Darkfield microscopy reveals WI-38 cellular extensions
conforming to the channels of the silicon chip substrate.   600X.

parallel to, and coincided with, these lines.   By darkfield
microscopy (Fig. 10) one could see many elongated cells and cell
processes extending in the pattern lines.   The scanning electron
micrographs show the L 929 mouse connective tissue fibroblasts
elongating parallel with the semiconductor lines (Fig. 11a).   It
appeared that the cell body more usually attached to the channels
of the pattern with the longer cell processes extending down the
trench (arrow).   At higher magnifications (Fig. 11b), fine
anchoring tendrils of the processes could be seen in abundance
reaching across the pattern lines.   Fig. 12a shows two cell
processes extending from the cell body conforming to the channel
and Fig. 12b at higher magnification shows the fine
multidirectional tendrils emanating from the longer process.

B.   In the Presence of an Applied Electric Field
      1)   Silk-screened Voltage Slide

The application of an electric field to human fibroblasts under culture showed a tendency for the cells to migrate toward the cathode (Table 1 and Figs. 13a,b). This was seen when both 1.5 and 7.5 volt fields were applied to these cells under culture on a Voltage Slide. The cell count increased across the slide; however, as the cathode was approached the cellular morphology was adversely affected. In the area immediately adjacent to the cathode (Fig. 13c), it appeared that the cells were damaged (by some type of streaming effect) and at the cathode itself, the cells appeared to have been completely disrupted.

At 12mm from the anode, under the different voltage conditions, ranging from zero or control electric field to 30 volts, considerable effects on the WI-38 cellular morphology could be seen (Figs. 14a-e). Under zero electric field, the spherical shape was retained by the cells. At 1.5 volts there were appearances of cell extensions (Figs. 14a,b). From 7 to 15 volts, a dramatic release of microexudative materials could be seen (Figs. 14 c,d) until at 30 volts, disruption of cells could be seen (Fig. 14e).

2)   Taped Voltage Slide for Silicon Chips.

The WI-38 cells cultured on silicon semiconductor substrates at 12 mm from the anode, showed a similar cellular response under the differing voltage conditions. Under zero electric field, (Fig. 15a) the cells retained their spherical shape and at 1.5 volts a flattening or spreading effect could be seen (Figs. 15b,c). At 7 volts the release of microexudative material was enough to obscure the presence of lines (Fig. 15c) and at 15 and 30 volts cell destruction was evident (Figs. 15d,e).

DISCUSSION

In this study we were able to show the directed attachment and growth of NCTC clone 929 (Strain L) mouse connective tissue fibroblast-like cells in oriented arrays on pattern lines of the SiO_2 device chips. The WI-38 fibroblasts did not elongate nor spread well, but appeared to adhere generally to the wider lines and could have been too large to deposit on the finer pattern

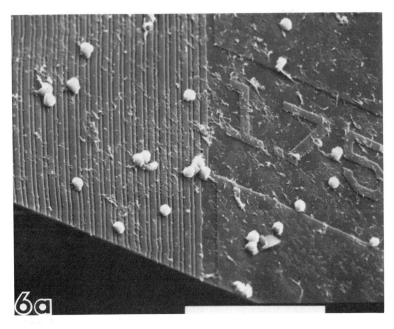

FIGURE 6.   a) Scanning electron micrograph of WI-38 cells randomly attached to silicon substrate. The pattern of the line width is indicated as 1.75 microns. 360X. Bar = 100 μm.   b) Scanning electron micrograph of WI-38 cells adhering to the silicon substrate under zero electric field but with fibronectin coating. 2400X. Bar = 10 μm.   c) Scanning electron micrograph showing thick bio-coating lining the channel (arrow). 18,000X. Bar = 1 μm.

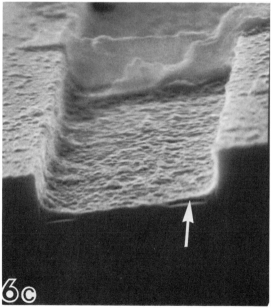

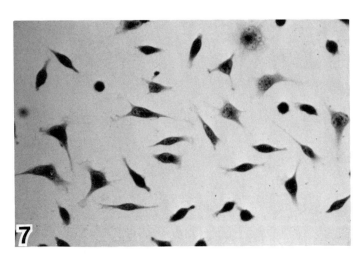

FIGURE 7. Light micrograph of L929 mouse connective tissue cells
shows cells spreading, elongating and having a random orientation
under zero electric field. 150X.

lines. In future experiments, this could be ascertained by
applying these types of cells to semiconductor substrates having
wider lines.

The effects of the applied potential in eliciting
microexudates from the WI-38 cells also deserves comment. The
secretion of a microexudate by a cell can replace serum-containing
media and have a great effect on cell adhesion and spreading (14).
Thus, the enhanced secretion of microexudate observed in our
studies from the application of the electric field could be a
factor responsible for better cell attachment.

Table 1 shows the influence of an electric field upon
definitive motion of WI-38 cells toward the cathode. Controls, in
the passive mode, where current was not applied, showed a
uniformity of cells across the slide and no apparent drift toward
the cathode.

It is well known that all mammalian cells, including the
connective tissue fibroblast-like cells NCTC L 929 and WI-38 used
in this study, carry a net negative surface charge at physiologic

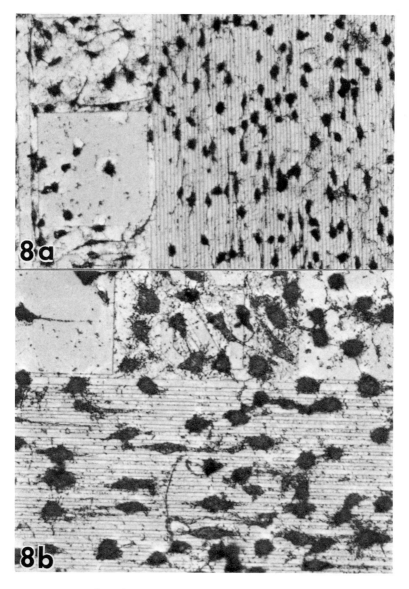

FIGURE 8.   a) Light microscopy showing L929 mouse connective tissue cells elongating along semiconductor lines. 120X. b) L929 cells at 240X. Note random attachment in areas where lines are absent.

FIGURE 9. a) Scanning electron micrograph of L929 mouse
connective tissue cells showing tendency to adhere to semiconductor
in orientation with line pattern. 1200X. Bar = 10 μm. b) At
higher magnification, note long cell extensions (arrows). 1600X.
Bar = 10 μm.

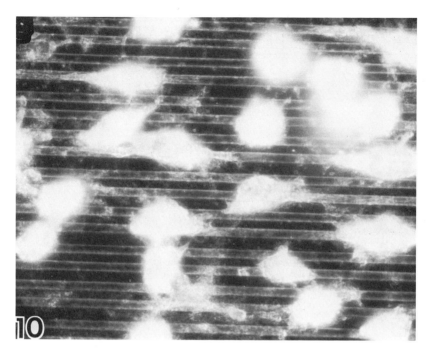

FIGURE 10.   L929 mouse connective tissue cells are shown extending in channels of the pattern by darkfield microscopy.   600X.

pH (15).   The effect of surface charge of substrate on adhesion can, for this reason, be quite dramatic (16). Cells can be made to adhere very rapidly to any surface carrying a positive charge by means of coulombic attractions. Such attractions play a demonstrable, although minor, role in cell to cell and substrate adhesion (17). Altering the electric charge of the substrate can markedly affect adhesion and locomotion of fibroblasts (18). It is not unexpected, therefore, that application of an external electric field could have a marked effect on fibroblast adhesion, orientation, and locomotion on a substrate even in the presence of extracellular proteins.

Better attachment and orientation was achieved on uncoated substrates, especially when the L 929 cells were used.   (The distortion of interfacing proteins and diminished cellular

Table 1. MIGRATION OF WI-38 HUMAN FIBROBLASTS
IN AN APPLIED ELECTRIC FIELD

| AREA OF "VOLTAGE" SLIDE | CELL COUNT (per unit area) | |
ANODE	@ 1.5V	@ 7.5V
1	15	33
2	38	67
3	62	80
4	82	97
5	94	102
6	93	113
CATHODE		

No apparent migration of cells was observed on slides incubated in the absence of an applied potential.

attachment observed in the presence of an applied electric field led to discontinuation of the use of interfacing proteins in this study.) However, distinct migration of undisrupted WI-38 human fibroblasts was observed up to 7.5 V (Table 1; Fig. 13) before any adverse effect on morphology was noted in the cells adjacent to the cathode.

In the case of L 929 cells, although counts were not done, there is a significant geometrical epitaxial-like orienting effect of substrate line patterns on cell deposition which was observed (Figs. 8-12) in the absence of an applied electric field. The effects of applied electric fields on the attachment and migration of L 929 cells on uncoated silicon microelectronic substrate lines would be a matter of great interest as would the design of microelectronic substrates to accomodate cells having different functions, properties and dimensions.

FIGURE 11. a) Scanning electron micrograph shows L929 mouse
connective tissue cell elongating parallel with semiconductor lines
(arrows). 3200X. Bar = 10 µm. b) At higher magnification, note
the fine anchoring tendrils of the cellular extension. 20,000X.
Bar = 1 µm.

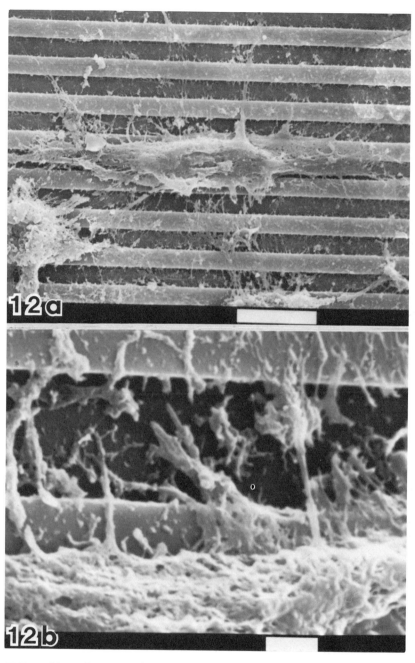

FIGURE 12. a) Scanning electron micrograph of L929 mouse connective tissue cell extending lengthwise in the trench and reaching across the pattern lines. 2000X. Bar = 10 μm. b) At higher magnification, cell tendrils can be seen readily multidirectionally across pattern lines. 13,000X. Bar = 1 μm.

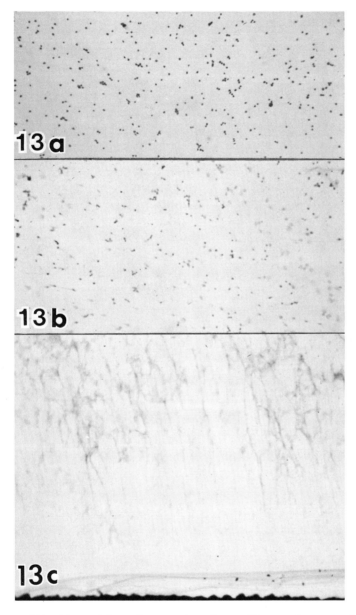

FIGURE 13. Light micrographs show effect of an applied electric field on WI-38 fibroblasts. a) Cells from separate areas nearer the anode. b) Near the cathode some of the WI-38 fibroblasts are flattened and have been disrupted. c) At the cathode, the cells appear totally disrupted. 20X.

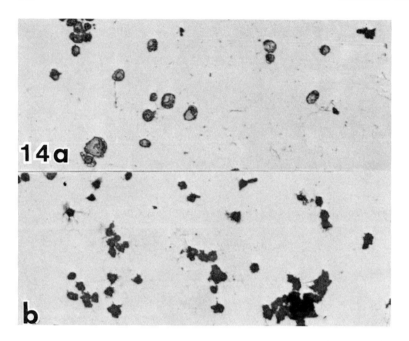

FIGURES 14 AND 15.   Light micrographs of WI-38 fibroblasts at the same distance from the anode but under different applied electric fields. a) Zero voltage, b) 1.5V, c) 7.0V, d) 15.0V, e) 30.0V. Note the release of microexudate material at higher voltage and disruption of cells at highest voltage. Figure 14 shows cultured cells on Voltage Slide. Figure 15 shows cells cultured on silicon semiconductor substrates. 120X.

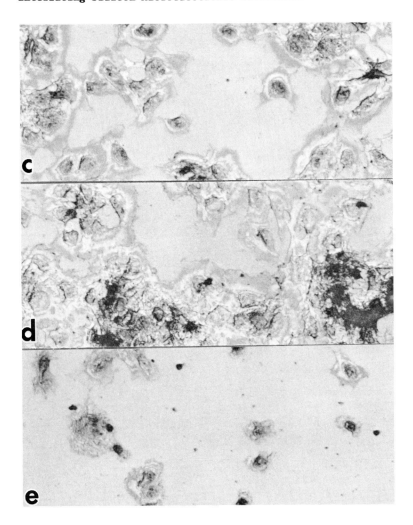

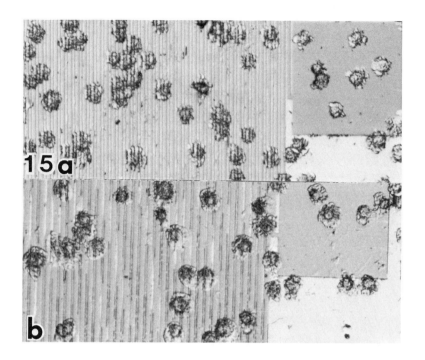

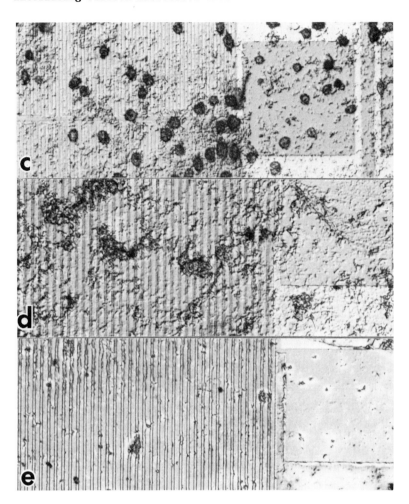

ACKNOWLEDGEMENTS

The authors wish to thank Peggy Yates for culturing the cells, David Chandler and John Mackenzie for their assistance in scanning electron microscopy, Larry Hanker for photography, and Cindy Grady and Pam Johnson for typing the manuscript. Supported in part by USPHS grant RR05333 and in part by grant Z3398 from the North Carolina Biotechnology center.

REFERENCES

1. F. Grinell, Int. Rev. Cytol. 53, 65-144 (1981).

2. A.C. Taylor, Exp. Cell Res., Suppl. 8, 154 (1961).

3. H.E. Kleinman, R.J. Klebe and G.R. Martin, J. Cell Biol. 88, 473-485 (1981).

4. J.H. McAlear and J.M. Wehrung, Biotechnical electron devices. In: Molecular Electronic Devices, F.L. Carter, ed. New York: Marcel Dekker, Inc. 1982, pp. 175-179.

5. J.S. Hanker and B.L. Giammara, Selective deposition of metals by biochemical reactions for electron device fabrication. In Molecular Electronic Devices, F.L. Carter, ed. New York: Marcel Dekker, Inc. 1982, pp. 181-194.

6. A. Cooper, H.R. Munden and G.L. Brown. Exp. Cell Res. 103, 435-439 (1976).

7. L.S. Darken and D.H. Lowndes, Appl. Phys. Letters 40, 954-956 (1982).

8. H.I. Smith and D.C. Flanders, Appl. Phys. Letters 32, 349-350 (1978).

9. S. Ingzar, Proc. Soc. Exp. Biol. Med. 17, 198-199 (1920).

10. P. Weiss, J. Exp. Zool. 68, 393-448 (1934).

11. N. Patel and M.-M. Poo, J. Neuroscience 2, 483-496 (1982).

12. L. Hinkle, C.D. McCraig and K.R. Robinson, J. Physiol. 314, 121-135 (1981).

13. J.C. Bean and G. Rozgonyi, Appl. Phys. Letters 41, 752-755 (1982).

14. D. E. Maslow and L. Weiss, Exp. Cell Res. 71, 204 (1972).

15. E.J. Ambrose, Progr. Biophys. Mol. Biol. 16, 241-265 (1966).

16. S. Gabor, T. Frits and Z. Anca, Int. Arch. Occup. Environ. Hlth 36:47-55 (1975).

17. L. Weiss. Cell Adhesion. Int. Dent. J. 28, 7-17 (1978).

18. Y. Sugimoto and A. Hagiwara, Exptl. Cell Res. 120:245-252 (1979).

16

Electron Beam Resists Produced from Monomer-Polymer Langmuir-Blodgett Films

G. Fariss, J. Lando and S. Rickert/
Department of Macromolecular Science,
Case Western Reserve UNiversity,
Cleveland, Ohio

Recent technological advances in very-large-scale integration are beginning to outstrip the technology used to produce these circuits. UV and optical lithography techniques are producing resist patterns with resolutions approaching the expected minimum. Application needs demand that this minimum be improved to achieve more features per unit area of circuit.

We have achieved improved resolution of resist patterns. Through the use of monomer-polymer multilayers, fine line resolution, produced by way of a novel computer-controlled electron beam technique, is now possible. The relative ease and high reliability of multilayer formation ensures uniform films of a thickness an order of magnitude less than that of spin-cast films. The high electron sensitivities of these films enables excellent degradation (or polymerization) upon exposure to an electron beam. Final pattern resolution, for both positive and negative resists, is an order of magnitude higher than conventional resist resolutions, offering possible improvement in circuit capability.

Morphological studies of these multilayer films provide unique information on domain size and structure. The ultimate morphology of these films is shown to be dependent on deposition conditions as well as the substrate used.

INTRODUCTION

Typical commercial resists, such as poly(methyl methacrylate) and poly(cinnamic acid), are deposited by precipitation from solution on a spinning wafer. Defect-free films are guaranteed only when the film thickness exceeds 1 μm[1]. At thicknesses of less than 1 μm, pinholes (poorly covered regions of the wafer surface) appear, providing films unacceptable for high resolution (submicron) patterns.

The introduction of thinner films into current resist production techniques would offer little advantage if UV or optical lithography were still employed. The wavelengths of these techniques (λ = 200-350 nm) are so large that line resolutions would be effectively limited by the wavelength instead of the film thickness. With electron lithography, however, it is possible to focus a beam of diameter as small as 10 nm onto the film, solving this problem.

The advantages offered by electron beam lithography do not solve the nagging problem caused by film thickness. The work of Hatzakis[2] and others still suffers from scattering of electrons, limiting the resolution to one-half of the film thickness. This problem exists regardless of the beam diameter used for reasons mentioned earlier. With the film thickness of about 1 μm required, maximum resolution would be 0.5 μm.

The Langmuir-Blodgett technique[3] of film deposition creates uniform defect-free films of thickness 3-100 nm, depending on the length of one molecule and the number of layers deposited. Thus, ultrathin films an order of magnitude thinner than conventional films can be produced. Through recent expansion of this technology, polymerizable materials are included in monolayer formers. This, in combination with electron beam lithography, suggests potential improved line resolution.

EXPERIMENTAL DETAILS

Material

The requirements for a material to form a stable monolayer are very rigid and become more so for the subsequent formation of multilayers[3,4]. α-octadecylacrylic acid (ODAA) fulfills these requirements, as well as the need for electron sensitivity.

Substrate

The deposition of multilayers was carried out on Si<111> chips coated with 130 nm of silicon oxide on one side. Gallium arsenide (GaAs) chips were also used, and both types were precut to 1.5 cm X 0.5 cm. The GaAs chips were cleaned by repeatedly dipping them into tetrachloroethylene and rubbing vigorously with Fisher lens paper. After all visible impurities were removed, the chips were placed in a mixture of two-thirds sulfuric acid and one-third 30% hydrogen

peroxide in water at 70°C for 5 min. The chips were then rinsed
thoroughly with doubly distilled water. A clean chip was declared
so if no visible impurities were present and the contact angle of
water with the surface approached zero. Clean chips were stored
between sheets of Fisher lens paper and unclean ones were cleaned
again.

Silicon wafers were treated for 5 min. with hexamethyldisila-
zane (HMDS; $[(CH_3)_3Si]_2NH$) to remove both surface water and hydroxy
groups by the following reactions:

$$2SiOH + HMDS = 2SiOSi(CH_3)_3 + NH_3$$

$$H_2O + HMDS = [(CH_3)_3Si]_2O + NH_3$$

Deposition Environment

All depositions were carried out in an environmentally controlled dust-free room with a temperature of $23.3 \pm 0.5°C$. The film
balance used was a commercial MGW Lauda-Filmwaage preparative film
balance, designed for continuous isotherm measurement, with control
of the film pressure and the water temperature. The subphase consisted of doubly distilled tap water.

The passing of the substrate through the water surface was
accomplished with a pneumatic dipping device. The upward and/or
downward velocities of the device could be controlled physically,
with speeds ranging from 1 to 1000 mm min^{-1}. The pneumatic dipper
offered advantages over electric-motor-driven dipping devices in
that movement was fluid and no hesitations were present at any time.

Formation of Monolayers and Multilayers

All solutions were deposited dropwise onto a clean water surface via a micropipette. Drops typically measured 10 μl in volume
with monomer concentrations in the range of 0.5 mg ml^{-1} in heptane.
Isotherms were always run before deposition procedures, and were
very reproducible because of accurate volume delivery.

A typical isotherm[1] of ODAA reaches a local maximum at a
surface pressure of 46 dyn cm^{-1}. Depositions were therefore usually
carried out under a surface pressure of 36 dyn cm^{-1}.

Other dipping conditions were determined by experimentation.
Typical values for deposition were a temperature of 18.2°C, a
dipping speed of 4 mm min^{-1} (upward and downward) and a subphase pH
of 3.5.

Resist Patterning

A novel computer-controlled electron beam lithography technique
was used to pattern the ODAA resist. A JEM JEOL-100B electron

microscope was coupled with a VC-100 digital computer to allow complete control of the lithographic process.

The sample holder was made out of light gauge brass and made to hold the chip between clips while still providing enough pressure by the clips to secure the chip. The holder was designed to attach to the microscope sample entry unit and was inserted into the microscope from there.

The lithographic exposure could result in either polymerization or degradation of the ODAA resist, producing either a negative or a positive resist respectively. The type of resist obtained depended upon the treatment prior to introduction of the sample into the microscope. All films were deposited pure, as monomers, with no sensitizers added. Initial results illustrated that, when the film received a small dose of irradiation from a small UV source, it polymerized slightly. This slight polymerization provides enough long-range order to render the ODAA film stable in the microscope chamber. Any exposure to electron irradiation of such a film produced a negative resist. Treatment of a multilayer film with high doses of irradiation (up to 4.8 Mrad) essentially completely polymerized the film. A positive resist results upon exposure of this type of film to the electron beam.

Post-pattern treatment of these films is usually necessary in order to dissolve away any soluble fractions of the film. When it was required, the films were placed in ethanol for a predetermined time to develop the pattern and make it sharper.

Deposition Variables

The quality of the film deposited onto the substrate is dependent upon the conditions under which the deposition was done. In all cases, the type of deposition obtained was the alternating Y structure. The effect of the dipping conditions on the number of domains, their size and orientation, and also the ease of film deposition, was studied.

Morphology

The morphology of the final multilayers is determined by the spreading conditions of the monolayer. It has been shown that domains in the monolayer continue to grow until the solvent has completely evaporated. These domains are a general phenomenon of all multilayers. By controlling the evaporation rate, the domain size can be varied as desired[5]. Once the evaporation is finished, the domain size in the monolayer remains the same.

Deposition of the monolayer onto a substrate has a detrimental effect upon the domain size. In most cases, passing the substrate through the monolayer surface disrupts the domains and breaks them

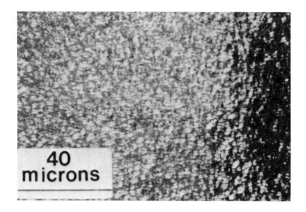

FIGURE 1. Differential interference contrast optical micrograph of
ODAA on silicon. The domain sizes averaged 2 μm x 1 μm and the
domains exhibited little orientation.

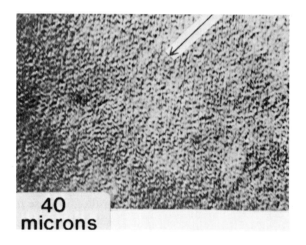

FIGURE 2. Differential interference contrast optical micrograph of
ODAA on GaAs. The domain sizes averaged 1.5 μm x 0.75 μm. The
higher degree of orientation (lines) indicates a possible epitaxial
effect.

apart. Multilayer domains are therefore probably smaller than those
in the original monolayer. This disruption of domains is affected
by the dipping conditions.

Substrate Dependence

Little difference could be seen visually between deposits of
ODAA on Si<111> and on GaAs<110>. In both cases the deposit was
smooth and uniform and no problems occurred with the final film.

Magnified views showed that the films differed significantly
on the microscopic level (Figs. 1 and 2). Differential interference
contrast optical microscopy (Nomarski method) is a relatively new
method in the study of thin films. The interference method is
superior to phase contrast as only the surface layers of a specimen
are examined and surface reliefs are possible, similar to scanning
electron microscopy. Its main use has been in the field of metallurgy for studying grains in metal surfaces.

The fundamental domains in a monolayer can be thought of
perfect crystals. The Nomarski method yields a wealth of information when applied to the study of multilayers.

Multilayer domains on silicon consist of anisotropic regions
about 2 μm x 1 μm in size. The domains are seen to exist throughout
the film, not just at the external portions of the film. The aspect
ratio of the domains is 2, which is expected, for ODAA is a paraffin-
like material and paraffin has an aspect ratio of 2 because of its
unit cell. The domain packing is more or less biaxially alig
This alignment is not characteristic of any epitaxial effect between
film and substrate. Rather, this biaxial packing is merely a consequence of any dense organization of anisotropic rods.

The same biaxial alignment exists in multilayers on GaAs(Fig. 2).
However, the alignment is much more ordered in this case, as indicated by the arrow in Fig. 2. This rather high degree of orientation indicates a possible epitaxial effect between GaAs<110> and
the ODAA film. The domain size has now been decreased (1 μm x 0.5 μm)

Temperature

Of the operational variables present in the Lauda balance,
temperature would be expected to have the greatest impact on the
final film morphology. Monolayers have been known to undergo large
changes in viscosity with slight changes in temperature. This
change in viscosity leads to a change in the isotherm of the monolayer[1], ultimately affecting deposition.

The quality of film deposition is highly dependent on the
fluidity of the monolayer (Table I). At lower temperatures, higher
viscosities cause film fracture as the flexibility of the monolayer
is removed. Higher temperatures lead to a reduction in viscosity,
effectively reducing the stability of the monolayer. Smaller domain

sizes result, because of a lack of stability in the film. At temperatures in between, domain characteristics appear to be at an optimum.

Surface Pressure

The orientation was affected little, if any, by varying the
the surface pressure of deposition. Domain size decreasing pressure
in increments of 2 dyn cm^{-1} from 37 dyn cm^{-1} to 31 dyn cm^{-1}, the
domain size remained roughly the same, but the number continued to
decline. This could easily be seen visually, as less film was
deposited at lower surface pressures (Table II).

pH of the Subphase

The only true difference in deposition through variation in the
subphase pH occurs at a pH of about 5.5, which is related to the
extent to which the ODAA is ionized. Dissolution of the ODAA monolayer tends to occur if the pH of the subphase is higher than 5.5.
In these cases, we observed that dissolution led to smaller domain
sizes.

Dipping Speed

Any monolayer deposited onto a substrate would be expected to
decrease in quality and quantity as the dipping speed increases. At
higher dipping speeds, the monolayer does not have enough time to
flow and to deposit well. Rather, pinholes form as the increased
pressures caused by the dipping speed fractures the film. The
domains that are deposited are quite small after being fractured
during deposition. As the dipping speed is decreased, these pressures are easier to overcome and the domain size becomes larger
(Table III) in a defect-free film. This relationship continues
until a critical dipping speed is reached, below which no change
in film morphology can be detected. For ODAA, this dipping speed
appears to be about 4 mm min^{-1}.

Positive Resists

Initial problems in the resist area were concerned with
obtaining films of sufficient quality for high resolution work.
Such films would need to be nearly defect-free at a thickness much
less than 1 μm.

We decided to concentrate on films composed of 20 monolayers of
ODAA, which would make the films about 69 nm thick. These films
were deposited at 36 dyn cm^{-1}, a subphase pH of 3.5, a dipping speed
of 4 mm min^{-1} (both upward and downward) and a temperature of 18.2°C.
Typical early films (Fig. 3) contained pinholes due to the defects
present in the substrate surface prior to dipping. Hydrophobic areas
of the substrate surface lead to tail attachment of the monolayer

TABLE 1. Temperature Effect on Crystal Size

Temperature (°C)	Domain Size (microns)
7.0	(unstable monolayer)
9.8	1.5 x 0.8
13.2	2.0 x 1.0
16.0	2.0 x 1.0
20.2	1.5 x 0.8
22.9	1.0 x 0.5
26.0	1.0 x 0.5

TABLE 2. Surface Pressure Effect on Crystal Size

Surface Pressure	Domain Size (micron)	% Coverage
25	1.0 x 0.5	40
27	1.0 x 0.5	50
29	1.0 x 0.5	65
31	1.0 x 0.5	85
33	1.5 x 0.8	95
35	2.0 x 1.0	100
37	2.0 x 1.0	100

TABLE 3. Dipping Speed Effect on Crystal Size

Dipping Speed (mm/mn)	Domain Size (microns)
2.6	2.0 x 1.0
3.9	2.0 x 1.0
4.9	1.5 x 0.8
7.7	1.3 x 0.7
10.7	1.3 x 0.7
16.2	1.0 x 0.5

1.9
microns

FIGURE 3. Example of a pinhole-containing multilayer film. The
pinholes range in size up to 0.5 μm, leading to a film unsuitable
for submicron lithography. The high electron sensitivity of ODAA
is illustrated by the dark square at the bottom of the figure.
This area was focused on for 3 s and in that time underwent a high
degree of degradation.

while no molecules will deposit during the first immersion in the
areas of hydrophilicity. This phenomenon causes stresses in early
monolayers resulting in poorly covered regions of the wafer surface.
The detrimental effect of these pinholes can be seen in the fuzzy
appearance of the degraded lines in Fig. 3. Pretreatment of the
silicon surface with HMDS essentially eliminates these pinholes,
providing ample films for submicron work.

The degradation of a polymer film by the electron beam renders
the exposed material soluble in a developer, producing a positive
resist. Positive patterning of the ODAA resist was accomplished
after varying the operational parameters in a concerted manner to
obtain optimum results (Fig. 4). A beam with an effective diameter
of 250 nm was used to draw these 300 nm wide lines, i.e. there was
a 20% Compton backscattering effect.

Improving the resolution from 300 nm was performed at higher
magnifications, for the magnification controlled the beam diameter.
50 nm resolution lines were produced (Fig. 5) at a magnification of
10,000X. At this resolution, the electron sensitivity of the ODAA
multilayers was determined to be in the range of 10^{-6} C cm^{-2}, making

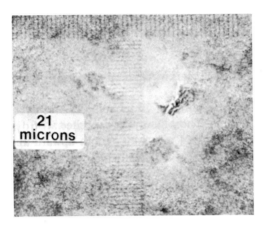

FIGURE 4. Positive resist pattern of ODAA. This pattern was
written with an electron beam of effective diameter 250 nm. The
resolution of the lines is 300 nm (resulting in a 20% backscattering
effect). No development was necessary.

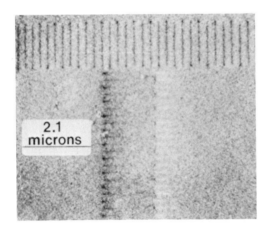

FIGURE 5. Positive resist pattern of ODAA with a resolution of 50
nm. The effective beam diameter used was 25 nm, with a backscattering effect of 100%. No development was necessary.

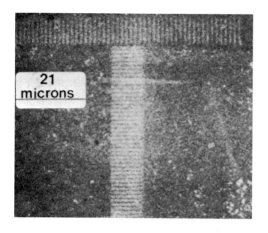

FIGURE 6. Negative resist pattern of ODAA with a resolution of
100 nm. This pattern was written with an electron beam effective
diameter 250 nm. The backscattering effect was 300%.

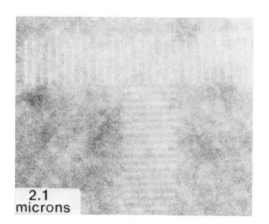

FIGURE 7. Negative resist pattern of ODAA with a resolution of
80 nm. The effective beam diameter was 25 nm. The backscattering
effect was 220%.

them much more sensitive than pure conventional resists. The resolution obtained is slightly better than that obtained in previous work with multilayer films[6].

Negative Resists

Polymerization of the monomer resist material (negative resist) by the electron beam proved to be more difficult than positive resist work. The length of exposure during the pretreatment period had to be just long enough to polymerize the film slightly. If the pretreatment period was too short, the required exposure in the electron microscope became too long to be practical. However, if the films were polymerized to an exceedingly high degree before exposure to the electron beam, degradation began.

The best pretreatment results were achieved through exposure of the film to 250 mW cm^{-2} of radiation from UV lamp for 1 h at a distance of 10 cm. Once the proper exposure was determined, negative resist patterning became rather simple (Fig. 6). Typical resolutions obtained were 1000 nm. Shorter exposure times than positive resists were needed in order to obtain this resolution.

The poorer resolution obtained for the negative resist compared with that for a positive resist at the same magnification is an indication of the Compton backscattering problem. The patterns in Figs. 4 and 6 were written with a beam of effective diameter 250 nm. The high electron sensitivity of the monomer led to wider lines as the material was polymerized by the backscattered electrons. While the resolution was decreased somewhat in the positive resist, the polymer was not sensitive enough to react on exposure to the backscattered electrons as extensively as the monomer.

This factor holds true even at higher resolutions (Fig. 7). The best negative resist results have typically been 80 nm, compared with 50 nm normally obtained for positive resists. Since an electron beam of effective diameter 25 nm was used, we see that the back scattering effect is still greater for negative resists than for positive resists, though not by as much as before.

CONCLUSIONS

The need to develop a new resist capable of submicron lithography has led to several advantageous innovations related to resist technology.

Quality multilayer deposition onto an SiO surface is possible through pretreatment of the surface with HMDS. If no pretreatment is done, pinholes about 1 μm in size appear in the films, which are thus inadequate for high resolution work. Electron micrographs show that after treatment hydrophilic protions are eliminated and that excellent virtually defect-free films are formed on the treated surfaces.

The final morphology of multilayer films is dependent upon the dipping conditions as well as the spreading conditions. Manipulation of the dipping conditions may prove useful in obtaining a certain required film morphology. Domain size and orientation are also possibly affected by the substrate used for deposition, with possible epitaxial effects.

Langmuir-Blodgett films of ODAA are excellent electron beam resists, offering several advantages over conventional and other electron beam resists.

(a) Higher resolutions are obtained compared with conventional resists.

(b) Because the ODAA multilayers exhibited an electron sensitivity in the range of 10^{-6} C cm^{-2}, no sensitizers are needed as are usually required with conventional UV resists.

(c) Both positive and negative resists can be made from the same material, while normal resists require special treatment to become one or the other.

(d) A lower energy electron beam can be used. The high contrasts normally required in high resolution lithography are then unnecessary.

(e) Development of such a positive resist is not necessary, though it is usually advantageous; direct electron beam removal of material can be observed.

REFERENCES

1. A. Garito, K. Hayes, K. Desai, M. Filipkowski and S. Rickert, to be published.
2. M. Hatzakis, J. Vac. Sci. Technol., 16(6)(1979) 1984.
3. G. L. Gaines, Jr., Insoluble Monolayers at Liquid-Gas Interfaces, Wiley-Interscience, New York, 1966.
4. G. L. Gaines, Jr. J. Colloid Interface Sci., 62(1)(1977) 191.
5. D. Day and J. Lando, Macromolecules, 13(1980) 1479.
6. A. Barraud, C. Rosilio and A. Ruaudel-Teixier, Solid State Technol., 22(8)(1979).

17

The Ultimate Spatial Resolution of Electron Beam Lithography

David C. Joy / Bell Laboratories, Murray Hill, New Jersey

Molecular electronic devices will require individual interconnects, and arrays of conductors, with dimensions that are measured in nanometers rather than in the microns now current in the semiconductor industry. Of the technologies now available for microfabrication, the one that offers most promise for use at this size scale is that of electron beam lithography. Although the fabrication of structures at the nanometer scale has been demonstrated in a few special materials [1,2] the speed of writing in such cases is far too slow to permit their use in practical situations. On the other hand, e-beam lithography using organic resists is very rapid, but recent experimental results [3,4] seem to indicate that the minimum line width is only of the order of 100 A. The purpose of this paper is to examine, through the use of a Monte Carlo simulation and an idealised specimen geometry, whether there is an intrinsic limit to the minimum line width that can be produced in a conventional positive resist such as poly-methyl methacrylate (PMMA), to deduce the exposure conditions which would allow this resolution to be achieved, and to attempt to identify ways of designing resist materials which would permit still finer lines to be produced.

THEORY

The cross-section profiles of lines produced by the exposure and development of a positive resist follow the contours of equal energy deposition in the resist. The line width achieved under some specified exposure condition can thus be predicted if the energy deposited from the incident beam can be calculated and if the response of the resist to electrons is known. The simplest assumption that could be made in such a calculation is that exposure occurs solely as the result of energy deposited by the primary electrons as they are elastically scattered through the resist layer. Under the conditions currently employed for commercial e-beam lithography (i.e low energy electron beams with a probe size of the order of half a micron, and thick resist layers on a bulk substrate) this approximation is a reasonable one since the scale of the interaction is determined mainly by the volume over which backscattered electrons emerge from the substrate and travel back to the entrance surface. The conditions that would be used for the fabrication of conductors for molecular devices might, however, be expected to be very different to these. In particular it would seem necessary to use a thin resist layer and high accelerating energy, to reduce scattering of the incident beam in the resist, and to thin, or even remove entirely, the substrate to minimise backscattering. Under such conditions the assumption made above leads to the conclusion that the interaction volume should tend to a radius that varies as about $t^{3/2}/E_0$, where t is the sample thickness and E_0 is the incident beam energy[5]. Thus a line of arbitrarily small width could be obtained in a thin enough resist and at a sufficiently high energy. Experimentally however it is found that even under optimally chosen conditions the line width reaches a minimum value which, to first order, stays constant independent of either the beam energy or resist thickness.

It has been suggested that this result might be due to such effects as mechanical instabilities in, or the granularity of, the resist, but there is little direct evidence to support this contention. Alternatively it has been proposed[3,4,6] that the resolution is limited by the action of secondary electrons. It is this proposition which is examined here, since there are at least two reasons for believing that secondary electrons might play a significant role in the lithographic process. Firstly, PMMA can be exposed by electrons with energies as low as 5eV, so any theory which ignores the effects of secondaries is likely to be in error. Secondly, in the usual Bethe approximation, the rate of energy deposition from an electron (eV/cm of path length) depends inversely on the energy of the electron, so a 1 keV secondary is depositing energy at about 100 times the rate of the 100 keV primary electron which produced it. Since the average mean free path of a secondary is tens of nanometers it is clear that the production of even a few percent of such electrons could have a significant effect on the magnitude and distribution of energy deposition in the resist.

A Monte Carlo simulation was developed to investigate the consequences of considering secondary electron production in lithography. The cross-section for secondary production was the classical one given by Evans[7] for a knock-on collision:

$$\frac{d\sigma}{d\Omega} = \frac{\pi e^4}{E^2} \left\{ \frac{1}{\Omega^2} + \frac{1}{(1-\Omega)^2} \right\} \tag{1}$$

where e is the charge of the electron, E is the energy of the primary electron, and $\Omega.E$ is the energy of the secondary produced. This expression assumes that the primary and secondary electrons are indistinguishable, thus Ω is limited to 0.5. The dynamics of the knock-on collision are shown in figure (1). After the collision the electron with highest energy is travelling at an angle α to the original trajectory where:

$$\sin^2 \alpha = 2 \, \Omega/(2 + t - t\Omega) \tag{2}$$

and t is the energy E of the primary electron normalised by its rest mass (511 keV). The electron with lower energy travels away at an angle β where:

$$\sin^2 \beta = 2 \, (1-\Omega)/(2 + t\Omega) \tag{3}$$

The significance of these equations can be appreciated by considering the example of a 100 keV electron producing a 2 keV secondary (i.e $\Omega = 0.02$). α is then about 1 degree, but β is nearly 80 degrees. The production of secondary electrons therefore provides a mechanism for the lateral transfer of energy relative to the incident beam direction.

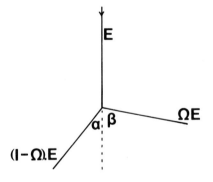

FIGURE 1   The scattering dynamics of a knock-on collision which produces a secondary of energy ΩE.

The energy loss dE along a segment of path length dS of the primary electron is

$$-\frac{dE}{dS} = \sum_i N_i Z_i \int_{\Omega_c}^{0.5} E \cdot \Omega \cdot d\sigma \cdot d\Omega \frac{}{d\Omega} \qquad (4)$$

where N_i is the number (per unit volume) of atoms of atomic number Z_i. The lower limit of integration Ω_c is necessary to avoid an infinite cross-section at zero energy, and it also removes from consideration secondaries of low energy which contribute little to the total energy deposition but take up large amounts of computer time. It is found experimentally that the computed energy doses do not depend sensitively on the choice of Ω_c, since lowering the value will increase the number of events considered but reduce their average energy. Here a value of 100 eV (i.e Ω_c = 0.001 at 100 keV) was used, and the secondaries were usually tracked until their energy fell below 40 eV. The secondaries were assumed to lose energy with the usual Bethe cross-section

$$-\frac{dE}{dS} = \frac{2\pi e^4}{E} \sum_i N_i Z_i \frac{\ln(1.166\ E)}{J} \qquad (5)$$

where E is the instantaneous energy, and J is the mean ionisation

$$J = (9.76Z + 58.5\ Z^{-0.19})eV \qquad (6)$$

This model of secondary electron production was incorporated into a single scattering Monte Carlo simulation. The fractional yield of secondaries was determined by the ratio of the total mean free path (i.e elastic plus inelastic) to the elastic mean free path for the incident electron, and was typically between 1 and 5%. When during the simulation of a primary trajectory a random number call determines that a secondary electron has been produced as the result of a knock-on collision, then tracking of the primary is suspended and the secondary is tracked until it leaves the sample or falls below the minimum energy. Tracking of the primary is then resumed until it also leaves the specimen. For the results discussed below typically 250,000 primary trajectories were computed in each case. The program was run on an APPLE II+ computer equipped with a hardware arithmetic processor.

RESULTS

For the purposes of investigating the spatial resolution limit in e-beam lithography it is convenient to consider the idealised case of a thin resist layer which is free standing, or supported on a substrate of negligible back scattering power. As briefly discussed later this is a valid simplification since it can be shown that at high beam energies and small probe diameters the effect of even a bulk substrate is minimal. Consider the example of a 1000A film of PMMA irradiated with a 20A probe of electrons. Figures (2) and (3) show typical primary and secondary trajectories for an incident beam energy of 100 keV. The primary trajectories show a conical envelope, with a radius which increases steadily with depth in the layer. The secondaries, on the other hand, show a distribution which is almost cylindrical about the incident beam direction suggesting that the energy deposition from the secondaries should be essentially independent of depth.

This is confirmed by the plot of absorbed energy shown in figure (4) for these conditions. The contours of equal energy dose are seen to be nearly parallel to the beam axis, with little difference in the energy absorbed at corresponding points on the entrance and exit surfaces. This is very different to the result predicted by the more usual theory which considers only the primary electrons. As shown by the

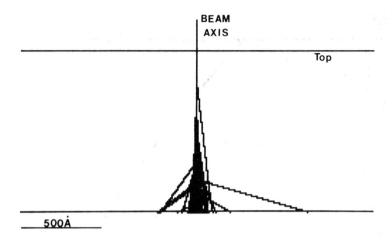

FIGURE 2   Typical primary electron trajectory plots in 1000A of PMMA
exposed by a 100 keV beam.

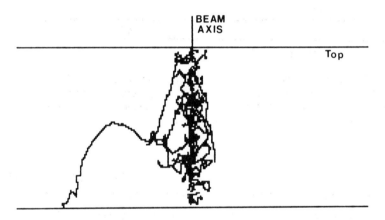

FIGURE 3   Typical secondary electron trajectories   in 1000A of PMMA
exposed by a 100 keV beam.

dotted lines on figure (4) this predicts an absorbed energy profile
which broadens monotonically with increasing depth. The contribution
from the secondaries is seen not only to transform the shape of the
energy profiles, but also to give a significant increase in the actual
energy deposited in the sample at any point. Equivalent computations at

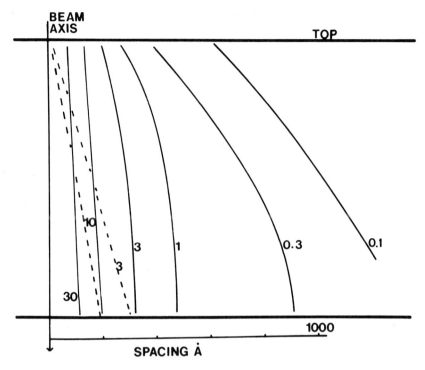

FIGURE 4   Contours of equal energy deposition in 1000A of PMMA exposed by 20A gaussian probe at 100 keV. The dotted lines are the equivalent contours for a simulation ignoring the effect of secondary electrons. The contours are in units of 10^{15} eV/cm^3/electron.

30 and 50 keV generate profiles which are very similar in general shape and predicted energy deposition to that found at 100 keV, indicating that the energy deposited within a few hundred angstroms of the beam axis is relatively insensitive both to the beam energy and depth in the sample. Since the magnitude and spatial distribution of the energy deposition from the primary beam varies widely over this same range this result confirms that in the region close to the beam axis the energy dose is dominated by the contribution of the secondary electrons.

For practical applications the resist will usually be exposed in a line pattern, rather than by a static probe. Figure (5) plots the energy

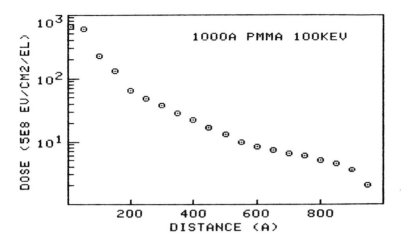

FIGURE 5 The absorbed energy profile at the exit surface of a 1000A film of PMMA exposed by a 20A line source at 100 keV. The energy units are 5.10^8 eV/cm^2/electron.

deposition at the exit surface of the 1000A PMMA film for exposure by a 20A line source incident at 100 keV. The energy absorbed shows a rapid rise as the distance from the beam axis decreases, but at spacings below about 75 angstroms the slope of the line drops rapidly and the plot becomes almost horizontal. The line width attainable under these conditions can be found by applying the threshold energy density criterion of Greeneich and Van Duzer[8] which assumes the lower solubility limit of PMMA to be 10^{22} eV/cm^3. For any given incident dose the resultant line width, after development, will then be defined by the extent of that region lying above this critical dose value. It is evident that the line width will fall rapidly as the dose is reduced, e.g changing the exposure from 10^{-7} C/cm to 10^{-8} C/cm will lower the total line width from about 1200 to around 500 angstroms. Eventually, at a dose of around 2.10^{-9} C/cm, only the plateau region of the profile will lie above the threshold limit. Because the slope of the profile in this region is very low any dose big enough to cause exposure will raise the entire plateau above the threshold to give a total line width of about 100 to 120 A at the minimum dose. Since it has already been shown above that the absorbed energy profile varies only slightly with either depth in the resist layer or beam energy it is thus reasonable to infer that this result represents the best performance of which this resist is capable. The values of optimum exposure, and finest line width, derived here are in excellent agreement with the results of Broers[3,4] taken under comparable experimental conditions indicating that the simulation is correctly representing the physics of the situation.

Further evidence that it is the lateral spread and energy deposition of the secondaries which sets the minimum feasible line width can be found by considering the case of a resist layer so thin that essentially zero scattering of the primary beam occurs within it. Figure (6) plots the absorbed energy deposition at the exit surface of a 250A, free standing film of PMMA, again exposed by a 20A line source. At 100

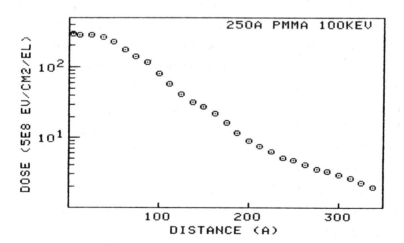

FIGURE 6   Absorbed energy profile at exit surface of 250A film of PMMA exposed by 20A line source at 100 keV. Energy units are 5.10^8 eV/cm^2/electron.

keV the resist thickness is now only about 0.1 of the mean free path, so scattering of the incident beam is very small. Nevertheless, as can be seen, the peak energy absorption is still about 30% of that for the 1000A film and this value stays constant out to a spacing of about 40 to 50 A from the axis before starting to fall rapidly away. Thus even in this highly idealised example the minimum line width and the optimum exposure dose would still be of the same order as that found in the thicker, and physically more realistic film.

DISCUSSION

The computation described above shows that the line width limit in e-beam lithography is dominated by the secondary electrons and their intense and highly localised contribution to the energy deposition. Because of the mechanics of the knock-on process which generates the secondaries, and the scattering that they subsequently undergo, the energy deposited by the secondaries is essentially independent of depth in the resist layer until the scattering of the primary beam becomes comparable with the extent of the secondary distribution. Consequently both the minimum line width (which is determined by the intrinsic width of the secondary distribution), and the optimum exposure dose (which is a function of the energy deposition from the secondaries), will be fairly constant over a wide range of accelerating voltages and resist thicknesses. These calculations lead to the conclusion that the minimum line width cannot be improved either by manipulating the electron-optical conditions or reducing the resist thickness. As a corollary, however, it can be noted that this same effect also makes it possible to achieve comparable resolution even with the resist supported on a bulk substrate. This is because, under the experimental conditions assumed here of a high energy (greater than 30 keV) beam, the volume over which the backscattered electrons emerge has a radius of many microns. Consequently the absorbed energy density contribution (ev/cm^3) from the backscattering is very low compared to that

produced by the secondaries around the incident beam axis. By carefully
choosing the incident exposure dose it is thus possible to ensure that
only in the secondary electron region is the dose above the threshold
value[9]. The optimum exposure dose, line width and the form of the
proximity correction predicted by the theory are again in good agreement
with recent experimental data[10]. However the possibility of
fabricating 100A structures on a bulk substrate is rapidly lost once the
incident probe size becomes comparable with the secondary electron
volume i.e about 200 to 300 A, because this delocalises the secondary
energy deposition so reducing the differential between the it and that
due to the backscattered electrons.

Instruments are already in existence which are capable of generating
a probe which fully meets all the parameters (e.g 50 or 100 keV energy
and a probe diameter in the range below 100A diameter) needed to expose
the narrowest possible line in a resist such as PMMA. Since it also has
been shown here that reducing the thickness of the resist layer below
about 1000A (at 100 keV) will not lead to any improvement in line width
it is necessary to ask what factors might be changed to produce an
improvement in resolution. The most direct solution is to find materials
in which the initial step in the exposure process cannot be initiated by
low energy electrons. This is demonstrated in figure (7) which plots
energy deposition in a 250A film of PMMA-like material irradiated with a
20A line source at 100 keV, but making the arbitrary assumption that
electrons below some cut-off energy no longer contribute to the
exposure. As the cutoff energy is increased from the usual low value
(dotted line) to first 200eV (crosses), then 1 keV (squares) and 5 keV
(circles) the peak energy deposited in the film falls, the slope of the
distribution increases and the average width of the distribution
decreases. If this material were presumed to have the same type of
threshold behaviour as PMMA then clearly the higher the energy cutoff
the narrower the minimum line width that could be achieved. This

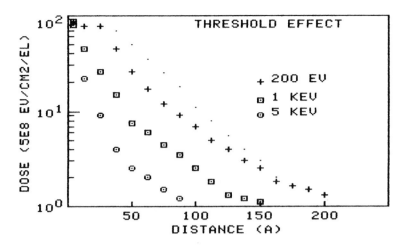

FIGURE 7   Variation of absorbed energy in 250A PMMA film, irradiated with a 20A line source at 100 keV, assuming that only electrons with greater than 40 eV (dots), 200 eV (crosses), 1 keV (squares) or 5 keV (circles), contribute to the exposure process.

improvement will, however, be bought at the price of writing speed since the amount of "usable" energy deposited in the specimen will be lower. It is possible that the reports of very high resolution lithography performed on such materials as NaCl[1], and sodium β-alumina[2], are examples of this because it is probable that in both cases the initial step in the exposure is the ionization of the sodium, an action which requires an electron to have in excess of 1keV of energy. At this energy level the energy deposition vs distance data (figure 7) has almost lost its flat-top and increases smoothly towards the axis. By a judicious choice of exposure dose it should thus be possible to fabricate a structure with a size limited only by the probe diameter, as was in fact observed. Similar kinds of improvements might perhaps also be obtained in organic materials in which exposure was obtained by direct sublimation or some other form of gross mass removal which requires more energy than could be obtained from the secondary elctrons. Clearly systematic investigation is required in this area to find materials which provide a reasonable compromise between resolution and writing speed.

ACKNOWLEDGEMENTS

The author is grateful to Dr. D.E.Newbury for many stimulating discussions, and much helpful advice, on Monte Carlo techniques.

REFERENCES

1 M.Isaacson and A.Muray, Molecular Electronic Devices, Marcel Dekker, New York, 1982, p165

2 M.E.Mochel, C.J.Humphryes, J.A.Eades, J.M.Mochel and A.M.Petford, Appl.Phys.Lett., 42, p392,(1983)

3 A.N.Broers, Proc. 38th Annual Meeting EMSA, Claitor's Press, Baton Rouge, 1980, p308

4 A.N.Broers, J. Electrochem. Soc., 128, p166, (1981)

5 J.I.Goldstein, J.L.Costley, G.W.Lorimer and S.J.B.Reed, Scanning Electron Microscopy 1977, SEM Inc, Chicago, 1977, 1, p315

6 K.Murata, D.F.Kyser and C.H.Ting, J.Appl.Phys, 52, p4396,(1981)

7 R.D.Evans, The Atomic Nucleus, McGraw-Hill, New York, 1955, p576

8 J.S.Greeneich and T. Van Duzer, IEEE Transactions on Electronic Devices ED-21, p286, (1974)

9 D.C.Joy, The Microscopy of Semiconductors, Institute of Physics, London, 1983, (In press)

10 H.G.Craighead, R.E.Howard, L.D.Jackel and P.M.Makiewich, Appl.Phys.Lett., 42, p38, (1983)

18

Macro-to-Molecular Connectors

D. J. Nagel/Naval Research Laboratory
Washington, DC

I. INTRODUCTION

The availability of complex electronic devices on the molecular
level could produce a revolution even greater than today's impact
of silicon-based microelectronics. Many technical barriers to
attainment and production of cheap, functional molecular devices
clearly exist. Identification and production of molecular groups
with the needed properties and their assembly into devices are
fundamental problems. Heat dissipation and connections to input/
output devices are two technological problems.

   This paper is concerned with the macro-to-molecular connections necessary for communication of information from humans to
molecular electronic devices. It is conceivable that electromagnetic energy could be employed to transmit information from the
large-scale to the molecular levels. That would be a scaled-down
version of a satellite communication system, with which data is

down-linked from earth orbit with radio wavelengths to local, sensitive receivers. However, molecular-to-macroscopic electromagnetic communication seems unlikely due to power requirements, so attention is restricted to physical electronic connections. Conductors with widths near 100 A are considered. That is, 100 A is considered to be at the molecular scale for this paper. It is conceivable that conductors 10 A wide may eventually be produced, although their maintenance against diffusion and their high resistivity would be problematic.

Macro-to-molecular level connections are of current interest for reasons other than potential information exchange with molecular computers. Probing of biological and chemical systems with microelectrodes in order to determine voltages or cation concentrations has a three-decade history. Alteration (e.g., injection into) and field-induced orientation of biological cells and molecules has also been accomplished. Physical, especially electronic, systems with submicron-wide conductors have been heavily studied in the past decade. Interest in molecular electronic devices and their connection to the macroscopic level integrates much past work in these diverse areas.

This paper has two thrusts. Firstly, the production, characteristics and use of microconductors are briefly reviewed (Section II), with emphasis on microelectrodes and microchannel plates produced by glass drawing techniques (Section III). Secondly, an idea on how arrays of microconductors can be made by use of the techniques employed to produce fiber optics is presented (Section IV) and potential uses of such "fiber electronics" are considered (Section V). This paper merely gives background information and factors relevant to a planned experimental program aimed at producing arrays of conductors as fine as about 100 A, and connecting them to ordinary electronic circuits. Fabrication and testing of fiber electronics will be described at the appropriate time.

II. PRODUCTION, CHARACTERISTICS AND USES OF MICROCONDUCTORS

Microelectrodes and fine conductors have been produced by a variety
of techniques. Linear conducting structures with diameters of a few
microns or less can be produced by direct growth, by etching for
diameter reduction, or by drawing-down to achieve small lateral
dimensions. The resulting structures are characterized according
to the number of fine points on a single electrode, the number of
electrodes in an array, the spatial ordering (relative placement) of
multiple electrodes and whether or not the individual conducting
paths are isolated from each other electrically. Production techniques and microelectrode characteristics are enumerated in Table I
and discussed in the remainder of this section. Ordered arrays of
isolated electrodes are of most interest for macro-to-molecular
connectors.

Micron-sized carbon conductors can be grown by thermal decomposition of polymeric materials such as rayon. Pyrolysis near

Table I - Production Techniques and
Structural Characteristics of Microelectrodes

PRODUCTION TECHNIQUE	ELECTRODE CHARACTERISTICS				
	Single Electrodes	Multiple Points	Multiple Electrodes	Ordered Arrays	Electrodes Isolated
Fiber Growth	Yes	No	-------Possible------		
Chemical Etching	Yes	No	-------Possible------		
Ion Etching	Yes	Yes	No	–	–
Orientation-Dependent Etching	Yes	Yes	No	–	–
Lithography	Yes	Yes	Yes	Yes	Yes
Drawing Metals	Yes	No	Yes	Yes	No
Drawing Glasses	Yes	No	Yes	Yes	Yes

$1000^{\circ}C$ followed by graphitization at $2000-3000^{\circ}C$ yields high-con-
ductivity fibers with diameters as small as 5 µm. Alternatively,
catalytic pyrolysis of benzene and hydrogen near $1100^{\circ}C$, followed
by graphitization at $2800^{\circ}C$, also yields conductive carbon fibers.
Production, characteristics and electrochemical uses of graphitized
fibers have been reviewed (1).

Single crystal whiskers of a wide variety of materials have been
widely studied because of their high internal perfection and their
value as strengtheners in composite materials (2). They can be
produced by a wide variety of techniques, including ion-bombard-
ment (3), stress-induced growth, vapor phase reactions (chemical
vapor deposition), electrolytic deposition, growth from melts
(including directional solidification of binary eutectics) and
growth from solution (4). Whisker diameters less than 500 A occur
for some materials (5), although diameters in the 1-10 µm range
are most common (2). The finest point which can be grown on a
whisker does not appear to be known, but equivalent radii significantly less than 500 A may exist for particular materials.

Single, finely-pointed conductors can also be made by etching
of the near-tip region of larger-diameter metals, alloys or selected
compounds. Recipes for preparation of fine points on a wide variety
of materials have been developed, including useful electrolytes
and etching voltages (5,6). Tip radii as small as about 100 A can
be achieved for some materials, such as tungsten (6). Sharply-
pointed conductors are employed as samples in field-ion microscopy
for studies of atomic structure and motion. After coating with
lacquer (7) or glass (8) insulation, except near the tip, metallic
points have been used as probes of biological systems.

Individual conductors prepared by the growth or etching techniques outlined above are of limited use for macro-to-molecular
connections. However, it should be possible to arrange the tips of
fine-scale conductors in three-dimensional arrays and to insulate
them from each other. Micromanipulation techniques are readily

available with positional resolution of less than 1 μm. Position
control on the 100 A level is achievable with commercial piezo-
electric devices. Sub-Angstrom translation control has been
demonstrated with an x-ray interferometer for positional readout
(9). With such modern micropositioning techniques, individual
conductors, e.g., etched tips, could be placed relative to each
other, possibly with control at a level approaching 100 A. Electrical contact might provide a reference for relative positioning.
It could be necessary to use a scanning electron microscope (with
a large sample region) to observe the microelectrodes during positioning. Possibly, a wax or epoxy could be employed to maintain
the relative positions of the points during use. Laborious construction and potential position maintenance problems are drawbacks
to fabrication of electrode arrays from individual conductors.

Arrays of fine conducting points can be produced by other
etching techniques. Bombardment of crystals, either by an ion
beam or by ions from a plasma, can result in sharply-peaked cones
due to sputter etching of the surface (10,11). The mechanisms for
production of cones include sputtering, lateral movement of atoms
along the surface and whisker growth, with the relative importance
of these processes for specific targets as yet undetermined (3).
Cones are commonly about 10 μm high and can have tips less than
500 A in diameter. Ion-etched cones usually occur in irregular
patterns, although they are sometimes quite regular in height and
surface distribution (12). Highly-ordered arrays of uniform
pyramids with fine tips can be produced by lithographic patterning
and orientation-dependent-etching (ODE) of silicon (13). ODE
results from use of chemicals which attack particular planes in a
single crystal much faster than others. As indicated in Table I,
ion and orientation-dependent etching produce multiple points on
one conductor (electrode) but not multiple isolated conductors.
They may find use for connection of a single macroscopic terminal
to multiple micro-level points. However, they are not considered
further here.

Lithographic methods are the most recent and versatile techniques for production of fine conductors. All three of the major approaches to submicron lithographic exposures (x-ray, electron-beam and ion-beam) have been used to produce lines with widths in the 200-500 A range. Single, branched and isolated multiple conductors can be produced in arbitrary arrays on flat surfaces. Such arrays have been employed as substrates in neurological studies (14). Structures with one to four electrodes for probing cells have been fabricated by integrated-circuit technology. Electrode tip diameters as small as 2 μm and interelectrode separations down to 10 μm were achieved (15,16).

Single very-fine conductors have also been produced for physical studies in microelectronic devices by using the processing methods developed for semiconductor production. Semiconductor wires with cross sections of 200 A x 200 A were formed in GaAs under the gates of metal-oxide-semiconductor field-effect transistors (17,18). Electron transport and optical properties of such quantum-well structures have been studied. Single metallic conductors which have been produced by lithographic, deposition and etching techniques used for microcircuit production. A novel "step lithography" technique was employed to produce microconductors with triangular cross sections (19,20). Areas equivalent to squares 200-300 A on a side were produced for the study of electron localization at low temperatures (21,22). A "step edge shadowing" technique (23) was used to make conductors with rectangular cross sections as small as 200 x 220 A (24). The pure metal (e.g., Pt) and alloy (AuPd) microconductors produced by these methods have been used for electron localization and noise studies.

The remaining microelectrode production techniques listed in Table I involve diameter reduction by drawing of metallic or glassy materials. The work with composite metals was motivated by physical studies such as strength enhancement and magnetism. Fine

conductors embedded in metal matrices are shorted electrically.
Their superconductivity has been studied as drawn. Alternatively,
the matrix can be etched away for use of individual filaments.
Various starting materials have been employed for production of
metallic composites with very fine filaments. Directional solidification of eutectic alloys can produce structures with diameters
near 0.5 μm (25,26). Subsequent drawing of such materials has
yielded filaments with lateral dimensions of 100 A (27). Powder
metallurgy or quenching of a molten alloy into a two-phase region
also give suitable starting materials for diameter reduction by
drawing (28,29). Filaments with thicknesses of 100-200 A result
after cross-sectional area reduction of 99.999%. An excellent
review of preparation and the mechanical, electrical and magnetic
properties of filamentary composites was published recently (30).

Metal drawing techniques have also been employed to produce
single isolated filaments. Platinum wires as thin as 800 A were
made by drawing 0.5 mm diameter Pt wire encased with Ag (31). The
Ag was chemically etched away to produce the isolated Pt filaments
microconductors, which had highly disordered internal structure.
Large increases in the low-temperature (<20 K) resistivity of such
Pt filaments 2000 A or smaller was observed. Numerous magnetic
filaments in a non-magnetic matrix have been produced by a repeated
drawing technique. Approximately 2×10^6 Fe filaments 4000 A in
diameter were produced in a copper alloy matrix by repeated drawing,
bundling, drawing, etc. (32). The same method was used to produce
almost 10^7 superconducting Nb wires, some as fine as 200 A diameter, in a copper matrix (33).

Production of microconductors by drawing of glasses, the last
entry in Table I, is most germane to the fabrication of fiber-elec-
tronics. Hence, the next section is devoted to reviewing production
of hollow and metal-filled microelectrodes. A discussion of related
aspects of the production of electron multipliers, called microchannel plates, is included.

III. GLASS MICROELECTRODES AND MICROCHANNEL PLATES

Fine glass structures have been drawn in single, coaxial, side-by-
side ("double barreled") and multiple configurations. A saline
solution, such as 3 Molar KCl, is commonly employed as the conductor, although metal-filled glass microelectrodes have been produced also. Electrical and chemical measurements have been made
in biological and chemical systems with glass microelectrodes.
Various configurations, conducting media and uses will be enumerated in the following paragraphs.

Hollow glass tubes are quite easily drawn into microcapillaries. The use of electrolyte-filled electrodes in cellular research was well-developed already 25 years ago (34). Production
and use of hollow microelectrodes has been reviewed (35,36).
Openings at the microelectrode end in the 2000 A (37) down to 1000
A (38) range have been produced. These compare with cell dimensions which are commonly about 10 μm. The electrical properties
of hollow, solution-filled microelectrodes are summarized (39).
Rather large ($\sim$20 μm) ion-selective electrodes have been employed
to study distributions of H^+, Na^+, K^+ and Cl^- ions in giant ($\sim$2000 A
diameter) barnacle cells (40). Intracellular ion activities have
been measured with ion-exchanger microelectrodes (41). Hollow
microelectrodes have also been employed for microinjection into
cells (42).

Coaxial (38,40,41) and double-barreled (34,43) electrodes
allow separate application of a potential and sensing of its
effects. The side-by-side microelectrodes are drawn, in a fashion
similar to production of single microcapillaries, from two fused
glass tubes. Electrical and thermal characteristics of double-
barreled microcapillaries are summarized (44). Multiple-bore
microcapillaries with 19 openings within a 100 μm outside-diameter
structure have been produced to permit use of separate conductors
in electro-biological measurements (45). Multiple microcapillary

filaments, also 100 μm in overall diameter but with seven openings, have been made as strength members in composite materials (46).

Metal-filled glass microelectrodes can be produced by insulating a sharply pointed metal rod except near the tip (7,8), or by drawing of glass tubes filled with a metal or alloy. Sixty years ago Taylor produced glass-coated wires of Pb, Sb, Bi, Au, Ag, Cu, Fe, Sn, Tl, Cd, Co, Ga, In and some of their alloys with metal diameters down to about 2000 A (47). Gallium, indium and many of their alloys have low melting points and will wet glass. For example, a 50-50 alloy of In and Sn, which melts at $110^{\circ}C$, was employed to produce microelectrodes with 2-4 μm tip diameters for studies of cellular discharges (48). Oxygen concentration in cat muscle was measured with 2 μm diameter microelectrodes filled with Woods metal (49). In both these cases, the low-melting metal was electroplated with gold at the electrode tip. Production and use of metal-filled microelectrodes has been reviewed (35,36). Equivalent circuits for such microconductors and their associated electronics have been studied (50).

A microchannel plate (MCP) is a glass structure, typically 2-4 cm in diameter and about 1 mm thick. MCPs consist of numerous fine holes across the 1 mm plate thickness (the microchannels) which are typically about 15 μm in diameter with 20 μm center-to-center spacing. The channels are coated internally with a semiconducting layer having a high secondary-electron production coefficient. The plate faces are metallized. With typically 1000V across the plate, an electron cascade within a channel gives an electron multiplication of about 10^3. The initial electrons can be due to absorption of photons, electrons or ions incident on the plate face. MCPs have many uses as electron multipliers and image intensifiers (51).

The production of MCPs has much in common with the making of fiber electronics (FEs) to be considered in the next section. Hollow glass tubes, soft-glass-filled glass tubes or metal-filled tubes are drawn, bundled, drawn and bundled (52). The process is

similar to the Levi process for production of metallic composites
(32). Since hollow channels are required in the end, the glass or
metal filling is removed chemically after transverse cutting to
form the disc-shaped plate. The major differences between MCPs
and potential FEs are (a) MCPs have hollow semiconducting paths
while FEs will have solid metallic conductors, (b) the final MCP
channel is about 10^3 larger than the planned FE microconductor,
and (c) lack of transverse sectioning in the FE, since continuous
conductors are desired.

IV. POTENTIAL FIBER ELECTRONICS

The central idea discussed in this section involves a hybrid of the
techniques for bundling and repeated drawing of metal-filled glass
tubes mentioned in the last section and modern techniques for production of fiberoptics (53). The process is indicated in Figure 1.
Initially, a 2 mm outside diameter glass tube with a metal-filled
100 μm core is bundled with about 20 2 mm diameter rods made of
the same glass and sintered to produce a "preform" about 1 cm in
diameter. This is inserted in a machine ordinarily employed to
draw fiber-optics instead of the usual optical preform (which
consists of a high-index glass clad with a low-index glass).
Pulling a 1 cm diameter preform produces a 100 μm diameter fiber.
That is, a diameter reduction of 100X is typical of fiberoptics
production. At this stage the conducting core is 1 μm (10,000 A)
in diameter. Then a new preform is assembled from hundreds of 100
μm diameter fibers with and without metallic cores arranged in the
desired pattern. Again, sintering to produce the 1 cm diameter
preform and drawing are performed. The second 100X reduction
would yield 100 A diameter conductors in a fiber, the ends of
which are 1 μm in diameter in the remaining undrawn ends of the
preform.

Several problems with production of continuous fiber electronics on the 100 A level are envisioned. Compatibility of metal

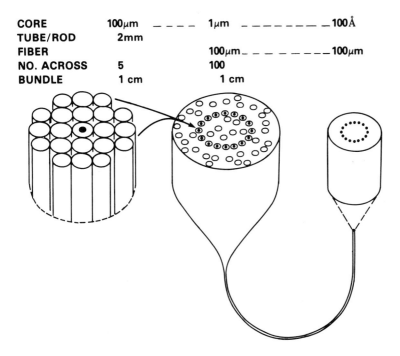

CORE	100μm	_ _ _ _	1μm	_ _ _ _ _ _ _ _ 100Å
TUBE/ROD	2mm			
FIBER			100μm_ _ _ _ _ _ _ _ _100μm	
NO. ACROSS	5		100	
BUNDLE	1 cm		1 cm	

Figure 1. Diameters (table) and geometry (schematic) for preparation of fiber electronics by a four-step process: assembly (1)
and drawing (2) of the initial bundle containing one metallic
core (shown on left), and assembly (3) and drawing (4) of the
final fiber containing multiple microconductors in a predetermined
array geometry (center and right, with 16 metallic paths shown).

and glass will be critical to attainment of continuity on that

scale. It has already been shown that various metals and alloys

will work to the micron diameter level (47,49), which is essen-

tially the first of the two drawing stages for fiber electronics.

Possibly, even with surface wetting, the shear forces produced

during drawing will disrupt the conductor at some level. Metals

and alloys which have little volume change upon melting or solidi-

fication are available. Their use, and very slow drawing to allow

for stress relaxation, may be necessary to achieve the finest

conductor diameters.

    We now turn to the problem of connecting fiber electronics to

exterior circuits. The approach is sketched in Figure 2. Embedd-

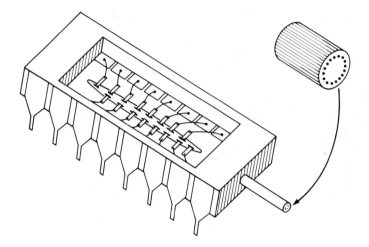

Figure 2. The large end of an electronic fiber sectioned obliquely
and mounted in a conventional microcircuit package for connection
to standard electrical circuits.

ing the large end in plastic and obliquely sectioning it at a
shallow angle will expose the 1 μm diameter conductors over
several microns. Polishing and lithographic production of contacts, including plasma etching of superficial oxides prior to
deposition of the (probably gold) contacts, is envisioned. Microchip amplifiers (not shown in Figure 2) could be produced on the
oblique surface for signal amplification. Then mounting in a
standard microchip package and routine connection with fine gold
wires would complete the connection. It is noted that in current
chips the scale of the pads and the finest features (e.g., gates
in MOS devices) are in a ratio of about 100:1 (100 μm to 1 μm).
Fiber electronics would give a second ratio of 100:1 between the
lithographically-accessible micron scale and 100 A molecular scale.
That is, a combination of two technologies, lithography as currently practiced and fiber-drawing with metal rather than glass-filled
preforms, is advocated to achieve macro-to-molecular connections.
     The final step is terminate the fine end of the fiber elec-
tronics bundle in such a way that it is useful. A simple scheme is

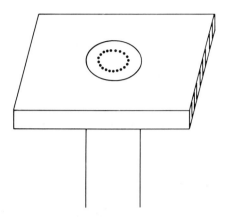

Figure 3. One method to terminate the small ($\sim$100 μm diameter)
end of an electronic fiber by embedding in plastic, and cutting
and polishing the fiber transverse to its axis. Plasma etching
and electroplating may be used to prepare the fine conductors
($\sim$100 Å) for use. The plastic could form a substrate during use,
or be removed for insertion of the fiber into biological or chemical environments.

indicated in Figure 3. Embedding the fiber in plastic, transverse

sectioning, polishing and etching would yield an array of exposed

conductors. Use of a dry (plasma) etch which would selectively re-

move the plastic and leave protruding metal is envisioned. Elec-

trochemical oxide removal and electron deposition of gold or an-

other metal would ready the termination for use. The critical

problem of contact between the fine end of a fiber electronic con-

nection and electronic devices on a molecular level is not given

due consideration here. A similar problem will have to be solved

if any dense, two- or three-dimensional electronic devices are to

be electrically connected to the macroscopic world. Chemical

attraction to the material at the fiber electronic termination and

self-assembly may play roles. Whatever the progress on this front,

fiber electronic arrays may have uses independent of communication

with molecular electronic devices. These are considered in the

next section.

V. USES FOR FIBER ELECTRONICS

Possible applications in biology, chemistry and physics are discussed in this section. They primarily depend on the very fine-
scale conductors expected in fiber electronics, although the
availability of numerous parallel channels which can be separately
biased might also be exploited. Potential uses of fiber electronics as sensors in macro-to-micro-to-macro circuits are outlined at the end of this section. A possible role for fiber
electronics in microelectronics and computer circuitry is noted.

Biological applications of microelectrodes have already been
noted (34-45). A more recent survey of uses (54), and a news article on electrical probing of nerves (55), are also available.
Microterminations at the end of an electronic fiber could be used
to orient as well as excite and probe cells. Their structure and
composition alone provide a substrate with landmarks for orientation (56). Further, appropriate biasing of the individual conductors would produce surface fields and currents of use in affecting
cellular orientation and structure. Such fields have often been
used to orient fine fibers suspended in liquids (57). Oriented
fibers could play a role in influencing the attachment and geometry of cells. In any event, the availability of numerous,
individually-controllable microelectrodes would be useful for
applying voltages to and sensing currents in cells.

Chemical uses of microconductors have been less numerous than
applications in biology, although electrochemistry is fundamental
to several cellular uses (40,41). The measurement of noise associated with chemical reactions is an especially attractive use of
fiber electronics. For small systems, a single molecular event
can lead to a measurable electrical fluctuation. Nucleation and
growth of crystals, pore formation in lipid bilayers, and electro-
catalysis are examples of stochastic effects (58). Passive film
breakdown in electrochemical cells is another (59). By using
electrode separation as a variable, information on the spatial

extent of electrochemical fluctuations could be obtained with a terminated electronic fiber. Use of fiber electronics for chemical sensors also deserves study.

Applications of fiber electronics in physics would include noise studies. For example, 1/f noise was recently studied (60) with microconductors made by step-edge lithography (19,22). Studies of electrical conductivity and electron localization in fine conductors have been popular recently (19-24, 61,62). It seems likely that fiber electronics would also be useful for such work, which would in turn produce information on the characteristics and utility of fiber electronics.

The chemical and physical behavior of arrays of fine-scale conductors might make them useful as sensors. The development of micro-chemical sensors is a new and rapidly-developing area of study (63). Fiberoptics have been employed as sensors for electromagnetic and acoustic waves, magnetic fields, ionizing radiation, acceleration, temperature, pressure, and a wide variety of other variables (64). Some of these uses of optical devices may carry over to fiberelectronics, with no need for light generation and detection. That is, the application of electric or magnetic fields or a mechanical stress (strain) to an electronic fiber might alter its conductivity in a way uniquely dependent on the applied field or stress. The presence of a magnetic conductor within the fiber electronic should produce interesting interactions with externally-applied magnetic fields. A macro-to-micro-to-macro circuit for such applications is indicated in Figure 4. Production of such a circuit would clearly require preservation of large diameters on both ends of the electronic fiber.

A fiber electronics circuit, such as shown in Figure 4, could serve as a basis for a wholly new type of microelectronics. Application of conduction-altering electric or magnetic fields to the entire fiber would be analogous to the action of the gate in a MOS field-effect transistor. That is, a conducting structure outside

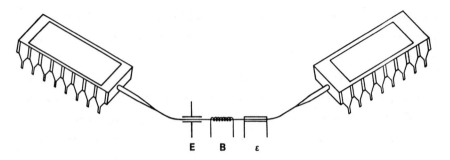

Figure 4. Schematic of a fiber electronic circuit in which conductivity would by modulated by application of electric (E) or
magnetic (B) fields, or a mechanical or acoustical strain (ϵ).

of and transverse to the electronic fiber (possibly in a helical
geometry) could simultaneously affect signal transmission in all
conducting paths within the electronic fiber. Such a capability
could be useful in signal processing, or for connecting microprocessors in a parallel-processing computer architecture.
Altering the topology by twisting the electronic fiber could
expand the range uses.

A major concern for fine-scale semiconductor and molecular
electronic devices is interaction between circuit elements due to
leakage, capacitances and fields in general (65). Fiber electronics, with their known geometry and variable scale (e.g.,
conductor size and separation) could prove useful for experimental
assessment of microcircuit interference. Measurements as a
function of signal frequency and strength could be used to test
models of correlated electrical behavior. Also, short pulses of
ionizing ultraviolet or x-radiation might provide transient
inter-conductor shorts with interesting effects.

Glass-covered metallic fibers could find uses in plasma physics which do not involve their ordinary electrical conductivity.
Plasmas can be created and heated to temperatures near 10^7K by
10^{-7} sec megampere electrical currents. In such work, fibers
about 25 μm in diameter are stretched across the cathode-anode gap

prior to the discharge (66). Drawing of metal fibers within glass
and subsequent chemical removal of the glass could produce micron-
scale fibers for discharge heating. Metal-in-glass fibers might
also be useful targets for plasma generation and heating by very
high power laser pulses. Here, multimillion-degree temperatures
are achieved in times of 10^{-8} to 10^{-11} sec. Linear targets containing at least two elements are of interest for x-ray laser
research (67). One of the elements in the glass, excited by a
long-wave-length laser pulse focused to a line, could pump a
second element in the core for such work.

This listing of potential uses for fiberelectronics is meant to
be illustrative. It seems likely that, once they are available,
electronic fibers would find a variety of applications in research
and possibly in industry.

VI. DISCUSSION

Production and use of fine conductors has had a long history.
Drawing of single and multiple microelectrodes for biological
probes gained momentum about 30 years ago. Etching of single fine
points for atomic-scale microscopy was an active area about 20
years ago. Production of microconductor arrays by lithographic
and related techniques achieved wide interest in biology and,
especially, physics in the last five years. Many opportunities
remain for exploitation of lithographically-produced macro-to-
molecular conductors. For example, tapered conductors, and
embedding with transverse sectioning of microcircuits as in Figure
3, should be useful. The aim of this paper has been both to
review earlier approaches to production of microconductors, and to
propose a potentially straightforward way to make ordered arrays
of connectable microelectrodes in electronic fibers.

The fabrication of continuous metallic conductors from the 1 μm
to the 0.01 μm (100 A) scales will not be sufficient to insure the
practicality of fiber electronics. High electrical resistance, even
in the absence of electron localization effects, may be a major

limitation. The resistance, R in ohms, of a conductor is given by
$R = \rho L/A$, where the resistivity ρ is typically 10^{-5} Ωcm. Hence,
for a length L of 1 cm and area A of $(100\ A)^2 = 10^{-12}$ cm^2, R=10 MΩ.
If a voltage drop of 10V is acceptable, the current will be 10^{-6}
amps, and the power dissipation 10^{-5} watts/cm. Interdiffusion and
surface effects may increase ρ by a factor of 10 or more. Experimental determination of the performance characteristics of very
fine, glass-embedded conductors is needed.

Ionic conductors (glasses or compounds) could be used instead
of metals in electronic fibers if high conductivity and fast
response are not needed. Such structures could be employed for
basic studies of ionic conductivity and for chemical (ion exchange)
studies. The influences of fields and pressure on ionic motion
might produce interesting effects.

Hybrid fiber optics and electronics are possible. That is,
few micron-diameter, single-mode optical conductors could coexist
in a single structure with fine-scale electrical conductors. The
termination at the large-scale end (Figure 2) would have to include
one or more optical couplers, while termination at the fine-scale
end (Figure 3) could remain simple. Electrical probing of optic-
ally-produced changes may be possible on a molecular level with
hybrid fiber optics and electronics. Complete fiber electro-optic
circuits are also envisioned (as in Figure 4).

The relationship of the production and study of fine conductors to work with other microstructures can be noted. Thin films
are small in one dimension. They can be produced by a wide variety
of techniques, such as evaporation, sputtering and chemical vapor
deposition. Their study has a long history and they are very
important commercially. Very thin (<100 A) single layers and
multilayers are now receiving heavy research attention. Fibers,
small in two dimensions, are also important in both science and
technology. Very fine (<100 A) conductors were discussed in this
paper. Powders and other particles, small in all three dimensions

are also broadly important. The production of "few"-atom (<100 A diameter) clusters by condensation or sputtering, and study of their physical characteristics, is now an active research area. During one era, from the development of the optical microscope through improvement of electron microscopes, a major thrust in science was to see structures on a very fine scale, even to imaging of molecules and atoms. Now, research is producing ways to build ordered structures on a scale reaching to the molecular level.

ACKNOWLEDGEMENTS

This paper resulted from a chance conversation with F. L. Carter. His stimulus is greatly appreciated. Helpful conversations with T. W. Barbee Jr., J. Bevk, J. Buchanan, J. Costa, M. Fatemi, D. U. Gubser, J. S. Murday, D. L. Nelson, M. C. Peckerar and G. H. Sigel Jr., and comments on the manuscript by J. V. Gilfrich, are also acknowledged with pleasure.

REFERENCES

1. J.S. Murday and D.D. Dominguez in E. B. Yeager, et al. (Editors), Membranes and Ionic and Electronic Conductors, Vol. 83-3, Electrochemical Society, Pennington, NJ (1983) p. 359.

2. A.P. Levitt (Editor), Whisker Technology, Wiley-Interscience, New York (1970).

3. G.K. Wehmer, Appl. Phys. Letters, 43, 366 (1983).

4. T. Gabor and J.M. Blocher, Jr., J. Appl. Phys., 40, 2696 (1969).

5. J.J. Hren and S. Ranganathan (Editors), Field-Ion Microscopy, Plenum Press, New York, 1968, p. 231.

6. K.M. Bowkett and D.A. Smith, Field-Ion Microscopy, North-Holland Publ. Co., Amsterdam, 1970, p. 221.

7. D.H. Hubel, Science, 125, 549 (1957).

8. C. Guld, Med-Electron. Biol. Engr., 2, 317 (1964).

9. R.D. Deslattes, Appl. Phys. Letters, 15, 386 (1969).

10. B. Navinvek, Prog. in Surface Science, 7, 49 (1977) and references therein.

11. O. Auciello, J. Vac. Sci. Tech., 19, 841 (1981).

12. J.L. Whitton, L. Tanović and J.S. Williams, Applications of Surface Science 1, 408 (1978).

13. D.J. Nagel, K.E. Bean and R.K. Watts, Nucl. Instr. and Methods 172, 321 (1980) and references therein.

14. G.W. Gross, A.N. Williams and J.H. Lucas, J. of Neuroscience Methods 5, 13 (1982) and references therein.

15. K.D. Wise, J.B. Angell and A. Starr, IEEE Trans. on Bio-Med. Engr., BME-17, 238 (1970).

16. K.D. Wise and A. Starr, Proc. 8th Intl. Conf. on Medical and Biological Engr., 1969, Section 14-5, quoted in L.A. Geddes, Electrodes and Measurement of Biological Events, Wiley-Interscience, New York, 1972, p. 144.

17. H. Sakaki, Jap. J. Appl. Phys., 19, L735 (1980).

18. P.M. Petroff, A.C. Grossard, R.A. Logan and W. Wiegman, Appl. Phys. Letters 41, 635 (1982).

19. M.D. Feuer and D.E. Prober, Appl. Phys. Letters, 36, 226 (1980).

20. D.E. Prober, M.P. Feuer and N. Giordano, Appl. Phys. Letters, 37, 94 (1980).

21. N. Giordano, W. Gilson and D.E. Prober, Phys. Rev. Letters, 43, 725 (1979).

22. J.T. Masden and N. Giordano, Physica 107B, 3 (1981).

23. D.C. Flanders and A.E. White, J. Vac. Sci. Tech., 19, 892 (1981).

24 A.E. White, M. Tinkham, W.J. Skocpol and D.C. Flanders, Phys. Rev. Letters 48, 1752 (1982).

25. J.D. Livingston, J. Appl. Phys., 41, 197 (1970).

26. M.J. Salkind, F.D. Lemkey and F.D. George in reference 2, p. 343.

27. G. Frommeyer and G. Wasserman, Acta Metallurgica, 23, 1353 (1975).

28. J. Bevk, M. Tinkham, F. Habbal, C.J. Lobb and J.P. Harbison, IEEE Trans. on Magnetics, MAG-17, 235 (1981).

29. J. Bevk, J.P. Harbison and J.L. Bell, J. Appl. Phys. 49, 6031 (1978).

30. J. Bevk, Ann. Rev. Material Science, 13, 319 (1983).

31. A.C. Sacharoff, R.M. Westervelt and J. Bevk, Phys. Rev. B, 26, 5976 (1982).

32. F.P. Levi, J. Appl. Phys., 31, 1469 (1960).

33. H.E. Cline, B.P. Strauss, R.M. Rose and J. Wulff, J. Appl. Phys. 37, 5 (1966).

34. D.W. Kennard in P.E.K. Donaldson (Editor), Electronic Apparatus and Biological Research, Academic Press, New York, 1958, p. 534.

35. K. Frank and M.C. Becker in W.L. Nastuk (Editor), Physical Techniques in Biological Research, Vol. 5, Electrophysical Methods, Part A, Academic Press, New York, (1964).p. 22.

36. M. Lavallee, O.F. Schanne and N.C. Hebert (Editors), Glass Microelectrodes, Wiley and Sons, New York, (1969).

37. P.G. Kostyuk, Z.A. Sorokina and Yu.D. Kholodova in reference 36., p. 322.

38. T. Tomita in reference 36, p. 124.

39. O.F. Schanne, M. Lavallee, R. Laprade and S. Gagné, Proc. IEEE 56, 1072 (1968).

40. J.A.M. Hinke in reference 36, p. 349.

41. F.W. Orme in reference 36, p. 376.

42. T.K. Chowdhury in reference 36, p. 404.

43. E. Coraboeuf in reference 36, p. 224.

44. S. Rush, E. Lepeschkin and H.O. Brooks, IEEE Trans. on Bio-Med. Engr., BME-15, 80 (1968).

45. V.A. Vis, Science, 120, 152 (1954).

46. R.A. Humphrey in L.J. Broutman and R.H. Krock (Editors) <u>Modern Composite Materials</u>, Addison-Wesley, Reading (1967) p. 330.

47. G.F. Taylor, Phys. Rev. <u>23</u>, 655 (1924).

48. R.M. Dowben and J.E. Rose, Science, <u>118</u>, 22 (1953).

49. W.J. Whalen in reference 36, p. 396.

50. D.A. Robinson, Proc. IEEE, <u>56</u>, 1065 (1968).

51. B. Leskovar, Physics Today (Nov 1977) p. 42.

52. D. Washington, V. Dushenois, R. Polaert and R.M. Beasley, Acta Electronica, <u>14</u>, 201 (1971).

53. R.E. Jaeger, A.D. Pearson, J.C. Williams and H.M. Presby in S.E. Miller and A.G. Chynoweth (Editors) <u>Optical Fiber Telecommunications</u>, Academic Press, New York (1979), p. 263.

54. R.F. Thompson and M.P. Patterson (Editors) <u>Bioelectric Recording Techniques</u>, Part A - Cellular Processes and Brain Potentials, Academic Press, New York (1973).

55. J.A. Miller, Science News, <u>123</u>, 140 (1983).

56. B.L. Giammara and G.A. Rozgonyi, these proceedings.

57. A. A. Winer and H. M. Woodrooffe, U. S. Patent No. 3,497,419 (Feb. 1970).

58. M. Fleischmann, M. Labram, C. Gabrielle and A. Sattar, Surface Science, <u>101</u>, 583 (1980).

59. V. Bertocci and J. Kruger, Surface Science, <u>101</u>, 608 (1980).

59. D.M. Fleetwood, J.T. Masden and N. Giordano, Phys. Rev. Letters, <u>50</u>, 450 (1983).

61. N. Giordano, Phys. Rev. <u>B22</u>, 5635 (1980).

62. D.J. Thouless, Physica, <u>109/110B</u>, 1523 (1982).

63. H. Wohltjen, Anal. Chem., <u>56</u>, 87A (1984).

64. T.G. Giallorenzi, J.A. Bucaro, A. Dandridge, G.H. Sigel, Jr., J.H. Cole, S.C. Rashleigh and R.G. Priest, IEEE J. of Quant. El., <u>QE-18</u>, 626 (1982).

65. R.O. Grondin, W. Porod, C.M. Loeffler and D.K. Ferry, these proceedings.

66. P.G. Burkhalter, C.M. Dozier and D.J. Nagel, Phys. Rev. A 15, 700 (1977).

67. D.J. Nagel in C. Bonnelle and C. Mande (Editors) Advances in X-Ray Spectroscopy, Pergamon Press, Oxford (1982) p. 371.

19

Some Thoughts on Molecular Electronic Devices and Chemical Sensors

Hank Wohltjen / U.S. Naval Research Laboratory, Code 6170, Washington, DC

An interesting application of the molecular electronic device (MED) concept is chemical sensing. In this very brief presentation, I would like to make three points which describe the thinking underlying some of the chemical microsensor research being conducted at NRL. First, chemical microsensors offer many immediate opportunities for incorporating the MED concepts being discussed at this workshop. Second, the development of MED devices will require a systematic reduction in the dimensionality of the materials being used to fabricate the devices. Third, as the size of a physical system is reduced to very small dimensions, the physical phenomena which dominate the behavior of the system can change dramatically. Some of these concepts have been elaborated on by Hirschfeld and others but it is useful to present them together in the context of this workshop (1,2).

The objective of the chemical microsensor research being
conducted at NRL is to develop new methods for selectively detecting very low concentrations of vapor phase contaminants in the
atmosphere. Basically we are trying to develop an analog of the
nose; a device occupying a few cubic centimeters of space which is
capable of detecting a wide array of vapors at the parts per
million concentration level and below. While we are a long way off
from being able to match the nose in its capabilities, the biological example gives us considerable encouragement that our goal
does not violate fundamental thermodynamic considerations. In
fact, the nose offers many insights into how we can build practical
vapor sensor systems. For example, it is known that the nose
possesses a number of selective receptors which generate a characteristic pattern in the brain whenever a particular odor is
encountered. We, too, are attempting to use this multiple sensor
array approach combined with computerized pattern recognition to
identify vapors.

In general, all chemical microsensors consist of two elements:
a chemically selective coating which interacts with the chemical
species to be detected and a microfabricated physical probe device
which monitors some property of the coating and provides an
electronic output signal (Figure 1). The device technologies
presently being investigated as physical probes for chemical
microsensor applications include surface acoustic wave (SAW)
devices which are extremely sensitive to mass or elastic modulus
changes in the coating film; optical waveguide spectrometers which
respond to variations in optical density, refractive index,
thickness, or scattering characteristics of the coating; and a
microelectrode array (called a Chemiresistor) which can sense small
changes in the transport of electrons through a thin,
semiconductor coating. A typical probe device presently has an

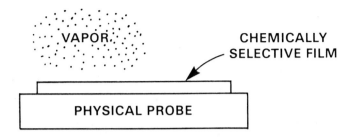

PROBE DEVICE	SENSITIVE PROPERTY
SURFACE ACOUSTIC WAVE DEVICE	MASS, MODULUS
OPTICAL WAVEGUIDE	OPTICAL DENSITY, REFRACTIVE INDEX
MICROELECTRODE ARRAY	ELECTRICAL CONDUCTIVITY

1. Generalized chemical microsensor device. Variations in the
    characteristics of the coating caused by chemical interactions
    are monitored by a microfabricated physical probe device.

active area on the order of a few square millimeters or more and is

fabricated using microlithographic techniques. Devices having

active areas a thousand times smaller (e.g. 10^{-6} cm^2) are

presently possible. Thus, chemical microsensor technology may

offer attractive methods for monitoring physical changes in

extremely small quantities (e.g. 10^8 molecules) of candidate MED

material coated onto the device surface. As device technology

improves it is tantalizing to consider extremely small sensors

(e.g. 10^{-10} cm^2 coated with 10^4 molecules) which might

respond to a single molecule of a vapor of interest. Such

sensitivity is not unknown in nature. Butterflies are attracted to

a romantic rendezvous by a few pheromone molecules thoughtfully

provided by the opposite sex.

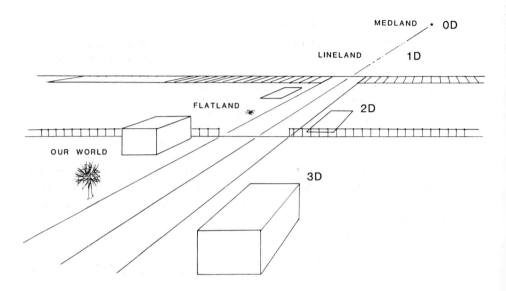

2.   The road to MED land involves successive reductions in the
     dimensionality of the materials used to fabricate the device.
     Practical systems will require successive increases in the
     dimensionality of the interconnected devices.

Clearly, the devices described previously are not MEDs but
they might be in the future. The selective coatings are extremely
thin, typically 1000 Angstroms or less in thickness. They are
frequently deposited using the Langmuir-Blodgett film transfer
technique. It is expected that as the probe device sizes shrink,
so will the coating thickness. Soon, two dimensional monolayer
films will be used for the chemically sensitive coatings. This
represents an important step since we will have reduced the
dimensionality of our device from three dimensions to two.
Conceptually, this means we are working with a flat array of
individual molecules. An individual MED is obtained after reducing
the two dimensional ensemble of molecules down to a one dimensional
line of molecules and from there, down to the zero dimensional
(conceptually, not physically) MED. It is doubtful that the

THE RELATIVE IMPORTANCE
OF VARIOUS PHENOMENA
IN MICRO SPACE

<u>MORE IMPORTANT</u>   <u>LESS IMPORTANT</u>

ELECTROSTATICS   GRAVITY

SURFACE DIFFUSION   FLOW

TUNNELING   INERTIA

VAN DER WAALS FORCES   INDUCTANCE

ETC.

3.   Careful attention should be paid to the scaling laws which
     govern the phenomena of importance to MEDs.

isolated MED will be of much use to anyone. Rather they will be

combined and interconnected in some systematic way, first in one

dimension, then in two dimensional planes, and finally in complex

three dimensional systems. Thus, the road to MED land (Figure 2)

will likely be marked by progressive reductions in the dimen-

sionality of the materials used to make the devices, followed by a

progressive increase in the dimensionality of the interconnected

MEDs used to make a useful system.

Finally, MED designers must be sensitive to the relative

importance of physical phenomena as one reduces the dimensionality

of a system. Phenomena which are important in the macroscopic

three dimensional world, such as gravity, are totally irrelevant at

small dimensions. In fact, fields of all kinds (e.g. magnetic,

electric, thermal, etc.) which are external to the system have a
negligible effect. Rather, phenomena such as surface diffusion,
electrostatic forces, tunneling and others, whose magnitude scales
with some inverse power of distance, tend to dominate the behavior
of the system. Careful consideration of the physical scaling laws
which describe various phenomena will undoubtedly yield many fresh
approaches to MED design.

REFERENCES

1.   E.A. Abbot, "Flatland: A Romance of Many Dimensions", Fifth
     edition, Harper & Row, New York, 1983.

2.   T. Hirschfeld, "Microstructures and Microinstrumentation",
     presented at the 185th National ACS Meeting, March 22, 1983,
     Seattle, WA.

IV
Two-Dimensional Approaches

20

Self-Organizing Molecular Electronic Devices?

Hans Kuhn/Department Development of Molecular Systems
Max-Planck-Institute for Biophysical Chemistry
(Karl-Friedrich-Bonhoeffer-Institute) D 3400 Göttingen-
Nikolausberg
Federal Republic of Germany

The search for methods to construct molecular devices and
the search for ways how first simple living systems could
have originated are strongly related. Finding ways to
construct simple machinery of molecular size is helpful
in finding possibilities in the origin of life, and considerations on selforganization of matter should stimulate the search for future possibilities of making molecular devices.

1. Monolayer Organizate Construction, Energy and Electron
Transfer

The first and simplest molecular devices were obtained
by assembling monolayers, using the Langmuir-Blodgett
technique in a new way [1]. Fig. 1 gives simple examples

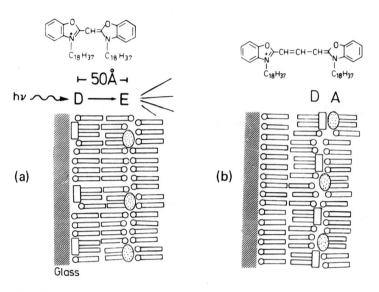

FIGURE 1.   Monolayer assemblies for studying energy trans-
             fer (a) and electron transfer (b).

of organized systems of molecules. A molecule D absorbs
a UV light quantum, the energy is transferred to a molecule E (absorbing at the fluorescence wave length of D),
which then emits a yellow light quantum (Fig. 1a). This
cooperation of the two molecules takes place at 50 Å
distance. The two different chromophores can be placed
at fixed positions by assembling monolayers. A mixed monolayer of dye D and arachidic acid is produced at the
water surface. The monolayer is deposited on a glass
slide by lifting the slide which has been immersed in the
water. A monolayer of dye E in a fatty acid matrix is
produced at the water surface and this layer is deposited

on top of the first layer by dipping in the glass slide. A monolayer of a fatty acid produced at the water surface is deposited on the slide by lifting it. This layer is necessary to stabilize the molecular architecture of the slide stored in room air. In this case the dye chromophores D and E are kept separated at 50 Å distance by the hydrocarbon parts of the two layers.

The electron transfer from an excited dye molecule D to an acceptor A can be investigated in monolayer assemblies by measuring the fluorescence of the dye which is quenched in the presence of the electron acceptor in the arrangement of Fig. 1b. However, if the electron acceptor is separated from the dye by two fatty acid layers, the fluorescence of D is unquenched. Electron transfer (in contrast to energy transfer) is inefficient at 50 Å distance between dye and acceptor. If the acceptor is separated from the donor by only 20 Å by a fatty acid ester interlayer a 50 % quenching of the fluorescence is observed.

An appropriate molecule with π-electrons can be incorporated in the interlayer acting as a molecular wire. The rate of electron transfer from the excited dye to the electron acceptor is enhanced by a factor of 3.

Energy transfer and electron transfer taking place at distances below 100 Å are of interest for switching visible light in a range not restricted by the wavelength of light.

2. Signal Transfer Across Monolayer Assembly

Fig. 2 shows some types of arrangements studied by the monolayer assembly technique [2].

a) Two step energy transfer with the third dye absorbing at the fluorescence wavelength of the second;

b) Energy transfer from D to photochromic acceptor dye E (a merocyanine) which is transformed to a spiropyrane S by a 545 nm light flash and recovered by exciting the spiropyrane with a 366 nm flash. Only E acts as acceptor in the assembly. Therefore the fluorescence of D can be turned on and off by transforming E into S and S into E.

c) Energy transfer from proton sensor D to acceptor E. Dye D (the basic form of a coumarine) is fluorescent and acts as energy donor; dye DH^+ is non fluorescent; it does not transfer energy to E.

d) Electron donor dye D and electron acceptor A combined with another dye D' at 50 Å distance. D' is excited continuously and fluoresces. However, if D is excited by a flash it transfers its electron to A, the absorption band of A^- thus obtained overlaps with the fluorescence band of D and therefore A^- acts as energy acceptor of D'; the fluorescence is quenched. The operation of this switch depends on the exact position of each component molecule. D and A must be in contact and D' must be at 50 Å distanc At a larger distance energy transfer would not be possible anymore; at a smaller distance, the electron transfer from D' to A would take place.

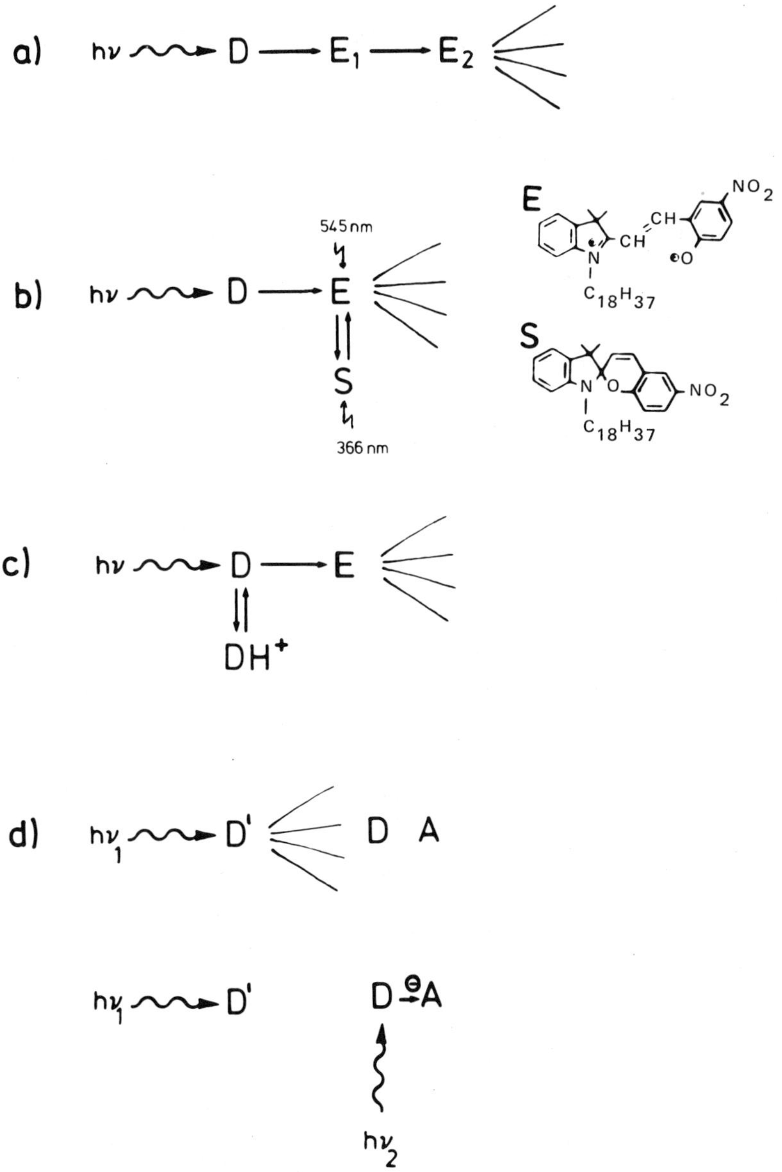

FIGURE 2. Arrangements of fluorescent donor (D) and
energy acceptor (E), or electron acceptor (A)
a) two step-energy transfer
b) arrangement with photochromic acceptor E
c) arrangement with proton sensing donor D
d) arrangement for switching fluorescence of
D' by electron transfer from D to A.

3. Visible Light Signal Transfer in the Submicrometer Range

It is of interest to demonstrate experimentally signal transfer in a range below the wavelength of light [3]. A small Platinum disc is inbedded in an epoxy-resin (Fig. 3a). The film thus obtained is contacted with a dye layer which has been deposited on a thin polystyrene film supported by a glass slide. The dye is excited until it has bleached. No bleaching takes place at the portion contacted with the platinum disc due to the energy transfer from the dye to the platinum. The dye monolayer is decorated with silver and observed in the electron microscope. A resolution of 500 Å is observed.

For a submicrometer-switching with visible light a 1000 Å light source is required. Such a source has been made by U. Ch. Fischer in this laboratory by covering a hemisphere of glass with a Tantal/Tungsten mirror with a 1000 Å hole (Fig. 3b). Another mirror with a ten times larger hole is made on a glass slide. By moving the glass slide along the hole (using a piezo quartz) a signal is obtained with a 1000 Å resolution.

4. Signal Transfer Within Monolayer

Signal transfer within a monolayer is of interest for future possibilities to connect submicrometer light switches. Monolayers of J-aggregating dyes are of particular interest for exciton, electron and hole migration. In a monolayer of N,N'-dioctadecyl-oxacyanine and octadecane of molar mixing ratio 1:1 the chromophores are in a com-

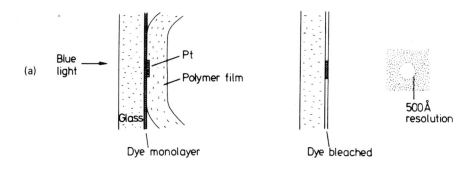

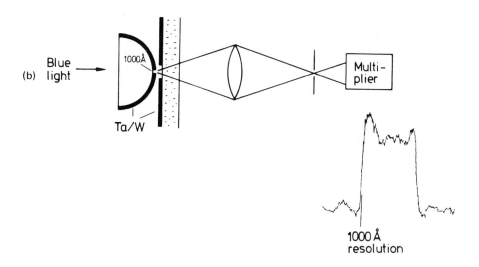

FIGURE 3.   Signal transfer in the submicrometer range
a) dye monolayer in contact with original
(Platinum disc) and separated after dye has
been bleached. Image developed by decoration.
Bleaching suppressed by energy transfer from
dye to Platinum.
b) 1000 Å light source and signal with 1000 Å
resolution.

pact brick-stone work-like arrangement and therefore strongly coupled. The dye fluoresces blue. By incorporating an exciton trap (the dye with two S instead of two O atoms) the blue fluorescence is quenched and a green fluorescence occurs even if only one molecule of the trap is among 10 000 molecules of the host, corresponding to an exciton motion over distances of the order of 1000 $\overset{\circ}{A}$.

The trap can be incorporated in a monolayer deposited on the monolayer of the J-aggregate. In this case it is somewhat less effective than in the case of being incorporated in the layer of the J-aggregate; the exciton is trapped within a distance of 200 $\overset{\circ}{A}$, and the trap can be an energy acceptor or an electron acceptor [4]. Arrangements of an electron donor, a J-aggregate and an electron acceptor can be investigated [5], and the results can be interpreted by assuming that the exciton is trapped at the acceptor because the excited electron is captured. It may fall back into the hole in the adjacent dye molecule in the aggregate, but there is a certain chance that this hole has already been filled due to hole migration through the aggregate. In this case the donor will stay oxidized and the acceptor reduced. This has been experimentally justified.

5. Information Storage in Monolayer

J-aggregates might be useful for information storage. The aggregates consist of two dimensional microcrystals that absorb light polarized in a certain direction and that do not absorb perpendicular to that direction. The absorption band is narrow and so high that 40 % of the in-

cident light is absorbed by a single monolayer in the
maximum. The two dimensional crystals can be oriented and
the orientation can be fixed, and thus it seems of interest to use such two dimensional crystals for information
storage. The size of the crystals can be varied widely by
variing the conditions of preparation. The chromophores
of dye molecules incorporated in a fatty acid monolayer,
present in the monomeric form, can be oriented by depositing the layer on a gypsum crystal, transferring it to
the water surface, and picking it up by a glass plate.
The orientation of the chromophore axes is indicated by
the dichroism of the layer. The direction of the chromophore of each dye molecule represents an information and
therefore an information storage at the molecular level
seems achievable.

6. Copying Information at Molecular Level, Separating Copy from Original

A dye monolayer can be deposited on a glass slide covered by a layer with oriented chromophores and the chromophores of this second dye are oriented in the direction
of the chromophores in the first layer. This indicates
sandwich dimer formation between the dyes in the two
layers. The layers can be separated, and the dichroism
observed in the two complementary monolayers is essentially unchanged indicating the possibility of copying information stored at the molecular level [6].

7. Monolayer Organizate Construction and Ideas about the Origin of Life

In constructing organized arrangements of molecules with
the monolayer techniques it is important to control con-

ditions exactly, to plan the intended molecular architecture carefully and to check the quality of the assembly
by measuring effects that depend strongly on the architecture, such as energy and electron transfer, tunneling
of electrons, spectroscopic effects indicating chromophore interactions. Considering the skill necessary for
constructing even the most simple molecular machinery it
seems obvious that the living machinery cannot simply result from a selforganization in a homogeneous or quasi
homogeneous prebiotic soup as usually assumed. Building
machinery needs a highly specific stimulus; such a stimulus might have been given on a prebiotic earth in very
particular regions where an appropriate set of conditions
was present by chance. A model pathway starting with reasonable conditions that lead to systems with the essential properties of living systems can be given [7]. In the
following we discuss a step in that pathway that might be
of interest in developing molecular devices, the emergence of a simple translation machinery.

   Let us first consider a short nucleic acid strand
that replicates under appropriate conditions. The temperature rises. Strand and replica separate, and in a periodically changing environment a periodic transition
occurs between phases where strands replicate and phases
where strands separate and then assume conformations related to the nucleotide sequence. We consider hairpin
conformations that must occur sooner or later under such
conditions. The replica of a hairpin (the (-) strands is
again a hairpin; the legs of the hairpin are twisted into a double helix and hairpins aggregate, since they can
interlock as indicated in Fig. 4a and it can be seen at

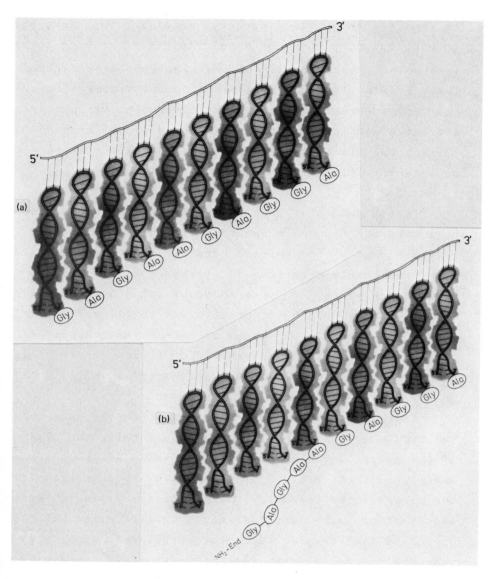

FIGURE 4. Assembler inducing aggregation of hairpin strands. (+) strands carrying glycine, (-) strands carrying alanine (a); formation of polypeptide by interlinking amino acids (b).

molecular models and by computer modelling that an aggregate of hairpins bound by base triplets to an open strand
(assembler) should originate fast and should be particularly stable. If an error appears in the replication process and the replicate is no more a correct hairpin or a
correct assembler it does not match, diffuses away and a
correct copy is introduced instead. Thus aggregation acts
as an error filter.

In the molecular model an activated amino acid can
be introduced in such a way that it binds specifically
to the 3'-end of the hairpin strand. From a detailed study of the steric relations it follows that two amino
acids (such as the two most abundant amino acids glycine
and alanine) should bind specifically to (+) and (-)
strand respectively. In the aggregate peptide bonds between adjacent amino acids should be formed resulting in
the formation of a polypeptide strand (Fig. 4b).

Hairpin (+) and (-) strands differ in the base in
the middle of the triplet in the loop forming the head
of the hairpin. The sequence of the amino acids in the
aggregate is therefore correlated with the sequence of
the bases in the assembler. This arrangement would represent a simple machinery that would translate the base-se-
quence in the assembler into the corresponding amino acid
sequence in the polypeptide. The model gives a plausible
explanation for many constructional details of the translation machinery in biosystems.

This particular model step concerning the origin of
life might be of interest in constructing molecular de-

vices. Another more general aspect of the origin of life
should also be considered in this connection. As mentioned in the beginning of this section, the emergence of
simple machinery is restricted to a very special environment with a very special program of periodical temperature changes that drives a complex sequence of processes.
Later the evolving systems get more and more complex,
since survival in regions with not exactly these special
conditions requires greater complexity but then offers
selectional advantages: The new and more elaborate system
does not have to compete with the others. In this way the
systems get increasingly independent on the initial environment by getting increasingly complex. The process of
liberation from the special sequence of environmental influences present in the original region results, sooner
or later, in the evolution of selforganizing systems,
where selforganization in the present context --organi-
zation according to an intrinsic program in contrast to
organization forced by a specific complex sequence of environmental events-- is a consequence and not a prerequisite of the evolution of adaptable organized molecular
systems.

8. Origin of Life and Ideas about Selforganizing Molecu-
lar Devices

In biosystems the genetic information is stored in a linear array. For molecular devices a two-dimensional storage of the blue print of the machinery seems more appropriate. According to sections 5 und 6 it seems feasible
to store information at the molecular level in a monolayer and to copy that information. Appropriate molecules

may adsorb at the corresponding charge pattern, chemically bind and by sequential adsorption of very special interlocking molecules a well planned complex arrangement
might then be obtained. One may speculate about obtaining by self-assembly a system of switches and molecular
wires as the translation product of the information
stored in the monolayer.

A difficulty in this approach is not only the necessity of developing appropriate molecules that are precisely interlocking, but also to write down the blueprint
on the monolayer at the molecular dimension. It should be
difficult to orient individually single small molecules
(such as dye chromophores) representing that blue print.
Also in biosystems meaningful information is not written
down at the molecular level (writing in the sense of
feeding a memory occurs at the multicellular level in the
nerve system), but genetic information evolves: a meaningless sequence transforms into a meaningful sequence
in the Darwinian process by multiplication, variation and
selection many times repeated.

One may speculate on applying this evolutionary
principle to insert meaningful information into a monolayer at the molecular level, i. e. to translate some
initial pattern in a monolayer by binding appropriate molecules, to select among translation products, to copy
the pattern, introducing sometimes a copying error, and
to repeat this many times. This seems rather complicated.

Another possibility would be to write the blue print
of the molecular device with an electron beam on a mono-

layer in a resolution of say 50 Å and then to adsorb
appropriate building blocks of some complexity that would
cooperate according to the pattern constituting the blue
print of the molecular device. The building blocks may be
complex molecules or well planned aggregates similar to
the aggregates discussed in section 7, that is aggregates
consisting of a strand with a planned sequence and adap-
ter molecules that would arrange according to that sequence. These adapters would carry appropriate functional
components such as switching and conducting elements (instead of amino acids as in section 7).(The binding of
protein molecules to 30 - 40 Å wide contamination lines
produced by the electron beam has been demonstrated in
this laboratory by Zingsheim [8] (a carbon film covered
with an insulin monolayer was used, and the lines were
made with a 60 kV scanning transmission electron microscope)).

By combining chemical synthesis with electron beam
surface modification this speculative approach attempts
to avoid the difficult problem of writing the blue print
of the molecular device in molecular resolution.

The chemist is challenged to construct organized systems of molecules by using the classical methods of synthesis but having in mind a new aim: to construct devices of molecular size as tools. Such entities would then
be as much objects of synthesis and products of inventive
faculty as the classical chemist's single molecule. Monolayer assembly design should be seen in connection with
this fundamental problem. The synthesis of molecules for
the purpose of being components of designed assemblies

should be a fascinating and promising new field in preparative chemistry [9].

REFERENCES

1. H. Kuhn, Pure Appl. Chem., 11, 345 (1965);
    H. Kuhn, D. Möbius and H. Bücher, Physical Methods of
    Chemistry, Arnold Weissberger and Bryant Rossiter eds.
    John Wiley & Sons, Inc., 1972, Vol. I, Pt. 3B;
    D. Möbius, Acc. Chem. Res., 14, 63 (1981);
    H. Kuhn, Thin Solid Films, 99, 1 (1983) (Literature
    Surway).

2. D. Möbius, Colloids and Surface in Reprographic Tech-
    nology, M. L. Hair and M. D. Croucher, American Chemi-
    cal Society, 1982, p. 93.

3. U. Ch. Fischer and H. P. Zingsheim, Appl. Phys.
    Lett., 40, 195 (1982); J. Vac. Sci. Technol., 19, 882
    (1981).

4. D. Möbius and H. Kuhn, Israel J. Chem., 18, 375
    (1979).

5. Th. L. Penner and D. Möbius, J. Am. Chem. Soc., 104,
    7407 (1982).

6. D. Möbius, Acc. Chem. Res., 14, 63 (1981); Photoche-
    mical Conversion and Storage of Solar Energy, J. Raba-
    ni Ed., The Weizmann Science Press of Israel, 1982,
    Part A;
    H. Kuhn, Pure & Appl. Chem., 53, 2105 (1981).

7. H. Kuhn and J. Waser, Angew. Chem., Int. Ed., 20,
    500 (1981); in Biophysics, Springer, New York, 1983,
    p. 830.

8. H. P. Zingsheim, Ber. Bunsenges. Phys. Chem., 80,
    1185 (1976); Scanning Electrons Microscopy Proceedings
    of the Workshop on Analytical Electron Microscopy,
    IIT Res. Inst., Chicago, 1977, Vol. I, p. 357.

9. H. Kuhn, in McGraw-Hill Yearbook of Science & Tech-
    nology, New York, 1977, p. 69.

21

Measurement of Nanosecond Dipolar Relaxation in
Membranes by Phase Sensitive Detection of Fluorescence

Joseph R. Lakowicz, Richard B. Thompson,
and Henryk Cherek / University of Maryland
School of Medicine, Department of Biological
Chemistry, Baltimore, Maryland

ABSTRACT

In this report we describe measurements of the time-dependent
orientation of polar molecules around the excited state of a fluorescent
molecule, a process we call solvent or dipolar relaxation. This process
results in partial dissipation of absorbed energy, and may thus be of
importance in molecular electronic devices which are triggered by light
absorption and which are intended to store information in the excited
state energy levels. We used phase-modulation fluorometric techniques
to measure the rates of dipolar relaxation around the fluorophore 2-(p-
toluidinyl)-6-naphthalene sulfonate (TNS). We found at low or high
temperatures in glycerol that solvent relaxation was either absent or
complete, respectively, prior to emission by the fluorophore. This is due
to the change in viscosity of the glycerol. At intermediate
temperatures, dramatic changes in the phase and modulation lifetimes

across the emission band were seen that could only have been caused by an excited state reaction. We were able to approximately measure the emission spectra of the relaxed and unrelaxed states by use of a new technique, phase sensitive fluorescence spectroscopy. The phase sensitive spectra and measured phase and modulation values were used to calculate the relaxation rates, which displayed a linear Arrhenius temperature dependence. We carried out a similar analysis for relaxation around TNS when it was bound to the lipid-water interface of phosphatidylcholine lipid bilayers. We found that the different phosphatidylcholines examined displayed activation energies that were similar among themselves, but less than in glycerol. The phase state of the acyl side chain region of the bilayer had no effect on the relaxation rate. The implications of these results with respect to molecular electronic devices are discussed.

INTRODUCTION

Molecular electronic devices have been contemplated that employ a variety of means to convey information, including solitons, phonons, and photons. Among the more promising approaches is that using photons or electrons to transfer information within Langmuir-Blodgett films. The work reported here describes a phenomenon of potential importance in the design and operation of such devices. We describe and quantitate a process wherein part of the energy of an excited fluorophore is lost-namely, solvent relaxation. Clearly, this process and in fact many other dissipative processes such as emission and radiation - less decay will have to be well controlled before high efficiency systems employing photons may be constructed.

Dipolar or solvent relaxation is a common phenomenon which accompanies fluorescence and it is the source of much of the sensitivity of fluorescence emission spectra to solvent polarity (1,2). In particular, many fluorophores display an increasingly red-shifted emission as solvent polarity is increased. The physical origin of this effect is that when the fluorophore absorbs a photon, its dipole moment will typically increase substantially. In response to this, solvent (or other surrounding) molecules will realign their dipoles with the new, stronger excited dipole of the fluorophore. The energy that the fluorophore loses in realigning the solvent creates a shift in its emission to lower frequency which may be estimated from the Lippert Equation:

$$hc \ \Delta\bar{\nu} \ = \ \frac{2\Delta f}{a^3} \ (\mu^* - \mu)^2 , \tag{1}$$

where $\Delta\bar{\nu}$ is the emission frequency shift attributable to relaxation, Δf is the orientation polarizability of the solvent, a is the radius of the cavity in which the fluorophore resides, and μ^* and μ are the dipole moments of the fluorophore in the excited and ground states, respectively.

In fluid solvents with low viscosities, the realignment process takes just picoseconds, and is thus complete before the average time of fluorescence emission - typically some nanoseconds. In more viscous solvents, or at the lipid-water interface of a membrane, the rate of dipolar relaxation is slower, and comparable to the rate of fluorescence emission. Such a system exhibits interesting effects when examined with time-resolved (pulse) fluorescence methods (2,3) or, in our case, phase-modulation fluorescence techniques (4-6). These effects are best understood by use of a Jablonski Diagram (Figure 1).

After excitation, E(t), and the extremely fast internal conversion to the lowest excited singlet state, the fluorophore is in what we call the unrelaxed or F (for Franck-Condon) state. This state decays via two routes: the ordinary emissive decay with rate = $1/\tau_F$, and via solvent relaxation to the spectrally different relaxed state, R. The relative

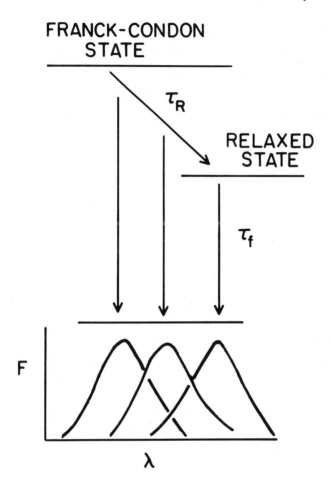

FIGURE 1.   Jablonski diagram of solvent relaxation.

intensity from the two states is a function of the two rates; when they
are comparable, the steady-state spectrum is the broadened one indicated by the dashed line. If impulse response data are used to generate
time-resolved emission spectra, and if relaxation is a pseudo-continuous
process resulting from many solvent-fluorophore interactions, the time-
resolved emission spectra shift gradually to longer wavelengths.

TNS is a fluorophore which is sensitive to solvent polarity and which localizes at the lipid-water interface region of membranes (7-9). Previously, time-resolved emission spectra of TNS-labeled membranes were used to determine the spectral relaxation rates (2,3). In contrast to the earlier time-resolved studies of spectral relaxation, we used the techniques of phase-modulation fluorometry and phase sensitive detection of fluorescence (10,11). Using the former method we show that time-dependent processes, and not spectral heterogeneity, are the origin of the wavelength-dependent fluorescence decays of TNS-labeled membranes. We then used phase sensitive detection of fluorescence to directly record the emission spectra of the relaxed and the unrelaxed states (6). Using these data the spectral relaxation times may be calculated.

THEORY

The analysis of excited state reactions (solvent relaxation is an example) by phase-modulation fluorescence spectroscopy is the subject of another paper in this volume (12) and other publications (4-6). We describe here the features of the theory relevant to solvent relaxation.

1. For a sample whose fluorescence intensity decays as a single exponential, the lifetimes measured by phase shift (τ_p) and demodulation (τ_m) will be equal. For a heterogenous group of fluorophores, $\tau_m > \tau_p$ or equivalently, $m/\cos \phi < 1$ where ϕ is the phase shift of the emission and m is the extent to which the emission is demodulated. For fluorescence which is the result of an excited state reaction (such as a relaxed fluorophore in the "R" state), $\tau_p > \tau_m$ and $m/\cos \phi > 1$.

2. When solvent relaxation occurs at a rate comparable to emission, the apparent phase and modulation lifetimes vary across

the emission spectrum. In particular, they increase with the emission wavelength, becoming larger as the R state predominates.

3. The relaxation rate of the solvent, k, may be calculated if we measure the intensities ($I(\lambda)$) and demodulation factors (m) of the F and R states, and the decay rate Γ (6):

$$\frac{I_F(\lambda)}{I_R(\lambda)} = \frac{m_F}{m_R} \frac{\tau_s}{\tau_o} = \frac{m_F}{m_R} \frac{\Gamma}{k} \qquad\qquad 2)$$

In this equation τ_s is the solvent relaxation time and $\tau_o = \Gamma^{-1}$ is the lifetime unaffected by solvent relaxation. To measure these values we assume that emissions from the F and R states predominate on the blue and red edges of the spectrum, respectively.

The intensities of the F and R states are measured using a new method: Phase Sensitive Fluorescence Spectroscopy. Since this technique has been described in some recent publications (1,10,11) we shall only discuss it briefly. This method allows measurement of the individual spectra of two components in a mixture, assuming their lifetimes differ. The modulated output of the phase fluorometer is passed to a lock-in amplifier, which gives a DC output proportional to the cosine of the phase angle difference between the components and the arbitrary reference of the lock-in. Practically speaking, one uses the method in a subtractive mode: the phase angle of the amplifier is adjusted to be 90° out of phase with the emission from one component. This entirely suppresses its emission, leaving the emission of the other component (Figure 2). Therefore, to measure the emission spectrum of the F state, the emission from the R state is suppressed by adjusting the phase angle of the lock-in amplifier to be 90° different from the phase of the red edge emission, which is dominated by the emission from the R

PHASE SUPPRESSION

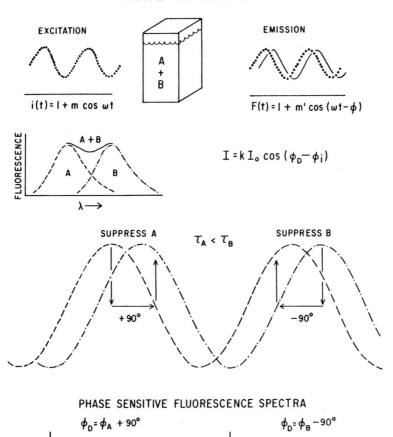

FIGURE 2.   Conceptual description of phase sensitive detection of fluorescence.

state. The converse operation is performed to measure the emission from the R state.

Finally, it is necessary to mention some caveats associated with our approach. We assume that emission from the F and R states predominates on the blue and red edges of the spectrum, respectively. Obviously, this is a good assumption only if the spectra are overtly shifted. Also, we used a two-state model to describe solvent relaxation. Clearly, this process may be multistep (13) or continuous (14). Finally, we assumed that the emissive rate of the fluorophor does not change as the solvent relaxes.

MATERIALS AND METHODS

Labeled vesicles composed of synthetic phosphatidylcholines were prepared as described previously (17). Detailed procedures for measurement of lifetimes (1,15) and phase sensitive spectra (10,11) were also recently described. All phase angles and demodulation factors were measured relative to the reference fluorophore p-bis[2-(5-phenyloxazolyl)]benzene in ethanol with a reference lifetime of 1.35 ns (15). This procedure minimizes the artifacts which appear when scattered light is used to estimate the phase and modulation of the incident light. These artifacts may originate with non-homogeneous modulation of the incident light by the ultrasonic modulator (E. Gratton, personal communication). The effect of Brownian rotation on the measured lifetimes was avoided by using vertically polarized excitation and an emission polarizer oriented 54.7° from the vertical direction (16). The measured phase angles (ϕ) and demodulation factors (m) were used to calculate the apparent phase (τ_p) and modulation (τ_m) lifetimes using

$$\tan \phi = \omega \tau_p \tag{3}$$

$$m = (1 + \omega^2 \tau_m^2)^{-\frac{1}{2}} \tag{4}$$

These apparent lifetimes are convenient because they are more familiar than are the values of ϕ and m.

The rotational correlation times (τ_c) for TNS in glycerol were calculated using the Perrin equation

$$\frac{r_o}{r} = 1 + \tau_o / \tau_c \tag{5}$$

where r_o is the fluorescence anisotropy in the absence of rotational diffusion, r is the steady state anisotropy, and τ_o the fluorescence lifetime. Using our experimental conditions we found $r_o = 0.357$.

RESULTS

Fluorescence Spectral Properties of TNS in Glycerol

Interpretation of the data for TNS-labeled vesicles is facilitated by examination of the data for TNS in glycerol. This particular solvent is useful because by variation of temperature its viscosity can be adjusted over a wide range, thereby changing the rate and extent of solvent relaxation. Steady state emission spectra measured at various temperatures are shown in Figure 3. At low temperature (-55°C) the emission maximum is near 410 nm. As the temperature is increased the emission spectrum shifts progressively towards longer wavelengths. This shift is essentially complete by 30°C (Figure 3). Such temperature-dependent spectra are a clear indication of solvent relaxation. This solvent relaxation is time-dependent, and is expected to alter the apparent fluorescence lifetimes, especially when measured at discrete wavelengths.

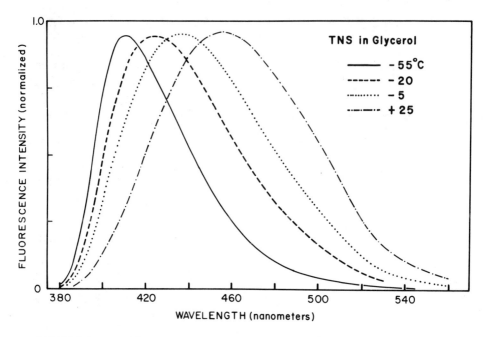

FIGURE 3.   Fluorescence emission spectra of TNS in glycerol.

Apparent fluorescence lifetimes for TNS in glycerol, measured at various temperatures and emission wavelengths, are shown in Figure 4. In the presence of time-dependent solvent relaxation the intensity at any given wavelength is not expected to decay exponentially. As a result the phase (τ_p) and modulation (τ_m) lifetimes, calculated from the measured phase angles (ϕ) and demodulation factors (m), are only apparent quantities. At the lowest ($< -50°$ C) and highest ($> 25°$ C) temperatures solvent relaxation is either much slower than emission or complete prior to emission, respectively. Thus, the emission is predominantly from a single state and is expected to decay mostly as a single exponential. This was observed at -52° C and +55° C. At these temperatures the phase and modulation lifetimes are essentially identical, and these values are nearly independent of emission wavelength. The residual wavelength-dependence seen at -52° C is probably a result of some solvent relaxation

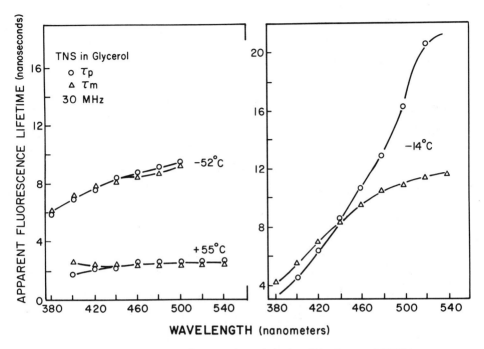

FIGURE 4.   Apparent phase and modulation lifetimes of TNS in glycerol.

at even this low temperature. In contrast, at intermediate temperatures, -14°C in this case, the apparent lifetimes are strongly dependent upon emission wavelength. Specifically, the apparent phase lifetime increases by 16 ns and the apparent modulation lifetime by 7 ns as the observation wavelength is increased from 380 to 520 nm. The origin of these wavelength-dependent lifetimes can be explained in an intuitive fashion. On the blue side of the emission the intensity is decaying by return to the ground state with a rate Γ and by relaxation with a rate k. At longer wavelengths the apparent lifetime is greater because one is selectively observing those fluorophores which have relaxed prior to emission.. These relaxed fluorophores display longer apparent lifetimes because the apparent phase lifetime is given by the sum of the phase angles of the F and the R states. Similar reasoning explains the longer apparent modulation lifetimes at the larger emission wavelengths (4,5,12).

The lifetimes seen for TNS in glycerol at -14°C (Figure 4) are also indicative of the nature of the relaxation process. In particular, both the apparent phase and modulation lifetimes decrease continuously at shorter observation wavelengths. Such results are consistent with the predictions of the continuous relaxation model of Bakhshiev et. al., (14), and with the measurements of Brand and co-workers (2) who used time-resolved measurements and found continuous relaxation in glycerol. If the relaxation was a two-state process, and if the spectral separation of states was adequate for selective observation of each state on the edges of the emission, then regions of constant apparent lifetime are expected and observed at the short and long emission wavelengths (4,5). Evidently, our data are not adequate for an unambiguous distinction between a continuous and a two-state process. However, we note that in instances where spectral separation is small, neither the time-resolved method nor the phase-modulation method is adequate to distinguish between the continuous and discrete models.

Conceivably, the wavelength-dependent lifetimes shown in Figure 4 could be due to the fluorophore being present in two different environments, with the fluorophores emitting at longer wavelengths having the longer lifetime. Fortunately, these cases can be distinguished by phase-modulation fluorometry. This requires measurement of both the phase angle and the demodulation factor. As stated in the Theory section, $m/\cos \phi < 1$ for the case of heterogeneity. This is clearly not the situation for the wavelength-dependent lifetimes of Figure 5. It is apparent that the $m/\cos \phi$ values on the red side of the emission exceed unity at those temperatures where the emission maximum is also the most temperature sensitive. The increasing values of $m/\cos\phi$, particularly in excess of 1.0, reflect an increasing contribution of the

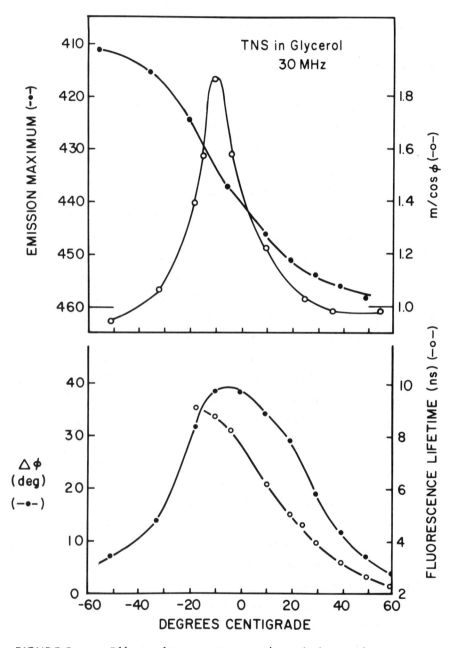

FIGURE 5.   Effects of temperature on the emission maximum,
m/cos φ (510 nm), lifetime and Δ φ for TNS in glycerol. The lifetime
was measured through a Corning 4-69 filter, which passed the entire
emission.

relaxed state to the total emission at the longer wavelengths. The phase difference between the red and blue sides of the emission ($\Delta \phi$) is also maximal near -10°C. These data may be regarded as typical of those expected for TNS in an environment for which the relaxation time is comparable to the fluorescence lifetime. In general, measurements at various temperatures reveal whether the spectral relaxation rate is faster or slower than the decay rate.

The relaxation times for spectral relaxation can be estimated from the phase sensitive fluorescence spectra (equation 2). Typical phase sensitive spectra for TNS in glycerol are shown in Figure 6. At low temperatures suppression on either side of the emission results in almost complete suppression of the entire emission spectrum. This occurs because at these temperatures the emission is from a single state and the phase angle is nearly constant across the emission spectrum (Figure 5). Similarly, at high temperature (50°C) suppression of either side of the emission also results in suppression of the entire emission. Of course, this is because at 50°C the emission is nearly all from the relaxed state. Contrasting results are found at intermediate temperatures (-35 to 25°C). Suppression on red and blue sides of the emission results in phase sensitive spectra which are similar to the steady state spectra seen at low and high temperatures, respectively (Figure 3). As the temperature is increased the intensity of the relaxed spectrum is increased, relative to that of the unrelaxed spectrum. We note that the phase sensitive emission maxima are not independent of temperature. This result indicates that either our simple procedure of red or blue suppression does not result in complete suppression of each state, or more likely, that the simple two-state model is not completely adequate to describe spectral relaxation of TNS in glycerol. Nonetheless, phase suppression does result in an approximate separation of these states.

We used the relative phase sensitive intensities at each temperature, along with the measured lifetimes and demodulation

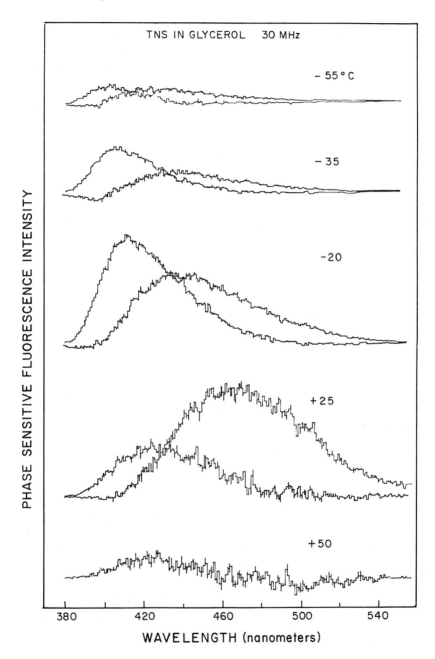

FIGURE 6.   Phase sensitive emission spectra for TNS in glycerol.
At each temperature the emission was suppressed on the blue and red
sides of the emission.

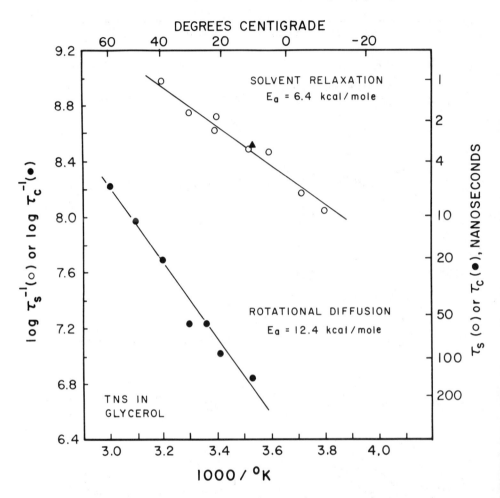

FIGURE 7.   Arrhenius plot for solvent relaxation and rotational diffusion of TNS in glycerol.   The triangle (Δ) indicates the spectral relaxation time measured by DeToma et. al., (2) using time-resolved emission spectroscopy.

factors, to calculate the solvent relaxation time (equation 2).   These calculated solvent relaxation times are shown as an Arrhenius plot in Figure 7.   Also shown are the rotational correlation times (τ_c) of TNS as calculated from the Perrin equation (equation 5).   Evidently, solvent relaxation occurs more rapidly than rotational diffusion, and with a smaller activation energy (6.4 versus 12.4 kcal/mole, respectively). This disparity is probably a result of the smaller molecular motions needed for spectral relaxation than for rotational diffusion. These results are in agreement with similar data obtained for a tryptophan derivative in a viscous solvent (6). In total these results indicate that the spectral properties of TNS are highly dependent upon its time-dependent interaction with the solvent. These same spectral properties for TNS, when bound to phospholipid bilayers, should reflect the dynamic properties of these membranes.

TNS-Labeled Phospholipid Vesicles

Representative steady state spectra of TNS bound to vesicles of dimyristoyl phosphatidylcholine are shown in Figure 8. Compared to the emission spectra of TNS in glycerol, only a small temperature dependent shift was observed. For TNS bound to the other lipid vesicles, dioleoyl- or dipalmitoyl-L-α-phosphatidylcholine, the temperature-dependent shifts were comparable or smaller. Such spectra were described previously in greater detail (17). The small extent of the spectral shifts suggests immediately that the rates of solvent relaxation in bilayers are not strongly dependent upon temperature. We used measurements of the apparent phase and modulation lifetimes, and phase sensitive emission spectra, to estimate the spectral relaxation rates for TNS bound to lipid bilayers.

Wavelength-dependent phase and modulation lifetimes, and m/cos ϕ values for TNS-labeled dimyristoyl phosphatidylcholine vesicles are shown in Figure 9. As was found for TNS in glycerol at intermediate temperatures, these lifetimes generally increase with increasing wavelength. For these vesicles this effect appears to be maximal near 20°C, which is comparable to the phase transition temperature of dimyristoyl phosphatidylcholine, 24°C. By comparison and analogy with TNS in glycerol, these results suggest that the spectral relaxation in

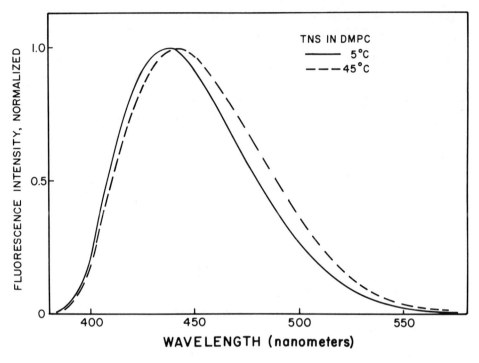

FIGURE 8.   Emission spectra of TNS-labeled vesicles of dimyristoyl
phosphatidylcholine.

these membranes is almost complete prior to emission at temperatures
above 35°C.

Phase sensitive spectra of TNS-labeled dimyristoyl phosphatidylcholine vesicles are shown in Figure 10. As for TNS in glycerol, the red
and blue suppressed spectra are comparable to those expected for the
unrelaxed and the relaxed emission of TNS, respectively. However, in
contrast to TNS in glycerol, the relative phase sensitive intensities do
not vary widely with temperature, and furthermore, both states are still
observable at the lowest (5°C) and highest (45°C) temperatures. These
results indicate that at these temperatures the membrane relaxation
rates are comparable to the fluorescence lifetime of TNS, and that the

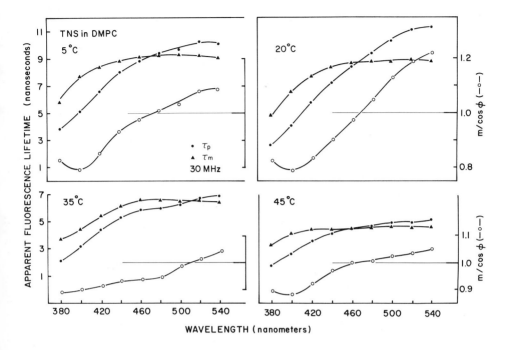

FIGURE 9.   Wavelength-dependent lifetimes and m/cos φ values for TNS-labeled dimyristoyl phosphatidylcholine vesicles.

ratio of the lifetime to the relaxation time does not vary greatly from 5 to 45°C (equation 2). Also, the phase sensitive spectral distributions do not vary greatly with temperature. This result probably indicates our simple two-state model is adequate to describe the data for the model membranes.

Similar phase sensitive spectra were obtained for TNS-labeled vesicles of both dioleoyl phosphatidylcholine and dipalmitoyl phosphatidylcholine. The relative phase sensitive intensities and spectral distributions varied only slightly with temperature. These results, along with the measured phase and modulation values, were used to calculate the rates of spectral relaxation in these bilayers (Figure

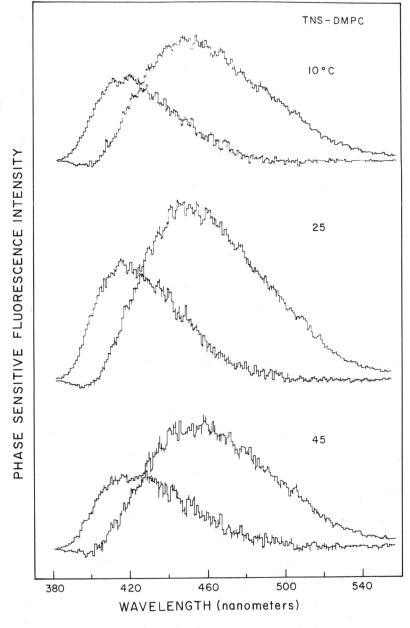

FIGURE 10.   Phase sensitive emission spectra of TNS-labeled
dimyristoyl phosphatidylcholine vesicles.

11). Perhaps surprisingly, but in agreement with previous studies (17), the rates are comparable among the various phosphatidylcholine vesicles. Furthermore, at the lipid-water interface region where TNS is probably localized, the spectral relaxation times are not strongly

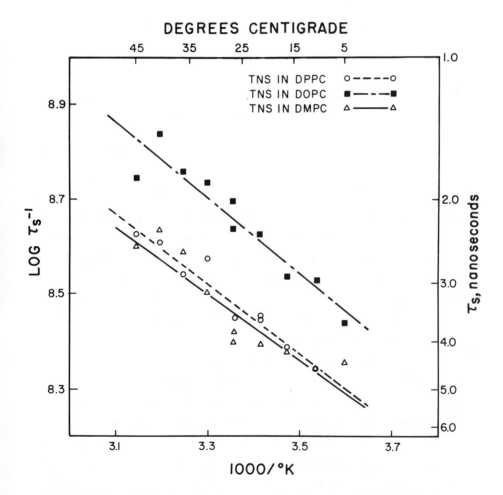

FIGURE 11. Arrhenius plot for spectral relaxation of TNS-labeled vesicles. DM PC, DO PC, and D PPC denote dimyristoyl, dioleolyl, and dipalmitoyl-phosphatidyl cholines, respectively.

TABLE I.   Activation Energies for Spectral Relaxation of TNS-
labeled lipids

Lipid	E_a (kcal/mole)
DO PC	3.7
DM PC	3.2
D P PC	3.2
TNS in Glycerol	6.4

affected by the phase transition which occurs in the acyl side chain region of these membranes (18).   We conclude that the dynamic properties of the polar head group region of these phosphatidylcholine vesicles are not subject to the dramatic phase changes found deeper in the bilayer. We note that only small molecular motions are needed for spectral relaxation (Figure 7 and the related discussion), and it is these smaller motions which appear to be mostly independent of the phase transitions of the membranes. The probability that only small molecular motions are the origin of spectral relaxation is also evident from the activation energies for spectral relaxation (Table I).   The activation energies are near 4 kcal/mole, and are relatively independent of the acyl side chain length.   We note that the apparent activation energies are comparable in magnitude to a few hydrogen bonds, which is reasonable if one considers the prevalance of such bonds in this region of the membrane.

DISCUSSION

We have shown how phase-modulation fluorescence spectroscopy, coupled with phase-sensitive detection, may be used to quantify the dipolar relaxation around a fluorophor in a viscous solvent and in phospholipid vesicles. While our results have obvious import in understanding membrane-mediated biochemical events, their relevance to the construction of molecular electronic devices may be less clear. Dipolar relaxation is a general phenomenon that will occur any time there is a localized change in dipole moment of a molecule in a non-rigid, somewhat polar medium. In particular, we have measured it at the lipid-water interface of phospholipid vesicles, which are not only model systems for cell membranes, but are very similar to Langmuir-Blodgett films as well. Hence, we feel it may be a non-trivial route of energy loss in such systems. Since they are likely to be complex, but utilize only small packets of energy (as little as one photon or soliton), molecular electronic devices must be very efficient. Therefore, measuring and controlling such processes may be crucial to constructing molecular electronic devices which depend upon excited states.

In a more philosophical vein, we should note that there is a certain biomimicry to using Langmuir-Blodgett films for molecular electronic devices. More precisely, virtually all biological phototransducers are housed in membranes; this includes chloroplasts, retinal rod outer segments, and photosynthetic bacteria. It is interesting to note that these biological systems have many of the properties we expect molecular electronic devices to have: they are small and may be densely packed, they are supremely efficient, and they are easy to make by the billion. Indeed, we may ultimately find that many of the structural motifs evolved in nature at the molecular level will find use in contructing molecular electronic devices.

AC KNOWLEDGMENTS

This work was supported by Grants PCM 80-41320 and PCM 81-06919 from the National Science Foundation and Grant GM-29318 from the National Institutes of Health. Correspondence should be addressed to J.R.L., who is an Established Investigator of the American Heart Association.

REFERENCES

1.   J.R. Lakowicz, Principles of Fluorescence Spectroscopy, Plenum, New York, 1983.
2.   R.P. DeToma, J.H. Easter, and L. Brand, J. Am. Chem. Soc., $\underline{98}$, 5001 (1976).
3.   J.H. Easter, R.P. DeToma, and L. Brand, Biochim. Biophys. Acta, $\underline{508}$, 27 (1978).
4.   J.R. Lakowicz and A. Balter, Biophys. Chem. $\underline{16}$, 99 (1982).
5.   J.R. Lakowicz and A. Balter, Biophys. Chem. $\underline{16}$, 117 (1982).
6.   J.R. Lakowicz and A. Balter, Photochem. Photobiol. $\underline{36}$, 125 (1982).
7.   G.L. Jendrasiak, and T.N. Estep, Chem. Phys. Lipids $\underline{18}$, 181 (1977).
8.   W. Lesslauer, J. Cain, and J. K. Blasi, Biochim. Biophys. Acta $\underline{241}$, 547 (1971).
9.   G. K. Radda, Biochim. J. $\underline{122}$, 385 (1971).
10.   J.R. Lakowicz and H. Cherek, J. Biochem. Biophys. Methods $\underline{5}$, 19 (1981).
11.   J.R. Lakowicz and H. Cherek, J. Biol. Chem. $\underline{256}$, 6348 (1981).
12.   J.R. Lakowicz and A. Balter, This Volume, p
13.   W. Rapp, H.H. Klingenberg, and H.E. Lessing, Ber. Bunsenges, $\underline{75}$, 883 (1971).
14.   N.G. Bakhshiev, Yu. T. Mazurenko, and I.V. Piterskaya, Opt. Spectrosc., $\underline{21}$, 307 (1966).
15.   J.R. Lakowicz, H. Cherek, and A. Balter, J. Biochem. Biophys. Methods, $\underline{5}$, 131 (1981).

16.   R.D. Spencer and G. Weber, J. Chem. Phys. 52. 1654 (1970).

17.   J.R. Lakowicz and D. Hogen, Biochemistry, 20, 1366 (1981).

18.   B.R. Lentz, Y. Barenholz, and T.E. Thompson, Biochemistry 15, 4521 (1976).

22

Chemical Communication Involving Electrically Stimulated Release of Chemicals from a Surface

Larry L. Miller and Aldrich N. K. Lau/ Department of Chemistry, University of Minnesota, Minneapolis, Minnesota

Consideration of molecular electronics usually presupposes that specific molecular entities, capable of performing electrical functions, can be built on surfaces. Work in our laboratories over the last several years is related to this concept in that we have developed techniques for preparing molecular surfaces on conducting solids. These new materials are used in solution as chemically modified electrodes. Their tailor-made molecular surfaces communicate electrically with the conductor and serve specific chemical functions. In this paper a brief review of these surface modification methods will be followed by a description of recent work in which we have developed an electrode which will release chemicals in response to an electrical signal. It is of particular significance when used as primitive analog of a synapse. Thus, it demonstrates how chemical communication might take place between spacially separated surface units.

In 1975 my coworkers described the first example of an electrode whose surface had been chemically modified to provide speci-

fic chemical properties.[1] In that example, a carbon surface was
modified by covalantly attaching a chiral amino acid derivative to
surface oxide sites. This electrode then had molecular surface
structure and it was used as part of an electrochemical cell to
perform an asymmetric electroorganic reaction. In other words, it
provided chiral selectivity, not otherwise available.

   Since then, many research groups have contributed to develop
methods for surface modification which provide molecular electrode
surfaces.[2] Those electrodes which hold electroactive groups
(capable of oxidation/reduction) have attracted the most attention.
It has been shown in a large number of cases that surface confined
electroactive species can communicate electrically with an underlying conductor-like carbon, platinum or tin oxide. It is known,
for example, that the electrochemical potentials of redox couples
are essentially the same if the species is in solution or attached
to a surface. Electrodes have been surface modified with monomeric
species by covalent attachment and by physical adsorption. Polymer
layers on surfaces have been extensively studied. These can be
formed by dip-coating or electropolymerization, for example. Ion
exchange polymers have been used to incorporate charged redox
species into the polymer film and polymers have been synthesized
holding pendant redox groups. A simple example is poly(-p-nitro)-
styrene.[3] A film of this polymer on platinum, used in a non-
aqueous solution, will be charged by injection of electrons into
the nitrostyrene units at about -1.5 V (SCE). This charging is,
naturally, accompanied by incorporation of counter ions from the
solution. If the potential is returned to values more positive
than -1.5 V the process is reversed and the film is neutralized.

   More recently, we have become interested in developing an
electrode which could be used to deliver biomedically useful
chemicals from its surface. Initial experiments have been directed toward the release of neurotransmitters.[4] This device is

constructed by coating an inert electrode with a suitably designed polymer. The polymer is strongly held to the solid surface of the electrode and the neurotransmitter is held covalently to the polymer backbone. Release of the neurotransmitter will only occur in response to a suitably negative electrode potential, which will cause cathodic cleavage of the covalent bond to the polymer backbone. If a microelectrode based upon this design could be properly developed, it would be useful for neuroscientists interested in testing the action of neurotransmitters and drugs at the single neuron level.

In the context of chemical communication these electrodes may serve as primitive analogs for the presynaptic terminal. The conduction of electrical signals in vivo involves chemical communication between neurons. At a synapse, transmitter substances are released from the presynaptic terminal in response to a change in the neuron's potential. These neurotransmitters diffuse across a small volume of solution, the synaptic cleft, and then are detected at receptor sites on the post-synaptic membrane. We have shown that a solid electrode, modified with a thin layer of a suitable polymer, will similarly respond to a change in potential to release the neurotransmitter, dopamine. The released dopamine can be detected at a nearby second electrode. In this way the primitive synaptic analog can be constructed.

In these experiments, a carbon electrode coated with an ultra-thin film of polymer I served as the delivery electrode. At sufficiently negative potentials, cathodic reduction cleaved the

amide bond of I, releasing the dopamine into a small volume of
electrolyte solution.

Experimentally, as shown in Figure I, a glassy carbon disk
electrode, set into Teflon, stood upright and typically 5 µl of
0.1 M KCl, pH 7 phosphate buffer, was syringed onto it. A second
carbon disk (0.08 cm^2) which had been coated with I (2.4 x 10^{-10}
mol) was lowered, using a micromanipulator, until a thin layer of
solution was in contact with both electrodes. Two micropipets
(tip diam. 3 µm) were inserted into this solution. Each pipet led
to a syringe. In one syringe barrel was a saturated calomel (SCE)

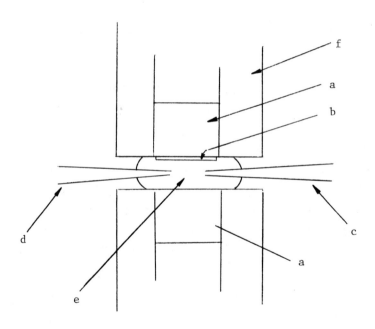

Figure 1 Thin layer cell: (a) glassy carbon disk; (b) polymer
coating; (c) glass micropipette filled with 3M KCl and connected to
a syringe containing 3M KCl and counter electrode; (d) glass micropipette filled with 3M KCl and connected to a syringe containing 3M
KCl and the reference electrode; (e) electrolyte (3 - 100 µl);
(f) teflon tubing.

reference electrode, in the other a platinum wire auxiliary electrode. As soon as the cell was assembled, the potential of the
coated disk was cycled from 0.0 to -1.0 V and back to 0.0 V. The
lower disk was then made the working electrode. Cycling the
potential of the lower disk from -0.1 to +0.4 V and back gave a
voltammogram identical to that of authentic dopamine in solution.
Sweeping the potential at 100 mV·s^{-1} gave an anodic peak at 0.16 V,
i_p^a = 0.25 μA, and, on the return half-cycle, a cathodic peak at
0.1 V, i_p^c = 0.20 μA. The presence of dopamine in solution was
confirmed by HPLC.

In a control experiment, the cell was assembled as usual, but
no current was passed. No dopamine was released. Indeed, the
potential could be made as negative as -0.6 V without causing
release. This observation is in concert with the voltammogram
observed for I, which was a cathodic peak at -0.99 V. In further
experiments the amount of dopamine released was quantitated under
a variety of conditions.

Thus, we have demonstrated for the first time that molecular
entities can be released from a surface in response to an electrical signal. This approach can be used for chemical communication
between spacially separated units.

REFERENCES

Journal:
1. B. F. Watkins, J. R. Behling, E. Kariv and L. L. Miller,
   J. Am. Chem. Soc., 97, 3549 (1975).

2. For leading references see:
   M. Fukui, A. Kitani, C. Degrand and L. L. Miller, J. Am. Chem.
   Soc., 104, 28 (1982).

   D. A. Buttry and F. C. Anson, J. Am. Chem. Soc., 104, 4824
   (1982).

   G. S. Calabrese, R. M. Buchanan and M. S. Wrighton, J. Am.
   Chem. Soc., 104, 5786 (1982).

J. S. Facci, R. H. Schmehl and R. W. Murray, J. Am. Chem. Soc. 104, 4959 (1982).

R. W. Murray, Acc. Chem. Res., 13, 135 (1980).

3. M. R. Van DeMark and L. L. Miller, J. Am. Chem. Soc., 100, 3223 (1978).

4. L. L. Miller, A. N. K. Lau and E. K. Miller, J. Am. Chem. Soc. 104, 5242 (1982).

23

Langmuir-Blodgett Films of Substituted Aromatic Hydrocarbons and Phthalocyanines

Richard A. Hann, William A. Barlow, James H. Steven and
Barbara L Eyres/Imperial Chemical Industries plc, Runcorn,
Cheshire, U.K.

Martyn V. Twigg/Imperial Chemical Industries plc, Billingham,
Cleveland, U.K.

Gareth G. Roberts/Department of Applied Physics and Electronics,
The University, Durham, U.K.

Presented by Donald J. Freed, ICI Americas, Wilmington,
Delaware

INTRODUCTION

Langmuir and Blodgett in their pioneering work showed that fatty
acids could be spread as monolayers on a water surface and then
deposited onto a solid substrate. Much subsequent work has been
concerned with modified fatty acid type materials, with one or more
long alkyl chains in the molecule. The application of this type of
compound in the molecular electronics area will be limited because
the long alkyl groups inevitably confer insulating properties on the
deposited layers.

    We wanted to obtain Langmuir-Blodgett (L-B) films with semi-
conducting properties in order to extend the range of usefulness of
the technique. Classical L-B materials have a hydrophilic headgroup,
which interacts with the water surface and a hydrophobic group
(usually incorporating alkyl chains) which ensures insolubility. We

(I)

have synthesised (1) a wide range of anthracene derivatives in order
to determine which structural features are necessary for good monolayer stability and multilayer deposition, while minimising the size
of attached groups which will inevitably dilute the semiconducting
properties of the anthracene.

We have also investigated L-B deposition of phthalocyanine
derivatives. Phthalocyanines are of particular interest because of
their strong absorption in the visible region of the spectrum and
because their electrical properties may be modified by changing the
central metal atom. This also acts as a centre for interaction with
other chemical species, thus allowing the electrical properties to
be modified by the chemical environment. Previous workers have been
unable to spread phthalocyanine monolayers on the Langmuir trough
because of the extremely low solubility of this type of compound.

They are non classical L-B materials in as much as they do not have clearly defined head and tail groups.

In a previous communication (2) we reported on the formation of multiple layers of phthalocyanines at the air-water interface and the deposition of these films onto substrates. In that work we used two approaches to obtain solutions. In the first approach we used dilithium phthalocyanine, which demetalated to phthalocyanine on the water surface. In the second approach we used tetra-t-butyl-phthalocyanine (TBP) (I), which is soluble in organic solvents (3). We now report on the formation of monolayers from the zinc and copper derivatives of (I), and the deposition of these onto substrates. These monolayers represent a significant advance in the controlled deposition of electrically active organic species.

EXPERIMENTAL

Surface-pressure/area isotherms were measured and Langmuir-Blodgett films deposited at 18°C on a constant perimeter trough designed and made by our group. All measurements on anthracene derivatives were carried out under a yellow light. The experimental procedures and precautions are described in ref. 4. Chemical synthesis of the anthracene derivatives is described in a supplement available on request from the authors.

ANTHRACENE DERIVATIVES

Earlier work (5) had indicated that anthracene derivatives of general formula (II) need a long alkyl chain (X $\geqslant C_{12}H_{25}$) for stability at the air-water interface. Because of the known (6) insulating properties of alkyl groups we were anxious to minimise the chain length, and therefore reinvestigated the stability of monolayers prepared from the series of anthracene derivatives of

(II)

formula (II) where $X = CH_2CH_2.CO_2H$ and $R \equiv C_nH_{2n+1}$ for $n = 4, 6, 8,$ 12.

Effect of Alkyl Chain Length

Surface-pressure/area isotherms were measured on a subphase of pH 6 after a saturation procedure whereby monolayers were repeatedly spread and compressed until a stable film was obtained. This procedure relies on the fact that marginally soluble molecules on the water surface dissolve much more rapidly at high surface pressures. The resultant isotherms (Fig. 1) show that in contrast to Stewart's work we were able to obtain stable condensed mono-layers for $n = 6$ and $n = 8$. However, under these conditions the compound with $n = 4$ did not show a condensed phase and could not sustain surface pressures above 20 mN m^{-1}.

Effect of Subphase pH

In order to further improve monolayer stability, we added cadmium chloride to the subphase as this is known (7) to insolubilise the film. A dramatic increase in stability was obtained by reducing the pH from the commonly used value of 6. Figure 2 contrasts the

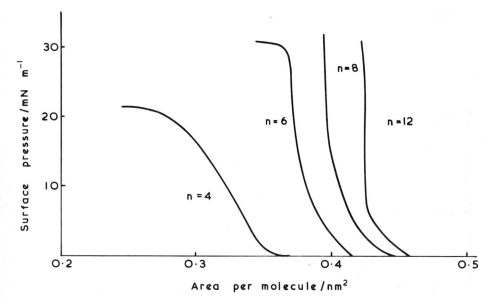

FIGURE 1. Surface-pressure/area isotherms of anthracene propanoic acids (II : R ≡ C_nH_{2n+1}; X ≡ $CH_2.CH_2.CO_2H$) for various values of n at a subphase pH of 6.

isotherms obtained with the anthracene derivative (n = 4) henceforth known as C4 anthracene at pH values of 4, 6 and 8. This clearly shows the improved stability at low pH values. The n = 12 derivative (line d) gives indistinguishable isotherms at all three pH values. C4 anthracene was selected for particular study because it was the shortest member of the series that we could stabilise on the water surface; the derivative with n = 3 gave unstable films under all the conditions we examined.

Effect of Varying Headgroup (X)

The structure of the head group has a profound effect on monolayer properties. In view of the stability of C4 anthracene we investigated the effect of head group variation on a range of compounds

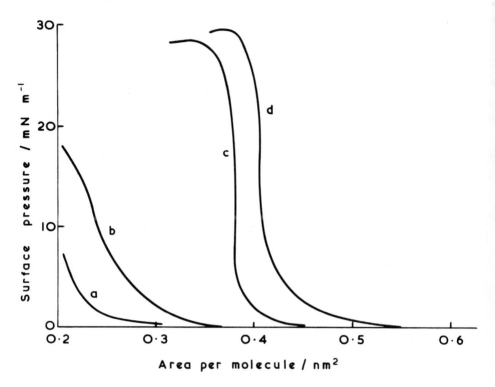

FIGURE 2. Surface-pressure/area isotherms of C4 anthracene at a
subphase pH of 8 (curve a), pH 6 (curve b) and pH 4 (curve c). The
isotherm of the long chain homologue (II : R ≡ $C_{12}H_{25}$; X ≡ $CH_2.CH_2.$
CO_2H) is independent of pH (curve d).

with n = 4 and varying X; the isotherms and structures of these are
illustrated in Fig. 3. It is worth noting several detailed
comparisons.

(a) As the chain length of the acid is reduced in the series
X ≡ $CH_2CH_2CO_2H$, CH_2CO_2H, CO_2H, the monolayer stability decreases
and no condensed phase is observed, probably because the reduction
in size of the hydrophobic group increases the molecule's
solubility. This is consistent with the poor isotherm observed for
the diacid (X ≡ $CH_2CH_2(CO_2H)_2$) which is expected to have increased
solubility by virtue of the second polar head group.

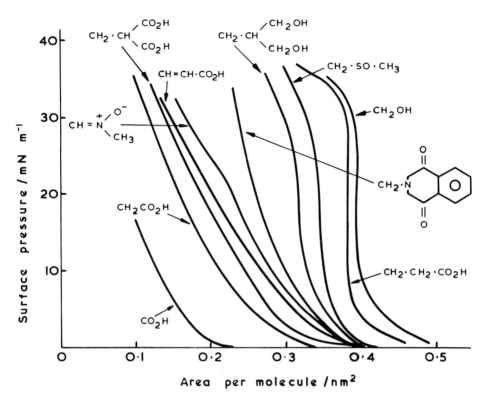

FIGURE 3.   Surface-pressure/area isotherms of anthracene derivatives
(I : R ≡ C$_4$H$_9$; X as indicated).

(b)   Introduction of unsaturation into the carbon chain
(X ≡ CH:CHCO$_2$H) gives a similar degradation of monolayer properties
and probably reflects the less hydrophobic character of unsaturated
carbon atoms.

(c)   The short-chain alcohol (X ≡ CH$_2$OH) has a well-defined
solid region and good stability. This is explained by the less
hydrophilic nature of the alcohol group relative to the carboxylic
acid group. The intermediate character of the diol (X ≡ CH$_2$CH(CH$_2$
OH)$_2$) reflects the balance between lower hydrophilicity and the
effect of introducing a second head group.

(d) Two other types of head group give compounds with promising monolayer properties. These are sulphoxide ($X \equiv CH_2SOCH_3$) and phthalimide ($X \equiv CH_2N:(CO)_2C_6H_4$).

Preparation of Deposited Multilayers

Previous attempts to prepare multilayers by deposition of alcohols have been unsuccessful (8) and we were similarly unable to prepare built up films from the alcohol ($II: R \equiv C_4H_9; X \equiv CH_2OH$) despite its exemplary monolayer properties. We have, however, deposited multilayers of C4 anthracene up to 500 molecules thick using a subphase of pH 4.5 and a cadmium chloride concentration of $2.5 \times 10^{-4}M$. The cadmium ions appear to act as bridges between the organic molecules; if the pH or cadmium concentration is too high then a relatively large number of cadmium ions are incorporated into the film, and it becomes very viscous on the water surface and very brittle. If the pH or cadmium concentration is too low the film lacks cohesion. Under the optimum conditions described above the ratio of cadmium ions to C4 anthracene molecules is approximately 1:4.

The x-ray structure obtained for the films (9) shows that the bulky anthracene nuclei dominate the packing. They are tilted with respect to the film plane, while the less bulky C_4H_9 groups are interleaved, so that the anthracene nuclei in successive layers approach each other more closely than they would in a classical L-B film structure. The headgroups ($CH_2.CH_2.CO_2H$) do not interleave because of the strong hydrogen bonding interactions between this part of the molecule. Gold electrodes have been evaporated on top of the films to allow capacitance measurements to be made. The measured capacitance values are consistent with the X-ray spacing of 1.22 nm per layer and the plot of reciprocal capacitance against number of layers is linear (9).

Of particular interest for molecular electronics are our observation (9,10) of anisotropic conductivity, electroluminescence.

and photoconductivity. The dark conductivity shows an anisotropy of 10^8 in favour of in-plane conduction. This reflects the relative closeness of anthracene rings within the plane of the film compared to the interlayer spacing, and vindicates our efforts to reduce the length of the groups X and R. The out-of-plane conductivity of the deposited film (10^{16} $\Omega^{-1}cm^{-1}$) is similar to that of single-crystal anthracene.

Photoconductivity shows a steep edge at 430 nm, which is consistent with the absorption spectrum of anthracene. Double-injection electroluminescence is observed in the thickest samples; electrons injected from the aluminium bottom electrode combine with holes from the gold top electrode to give initially a charge-transfer exciton which decays to an electronically excited anthracene molecule or bimolecular excimer, which emits blue light. The need to have relatively thick samples before EL can be observed is consistent with our investigations (11) of electroluminescence in evaporated anthracene films where the thinnest films (0.15μm) showed greatly reduced electroluminescence efficiencies relative to the thicker films.

PHTHALOCYANINE DERIVATIVES

The copper and zinc complexes of TBP (I) both form stable monolayers at the air-water interface, and we have built these up into L-B multilayers on glass and aluminium substrates.

Zinc TBP

The surface-pressure/area isotherm obtained for zinc TBP under a nitrogen atmosphere is shown in Fig. 4, showing that the close packed surface area/molecule is approximately 1.2 nm^2. In the presence of oxygen some decomposition takes place and lower surface areas are observed. We have investigated possible structures for the film using space-filling molecular models and a computer simulation program. The bulky t-butyl groups appear to dominate

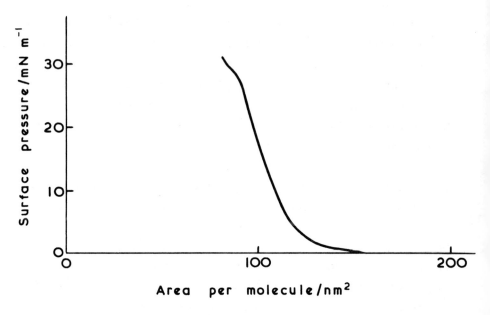

FIGURE 4.   Surface-pressure/area isotherm of zinc TBP.

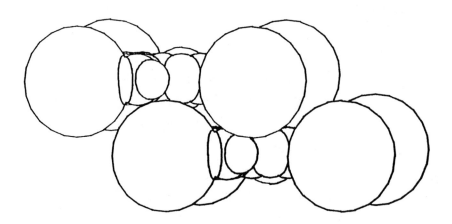

FIGURE 5.   Computer simulation of closely-packed metal TBP molecules. The t-butyl groups $(C(CH_3)_3)$ are represented as giant atoms.

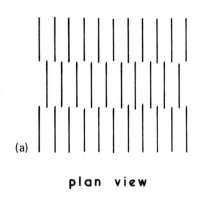

(a)

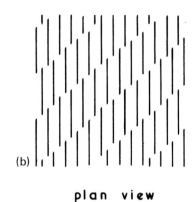

(b)

plan view   plan view

water
surface

FIGURE 6. Alternative packings of TBP molecules at the water
surface. The larger area packing (type a) corresponds to the
observed surface area.

the molecular packing, and a side elevation of two molecules packed
in this way is shown in Fig. 5, with the t-butyl groups $(C(CH_3)_3)$
represented as giant atoms.

Two close-packed arrays of this type of subunit can be
envisaged; one in which the plane of the molecule is tilted by
39° away from the water surface (Fig. 5a), and the other in which
the plane of the molecule is vertical to the water surface but
tilted to the shear line (Fig. 6b). The observed surface area of
1.2 nm^2 molecule^{-1} agrees closely with that expected for the first
arrangement (Fig. 6a).

We have deposited multilayers of zinc TBP on glass microscope
slides. The deposited layers show an optical absorption at 620 nm

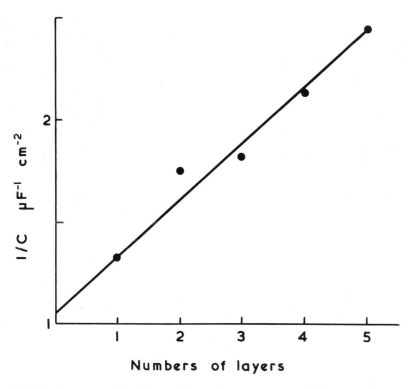

FIGURE 7. Plot of the reciprocal of the specific capacitance of zinc TBP multilayers as a function of the number of layers.

characteristic of the phthalocyanine chromophore. Multilayers have also been deposited onto aluminised glass slides; during this deposition process the angle of molecular tilt appears to decrease. This is indicated by our observation that the pickup ratio (reduction in area of monolayer on water surface − surface area of substrate being dipped) is approximately 0.5. This happens in a consistent way, and complete surface coverage with each monolayer is observed. The plot of reciprocal capacitance versus number of monolayers (Fig. 7) shows the excellent reproducibility achieved. As before, evaporated gold electrodes were used as the second plate of the capacitor.

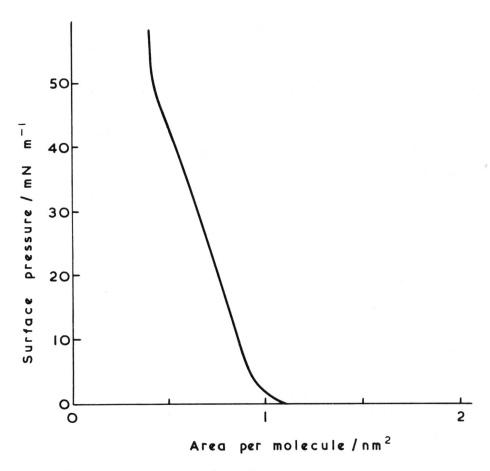

FIGURE 8.   Surface pressure/area isotherm of copper TBP.

Copper TBP

In contrast to zinc TBP, appears to be stable as a thin film in the
presence of air. We obtained the surface-pressure/area isotherm of
Fig. 8 after spreading the TBP from xylene solution. When chloro-
form was used as a solvent, evaporation appeared to be too rapid to
allow complete spreading. The observed surface area when spread
from xylene was 0.96 nm^2 $molecule^{-1}$, which is lower than observed
for the zinc derivative. It is, however, considerably greater than

would be expected for the alternative stacking of Fig. 6(b)
(0.75 nm^2 molecule^{-1}), and we suggest that the structure is probably
of Fig. 6(a) type with an increased angle of tilt (51°).

The built up multilayers on a glass substrate show an optical
absorption maximum at 616 mm, which is again consistent with a
phthalocyanine absorption. The shift in wavelength between the
copper and zinc complexes is probably due to the stacking
differences. In solution they both have an absorption maximum at
676 nm. We have observed conduction between gold electrodes
evaporated onto a monolayer of copper TBP deposited onto a glass
slide. The resistance is approximately ohmic and is modulated by
exposure to chemically reactive gases such as ammonia.

CONCLUSIONS

We have shown that it is possible to build up multilayers of
derivatives of the two best understood organic semiconductors,
anthracene and phthalocyanine. We have characterised the anthracene
films in some detail and have shown up some potentially useful
properties. The phthalocyanine work is at an earlier stage, but the
films examined are already showing promising properties.

ACKNOWLEDGEMENTS

We thank Steven Baker, Eric R. Cox, J. Alec Finney, and
Michael C. Petty for advice and assistance and David Baird for
computer modelling of the structure.

REFERENCES

1. J.H. Steven, R.A. Hann, W.A. Barlow and T. Laird, Thin Solid
    Films, 99, 71 (1983).

2. S. Baker, M.C. Petty, G.G. Roberts and M.V. Twigg, Thin Solid
    Films, 99, 53 (1983).

3. E.K. Lukyanets, J. Gen. Chem. U.S.S.R., 2770 (1971).

4. P.S. Vincett, W.A. Barlow, F.T. Boyle, J.A. Finney and
    G.G. Roberts, Thin Solid Films, 60, 265 (1979).

5. F.H.C. Stewart, Australian J. Chem., 13, 478 (1961).

6. F. Gutman and L.E. Lyons, Organic Semiconductors, Wiley-Interscience, New York, 1967.

7. G.L. Gaines, Jr., Insoluble Monolayers at Liquid-gas Interfaces, Wiley-Interscience, New York, 1966.

8. E.P. Honig, T.H. Th. Hengst and D. den Engelsen, J. Colloid Interface Sci., 45, 92 (1973).

9. P.S. Vincett and W.A. Barlow, Thin Solid Films, 71, 305 (1980).

10. W.A. Barlow, J.A. Finney, T.M. McGinnity, G.G. Roberts and P.S. Vincett, Inst. Phys. Conf. Ser., 43, 749 (1979).

11. G.G. Roberts, T.M. McGinnity, W.A. Barlow and P.S. Vincett, Thin Solid Films, 68, 223 (1980); Solid State Commun., 32, 683 (1979).

24

A New Method of Modulating Electrical Conductivity in Phthalocyanines Using Light Polarization

T.W. Barrett, H. Wohltjen, and A. Snow / Chemistry Division Naval Research Laboratory, Washington, DC

I.   INTRODUCTION

The phthalocyanines (Pc) (Fig. 1) are well known as gas adsorbers and useful as ambient vapor sensors (1-5), although their wide selectivity has seemed to make them unreliable for such use (6). A new method, reported here, involves circularly polarized light control of the Fermi level of a Pc-electrode system and possibly determines the selectivity range. Such control indicates the way to reliable sensing devices. The circularly polarized light-induced paramagnetic modulation of the Pcs in an inverse Faraday effect also indicates a possible use in future computer technology.

Circularly polarized light interaction with the Pcs in an oxygen or electron accepting atmosphere is possible from two points of view. On the one hand, magnetic circular dichroism studies (7-12) of the metal Pcs indicate resonance interaction

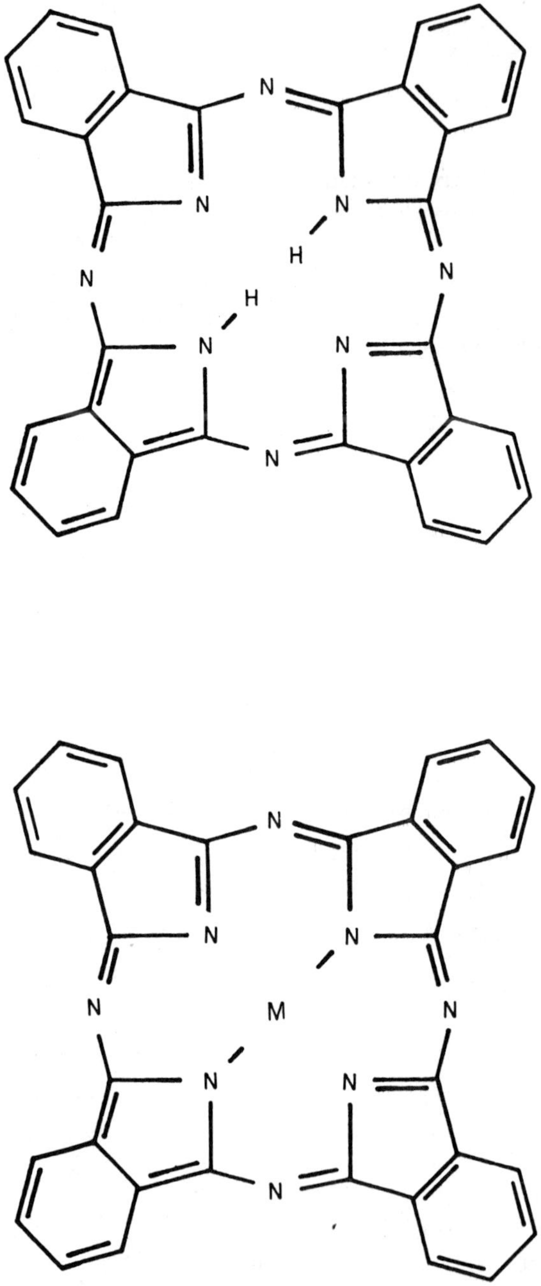

1.   H$_2$Pc, top, and metal-substituted Pc, bottom.

with the metal in the red (long wavelength) end of the spectrum. On the other hand, light interaction with dioxygen (O_2) adsorbed to the Pc ring occurs over most of the visible spectrum. In the present study, the two effects were dissociated by studying both metal-substituted and metal-free Pcs.

Apart from the circularly polarized light modulation effect, the subject of this paper, there is also the well-known photo-conductivity effect. The interaction of circularly polarized incident light with oxygen adsorbed to Pcs is of importance as this electron accepting oxygen molecule appears to be crucial in increasing the conductivity of Pc films by the formation of a p-type Pc semiconductor material (13,14). The photoconductivity effect in Pcs is also the result of adsorbed oxygen impurities (15). However, this better known photoexcitation process is due to adsorbed oxygen with singlet and singlet-triplet excitations (14) and the increase in conductivity involves a slower diffusion-related process. The singlet occurs only in metal-free Pc, and both the singlet and singlet-triplet in metal-substituted Pc, with both types of excitations equally dependent upon either adsorbed or on deeply trapped oxygen (14). The reversible thousand-fold increase in conductivity, which occurs on exposure to an oxygen atmosphere (16), is due to this slower diffusion-related process. The circularly polarized light effect, to be discussed here, is a direct light-molecule interaction resulting in a decrease in electron acceptors. This decrease can result in either an increase or a decrease in conductivity in a metal-semiconductor-metal (MSM) structure, depending on circumstances described below.

2. EXPERIMENTAL PROCEDURE

2.1 Sample Preparation

Copper and metal-free phthalocyanine compounds (Aldrich Chemical

Company) were purified by soxhlet extraction with tetrahydrofuran
followed by vacuum sublimation. Thin phthalocyanine films were
deposited on an interdigital microelectrode array by vacuum
sublimation ($<10^{-2}$ torr) at 340°C. The films were easily
visible to the eye and reasonably transparent to visible light.
Under an optical microscope, the electrodes were clearly visible
and the film had a uniform appearance with a slight graininess in
texture. This was smoothed by gently buffing with a clean soft
tissue.

2.2 Microelectrode Preparation

All conductivity measurements were performed using an
interdigital microelectrode array. The electrodes were
microfabricated on a polished quartz disk which was one-inch in
diameter. The array consisted of fifty pairs of gold "fingers"
(2000Å Au on 200Å Cr). Each finger was 25 microns wide. The
space between fingers was 25 microns and the overlap distance was
7250 microns. Phthalocyanine films (both Cu-substituted and
metal-free) were deposited onto the microelectrodes by
sublimation at a temperature of 340°C and a vacuum pressure of
$<10^{-2}$ torr. Film thickness was not measured but was estimated
to be less than 0.5 microns based on its optical density. The
resistance of a clean, dry microelectrode array was greater than
$3x10^{11}$ ohms.

2.3 Measurement Technique

The microelectrode array was clamped to an aluminum block and
sealed inside a 2-liter glass flask containing Drierite.
Electrical measurements were performed using a system interfaced
to an Apple II Plus computer. The measurement system consisted
of a selectable gain current-to-voltage converter, an A/D

converter, and a D/A converter. Bias voltage was applied to the microelectrode from the D/A under computer control. The resulting electrode current was converted into a voltage, digitized by the A/D and stored in the Apple II Plus. Thus, resistance versus time data could be obtained by operating at a fixed bias voltage and sampling the device current at periodic intervals. Current voltage curves could be obtained by scanning the D/A generated bias voltage and measuring the resulting device current.

The incident radiation was obtained from an Argon laser (for the 4765, 4880 and 5145Å wavelengths) and a HeNe laser (for the 6328Å wavelength). The laser beam was passed through a quarter wave plate set to $0°$ for linear polarization, or $±45°$ for right-left circular polarization. The light intensity was measured at all polarizations and found to be the same.

3. RESULTS

a. The resistance of thin phthalocyanine films can be modulated by the polarization (linear versus circular, or right versus left circular) in the presence of an electrode acceptor such as oxygen (Figs. 2 and 4).

b. The effect is more pronounced in the metal-free than in the CuPc film (i.e. ~16% modulation versus ~5%) (Fig. 2).

c. The I-V curves are slightly nonlinear and pass through the origin indicating some small deviation from ohmic behavior (Fig. 3). The slight hysteresis shown can be attributable to dielectric effects.

d. The I-V H_2Pc curves can exhibit major changes upon irradiation with circularly polarized light (Fig. 3F).

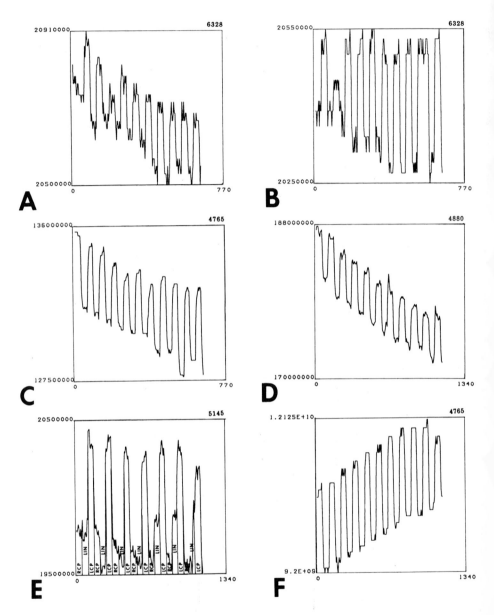

2. Light-induced modulation of (A-E) Cu-Pc and (F) H_2-Pc
conductivity (on the abscissa the total length is 2500s or 100s/
point; the ordinate is the resistance in ohms; increasing
conductivity is in the downward direction) where the polarization

e. O_2, or an electron acceptor, appears to be necessary for the effect, since the effect is not observed after lengthy (1-2 hour) exposure to dry N_2, He, or Ar atmospheres.

f. The modulation effect could not be observed with pulsed (7 nsec. pulse, 10/sec.) laser excitation, although the photoconductive effect was still obtained. As (i) the modulation effect is expected to be fast (<7 nsec.), (ii) the photoconductive effect is a slower diffusion process, and (iii) the data collection system integrates over 100 msec. during which the modulation effect lasts for at most 7 nsec., and the photoconductive effect considerably longer, this difference between the two effects is to be expected.

g. Circularly polarized light can either <u>increase</u> the electrical conductivity or <u>decrease</u> the electrical conductivity of the <u>same</u> Pc film, depending on the magnitude of the Fermi level which is set by both the ambient atmosphere and the bias voltage of the metal-semiconductor-metal structure. This result agrees with previous reports (17-21).

---

changes in the incident lasing line were made by cycling in the following order: (A) first linearly and then right circularly polarized incident light etc. (bias, 0.1 V; incident light wavelength and power, λ = 6328 Å and 5 mW); (B) first linearly and then right circularly polarized incident light etc. (bias, 0.1 V; incident light wavelength and power, λ = 6328 Å and 5 mW); (C) first left and then right circularly polarized incident light etc. (bias, 0.05 V; incident light wavelength and power λ = 4765 Å and 15 mW); (D) first left and then right circularly polarized incident light etc. (bias, 0.1 V; incident light wavelength and power, λ = 4880 Å and 50 mW); (E) first right circularly polarized, then linearly and then left circularly polarized incident light, etc. (bias, 0.2 V; incident light wavelength and power, λ = 5145 Å and 9 mW); (F) first left and then right circularly polarized incident light, etc. (bias, 0.5 V; incident light wavelength and power, λ = 4765 Å and 15 mW).

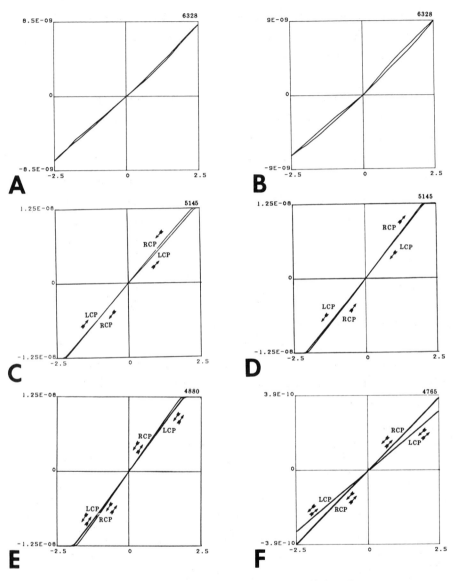

3. I-V relations (ordinate, resistance in ohms; abscissa, voltage) for (A) Cu-Pc (incident light wavelength and power, λ = 6328 Å and 5 mW; linearly polarized incident light), (B) Cu-Pc (incident light wavelength and power, λ = 6328 Å and 5 mW; circularly polarized incident light), (C), (D) Cu-Pc (incident light wavelength and power, λ = 5145 Å and 9 mW), (E) Cu-Pc (incident light wavelength and power, λ = 4880 Å and 50 mW) and (F) H_2-Pc (incident light wavelength and power, λ = 4765 Å and 15 mW): RCP, right circularly polarized incident light.

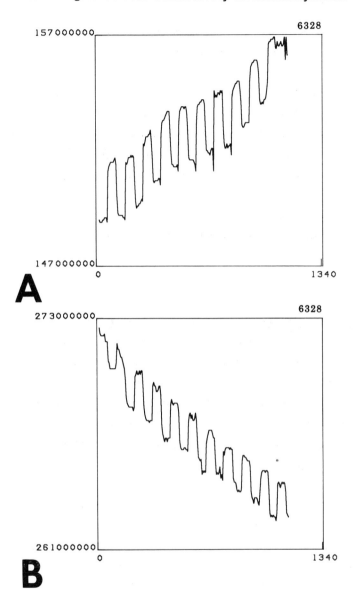

4. Light-induced modulation of Pc conductivity (ordinate, resistance in ohms; abscissa, voltage (incident light wavelength and power, λ = 6328 Å and 5 mW): (A) Cu-Pc with first linearly and then circularly polarized light incident, etc.; (B) Cu-Pc with first linearly and then circularly polarized light incident, etc.

h. The interaction of light and Pc film shows a chiral
preference. Left circularly polarized light has an
antagonistic action to right circularly polarized light (Fig.
2E). The direction of the response to a particular chirality of
circularly polarized light depends upon circumstances discussed
below. For example, depending on the magnitude of the bias
voltage with respect to V_{FB}, circularly polarized light can
result in either an increase (Fig. 4A) or a decrease (Fig. 4B) in
resistivity.

i. These results may be interpreted within the context of modern
semiconductor theory.

4. DISCUSSION

External Electron Acceptance by Paramagnetic Molecules or Ions

There is no such thing as a chemically or physically inert
solvent medium, and the same might almost be said of ambient
gases. Of particular importance in perturbing triplet states
are: oxygen, nitric oxide and other paramagnetic molecules or
ions (22). Other perturbing agents are: molecules consisting of
heavy atoms, e.g., I_3^-, and complexing agents with which the
solute can form charge-transfer complexes (22a). However, these
latter, if they are not paramagnetic, do not interact with the
effective magnetic field of circularly polarized light and will
not be considered here. Furthermore, paramagnetic salts forming
dative or coordination bonds with the Pc ring, such as $FeCl_3$,
increase conductivity greatly ($\times 10^{10}$) but the complex exhibits no
photoconductivity.

The effect of all of the above agents is to result in
external spin-orbit coupling. Oxygen, in particular, is well
known as a triplet state quencher (23). It is quite conceivable,

therefore, that oxygen adsorbed to a nonparamagnetic compound, such as metal free phthalocyanine, should couple its ground triplet state as an electron acceptor. This reaction is an S_1 → T_1 transition, in which the S_1 state is that of phthalocyanine, and the T_1 state, a dioxygen state. Terenin (24) called this "paramagnetic quenching" and reported many studies of the intermolecular effects of oxygen (25-29).

It has been suggested (22) that the enhanced spin-orbit coupling caused by oxygen is due to the inhomogeneous magnetic field present in O_2 acting on nuclear spins. According to the Tsubosmura-Mulliken model (30), the effects of oxygen on, e.g. hydrocarbons, are caused by the formation of transient charge-transfer complexes having an equilibrium constant of approximately zero. These complexes are known as "contact charge-transfer complexes" (31). The important point for the present discussion is that intermolecular charge-transfer adsorption can occur in the absence of a stabilization energy of the complex. The energy of the transition to the triplet state is borrowed from a charge-transfer transition.

An alternative interpretation involves the "exchange mechanism" of Dijkgraaf and Hoijtink (32-34). The energy for the transition to the oxygen triplet state is obtained, according to this theory, from the S_1 ← S_0 transitions (35), which, in the case under discussion, are those of the phthalocyanines.

The consensus of opinion is that both effects occur, but that shifts are observed more in line with magnitudes expected from the charge-transfer point of view (22). Finally, we may observe that although charge-transfer complexing with electron acceptor species which do not contain heavy atoms and which are

not paramagnetic produces much the same effects as does O_2, NO,
paramagnetic metal ions, and heavy atom solvents, i.e., spin-
orbit coupling, these do not offer the opportunity of light-
molecule switching, which is dependent upon the excitation of
triplet states in the electron accepting species. The circularly
polarized light-electron accepting molecule interaction, apart
from charge-transfer coupling, depends upon the populating of
magnetic (triplet) states, hence their absence precludes the
interaction.

Metal-Semiconductor Contacts

The MSM structure is basically two Schottky barriers
connected back-to-back (36). The many transport behaviors of the
MSM structure (37) can only be understood within the context of
two metal-semiconductor contacts, one of which is reverse biased,
and the other forward biased.

In this section, the metal-semiconductor contact is examined
with respect to an understanding of the MSM structure and from
the point of view of thermionic emission theory (38). The
subject is reputed to have begun in 1874 when Braun noted the
dependence of the total resistance on the polarity of the applied
voltage and on the detailed surface conditions (39). In 1931 the
transport theory of semiconductors was formulated based on the
band theory of solids (40); and in 1938 the metal-semiconductor
barrier was suggested to arise from stable space charges in the
semiconductor alone, without the presence of a chemical layer.
This suggestion is inherently the Schottky barrier model (41).

Regarding this barrier which forms at the metal-semiconductor interface, a first consideration is that the Fermi levels
in the two materials must be coincident at thermal equilibrium.

That is, if charge flows from the semiconductor to the metal,
then relative to the Fermi level in the metal, the Fermi level in
the semiconductor is lowered by an amount equal to the difference
between the two work functions. The work function itself is the
difference between the vacuum level and the Fermi level. The
work function for the metal is $q\phi_m$, and for the semiconductor is
$q(\chi + V_n)$, where $q\chi$ is the electron affinity, or the difference
between the conduction band E_C and the vacuum level, and qV_n is
the difference between E_C and the Fermi level. Finally, the
potential difference $q\phi_m - q(\chi + V_n)$ is called the contact
potential.

If the metal-semiconductor gap is small, then the difference
between the metal work function and the electron affinity of the
semiconductor, i.e., the Schottky barrier height, $q\phi_{Bp}$, for a
p-type semiconductor is:

$$q\phi_{Bp} = E_g - q(\phi_m - \chi),$$ (1)

where E_g is the bandgap energy.

The depletion layer is the resulting electric double layer
formed at the contact surface between a metal and a semiconductor
having different work functions. It arises because the mobile
charge density is insufficient to neutralize the fixed charge
density of donors and acceptors. The layer width, W, is
defined:

$$W = \sqrt{\frac{2\varepsilon_s}{qN_A}} \; (V_{bi} - V - \frac{kT}{q}),$$ (2)

where ε_s is the semiconductor permittivity; N_A is acceptor
impurity density (in the case considered here, this is O_2
adsorbed density); V_{bi} is the diffusion or built-in potential; q
is the magnitude of the electronic charge.

When the electric field is applied forming a charge at a
distance, x, across the metal-semiconductor contact, charge is
formed on the surface of the metal which is equivalent in
attractive force to that which would exist if an image positive
charge existed at a distance: -x.

This image force lowering, or Schottky barrier lowering, $\Delta\phi$,
which occurs when a field is applied, is defined as:

$$\Delta\phi \;\; = \frac{qE}{4\pi\varepsilon_s} \tag{3}$$

where E is the electric field. Thus, at high enough field, the
Schottky barrier is reduced and electron transport occurs.

The current transport in metal-semiconductor contacts is
mainly due to majority carriers. Apart from this Schottky
barrier lowering, there is quantum mechanical tunneling of
electrons through the barrier (in the case of heavy doping (high
levels of O_2 in the present instance) or operation at low
temperatures) and recombination effects. The total current of
the contact is:

$$J_{m\rightarrow s} \;\; = \int_{E_F \,+\, q\phi_B}^{\alpha} qv_x dn \tag{4}$$

where $E_F + q\phi_B$ is the minimum energy required for thermionic emission from metal to semiconductor, v_x is the carrier velocity in the direction of transport; and the electron density in an incremental range is given by:

$$dn = N(E)F(E)dE$$

$$= \frac{4\pi(2m^*)^{3/2}}{h^3} \sqrt{E - E_C \exp\ [-(E - E_C + qV_p)]} \tag{5}$$

where $N(E)$ and $F(E)$ are the density of states and the distribution function; m^* is the effective mass of the semiconductor; h is Planck's constant; and $qV_p = (E_C - E_F)$. Thus, this equation describes a process in which any (circularly polarized light-induced) change in the density of states or the Fermi level, influences conduction.

The barrier height of a metal-semiconductor system is thus determined by the metal work function and the surface states of the semiconductor. The latter are depleted by circularly polarized light in the experiments reported here. Fig. 5 shows a detailed energy band diagram for such a metal-p-semiconductor (phthalocyanine) contact (after Ref. 42).

In the case of an MSM structure with two metal-semiconductor contacts, current continuity requirements dictate that the electron currents across both barriers must be equal. The total current across the structure is made up of contributions from both electron and hole current and their ratio may be varied by varying the barrier heights of the two contacts (absorbed O_2 concentration in the case under consideration here). The types

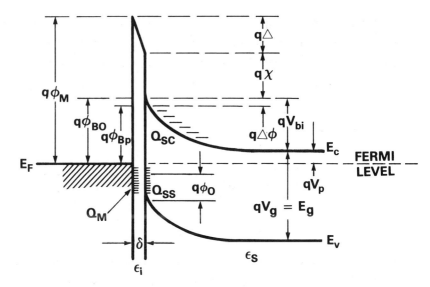

ϕ_M = WORK FUNCTION OF METAL
ϕ_{Bp} = BARRIER HEIGHT OF METAL-SEMICONDUCTOR BARRIER
ϕ_{BO} = ASYMPTOTIC VALUE OF ϕ_{Bp} AT ZERO ELECTRIC FIELD
ϕ_O = ENERGY LEVEL AT SURFACE
$\triangle\phi$ = IMAGE FORCE BARRIER LOWERING
$\triangle$ = POTENTIAL ACROSS INTERFACIAL LAYER
χ = ELECTRON AFFINITY OF SEMICONDUCTOR
V_{bi} = BUILT-IN POTENTIAL
ϵ_S = PERMITTIVITY OF SEMICONDUCTOR
ϵ_i = PERMITTIVITY OF INTERFACIAL LAYER
δ = THICKNESS OF INTERFACIAL LAYER
Q_{SC} = SPACE-CHARGE DENSITY IN SEMICONDUCTOR
Q_{SS} = SURFACE-STATE DENSITY ON SEMICONDUCTOR
Q_M = SURFACE-CHARGE DENSITY ON METAL

5. Metal-p-semiconductor contact (after Ref. 48).

of current-voltage relations obtainable fall between two limiting
cases depending on the relative barrier heights of the first and
second contacts. The two limiting cases are (1) the hole barrier
height at contact 1 is much larger than the electron barrier
height at contact 2; and (2) the hole barrier height at contact 1

is smaller than the electron barrier height at contact 2 (cf.
Ref. 37 for consideration of a metal-n-semiconductor-metal
device).

However, for all cases the current-voltage relations will be
linear for voltages between V_{RT} and V_B, where the reach-through
voltage, V_{RT}, is that applied voltage at which the sum of the two
depletion widths is exactly equal to L, the length of the
semiconductor; and V_B is the breakdown voltage. For very small
applied voltages and for large applied voltages, i.e. for applied
voltages on either side of this window, linearity is not to be
expected.

Data Analysis

The experiments reported here constitute an investigation
into the current-voltage relations of a symmetrical metal-
semiconductor-metal (MSM) structure - the semiconductor being the
Pc with adsorbed oxygen. The photovoltaic conversion is either
anodic or cathodic (17-20), depending on whether the potentials
applied to the MSM are positive or negative with respect to the
flat-band voltage (V_{FB}) (21). This flat-band voltage, V_{FB}, is
defined:

$$V_{FB} = \frac{q\ N_D\ W^2}{2\epsilon_s} \tag{6}$$

where N_D is the ionized impurity density (or density of holes in
the Pc created by adsorbed O_2), q is charge, $W = W_1 + W_2$, where
W_1 and W_2 are the depletion widths in the p-layer for the forward
and reversed biased barriers, respectively, and ϵ_s is the (Pc)
semiconductor permittivity. Taking V_{FB} as the bias voltage
across the total electrode device, for experiments reported $V_{FB} \sim$
0.02 - 2.0 volts.

If the applied voltage is at the flat band voltage, then the
Fermi energy level is equidistant between $q\phi_{Bn}$ and $q\phi_{Bp}$, where
ϕ_{Bn} and ϕ_{Bp} are the Schottky barrier heights on the n (metal,
gold in the present case) and p semiconductor (Pc, in the present
case) sides, respectively. Thus, the bias of an MSM structure
influences whether incident light enhances or decreases conductivity (see Result 7). Another influence is that of adsorbed
gases.

For example, hole and electron densities are:

$$p = N_V \exp \frac{(-E_F + E_V)}{kT}, \tag{7}$$

$$n = N_C \exp \frac{(-E_C + E_F)}{kT}, \tag{8}$$

where N_V and N_C are the density of states in the valence and
conduction bands, respectively. The conductivity at a p-n or n-p
junction is:

$$\sigma = q(\mu_p p + \mu_n n) \tag{9}$$

where μ_p is the mobility of the holes and μ_n is the mobility
of the electrons, or

$$\sigma = q(\mu_p N_V \exp \frac{(-E_F + E_V)}{kT} + \mu_n N_C \exp \frac{(-E_C + E_F)}{kT}). \tag{10}$$

Thus, a variation in E_F influences conductivity.

On gas adsorption, the mobilities μ_p and μ_n are not expected to change, in agreement with an earlier analysis (43,44). Furthermore, competition with the O_2 molecule for adsorption sites on the Pc ring and on the metal in metal-Pc, and on the ring only in metal-free Pc, will not affect the density of states, but rather the number of occupied levels, i.e., hole density. Thus, gas adsorption on Pcs is expected to result in the modification:

$$p = N_V \exp \frac{(-E_F + E_V + \zeta)}{kT} \qquad (11)$$

where ζ is the gas adsorption dependent change in the valence band energy. A consequent decrease in conductivity is then expected using Eqn. (9).

The conductivity of the CuPc MSM could be modulated by constant wavelength lasers provided that the bias to ground was approximately at the V_{FB}. The modulations obtained varied from ~ 1 to ~ 5%. Both linearly-circularly and left-right circularly polarized light changes induced conductivity modulations. The conductivity could be increased or decreased by a specific polarization setting depending on whether the bias voltage was above or below the V_{FB}.

The conductivity of the metal-free Pc MSM was modified to a greater extent, ~ 16%, by a variation in polarized light, than was the CuPc MSM (Fig. 4).

The photoconductivity and the polarized light modulation of conductivity could be abolished by application of N_2, He, and Ar, indicating the requirement of O_2 for both effects. The two

effects could be dissociated by use of a pulsed laser with a
pulse time of 7 nsec. and a cycle time of .10 sec. While the
photoconductive effect still remained, the polarized light
modulation was not detected by the current measurement system
which had a slow response time (e.g. 1 sec.) indicating that the
modulation effect is of very short time constant in decay after
light removal.

Figs. 2A and B demonstrate the effect of linearity, as
compared with right circularly, polarized light on the resistance
of CuPc MSM structures. In the instances shown, circularly
polarized light increased the resistance of the MSM structure.
However, depending on the magnitude of the bias voltage with
respect to the V_{FB}, circularly polarized light can result in
either an increase (Fig. 4A) or a decrease (Fig. 4B) in
resistivity.

Figs. 2C and D demonstrate the effect of left, as compared
with right, circularly polarized light on the resistivity of CuPc
MSM structures, and Fig. 2E demonstrates the effect of right
circularly, linear, and left circularly polarized light. At the
beginning of the record, linearly polarized light decreases the
resistance more than does right circularly polarized light.
Progressively through the record, however, the relative effect of
linearly polarized light in diminishing the resistance becomes
less than that of right circularly polarized light. This
variability indicates vapor changes in (Eqn. (11)), and was
observed when the ambient atmosphere was changed.

Fig. 2F demonstrates the effect of left, as compared with
right, circularly polarized light on the resistivity of
metal-free Pc MSM structures. Whereas the resistance of the CuPc

MSM structures studied could be modulated up to ~ 5% of the
initial resistance after photoconduction was elicited, the
resistivity of the H_2Pc MSM structures, of comparable
semiconductor material thickness, could be modulated up to ~ 16%.
Given the same quantity of complex O_2 and degree of charge
transfer, this result might indicate that circularly polarized
light removal of singlet excitons (creating hole carriers in the
case of H_2Pc), requires less photons than does removal of the
singlet-triplet excitons (creating hole carriers in the case of
CuPc). A long wavelength photoconductive response is a property
of CuPc, but not of H_2Pc, indicating the preponderance of
singlet-triplet adsorption in CuPc due to spin-orbit coupling.
Two excitons derived from singlet-triplet adsorption would be
needed to create free carriers, while only one singlet exciton is
needed (14,45,46). Circularly polarized light removal of those
excitons created by adsorbed or bound oxygen requires more
photons in the case of the metal triplet excitons than in the
case of the Pc ring singlet excitons.

Figs. 3A and 3B demonstrate slightly nonlinear current-
voltage relations for CuPc MSM structures with incident linearly
(Fig. 3A) and circularly (Fig. 3B) polarized light. The
increasing applied voltage passes through a $\dot{V}_{FB}$ in each case,
going from $V_{RT} < V < V_{FB}$ through $V_{RT} < V_{FB} < V$ and returning,
where V_{RT} is the reach-through voltage. Theoretically, beyond
reach-through, the current increases exponentially with applied
voltage.

Fig. 3C and 3D demonstrate reverse effects of right and
left circularly polarized light, indicating that light effects
are relative to the V_{FB}, which is influenced by ambient vapors,
i.e. by ζ.

Linear changes in the current-voltage characteristics
elicited by right versus left circularly polarized light are
demonstrated in Fig. 3E and 3F. As incident circularly
polarized light are demonstrated in Figs. 3E and 3F. As
incident circularly polarized light is expected to deplete the
hole density (Eqn. 7), this linear relation indicates that the
density of states is modified by circularly polarized light
interaction, and modifying Eqn. (11):

$$p = (N_v - \eta) \exp \frac{(-E_F + E_V + \zeta)}{kT} \qquad (12)$$

where η is the circularly polarized light-dependent change in the
density of states. This result is in agreement with recent
evidence that applied field and current density influences the
position of the Fermi level (47).

Figs. 4A and 4B indicate the vapor-dependent changes in ζ,
while circularly polarized light is, at the same time, producing
changes in η. Interestingly, ζ is increasing in Fig. 4A and
decreasing in Fig. 4B.

The circularly polarized light-induced modulation of the
resistance of MSM structures is due to the removal of the
electron-accepting oxygen molecule. Drago and Corden (48) have
proposed a spin pairing model of oxygen binding in transition
metal systems, the key feature of which is that the metals have
one or more unpaired electrons in d orbitals with enough energy
to spin pair with the electron in the oxygen binding orbitals.
In the case of oxygen adsorption to the Pc ring, the same
mechanism could apply.

The adsorbed oxygen imparts optical sensitivity to the Pc
ring and the left versus right circularly polarized light
resistance effect differences indicate the presence of circular
dichroism. An inverse Faraday effect (49-51) in which circularly
polarized light, acting as an effective magnetic field, induces
paramagnetism in the irradiated sample, is a possible mechanism
mediating the removal of the oxygen from the sample.

This inverse Faraday effect involves the induction of
magnetization in a non-absorbing gas, solution or material
through which a polarized beam of arbitrary ellipticity is
passed. A resonance inverse Faraday effect may be obtained as
the incident light energy approaches the negative value of the
zero-field splitting energy of magnetic sublevels, quite
independent of electronic dipole transitions which are also
involved. In the present study, the spin resonance of the
dioxygen molecule being broad, permitted the use of incident
light of widely varying wavelengths in the visible.

The induced paramagnetism would populate states split in
zero field and hence remove electrons from the oxygen
spin-pairing required by the Drago and Corden (48) model,
returning them to the Pc ring or the metal. The holes would be
removed. Certainly such an inverse Faraday effect exists in the
porphyrins (52-54) - a species of molecules closely related to
the phthalocyanines.

The complicated interactions of light and the MSM system are
diagrammed in Figs. 6 and 7. Fig. 6A shows the charge
distribution under low bias, the field distribution and energy
band diagram for an MSM structure. Fig. 6B diagrams the result
of obtaining reach-through and the flat band condition.

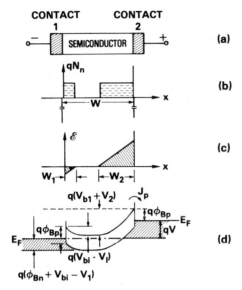

6a. Metal-semiconductor-metal (MSM) structure. (a) MSM with a
uniformly doped p-type semiconductor; (b) Charge distribution
made under low bias; (c) Field distributions; (d) Energy-band
diagram; (after Sze, et al. (37)).

Electrochemical studies have demonstrated control of V_{FB}
(17-21). The effect of vapors and voltage bias reported here is
indicated and the separate effect of circularly polarized light
on the flat band condition diagrammed.

The slight nonlinearity in the I-V curves, shown in Fig.
3, is attributed to the energy bands at the four conditions a-d
shown in Fig. 7A. During obtainment of the I-V curves, the
conditions a-d will be swept through.

Fig. 7B illustrates the effect of bias voltage and ambient
vapors on carrier concentrations at reach-through and at the flat

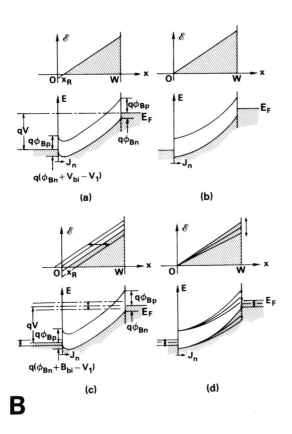

B

6b. Field distribution and energy-band diagrams. (a) At reach-through; (b) At flat band condition; (c) Effect of voltage bias and vapors on flat band condition; (d) Effect of circularly polarized light on flat band condition.

band condition. The effect of circularly polarized light is illustrated in Fig. 7C and the contrary effect of circularly polarized light, shown in Fig. 4, is illustrated in Fig. 7D and attributed to the possibility of either p or n majority carriers in the MSM structure.

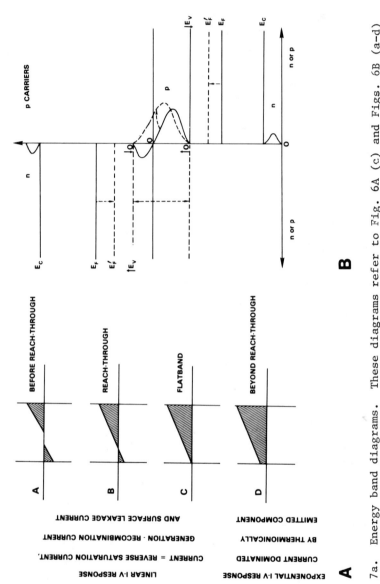

DETERMINANTS OF IV CHARACTERISTICS

EFFECT OF BIAS VOLTAGE AND VAPORS

7a. Energy band diagrams. These diagrams refer to Fig. 6A (c) and Figs. 6B (a-d) upper. (a) Before reach-through; (b) At reach-through; (c) At flat band; (d) Beyond reach-through.

7b. Effect of bias voltage and vapors on carrier concentrations; (a) At reach-through (straight line); (b) At flat band condition (hatched line).

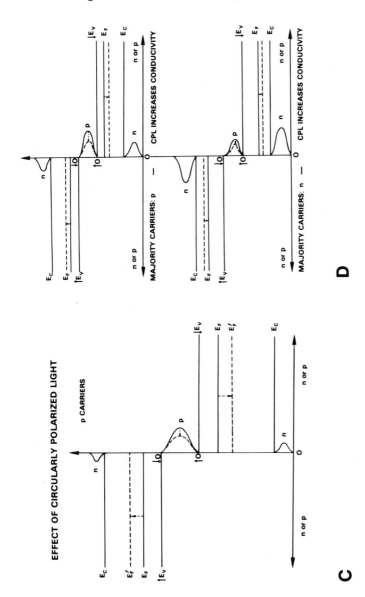

7c. Effect of circularly polarized light on carrier concentrations. (a) Before (straight line); (b) After circularly polarized light applied; (hatched line); p majority carriers

7d. Contrary effects of circularly polarized light. (a) p majority carriers; effect of circularly polarized light is to decrease conductivity; (b) n majority carriers; effect of circularly polarized light is to increase conductivity.

CONCLUSION

The electrical behavior of Pc MSM structures can be controlled by adjustments to the Fermi level achievable in a number of different ways including: (1) By gas adsorption which displaces an electron accepting, p semiconductor creating, species of molecule, such as O_2. (2) By changes in bias voltage which determines the state of the device with respect to reach-through and flat band conditions. (3) By conductance modulation through circularly polarized light irradiation, which removes the electron accepting adsorbed species of molecule, resulting in fewer p states. These results were obtained with an arbitrary thickness of phthalocyanine deposition and an arbitrary interelectrode spacing. It is to be expected that with thinner depositions and closer interelectrode spacings, even greater degrees of modulation will be obtained due to more extensive optical interaction.

REFERENCES

1. A.B.P. Lever, Adv. Inorg. Radiochem. 7 (1965) 27.

2. P. Bergveld, N.F. DeRoo and J.N. Zemel, Nature 273 (1978) 438.

3. J.N. Zemel, Sensors and Actuators 1 (1981) 31.

4. J.N. Zemel, B. Keramati, C.W. Spivak and A. D'Amico, Sensors and Actuators 1 (1981).

5. G.F. Gutman and L.E. Lyons, "Organic Semiconductors", Wiley, New York (1967).

6. A.W. Barendsz, C.A. van Beest and P.P.M.M. Wittgen, ms.

7. B.R. Hollebone and M.J. Stillman, Chem. Phys. Lett. 29 (1974) 284.

8. M.J. Stillman and A.J. Thompson, J. Chem. Soc. Faraday Trans. II, 70, (1974) 790.

9. M.J. Stillman and A.J. Thompson, J. Chem. Soc. Faraday Trans. II, 70 (1974) 805.

10. B.R. Hollebone and M.J. Stillman, J. Chem. Soc. Faraday Trans. II, 74 (1978) 2107.

11. K.A. Martin and A.J. Stillman, Can. J. Chem. Comm. 57 (1979) 1111.

12. C.H. Langsford, B.R. Hollebone and T. Vandernoot, Adv. Chem. Ser. 184 (1980) 139.

13. S.E. Harrison and K.H. Ludwig, J. Chem. Phys. 45 (1966) 343.

14. S.E. Harrison, J. Chem. Phys. 50 (1969) 4739.

15. G. Tollin, D.R. Kearns and M. Calvin, J. Chem. Phys. 32, (1960) 1013; D.R. Kearns, G. Tollin and M. Calvin, J. Chem. Phys. 32 (1960) 1020; D.R. Kearns and M. Calvin, J. Am. Chem. Soc. 83 (1961) 2110; D.R. Kearns and M. Calvin, J. Chem. Phys. 34 (1961) 2022.

16. J.M. Assour and S.E. Harrison, J. Phys. Chem. 68 (1964) 872.

17. H. Tachikawa and L.R. Faulkner, J. Am. Chem. Soc. 100 (1978) 4379.

18. F-R Fan and L.R. Faulkner, J. Chem. Phys. 69 (1978) 3334.

19. F-R Fan and L.R. Faulkner, J. Chem. Phys. 69 (1978) 3341.

20. F-R Fan and L.R. Faulkner, J. Am. Chem. Soc. 101 (1979) 4779.

21. C.D. Jaeger, F-R Fan and A.J. Bard, J. Am. Chem. Soc. 102 (1980) 2592.

22. S.P. McGlynn, T. Azumi and M. Kinoshita, "Molecular Spectroscopy of the Triplet State", Prentice-Hall, Englewood Cliffs, NJ (1969).

22a. E. Biatkowska and A. Graczyk, Organic Magnetic Resonance 11 (1978) 167.

23. G. Porter and L.J. Stief, Nature, 195 (1962) 991.

24. A.N. Terenin, Acta Physicochim. U.R.S.S. 18 (1943) 210.

25. A.V. Karyakin and A.N. Terenin, Izv. Akad. Nauk, SSSR. Ser. Fiz. 13 (1949) 9.

26. A.V. Karyakin and M.D. Galanin, Dokl. Akad. Nauk SSSR, 66 (1949) 37.

27. A.V. Karyakin, Izv. Akad. Nauk. SSSR, Ser. Fiz. 15 (1951) 556.

28. A.V. Karyakin and Y.I. Kalenichenko, Zh. Fiz. Khim. 26 (1952) 103.

29. A.V. Karyakin and A.N. Terenin, Dokl. Akad. Nauk, SSSR, 97 (1954) 479.

30. H. Tsubomura and R.S. Mulliken, J. Am. Chem. Soc. 82 (1960) 5966.

31. L.E. Orgel and R.S. Mulliken, J. Am. Chem. Soc. 71 (1957) 4839.

32. C. Dijkgraaf and G.T. Hoijtink, Tetrahedron Suppl. 2 (1963) 179.

33. C. Dijkgraaf, R. Sitters and G.J. Hoijtink, Mol. Phys. 5 (1962) 643.

34. G.J. Hoijtink, Mol. Phys. 3 (1960) 67.

35. J.N. Murrell, J. Am. Chem. Soc. 81 (1959) 5037.

36. S.M. Sze, "The Physical Basis of Semiconductors", 2nd Edition, (1981) Wiley.

37. S.M. Sze, D.J. Coleman and A. Loya, Solid State Electronics 14 (1971) 1209.

38. H.A. Bethe, MIT Radiation Lab. Rep. 43 (1942) 12.

39. F. Braun, Ann. Phys. Chem. 153 (1874) 556.

40. A.H. Wilson, Proc. Roy. Soc. London Ser. A, 133 (1931) 458.

41. W. Schottky, Naturwissenschaften 26 (1938) 843.

42. A.M. Cowley and S.M. Sze, J. Appl. Phys. 36 (1965) 3212.

43. B. Rosenberg, J. Chem. Phys. 36 (1962) 816.

44. T.N. Misra, B. Rosenberg and R. Switzer, J. Chem. Phys. 48 (1968) 2096.

45. P. Day and R.J.P. Williams, J. Chem. Phys. 37 (1962) 563.

46. P. Day and R.J.P. Williams, J. Chem. Phys. 42 (1965) 4049.

47. W. Mycielski, B. Ziolkowska and A. Lipinski, Thin Solid Films 91 (1982) 335.

48. R.S. Drago and B.B. Conden, Acc. Chem. Res. 13 (1980) 353.

49. P.S. Pershan, Physical Review 130 (1963) 919.

50. P.S. Pershan, M. Gouterman and R.L. Fulton, Mol. Phys. 10 (1966) 397; J.P. van der Ziel, P.S. Pershan and L.D. Malmstrom, Phys. Rev. Lett. 15 (1965) 190.

51. P.W. Atkins and M.H. Miller, Mol. Phys. 15 (1968) 503.

52. T.W. Barrett, Chem. Phys. Lett. 78 (1981) 125.

53. T.W. Barrett, in "Molecular Electronic Devices", Ed. F.L. Carter, Marcel Dekker, Inc., New York, NY (1982), p. 323.

54. T.W. Barrett, Chem. Comm. J. Chem. Soc. 1 (1982) 9, Comm. 789.

25

Conjugated Polymers as Electronic and Optical Materials:
Approaches Via Solid State Polymerization

Daniel J. Sandman, Gary M. Carter, Y.J. Chen, Sukant Tripathy, and Lynne A. Samuelson / GTE Laboratories, Inc., 40 Sylvan Road, Waltham, MA

INTRODUCTION

Molecular switching driven by any of several elementary physical excitations might be accomplished in organic polymers with conjugated backbones. Theoretical discussions of the electronic structures of polydiacetylenes

$$\begin{array}{c} R \\ | \\ \overline{(}C - C \equiv C - C\overline{)}_x \\ | \\ R \end{array}$$

(I, PDA) and polyacetylene, $(CH)_x$, lead to a description of these polymers as wide band one dimensional semiconductors (1,2). Thus, it might be anticipated that such polymers would have sufficiently high carrier mobilities and third-order suscepti-bility coefficients (3), $\chi^{(3)}$, that their study might lead to

prototype organic electronic and optical device concepts. The
experimental results on fully crystalline PDA are summarized
herein.

While experimental studies of currently available samples of
$(CH)_x$ and its derivatives have led to new conductive materials
with potential application in lightweight batteries and
suggestions of solitons in the electronic structure (2), these
materials are partially crystalline at best, and the need for
completely crystalline polyacetylenes has been recognized (4,5).

This paper reports some of the initial conceptual and
experimental work in our laboratories on solid state
polymerization of crystalline diacetylenic monomers and the
properties of materials derived from such processes.

POLYDIACETYLENES

The solid state topochemical transformation of macroscopic
1,3-diacetylene monomers (II) under thermal or radiative
excitation into fully crystalline polymers (III)

$$R-CH_2-C{\equiv}C-C{\equiv}C-CH_2-R$$

II

$$\begin{array}{c} RCH_2 \\ | \\ {\small(}C-C{\equiv}C-C{\small)}_x \\ | \\ CH_2R \end{array}$$

III

a $R = OSO_2-p-CH_3-C_6H_4$

b   R =

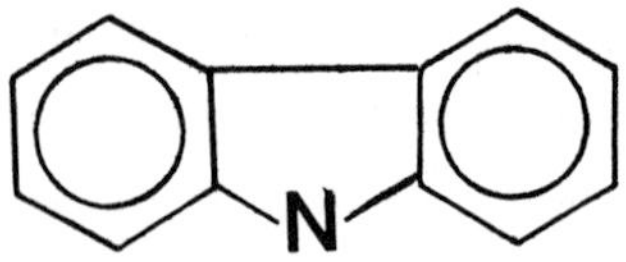

pioneered by G. Wegner and his collaborators (7) has by now been
studied in many laboratories throughout the world (8,9).

Colorless monomers such as the bis-p-toluenesulfonate of
2,4-hexadiyn-1,6-diol (PTS, IIa) and 1,6-di-N-carbazolyl-2,4-
hexadiine (IIb) are polymerized into single crystals with
metal-like reflectance. These polymers have absorption
coefficients greater than 10^5, and the role of excitonic states
in the visible absorption has been discussed (10).

There has also been considerable interest in the species
involved in the polymerization of diacetylenes. Carbenoid
species have been spectroscopically characterized as propagating
species in low temperature photopolymerization of PTS (6,9). The
monomer excited state may be represented as a diradical, DR, or a
bicarbene (11), (BC), and such species are

$$R-CH_2 \qquad\qquad R-CH_2$$
$$| \qquad\qquad\qquad\qquad |$$
$$\cdot C{=}C{=}C{=}C\cdot \quad\longleftrightarrow\quad :C{-}C{\equiv}C{-}C:$$
$$| \qquad\qquad\qquad\qquad |$$
$$DR \quad CH_2{-}R \qquad\qquad BC \quad CH_2{-}R$$

useful devices for discussion of extensions of the diacetylene
polymerization concept.

NONLINEAR OPTICS IN POLYDIACETYLENES
The linear index of refraction, $n(\omega)$ changes with the intensity
of the optical signal, I as follows:

$$n(\omega) = n_0(\omega) + n_2 I$$

The third order nonlinear index, n_2, contributes to
effects such as self-focusing, self-trapping, phase conjugation,
optical bistability, etc., which are fundamental to all-optical
signal processing applications. Recent discussions and

assessment of the prospects of optical switches and the optical computer have been given (12,13).

The third order nonlinear susceptibility coefficient $\chi^{(3)}$ is related to n_2 (14):

$$n_2 \equiv \frac{16\pi^2\chi^{(3)}}{C\varepsilon_1}$$

ε_1 is the dielectric constant of the nonlinear medium. Single crystals of PTS polymer were reported to have a value of $\chi^{(3)}$ (3ω) along the polymer backbone comparable to that of GaAs at 3ω below the absorption edge (15), and a large value of $\chi^{(3)}$ (ω) at 1.9 μm was also observed for poly -PTS crystals (16). Moreover, switching times of the order of 10^{14} sec. are deduced for poly-PTS (12,15). This subpicosecond switching time is ca. 10^{12} faster than estimates for the neuron (12).

While the studies of single crystal poly-PTS are critical to establish the scientific underpinnings of third order nonlinear optical phenomena in polydiacetylenes, they are not without limitations as one proceeds to consider realistic device configurations. For example, the crystals of poly-PTS grown to date do not have optically flat surfaces, and they have been studied by immersion in diiodomethane (CH_2I_2), which has a refractive index comparable to that of the polymer (16). We have noted that with polycrystalline films of poly-PTS about a micron in thickness that the absorbance loss (in air compared to the film immersed in CH_2I_2) in the near infrared due to reflectance and scattering is about 0.9 of an absorbance unit. Such films exhibit maximum absorbance at 6200 A and exhibit the additional structured noted in specular reflectance studies (6a).

Surfaces of improved optical quality may be obtained in
Langmuir-Blodgett multilayers of the polydiacetylene obtained
from $CH_3-(CH_2)_{16}C\equiv C-C\equiv C-(CH_2)_8-CO_2H$ (17). Films of this monomer
may be coated on silver gratings and polymerized to give a
structure which allows coupling of an incident laser photon to a
planar guided wave mode in the PDA waveguide or a surface plasmon
mode at the silver-PDA interface. The physics of the coupling
between radiation and confined modes (i.e., either guided waves
or surface plasmons) has been discussed (18).

Basically, one can relate the change in coupling angle $\Delta\theta$ (θ
is the angle with respect to the normal to the coupling plane)
with change in optical intensity in the mode, $\Delta\langle I\rangle$, to n_2 by:

$$N_2 \; \Delta\langle I\rangle = \Delta\theta \; \cos \theta$$

where $\langle \; \rangle$ indicates a suitable average of the optical intensity
over the mode volume. This technique allows one to obtain both
the sign (relative to $\chi^{(1)}$) and magnitude of $\chi^{(3)}$. By using
a surface plasmon mode, this technique yielded values of $\chi^{(3)}$
for Si and GaAs at 1 μm in satisfactory agreement with values
obtained with other techniques (14).

Fig. 1 shows the change in θ with incident beam intensity
at 6700 Å for a 5000 Å Langmuir-Blodgett multilayer of the
polymer derived from the diacetylene acid noted above. The shift
of θ to higher angles as the light intensity increases indicates
$\chi^{(3)}$ for the polymer is negative with respect to $\chi^{(1)}$ (14).
As the wavelength of the incident light decreases toward the
absorption edge of the polymer, $\chi^{(3)}$ is absorption enhanced.
Estimates of $\chi^{(3)}$ for the polymer are compatible with previous
reports (15,16) in crystals.

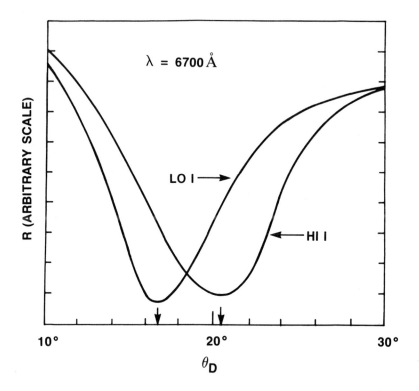

1. Reflectivity from a planar waveguide structure formed by
multilayers (total thickness ca 5000 A) of the PDA derived from
$CH_3-(CH_2)_{16}-C\equiv C-C\equiv C-(CH_2)_8CO_2H$ deposited on a metalized grating a
monolayer at a time from a Langmuir-Blodgett film balance. Loss
of reflection indicated coupling to the planar waveguide mode.
The shift in the angle of minimum reflectivity from LO I(low
intensity) to HI I(high density) is a measure of the nonlinear
index in the PDA film. θ_D is the angle the detector makes
relative to the input direction ($\theta_D = 2\theta_i$).

In addition to our work (17), other researchers have shown

interest in the assessment of Langmuir-Blodgett films of PDAs for

$\chi^{(3)}$ (3ω) and as waveguides (19 and 20, respectively).

POLYDIACETYLENES: ELECTRONIC PROPERTIES AND CHARGE-TRANSFER

While the metal-like reflectance of PDA is sufficient to stimulate inquiries into the electronic properties of these materials, this interest has been heightened by recent reports of room temperature electron mobilities in excess of 10^3 cm^2 V^{-1} sec^{-1} in the chain direction on the basis of pulse photoconductivity (21) in PTS and electron injection (22) and electroreflectance (23) studies of DCH polymer.

Hence, the extremely low dc conductivity of PDA crystals (6a) manifests extremely small numbers of carriers, and possible electronic application of such materials requires significant carrier concentrations. The issue of carrier creation by charge transfer thus arises.

Analogous to solids such as graphite and layered metal dichalcogenides whose interlayer forces are van der Waals, the interchain interactions in PDA crystals are also van der Waals and their structures might be "intercalated" by charge transfer reagents. Typically, such "doping" of PDA crystals does not occur (6c,24). Analogous to the partially crystalline $(CH)_x$, exposure of thin films and multilayers of PDA to halogen vapors reveals color changes and conductivity enhancements, but the observed conductivities do not exceed 10^{-7} (ohm-cm)$^{-1}$ and the structural and spectroscopic consequences of the halogen–PDA interaction were not reported (24,25). The polymerization of diacetylene itself with AsF$_5$ leads to a material with conductivity of 10^{-10} (ohm-cm)$^{-1}$ (26).

In view of the reports (6a,24), and our own experience with PTS polymer, that doping of PDA crystals does not occur, we were intrigued by the observation (27) that exposure of DCH polymer crystals to SbF$_5$ in pentane led to a dark colored material with a room temperature conductivity of 10^{-3} (ohm-cm)$^{-1}$. On the possibility that there might be a fundamental difference between DCH polymer and other PDA, we decided to interact DCH polymer with a variety of acceptors.

In perfluorohexane solution, SbF_5 interacts with thermally polymerized DCH to give antimony containing solids. By complete elemental analysis, a representative composition $[(DCH)_2 \ SbF_4]_x$ was obtained. The infrared spectrum of this material revealed all of the vibrations of pristine DCH polymer plus additional vibrations at 660, 300 and 280 cm^{-1}. These additional vibrations are not consistent with those reported for SbF_4 (28), but are compatible with an SbF_5 or SbF_6 species (29), hence the elemental analysis represents a mixture of Sb species. X-ray diffraction of these solids reveals only the pattern of DCH polymer. The solid state spectrum of thermally polymerized DCH polymer treated with SbF_5 is shown in Fig. 2. The salient feature is the broad band in the near infrared peaking at 1000-1100 nm associated with a partially oxidized PDA backbone. The remaining features in Fig. 2 may be associated with DCH polymer, and features at higher energies are associated with localized carbazole states, in agreement with the specular reflectance of this material (30).

In dichloromethane solution, $SbCl_5$ reacts with thermally polymerized DCH to give a material with the composition $(C_{29}H_{15}N_2)_{0.55} \ (SbCl_8)_{0.45}$ determined by elemental analysis. The 8:1 Cl:Sb ratio and the loss of hydrogen revealed by the elemental analysis suggest that ring chlorination and crosslinking of carbazole groups has occurred. Previously, an 8:1 Cl:Sb ratio has been observed on treatment of $(CH)_x$ with $NO^+ \ SbCl_6^-$. The infrared spectrum of our material is different from that of the DCH/SbF_5 material noted above, and absorption at 336 cm^{-1} is consistent with either a $SbCl_5$ or $SbCl_6$ species (32). X-ray diffraction of the material showed it to be amorphous.

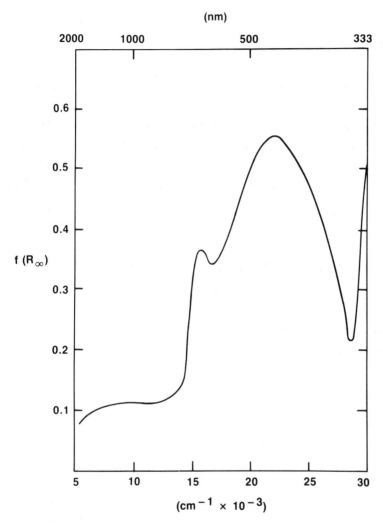

2. Diffuse reflectance of thermal DCH polymer treated with SbF$_5$ dispersed by 12% weight in NaCl.

.   The resistivity of the DCH polymer/SbX$_5$ samples discussed above is not less than 10^7 ohm-cm, and materials with lower resistivities were not obtained without loss of structural or compositional definition. Since pristine DCH polymer exhibits a resistivity greater than 10^{12}, the treatment with SbX_5 makes the material somewhat more conductive, but still not very high.

Since thermal polymerization of DCH proceeds heterogeneously to give a polycrystalline polymer with a fibrous texture (33), we exposed polymer crystals prepared by ^{60}Co-T-radiation of DCH monomer to $SbX_5(X=Cl,F)$. Examination of the materials resulting from this treatment by optical microscopy revealed they were inhomogeneous. This leads to the conclusion that the Sb moieties in the thermal DCH polymer reside in defects or vacancies.

Since neither Br_2 nor I_2 are taken up by the thermal DCH polymer, the question remains as to why the SbX_5 species are taken up. It is suggested that this is accomplished via initial ionization of the carbazole side chain, followed by reduction of the resultant carbazole-cation-radical by the conjugated backbone, as follows:

This side chain assisted ionization pathway is consistent with
the observed solid state ionization energies (I_c) of PDAs (34)
and carbazoles (35) with the former having the lower I_c.
Additionally, while Br_2, I_2, or SbX_5 can electron transfer
with a backbone such as PDA or $(CH)_x$, only SbX_5 can electron
transfer with a carbazole group. Ideally, the process of
"doping" via the side chain would not affect carrier mobility and
would be akin to modulation doping in semiconductor
superlattices.

SUMMARY AND ANTICIPATED DEVELOPMENTS
At present, the most straightforward approaches to completely
ordered polymers with conjugated carbon backbones involve (as
solid state processes), the polymerization of diacetylenes, the
polymerization of crystalline acetylenes, and inclusion
polymerization.

With respect to the latter process, exposure of acetylene
and phenylacetylene adducts of a cyclotriphosphazene to ^{60}Co-γ-
radiation did not lead to the relevant polyacetylenes (35).

Studies of the solid state reactivity of crystalline
acetylenes are in an early stage (37-40). It is apparent that
the strategy (38) of control of monomer structure of vinyl and
alkynyl systems via carboxylate salts or complexes may be adapted
to a broader range of nonbonded interactions. The studies of
Lando and Thakur (40) certainly fall into that pattern.

With respect to nonlinear optical phenomena in PDAs, we have
summarized above the successful coupling of laser light into a
PDA Langmuir-Blodgett waveguide structure. For studies of both
$\chi^{(2)}$ (9) and $\chi^{(3)}$ significant research opportunities exist in

both new PDA synthesis as well as improved processing techniques. With respect to $\chi^{(3)}$, time response measurements in the picosecond regime are clearly of interest.

The electronic properties and applications of PDAs remain an open question. It is easy to speculate that the conductivities of tightly packed PDAs may be markedly higher than those referred to above. At the present time, the prospects of PDAs for optical switching are somewhat more promising than for electronic switching, although the potential is clearly present.

ACKNOWLEDGEMENTS

M. Downey and J. Mullins provided the X-ray diffraction data cited and F. Kochanek and P. Martakos furnished FTIR spectra. Prof. B.M. Foxman provided access to the ^{60}Co source at Brandeis University. D.J.S. thanks Prof. G. Wegner for furnishing a copy of the Ph.D thesis of G. Schleier.

REFERENCES

1.   (a) A. Karpfen, J. Phys. C, $\underline{13}$, 5673 (1980), and
      reference therein; (b) J.L. Bredas, R.R. Chance, R. Silbey,
      G. Nicholas, and Ph. Durand, J. Chem. Phys., $\underline{75}$, 255
      (1981); (c) J.L. Bredas, R.R. Chance, R.H. Baughman, and R.
      Silbey, J. Chem. Phys. $\underline{76}$, 3673 (1982).

2.   S. Etemad, A.J. Heeger, and A.J. MacDiarmid, Ann. Rev.
      Phys. Chem., $\underline{33}$, 443 (1982).

3.   G.P. Agrawal, C. Cojan, and C. Flytzanis, Phys. Rev. B.,
      $\underline{17}$, 776 (1978).

4.   D.J. Sandman, J. Electronic Materials, $\underline{10}$, 173 (1981).

5.   D.J. Sandman in "Molecular Electronic Devices", F. Carter,
      ed. Marcel Dekker, New York, NY, 1982 p. 143 ff.

6. (a) G. Wegner, "Molecular Metals", W.E. Hatfield, ed.,
    Plenum Press, New York, 1979, p. 209 ff; (b) G. Wegner,
    Faraday Discussions Chem. Soc., 68 494 (1979); (c) V.
    Enkelmann, Die Angew. Makromol. Chem. 109/110, 253 (1982).

7. R.H. Baughman and K.C. Yee, J. Polymer Sci., Macromolecular
    Rev., 13, 219 (1978).

8. D. Bloor in "Developments in Crystalline Polymers", C.D.
    Bassett, ed., Applied Science Publishers, Englewood, NJ,
    Vol. 1, 1982, p. 151 ff.

9. A.F. Garito and K.D. Singer, Laser Focus, February 1982,
    p. 60.

10. M.R. Philpott, Ann. Rev. Phys. Chem. 31, 97 (1980).

11. (a) G. Wegner, Pure Appl. Chem. 49, 443 (1977);
    (b) H. Gross and H. Sixl, Chem. Phys. Lett., 91, 262
    (1982).

12. P.W. Smith, The Bell System Technical Journal 61, 1975
    (1982).

13. E. Abraham, C.T. Seaton, and S.D. Smith, Scientific
    American, 248(2), 85 (1983).

14. Y.J. Chen and G.M. Carter, Appl. Phys. Lett. 41, 307
    (1982).

15. C. Sauteret, J.P. Hermann, R. Frey, F. Pradiere, J. Ducuing,
    R.H. Baughman, and R.R. Chance, Phys. Rev. Lett. 36, 956
    (1976).

16. J.P. Hermann and P.W. Smith, Digest of Technical Papers,
    Eleventh International Quantum Electronics Conference,
    Boston, MA, June 23-26, 1980, p. 656.

17. (a) S.K. Tripatahy, G.M. Carter, Y.J. Chen, P. Cholewa, K.I.
    Lee, and D.J. Sandman, 164th American Chemical Society
    Meeting, Kansas City, MO, September 12-16, 1982, Polymer
    Preprints, 23 (2), 143 (1982); (b) G.M. Carter, Y.J. Chen,
    and S.K. Tripathy, Bull. Amer. Phys. Soc. 28, 492
    (1983).

18. G.M. Carter and Y.J. Chen, Appl. Phys. Lett., 42, 643
    (1983).

19. (a) F. Kajzar and J. Messier, Thin Solid Films, 99 109
    (1983); (b) F. Kajzar, J. Messier, J. Zyss, and I. Ledoux,
    Optics Commun. 45, 133 (1983).

20. F. Grunfeld and C.W. Pitt, Thin Solid Films, 99, 249
    (1983).

21. K.J. Donovan and E.G. Wilson, Phil. Mag., B44, 9 (1981).

22. W. Spannring and H. Bassler, Chem. Phys. Lett., 84, 54
    (1981).

23. L. Sebastian and G. Weiser, Phys. Rev. Lett. 46 1156
    (1981).

24. D. Bloor, C.L. Hubble and D.J. Ando, "Molecular Metals", see
    ref. 6a, p. 243 ff.

25. D.R. Day and J.B. Lando, J. Appl. Poly. Sci., 26 1605
    (1981).

26. P.J. Russo and M.M. Labes, J. Chem. Soc., Chem. Commun. 53
    (1982).

27. G. Schleier, Doctoral Dissertation, Faculty of Chemistry and
    Pharmacy, University of Freiburg, 1980.

28. M. Mehrain, B. Ducourant, R. Fourcade, and G. Mascherpa,
    Bull. Soc. Chim. France, 757 (1974).

29. (a) W.H.J. DeBeer and A.M. Heyns, Spectrochim Acta, 37A,
    1099 (1981); (b) A. Commeyras and G.A. Olah, J. Am. Chem.
    Soc. 91, 2929 (1969).

30. R.J. Hood, H. Muller, C.J. Eckhardt, R.R. Chance, and K.C.
    Yee, Chem. Phys. Lett. 54, 295 (1978).

31. S.C. Gau, J. Milliken, A. Pron, A.G. MacDiarmid, and A.J.
    Heeger, J. Chem. Soc., Chem. Commun., 662 (1979).

32. (a) H.W. Clark and B.I. Swanson, J. Am. Chem. Soc. 103,
    2928 (1981); (b) K. Olie, C.C. Smitskamp, and H. Gerding,
    Inorg. Nucl. Chem. Lett. 4 129 (1968); (c) I.R. Beattie,
    T. Gilson, K. Livingston, V. Fawcett, and G. Ozin, J. Chem.
    Soc. (A), 712 (1967); (d) H.A. Szymanski, R. Yelin, and
    L. Marabella, J. Chem. Phys. 47, 1877 (1967).

33. V. Enkelmann, R.J. Leyrer, G. Schleier, and G. Wegner, J. Mater. Sci. 15, 168 (1980).

34. (a) S. Arnold, J. Chem. Phys., 76, 3842 (1982); (b) A.A. Murashov, E.A. Silinsh, and H. Bassler, Chem. Phys. Lett. 93, 148 (1982).

35. P. Nielsen, A.J. Epstein, and D.J. Sandman, Solid State Commun., 15 53 (1974).

36. H.R. Allcock, W.T. Ferrar, and M.L. Levine, Macromolecules 15, 697 (1982).

37. A.O. Rohde and G. Wegner, Die Makromol. Chem. 179, 1999, 2013 (1978); (b) V. Enkelmann and O. Rohde, Acta. Crystallogr. B33, 3531 (1977).

38. B.M. Foxman and J.D. Jaufman, J. Polymer Sci. (C), Polymer Symposium, 70, 31 (1983).

39. D.J. Sandman and L. Samuelson, unpublished experiments.

40. J.B. Lando and M. Thakur, "Macroscopic Single Crystal of an Undoped Semiconducting Polymer", these proceedings.

26

Triplet Exciton Migration and Trapping in Molecularly Doped Polystyrene

Richard D. Burkhart and Augustine A. Abia / Department of
Chemistry, University of Nevada, Reno, NV

I. INTRODUCTION

The induction of electrical signals by the interaction of light
with matter is certainly a viable branch of electronics which has
many practical uses. Most applications of a light-generated
current flow may be represented by a diagram such as that shown
in Fig. 1 in which a charge separation occurs directly upon
photo-excitation. The resulting electrical potential is then
utilized for chemical or mechanical work.

An alternate way to accomplish such a task may be
represented in Fig. 2 in which a sensitizer molecule becomes
photoexcited and, as a result of exciton migration the energy is
transmitted to the donor and acceptor. Such a mechanism is
thought to operate in the chloroplasts of green leaves in the
initial stages of photosynthesis. In the following, a
description will be given of some recent work we have conducted
in an attempt to measure rates of exciton migration occurring

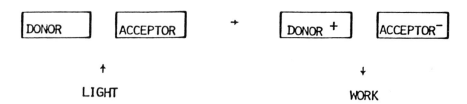

FIG. 1. Schematic diagram showing direct interaction of light
with matter to yield charge separation.

when molecules imbedded in plastic matrices are subjected to
photoexcitation. In particular, we will be interested in the
triplet electronic states which are produced and thus, the
triplet excitons.

The theoretical model most commonly used to describe the
transfer of triplet electronic excitation is that of Dexter who
proposed an electron exchange mechanism (1). In subsequent
studies Inokuti and Hirayama (2) applied this model to analyze
the triplet quenching experiments of Ermolaev (3) and, in the
process, they developed the relationship

$$n = \tau_o^{-1} \exp[\gamma(1 - \ell/\ell_o)] \tag{1}$$

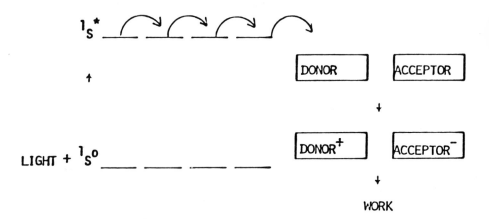

FIG. 2. Schematic diagram showing charge separation as a result
of interaction of light with a sensitizer followed by exciton
migration to a donor-acceptor site.

which has become known as the Dexter-Inokuti-Hirayama equation
(hereinafter to be called the DIH equation) wherein n is the
number of energy migration steps per unit time, τ_0 is the
triplet lifetime in the absence of quenchers and ℓ_0 is a
particular distance between donor and acceptor molecules such
that the probability of energy migration to a neighboring
molecule equals the probability of relaxation to the ground
state. The quantity represented by γ is $2\ell_0/\ell_b$ where ℓ_b is
an effective Bohr radius for a molecule and is related to the
spatial extension of its molecular orbitals.

A variety of uses have been made of the DIH equation.
Indeed, Inokuti and Hirayama were able to find ℓ_0 and ℓ_b
values for several different molecules thus yielding a
quantitative demonstration that triplet energy migration is a
relatively short-range process. For example, 1-bromonaphthalene
was found to have an ℓ_0 value of 13A. More recently, Morgan
and El-Sayed (4) have used the DIH relationship to investigate
the diffusive character of energy transfer among Eu^{3+} ions in
an amorphous solid.

In the present studies, the DIH equation plays a central
role for the interpretation of triplet migration rates among
dopant molecules present in polystyrene matrices. For example,
if a matrix is prepared having a solute concentration of 0.3M,
and if one assumes a completely random distribution of solute
molecules, then the nearest neighbor distance is given by (5) $\ell =$
$0.5539\ c^{-1/3}$ or 9.8A, c being the solute density. For 1,2-
benzanthracene as solute it has been estimated that $\ell_0 = 16$A and
that $\ell_b = 2$A. Also, it is known that $\tau_0 = 0.225$ sec. When these
numbers are inserted in eq. 1, n is calculated to be 2.2×10^3
steps per second.

In addition to this migratory activity, triplets may also
engage in radiative or radiationless deactivation to the ground
state as well as triplet-triplet annihilation to produce delayed
fluorescence. The specific rate constant of first order
processes is, of course, independent of any migratory events

which may be occurring, however, the specific rate constant for triplet-triplet annihilation is usually characterized as being dependent upon the frequency of encounters between potential reactants. For such processes the Smoluchowsky equation may be employed to evaluate the specific rate constant, that is

$$k_d = 4\pi R_o D N_A / 1000 \tag{2}$$

where k_d in the specific rate constant for the process in M^{-1} sec^{-1}, R_o is an encounter radius needed to insure reaction, D is the diffusion coefficient of potential reactants in cm^2/sec and N_A is Avagadro's number. Since, for a random flight process, $D = n\ell^2/6$ one may calculate $D = 3.5 \times 10^{-12}$ cm^2/sec for our example matrix. Then using $R_o = 9.8A$ one finds $k_d = 2.6 \times 10^3$ M^{-1} sec^{-1}.

The reason for working through this numerical example is to point out that the DIH equation, the Smoluchowsky equation and the theory of random flights, when used together in this way, are powerful tools for analyzing rates of triplet exciton migration. That is, in principle at least, one should be able to obtain k_d experimentally and to make a reasonable estimate of R_o. Thus, the diffusion coefficient for triplet excitons is available experimentally. Then, by using the random flight theory and the DIH equation ℓ_o and ℓ_b can be obtained from experimental values of the hopping frequency at different solute concentrations. Thus, in ideal cases, a rather complete determination of the migratory behavior of triplet excitons may be achieved.

An alternate approach is to evaluate the time dependence of the polarization of triplet luminescence following an excitation pulse of polarized light. Weber (7) has pointed out that the polarization, P, is related to the average number of migratory steps, $\bar{n}$, by the equation

$$P^{-1} - 1/3 = (P_o^{-1} - 1/3)(1 + \bar{n}) \tag{3}$$

where P_o is the polarization in the absence of migratory pro-

cesses. Using I_v and I_h to symbolize intensities of emission
through vertically and horizontally oriented polarizers,
respectively

$$P = (I_v - I_h)/(I_v + I_h) \tag{4}$$

where it is assumed that the excitation beam is vertically
polarized.

By combining these techniques of luminescence polarization
on the one hand, and utilization of the Smoluchowsky and DIH
equations on the other we may be able to characterize the
detailed mechanism of triplet migration in these plastic matrices
and to construct matrices to accomplish certain predetermined
tasks.

II. EXPERIMENTAL

The samples used in this work are prepared by dissolving the
desired quantity of polystyrene, usually 200 mg, and the dopant
species, usually from 1 to 20 mg, in a few ml of toluene. The
solution is placed in a glove box and purged with oxygen-free
nitrogen for about an hour after which the solvent is evaporated
away by heating the mixture at 120°C. This leaves a molten
solution consisting of polystyrene as solvent and, dopant
molecules as solute. The mixture is transferred to a preheated,
optically flat quartz plate where the molten mixture flows out
into a smooth film. Another plate is placed over the mixture to
form a sandwich which is then cooled to room temperature before
exposing it to air.

An alternate technique consists of allowing the molten
mixture to cool and harden and then to pulverize it to a powder.
It is then transferred to a 4mm O.D. quartz tube which has been
previously sealed to an outer 24/40 standard taper joint. When
the powdered sample has all been transferred, a vacuum stopcock
sealed to an inner 24/40 standard taper joint is attached to the
tube and closed to isolate the sample from the air. This

assembly is then removed from the glove box and attached to a
vacuum system where the sample tube is sealed off under vacuum
using a hand torch.

The spectroscopic equipment utilized consists of a
Perkin–Elmer MPF44–A spectrometer with phosphorescence attachment
and a locally constructed spectrometer which is especially useful
for examining both spectra and luminescence decays from the
plastic films. The custom built spectrometer has been described
previously (8,11) but it is worth repeating that the light
sources include both a conventional mercury–xenon lamp as well as
a variety of lasers obtained on loan from the San Francisco Laser
Center.

III. EXPERIMENTAL RESULTS

Fig. 3 is a spectrum of the delayed luminescence of 1,2-benzan-
thracene (BNZ) in a polystyrene film at ambient temperature. The

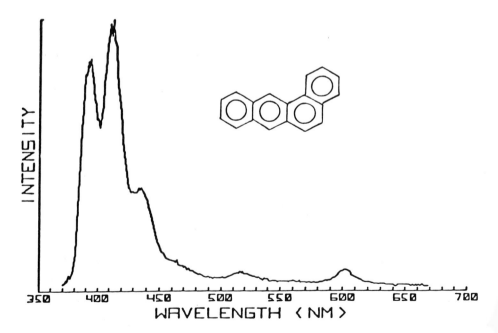

FIG. 3. Delayed luminescence spectrum of 1,2-benzanthracene in
a polystyrene matrix with delayed fluorescence (410nm), delayed
excimer fluorescence (520nm) and phosphorescence (605nm).

major features of the BNZ spectrum include delayed fluorescence
at 410nm, delayed excimer fluorescence at 520nm and a phosphorescence band at 605nm. The prompt luminescence spectrum consists
only of the fluorescence band. No prompt excimer fluorescence
has been observed from these films. It is clear that the BNZ
molecule displays a wide range of photophysical activity
providing many experimental access routes for the characterization of the migratory behavior of triplet excitons.

It was found that the phosphorescence decay exhibited
distinctly non-exponential behavior for sufficiently concentrated
matrices and at sufficiently short times following an excitation
pulse. A triplet lifetime of 225 msec for BNZ was finally
established for dilute samples as is seen in Fig. 4 where the
single exponential character persists for at least three
lifetimes (8). Furthermore, it was found that the

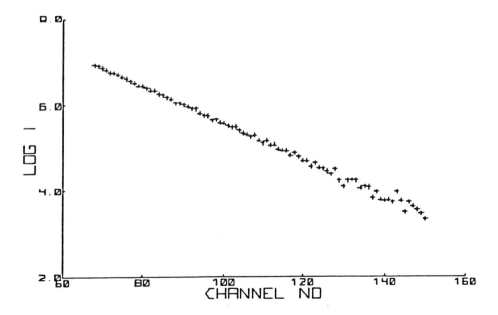

FIG. 4. Natural logarithm of phosphorescence intensity of
1,2-benzanthracene vs. channel number. Each channel equals
10 msec.

phosphorescence intensity is proportional to the first power of
the excitation intensity whereas the intensity of delayed
fluorescence emission depends upon the square of the excitation
intensity as shown in Fig. 5.   This, of course, is the sort of

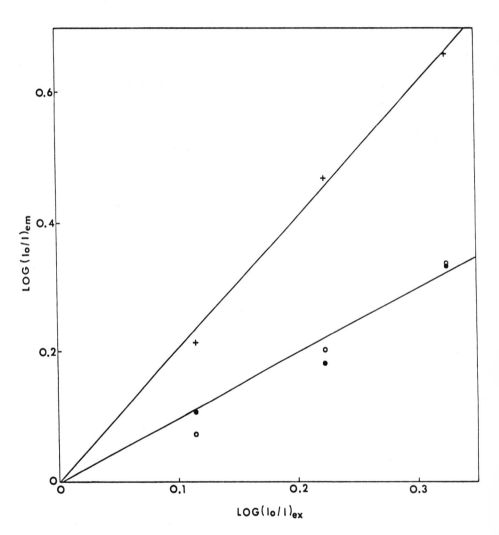

FIG. 5. Logarithm of the intensity of delayed luminescence
emission vs. the logarithm of the intensity of the excitation
light. Delayed fluorescence (+), delayed excimer fluorescence
(•), phosphorescence (o).

behavior expected if triplet-triplet annihilation is responsible for the delayed fluorescence.

Probably the most unexpected observation made on the BNZ-doped films was that the ratio, R, of delayed fluorescence to phosphorescence depends upon the square of the solute concentration, even after correcting for different rates of light absorption. This leads to a first major conclusion about the solute distribution in these films, namely that there is a propensity for BNZ molecules to form special solute pairs to which we have given the label "secluded pairs" (9). These are not excimer-like species but are simply two molecules separated by some distance r_p such that no third molecule is closer to either member of the pair than they are to each other. If either member of the pair is excited to the lowest triplet state either by direct excitation or as a recipient in an energy transfer process, there is a greater probability that this exciton will be transferred back and forth between the two members of the pair than there is that it will be transferred to a third molecule more distantly removed. If the ground state molecules which comprise the pair are labeled $^1S^0(1)$ and $^1S^0(2)$ and if the corresponding triplets are $^3S^*(1)$ and $^3S^*(2)$ then the trapping process may be represented by

$$^3S^*(1) + {}^1S^0(2) \;\rightleftarrows\; {}^1S^0(1) + {}^3S^*(2) \tag{5}$$

It should be emphasized that this secluded pair species, which we will symbolize by $^1S_P^0$ does not exist in any bound state as does an excimer species $^1E^*$ which in the non-bonded ground state will be symbolized as $^1E^0$ (10). Both $^1S_P^0$ and $^1E^0$ do have in common, however, a probable genesis in equilibrium processes occurring in the molten polystyrene at the time of film formation. That is the equations

$$2\,{}^1S^0 \;\rightleftarrows\; {}^1S_P^0 \tag{6}$$

$$\text{or} \quad 2\,{}^1S^0 \;\rightleftarrows\; {}^1E^0 \tag{7}$$

would represent these processes in the molten state and both
delayed fluorescence and delayed excimer fluorescence would be
expected to depend upon the square of the solute concentration as
is observed.

An even more interesting aspect of the solute distribution
has been found by a direct evaluation of the triplet decay using
triplet-triplet absorption measurements (9). In these studies,
the pool of triplets produced by an excitation pulse from a
nitrogen laser is subjected to an analyzing pulse by a visible
source at 485nm. A variable delay time was imposed between
excitation and analyzing pulses.

Since the molar absorptivity of BNZ triplets is known and
since the optical pathlength of the film could be deduced by
conventional spectroscopic measurements, these experiments on
triplet-triplet absorption make it possible to monitor the
absolute triplet concentration following an excitation pulse.
Using this method the data of Fig. 6 were obtained in the form

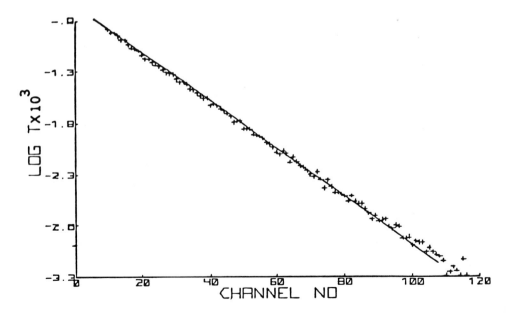

FIG. 6. Logarithm of the triplet concentration vs. channel
number. Each channel equals 5 msec.

of a first order kinetic plot over nearly three lifetimes
yielding a triplet lifetime of 215 msec which agrees quite well
with the earlier value of 225 msec obtained from the rate of
phosphorescence decay. At earlier time periods and with more
concentrated samples, there were significant deviations from
simple first order kinetics, however, and the data were analyzed
assuming concurrent first order and second order process. Such a
treatment is reasonable since the existence of triplet-triplet
annihilation has been confirmed. The kinetic equation is

$$-dT/dt = k_1[T] + k_2[T]^2 \qquad (8)$$

which may be integrated to yield

$$k_2 = \{((\,[T_o]/[T])\exp(-k_1t)-1\}/\{[T]_o[1-\exp(I-k_1t)]\} \qquad (9)$$

The results of these calculations are shown graphically in
Fig. 7.

Although the precision of these measurements is not excellent, there are two important features which stand out. The rate
constants are time dependent at times less than about 24 msec and
the long time values of k_2 are dependent on solute concentration.
Furthermore, the two most dilute samples yielded k_2 values on the
order of 10^3 M^{-1} sec^{-1}, which is the same order of magnitude
found in our earlier example calculation using random flight
theory, the DIH equation and the Smoluchowsky equation. Let us
now proceed a step further and calculate hopping frequencies of
triplet excitons.

The critical encounter radius required for triplet-triplet
annihilation has been estimated at 15A, however, all of the BNZ
matrices studied so far have concentrations such that the average
intermolecular separation distances were significantly less than
15A. Thus, to evaluate the diffusion coefficient from eq. 2,
R_o is set equal to ℓ, the average separation distance. The
results are summarized in Table 1.

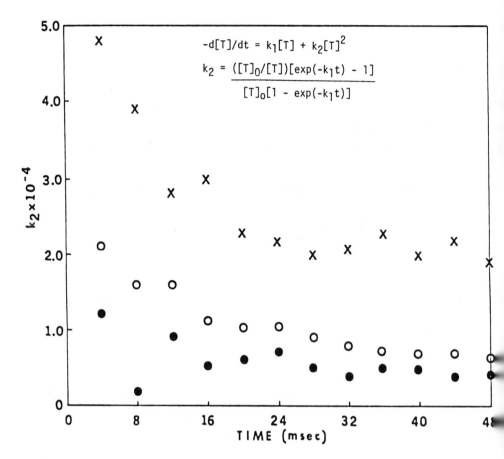

FIG. 7. Second order rate constants for triplet decay following
an excitation pulse. The solute concentrations are x (0.345 M),
o (0.260 M), ● (0.127 M).

If triplet exciton migration is to assume importance in the
construction of electronic devices, it will be necessary to
control the speed and direction of exciton flow. Although the
techniques for accomplishing this type of control are not
currently available, there is, at least, a rational approach for
determining migration rates based upon the kinetic-spectroscopic
approach used here for BNZ. In fact, the result suggests that
BNZ may be a good molecule to use as a test species in further
developments.

TABLE 1. Calculated Values of the Diffusion Coefficient, D, of Triplet Excitons, and the Exciton Migration Frequency, n, for 1,2-Benzanthracene in Polystyrene

Solute concentration	ℓ	$D \times 10^{12}$	$n \times 10^{-3}$	$k_2{}^a \times 10^{-3}$
(M)	(A)	cm^2/sec	sec^{-1}	$M^{-1} sec^{-1}$
0.127	13.1	4.0	1.4	4.0
0.260	10.3	9.0	5.1	7.0
0.345	9.4	28.0	19.0	20.0

[a] Taken from values of Fig. 6 at 40 msec.

One potential problem concerns the possibility that BNZ molecules are not, in fact, randomly dispersed in the polystyrene matrices but instead form clusters. The time dependence of the delayed fluorescence polarization given in Table 2 demonstrates the problem (11). Here the sample was excited with a vertically polarized laser source and the polarization, P, was determined at

TABLE 2. Experimental Values of the Delayed Fluorescence Polarization of 1,2-Benzanthracene in Polystyrene at Various Delay Times Following the Excitation Pulse.

Delay Time (msec)	Polarization
0.050	0.11
0.200	0.15
1.00	0.19
2.00	0.20
5.00	0.22
10.00	0.22
20.00	0.26

subsequent delay times following the excitation pulse. The fact
that the polarization increases with increasing delay time is an
indication that the environment of molecules emitting at long
times is different than at short times. In fact, it suggests
that the opportunities for exciton migration are much less for
the longer lived species. It seems reasonable to suppose,
therefore, that the longer lived species reside in less densely
populated regions than do their shorter lived counterparts.
Model calculations are currently underway to try to reproduce the
experimental time-dependent polarization. If clusters are being
formed in these matrices it may represent an opportunity as well
as a challenge since clustering suggests a tendency for these
molecules to assume some sort of ordered state. By suitable
modifications to the molecule or to the matrix, it may be
possible to influence the character of such ordered states.

REFERENCES

1. D.L. Dexter, J. Chem. Phys., 21, 836 (1953).

2. M. Inokuti and F. Hirayama, J. Chem. Phys., 43, 1978
    (1963).

3. V.L. Ermolaev, Opt. i Spektroskopiya, 6, 642 (1959)
    [English transl. Opt. Spectry. (USSR) 6, 417 (1959)].

4. J.R. Morgan and M.A. El-Sayed, J. Phys. Chem., 18, 3566
    (1981).

5. S. Chandrasekhar, Rev. Mod. Phys., 15, 1 (1943).

6. R.M. Noyes, Progr. React. Kinet., 1, 129 (1961).

7. G. Weber, Trans. Faraday Soc., 50, 552 (1954).

8. R.D. Burkhart, Chem. Phys., 46, 11 (1980).

9. R.D. Burkhart, J. Phys. Chem., 87, 1566 (1983).

10. General discussions of excimers may be found in J.B. Birks,
    Photophysics of Aromatic Molecules, John Wiley and Sons,
    New York, 1970, pp. 301-370.

11. R.D. Burkhart and A.A. Abia, J. Phys. Chem., 86, 468
    (1982).

27

Fast Optical Signals and Their Electrical Modulation from the Adsorbed Tetrasulfonated Phthalocyanines

B. Simic-Glavaski / The Chemistry Department and Case Center for Electrochemical Sciences, Case Western Reserve University, Cleveland, OH

INTRODUCTION

Optical devices and their optical bistability have received considerable attention in recent years (1) because their functional properties can be successfully employed in switches, optical memory elements, pulse shapers and several other devices. So far, most of the optical bistable devices have been exploiting non-linear refractive index characteristics of a medium in the resonator of the Fabry-Perot interferometer. However, the physical size of the interferometer is a serious obstacle in the effort to miniaturize optical devices.

This report describes optical properties of the adsorbed monolayer of water-soluble tetrasulfonated phthalocyanines, abbreviated TSPc, on silver substrates in contact with different electrolytes. A high intensity of the optical signals, their short duration (in the range of 10^{-13} seconds), and their

multilevel and multi-output response, which can be modulated by
electrical means, are achieved with the combined techniques of
Surface Enhanced Raman Spectroscopy (SERS) and cyclic voltammetry. The electrical modulation of the optical signals is
usually accompanied by a bistable hysteresis curve which is
caused by oxidized and reduced states of the adsorbed phthalocyanine molecule. Fundamental physical properties of the
phthalocyanine molecules, through their vibrational and intramolecular charge transfer mechanisms, have been employed and
experimentally demonstrated in the example to be discussed of the
adsorbed H_2-TSPc on a silver electrode. Other metallophthalocyanines studied, such as Fe-, Co- and Cu-TSPc, showed similar
behavior. On the other hand, the functionally versatile phthalocyanine molecules have shown the wide spectrum of their physico-
chemical properties as demonstrated in their electrical conductivity, photoactivity, catalytic activity. Naturally their
unusual properties have attracted considerable research interest
(2).

The molecule is composed of four pyrrole and benzene rings
which form greater inner and outer rings, respectively (see Fig.
1). The inner ring may contain two protons which form a
metal-free phthalocyanine H_2-Pc, or may have a four-coordinated
central metal ion which creates a variety of
metallophthalocyanines (M-Pc).

The Pc molecules are insoluble in aqueous media unless they
are sulfonated, as in the case of the tetrasulfonated (TSPc),
which structure is displayed in Fig. 1. The detailed study of
the M-TSPc (3) has shown that sulfonation has a minimal effect on
the overall molecular characteristics.

EXPERIMENTAL PROCEDURE

The metal-free tetrasulfonated phthalocyanine (H_2-TSPc) was
prepared by following the procedure described by Linstead and

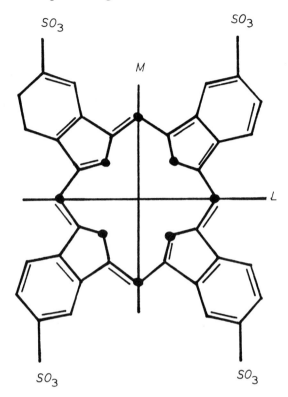

FIG. 1. The schematic structure of a tetrasulfonated phthalocyanine isomer. Nitrogen atoms are indicated with heavier dots.

Weiss (4). Then the H_2-TSPc was adsorbed on the silver
electrode.

The experimental set-up for the electro-optical studies of
the adsorbed H_2-TSPc on the Ag electrode is shown in Fig. 2.
The electrode interface with adsorbed H_2-TSPc was illuminated
with an argon ion laser line operating at 514.5 nm with output
power of about 50 mW. The incident angle was 78° relative to the
surface normal.

The scattered light is observed at 90° and analyzed by a
double monochromator (Spex 1400). A photomultiplier ITT FW130
and photon counting equipment were used as a detection system.

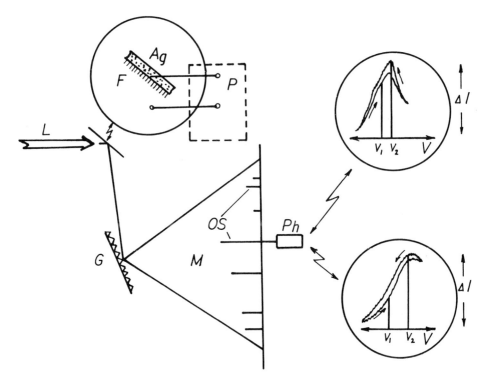

FIG. 2. Diagram of the electro-optical system. L is the laser,
Ag is the silver electrode with H_2-TSPc film F, P is the potential generator, M is the monochromator with the grating G, OS
are the optical signals, Ph is the photomultiplier, V is the
potential, and ΔI is the change of the optical signal.

RESULTS AND DISCUSSION

The laser excited interface of Ag|H_2-TSPc| electrolyte emits

scattered light in the form of the Surface Enhanced Raman

Scattering where the activated silver electrode acts as an

amplifier with a typical gain factor of about 10^6. A high

enhancement of the Raman signal by the activated silver electrode

is still the object of considerable theoretical and experimental

study and controversy (5). However, the SERS spectra obtained

from the adsorbed H_2-TSPc and other M-TSPc reliably reproduce

molecular vibrational properties of the macrocyclic species in

their solutions of solid phases (3).

It is also known that the intensity of the SERS spectra and their shapes are a function of the applied electrode potential (5).

SERS spectra obtained from the adsorbed H_2-TSPc on the silver electrode are shown in Fig. 3. The spectra are obtained for various electrode potentials and the intensities of the Raman lines show an appreciable change when the electrode potential is altered in discrete steps. The spectra in Fig. 3 were recorded

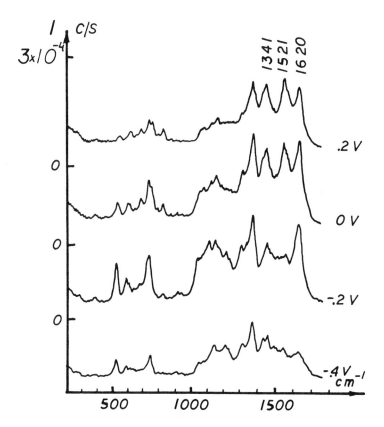

FIG. 3. Surface Enhanced Raman I_{HV} or I_{PS} spectra from the adsorbed H_2-TSPc on the silver electrode at various electrode potentials. The interface at pH $\simeq$ 1 was illuminated with the argon ion laser line at 514.5 nm at about 50 mW output power. Resolution is $\pm$ 2 cm^{-1}; scanning rate was 10 min per spectrum. Intensity I is in count/s.

with a relatively high resolution of the monochromator by using a
2.5 cm^{-1} band pass. A detailed analysis and the band
assignments of the SERS spectra obtained from the adsorbed
H$_2$-TSPc and other metallo-phthalocyanines are described
elsewhere (3). It was also shown (3) that molecular adsorption
on the Ag electrode is mainly due to a phys-adsorption mechanism,
without alteration of the molecular symmetry of the adsorbed
macrocyclic species which adsorb through the sulfo groups or an
edge-on configuration. The SERS spectra shown in Fig. 3 were
obtained with a non-resonant laser excitation and an additional
gain can be achieved by use of resonant excitation (3).

     For a particular Raman line, by increasing the band pass of
the monochromator to about 20 cm^{-1} in order to accommodate any
possible frequency shift of the Raman line as a function of the
applied potential, and by continuously varying the potential of
the electrode within the limits of the oxidation-reduction cycle,
one can obtain a continuously modulated Raman signal as a
function of the applied potential. The intensity change of the
Raman line as a function of the electrode potential is shown in
Fig. 4a for the Raman band at 1346 cm^{-1} obtained from the
adsorbed Fe-TSPc on the silver electrode in acid medium at pH1.
The Raman band at 1346 cm^{-1} is due to the N-C stretching mode in
the pyrrole ring (3). A cyclic voltammogram from the adsorbed
Fe-TSPc on the Ag electrode is shown also in Fig. 4b. Two waves
in the cyclic voltammogram indicate oxidation-reduction processes
of the macrocyclic ligand. Both the volt-Raman graphs and the
cyclic voltammogram are reversible and are not due to adsorp-
tion/desorption phenomena. The shape of the volt-Raman graph is
independent of the potential scanning rate up to 100 mV/s.
Volt-Raman graphs are characterized by a hysteresis. However, a
major portion of the hysteresis can be eliminated if the potential window of the electrode potential is limited to the values
of 0.2 V and the potential value of the more negative oxidation-
reduction voltammogram wave. This phenomenon of the hysteresis

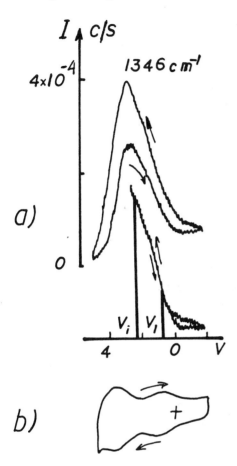

FIG. 4.   a) The volt-Raman graph for 1346 cm^{-1} Raman band obtained from Fe-TSPc adsorbed on the silver electrode in electrolyte of pH $\simeq$ 1.   The scan rate of the volt-Raman graph was 10
mV/s and the laser excitation was 514.5 nm lines. V_i and V_1
are the biasing potentials.   b) Cyclic voltammogram obtained
from the adsorbed Fe-TSPc on the Ag electrode with 100 mV/s
scan rate.   The arrows indicate direction of the scanning.
Potential V is versus SCE.

is interpreted as being due to the amounts of oxidized or reduced

states of the adsorbed macrocyclic molecule.   A relatively huge

change of the optical signal of about 40,000 counts is achieved

in a very narrow potential window of about 400 mV.   On the other

hand, depending on the biasing potential V_i of the silver

electrode and by switching the laser excitation of the interface
on and off, one can obtain short optical pulses whose duration is
related to the excited vibrational lifetime of the particular
Raman line.   Typical values of the vibrational lifetimes are
about $2-5 \cdot 10^{-13}$ s for the adsorbed TSPc.   The intensity of the
optical pulses will be within the wave envelope of the hysteresis
and will be dependent on the direction of the raising potential.

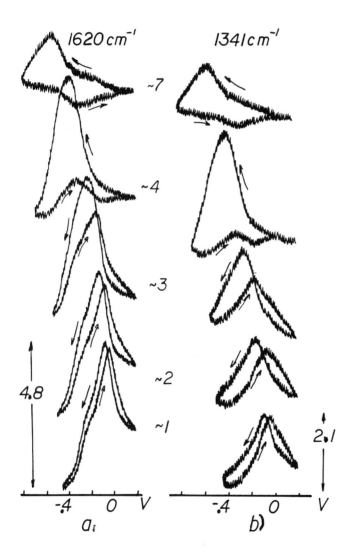

Volt-Raman graphs obtained from several Raman bands from the
adsorbed H_2-TSPc on the silver electrode are displayed in Figs.
5a, b and c. The pH of the supporting electrolyte is the
experimental parameter. Cyclic voltammograms are also shown in
Fig. 5d for a comparative analysis.

The Raman band at 1341 cm^{-1} is due to N-C stretching modes
in the pyrrole ring and the bands at 1521 and 1620 cm^{-1} are not

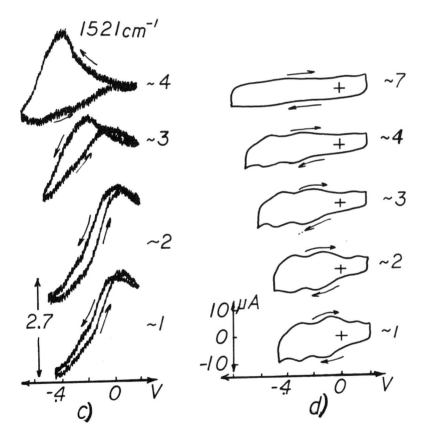

FIG. 5. Volt-Raman graphs for three Raman bands obtained from
the adsorbed H_2TSPc on the silver electrode for various pH
values. The intensity change of the optical signal is in 10^4
counts units. a) volt-Raman graph for 1620 cm^{-1} band; b) volt-
Raman graph for 1341 cm^{-1} band; c) volt-Raman graph for 1521
cm^{-1} band; d) corresponding cyclic voltammograms.

positively identified (3). Two distinct types of the volt-Raman graphs are noticed, ones with a pronounced maximum and others with a gradual change in intensity as a function of the oxidation-reduction cycle. The hysteresis in the volt-Raman graphs appears to be related to the pH of the supporting electrolyte and in electrolytes with higher pH values the volt-Raman graphs appear as elongated ellipses. A maximum in the voltammetry curve corresponds to the greatest rate of the change in the oxidation state of the adsorbed species and hence should correspond to a maximum in the first derivative of the volt-Raman graph. Moreover, the maxima in the volt-Raman graphs show a pH dependence of 60 mV/pH decade which is expected for the loss of a proton with reduction of the ligand. A comparative analysis of the vibrational mode at 1341 cm^{-1} and the shape of the volt-Raman graph suggests a high sensitivity to the oxidation state of the ligand and particularly to the delocalization of the electron in the pyrrole group. The more positive shift of the oxidation maxima in Figs. 5a and b most probably can be associated with additional protonation of the inner ring, which is characteristic for H_2-TSPc in acidic media (3,6). For pH values above 4, the oxidation maxima are more negative and the looped structure disappears. This phenomenon most probably reflects a change of molecular symmetry from D_{4h} to D_{2h} (6).

In conclusion, the experimental data obtained by SERS and cyclic voltammetry from the adsorbed H_2-TSPc on the silver electrode show a direct electrical modulation of the optical signals which are spatially resolved and of different intensities, and where the intramolecular charge transfer plays a significant role. Furthermore, the short duration of optical signals are limited by the vibrational lifetimes of the specific vibrational modes.

ACKNOWLEDGEMENT

The extensive study on catalytic properties of phthalocyanines by B. Simic-Glavaski, S. Zecevic and E. Yeager (see Ref. 3) has been stimulating and the assistance of Drs. S. Zecevic and E. Yeager is appreciated.

REFERENCES

1. P.W. Smith, E.H. Turner and P.J. Maloney, I.EEE J. Quantum Electron QE-14, 207 (1978).

2. A.B.P. Lever, in Advances in Inorganic Chemistry and Radio-chemistry, Eds. H.J. Emeleus and A.G. Sharpe, 7, 27 (1965).

3. B. Simic-Glavaski, S. Zecevic and E. Yeager, J. Raman Spec. 14 338 (1983), J. ACS 107, 5625 (1985).

4. R.P. Linstead and F.T. Weiss, J. Chem. Soc. 2975 (1950).

5. "Surface Enhanced Raman Scattering", Eds. R.K. Chang and T.E. Furtak, Plenum Press (1981).

6. L. Edwards and M. Gouterman, J. Mol. Spectrosc., 33, 293 (1970).

V
Enroute from Molecular Technology to Future Systems

28

Molecular Machinery and Molecular Electronic Devices

K. Eric Drexler / M.I.T. Space Systems Laboratory, Cambridge, MA

Like ordinary electronics and machinery, molecular electronics and molecular machinery promise to be closely linked in development and use. They will rest on a common technology base, the ability to assemble structures to complex, atomic specifications. Molecular machines can help assemble molecular electronics, molecular electronics can help direct molecular machinery, and both can be used in hybrid systems. They thus seem likely to emerge together, grow together, and influence each other strongly.

Modern microelectronic systems control macroscopic machines, despite mismatches in scale and power level. Macroscopic machines make microelectronic systems, despite the mismatch of scales and the need for great precision. Microelectronic technology has even brought a limited micromechanical technology (1), though current techniques cannot fabricate the range of moving parts found in ordinary machines.

Molecular machinery, in contrast, will bring a full range of mechanical functions, and will match molecular electronics in both power and scale, encouraging closer interfacing. Accordingly, molecular machinery seems important to the future of molecular electronics. Though molecular electronics can make impressive progress through chemistry, microtechnology, and biological-style self-assembly, advanced molecular electronics will likely be built by molecular machinery, because it will eventually provide the best tools for the job.

A PATH TO MOLECULAR MACHINERY

With the advent of protein design (2,3,4), it will become possible to design self-assembling protein systems, perhaps with the aid of connectors made from antibodies that bind to predetermined sites on the functional proteins (5). Protein systems could bind other molecules, holding them together to serve as circuit elements (6), or wielding them as tools to effect chemical reactions after the fashion of enzymes (2). As Table 1 indicates, proteins and other biomolecules provide examples of components serving as frameworks, moving parts, bearings, motors, and actuators. Since the function of a machine follows from the functions of its components, this indicates that complex, power-driven machinery can eventually be constructed on a molecular scale.

One class of protein machines could hold a molecular workpiece and bring molecular tools to bear, assembling reactive groups in a site-specific manner. Such machines could be programmed by molecular tapes, like early numerically controlled machine tools (programmed with punched tapes) or like the ribosome (programmed with mRNA). With a variety of tools under programmed control, such a machine could assemble a wide range of molecular structures to atomic specifications.

TABLE 1. Macroscopic and Molecular Mechanical Parts

Technology	Function	Molecular example(s)
Struts, beams, casings	Transmit force, hold positions	Microtubules, cellulose, mineral structures
Cables	Transmit tension	Collagen
Fasteners, glue	Connect parts	Intermolecular forces
Solenoids, actuators	Move things	Conformation-changing proteins, actin/myosin
Motors	Turn shafts	Flagellar motor
Drive shafts	Transmit torque	Bacterial flagella
Bearings	Support moving parts	Sigma bonds
Containers	Hold fluids	Vesicles
Pipes	Carry fluids	Various tubular structures
Pumps	Move fluids	Flagella, membrane proteins
Conveyor belts	Move components	RNA moved by fixed ribosome (partial analogue)
Clamps	Hold workpieces	Enzymatic binding sites
Tools	Modify workpieces	Metallic complexes, functional groups
Production lines	Construct devices	Enzyme systems, ribosomes
Numerical control systems	Store and read programs	Genetic system

These structures could themselves serve as machine parts having properties (strength, thermal stability, etc.) not found in the coiled, hydrated polypeptide chains of protein. Engineers could use such components in second-generation machines, including machines for assembling reactive groups to workpieces; with proper design, they could assemble atoms into virtually any stable pattern (2). Machines with this capability may be called assemblers.

In 1959, Richard Feynman outlined an alternative path to this capability, based on the use of larger machines to build successive generations of smaller machines (7). He stated, "...It would be possible (I think) for a physicist to synthesize any chemical substance that the chemist writes down...How? Put the atoms down where the chemist says, and so you make the substance."

Where technology based on bulk processes is often limited by fabricated difficulties, technology using assemblers will be limited chiefly by physics and by our ability to design things. Assemblers will naturally find use as a direct and versatile means for building molecular devices and systems. Regardless of the assembly techniques used in early MED, construction of molecular machinery to aid future assembly will rapidly become attractive. For the present, the prospect of assemblers able to build almost any stable configuration of atoms should encourage theoretical work on MED to consider a broader range of structures than might otherwise be of interest.

SCIENCE AND ENGINEERING

In pursuing theoretical work on the future of molecular technology, it is useful to distinguish between two modes of discussion. Exploring the future of molecular technology can involve scientific speculation. Where a physical system is complex, involving (for example) cooperative quantum-mechanical effects, one may be unable to calculate the system's behavior;

statements regarding it may then be speculative. Predictions of
high-temperature superconductive behavior (8) fall in this
category--we do not yet know whether other effects will disrupt
the calculated effects.

Discussion of the future of molecular technology can,
however, avoid scientific speculation. Where a technology has
close chemical, biochemical, or macro-mechanical analogies, or
where motions can be calculated by Newtonian mechanics,
possibilities can be firmly established. Certain statements may
then be described, not as scientific speculation, but as
engineering projection. The price of limiting discussion to
projection, of avoiding speculation, is that one must omit many
interesting ideas, some of which may well prove true. The
reward, however, is that by using solid, dull bricks one can
build a firm structure of greater size. Where speculation piled
on speculation rapidly weakens, firm projections can often be
combined to form the basis of further projections.

An historical example illustrates this. In the 1930's (and
before), principles known to rocket designers showed the
possibility of flight into space. Indeed, knowledge of the
strength of materials and the energy content of liquid
propellants guaranteed its essential feasibility. Knowing that
orbit could be achieved, it was clear that space stations could
be built and ships assembled to reach the Moon and beyond. These
projections of possibility were correct, though the historical
path to the Moon differed from anyone's detailed speculations.

In a similar fashion, the principles of chemistry and
molecular biology now show the possibility of certain classes of
molecular machinery. Assemblers form one such class. Their
mechanisms would be analogous to those of ribosomes, enzymes,
ordinary chemical reactions, and existing molecular machinery.
They, too, will open a new frontier: they will make possible a
technology limited not by fabrication problems, but instead by
the laws of physics and by our ability to conceive and design.

SOME NON-ELECTRONIC APPLICATIONS OF ASSEMBLERS

The rewards from miniaturization of electronics indicate the
value of molecular electronics: this, in turn, shows the value of
one application of assemblers. Other applications, however, seem
comparably important.

Since assemblers could build virtually arbitrary molecular
structures, systems based on assemblers could, if properly
supplied and programmed, build copies of themselves. Like
ribosomes in bacteria, assemblers could form the core of
replicators; also like ribosomes, they could be made cheaply in
bulk through replication. With inexpensive assemblers, many
products could be made for energy costs comparable to those of
biological materials (2) such as bacteria and wood.

This makes non-electronic products of interest. Diamond,
for example, consists of properly arranged carbon atoms and
should be inexpensive to manufacture with molecular machinery;
with a tensile strength of about 50 giganewtons per square meter,
it has roughly 50 times the strength to mass ratio of the
aluminum used to build the space shuttle. In an atomically
tailored fibrous composite, diamond could yield an exceedingly
strong, tough engineering material.

Unlike passive structural materials, muscles can contract by
means of molecular machines (the actin/myosin system) that pull
fibers past each other. Analogous artificial molecular machines
presumably could draw energy from various electrical or chemical
sources and could exceed the strength and operating temperature
of muscle proteins; the fibers they pull could have great
strength. Strong, rugged muscle-like materials thus seem
possible.

Assemblers could also make optical fibers with extremely low
absorption, scattering, and dispersion. Detailed control of
atomic patterns eliminate absorption caused by impurities, and
use of a crystalline structure could eliminate scattering caused

by the inhomogeneities inherent in glassy materials. Finally, free use of concentric layers of materials with different dispersion relationships could permit the dispersion curve to be more accurately flattened in a particular frequency band.

These examples describe but a few of the materials properties and device characteristics that could be improved through use of assemblers (novel devices will be made possible as well). Indeed, it is difficult to think of a form of hardware that could not be improved by removing fabrication constraints, that is, by making possible more nearly optimal arrangements of atoms. Assembler technology will bring a fundamental revolution reaching far beyond the field of molecular electronics; the related technology of molecular machinery, however, will be integrated with MED in important ways.

ASSEMBLERS, MED, AND MMD

Modern microelectronic systems function through essentially electronic effects, rather than through mechanical, ion displacing effects. Although ion motion can affect electron behavior (through phonon-electron scattering, polaron formation, etc.), these devices rely chiefly on electronic degrees of freedom.

As currently used, "molecular electronics" covers both essentially electronic devices, such as switches based on tunnelling (9), and devices involving substantial mechanical degrees of freedom (10). Since effects such as tautomeric conformational changes and soliton motion (11) involve mechanical degrees of freedom, and since assembly of molecular devices is an inherently mechanical process, proposals for molecular electronics already involve a measure of molecular mechanics.

Molecular machinery can help build purely electronic devices. Assemblers can make precise connections between conducting one-dimensional polymers and switching groups,

providing a general solution to the problem of connecting
molecular electronic devices to form molecular electronic
circuits. Further, they can improve electronic devices which
remain stylistically similar to the conductor/semiconductor/
insulator structures now used.

Examples of the latter might involve very thin, crystalline
wires with smooth surfaces, replacing surface scattering with
specular reflection, eliminating one source of increased
resistance in thin wires. In semiconductors made with similar
precision, doping atoms could be placed in regular arrays to
tailor scattering properties. Further, use of materials with
large bandgaps (such as sodium chloride, with a bandgap over 7.5
eV) should permit use of thinner insulation while keeping
tunnelling currents within acceptable bounds--the use of such
materials in stable devices will be aided by the ability,
implicit in the abilities of assemblers, to hermetically seal
circuits. Finally, circuits could be arrayed in three dimensions
(to shorten signal transmission times) and could be interwoven
with a branched system of tubes, circulating cooling fluid in a
pattern based on the principles of heat and mass transfer
underlying the artery-capillary-vein system in the mammalian
circulatory system.

MED and MMD will share similar scales and similar
characteristic energies. The energy required to move an arm on a
molecular assembler by a nanometer in a nanosecond, for example,
would be about an electron volt--also a reasonable energy for
signal processing in molecular electronics. Electronic changes
could be coupled to mechanical motions either through direct
application of electrostatic forces, or through rearranging bonds
and changing molecular conformations. Like electronic devices,
molecular machines lack wear mechanisms (though damage mechanisms
remain); further, they can operate as swiftly as today's
microcircuits (if not as swiftly as tomorrow's molecular
circuits).

These facts make molecular machinery and electronics extraordinarily compatible. Although differences of scale and energy preclude direct and intimate interfaces between, say, macroscopic robot arms and microcircuits, ordinary electronics and machinery are nonetheless combined; greater compatibility may encourage more intimate hybrids among the molecular analogs of macroscopic technologies.

In one class of hybrid system, molecular machinery could be used to reconfigure circuitry that would then compute by means of faster, purely electronic effects. If one identifies mechanical systems with ionic degrees of freedom, then a simple example of this would be any use of a conformation change to reconfigure a molecular circuit that then relies on electronic switching to perform actual computation.

Software has long been used to generate software, through compilers; recently, software has been used to generate hardware designs through so-called silicon compilers. Future systems might incorporate both molecular-scale manipulators and building-block circuit elements, using a "circuit compiler" to direct assembly of custom logic circuits to relieve bottlenecks that might appear in a repetitive computation with more general-purpose circuits. Full customization might involve assemblers using reactive molecules as building blocks; swifter construction (but slower computation) would result from use of prefabricated gates, wires, and so forth. The small scale of molecular technology would allow such compilation to hardware inside a user's machine, in response to a user's programs.

This sort of integration of MMD and MED can also ease problems of device failure resulting from radiation. As has been noted, redundancy and error correction will be needed to ensure system reliability (12); repairs cannot substitute for this, since they can neither compensate for circuit noise problems nor negate the effects of damage in mid-computation. Repairs will be useful, however, to prevent accumulation of damage. This might

be achieved by means similar to those proposed above for circuit
assembly and reconfiguration; the ability to assemble and
disassemble a system, combined with knowledge of the structure
desired, implies the ability to perform repairs.

MECHANICAL COMPUTERS AND PERIPHERALS

Mechanical elements could themselves perform computation and data
storage, after the fashion of Charles Babbage's mechanical
computer design of the mid-1800's, or of punched-tape systems and
mechanical desk calculators. In speed, mechanical devices will
almost certainly be inferior to electronic devices; in
data-storage density, and perhaps gate density, they may well
prove superior. From a conceptual point-of-view, they offer the
advantage of relative simplicity. Since MMD's can often be
adequately described by Newtonian mechanics (and statistical
mechanics), it becomes relatively easy to set approximate lower
bounds to their capabilities--and thus to the capabilities of
molecular computational systems--while avoiding speculation.
Such lower bounds can then serve as points of departure for
further projections.

A simple calculation provides a lower bound on information
storage density. Given abilities less general than those of
assemblers, a device could add and remove R-groups to a
polyethylene molecule. Assume, for the sake of concreteness,
that the side groups are hydrogen and fluorine; in a
properly-oriented, partially-fluorinated polyethylene molecule,
R-groups can then store two bits per carbon atom. Such molecules
could serve as tape memory, analogous to punched paper tapes or
RNA; bits could be read by sensing the size of the R-group with
a mechanical probe. Approximating the bulk density of this
polymer as the average of those of polyethylene and
polytetrafluoroethylene, and allowing a factor of three for the
volume of reel, drive, and read/write apparatus, tape memories
could store over 15 bits per cubic nanometer.

Access times may also be estimated. Tape speed could be comparable to that in macroscopic devices; to first order, energy densities and dynamic tape stresses remain constant regardless of scale. If a reasonable tape length is a micron and a reasonable speed is a meter per second, then maximum access times would be about a microsecond. At room temperature, the kinetic energy per atom in the moving tape would be on the order of 1/10,000 kT. Such a tape drive would store about 16K bits, and be roughly 10 nanometers across.

In connection with rotating parts, an upper bound can also be set on the energy dissipated in a sigma-bond bearing. Consider hindered rotation: in the unrealistic worst-case, the energy dissipated per rotation would be the amplitude of the potential function times the number of peaks (in practice, of course, energy would be recovered in traveling from peak to valley). For ethane, the rotational barrier is about 12.0 kJ/mole (13), and the molecule has threefold symmetry; the resulting upper bound on energy dissipation is about 0.4 eV/rotation. This is substantial, and even with a more realistic model for energy dissipation, the associated torque may be too large. Fortunately, separating the two methyl groups in ethane with a pair of triple-bonded carbon atoms (to make dimethylacetylene) reduces the rotational barrier to a listed value of 0.0 kJ/mole (13); even taking this as 0.05 kJ/mole, both energies and torques would be low, resulting in good bearing properties. Some carbon-carbon sigma bonds also have rotational barriers listed as 0.0 kJ/mole.

MECHANICAL RAM AND RANDOM LOGIC

A mechanical computer could transmit information by pushing and pulling sliding rods of molecular scale; ones and zeros could be represented by clamped and unclamped rods. With this approach one can design the moving parts of a random access memory. Some

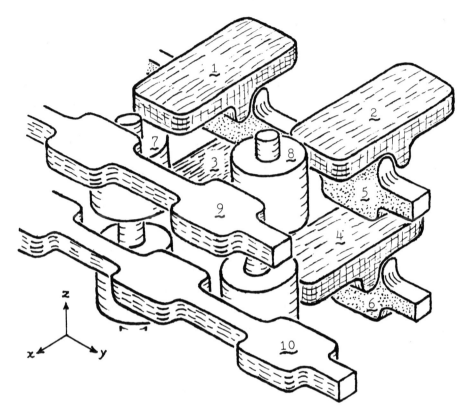

FIGURE 1.   Schematic sketch of four mechanical random-access
memory cells. See text for discussion.

detail will be given here, to provide a more concrete image of
how molecular machinery and mechanical computers might work.

Picture a random-access memory as a block holding parallel
planes of memory cells. Outside the block, mechanical linkages
perform binary decoding, addressing (unclamping) particular rods
reaching into the interior of the block. Figure 1 is a schematic
sketch of one approach to building memory cells inside such a
block, omitting atomic detail for clarity. It shows four cells,
storing four bits. Supporting frameworks are also omitted for
clarity, but constrain each moving part to (essentially) one
mechanical degree of freedom. The sliders (labeled 1, 2, 3, and
4) can move back and forth in the x direction, the write-protect

rods (5 and 6) and the row-selector rods (9 and 10) in the y direction, and the read/write rods (7, 8) in the z direction.

Operation of the memory depends on rod motions; bits are encoded by slider positions. The selector rod determines which row the read/write rods respond to: if it is in the position shown by 10, the read/write rods can move regardless of slider positions; if it is in the position shown by 9, then a forward slider (such as 1) suffices to clamp the read/write rod, blocking motion of the knob past the resulting constriction. The gap between 9 and 2, however, remains wide enough to allow the read/write rod to move upward. Thus, the response of a set of read/write rods to upward tugs depends on the row selected and on the positions of the sliders along that row.

To change slider positions in a row (to write), the corresponding write-protect rod (such as 5 or 6) is moved away from the position shown, placing its knobs in the spaces between the sliders and permitting them to move. If the sliders are biased to move forward unless pushed back (perhaps by the pressure of a few trapped gas molecules), all will then move to the "clamping" position, and could be locked there by returning the write-protect rod to its illustrated position. If the row-selector rod is in the activated position (like 9), however, and the write-protect rod is in the unlocked position (unlike 5), then lifting a read/write rod (like 8) will push the corresponding slider back; it can then be locked in this position. Thus, manipulation of the three sets of rods can both change and determine the position of any given slider, permitting the system to function as a random-access memory.

This scheme is abstract; it could be implemented with rods of tensioned steel cable linking cast-iron blocks. A plausible molecular implementation, however, would use tensioned carbyne chains as the rods. Carbyne, a pure carbon polymer with alternating single and triple bonds, seems the best material to use for thin, longitudinally-stiff rods. The knobs on the write-

protect and row-selector rods could then be phenyl groups, and
those on the read/write rods bicyclooctane groups; the sliders
might be halogenated 9,10-dihydroanthracene molecules. Indeed,
in all cases certain hydrogen atoms would be replaced with bulker
groups (such as fluorine, chlorine, or methyl moieties) to
provide the proper steric hindrances. This approach results in a
RAM cell some 1.4 by 1.4 by 2.5 nanometers on a side, or roughly
5 cubic nanometers: it is thus some 75 times less dense than the
tape memory described above, but offers faster access times (see
below).

Memory cell volumes can also be estimated through comparison
to present memory chips. As a crude approximation based on round
numbers, take a typical line width as three microns and the
corresponding molecular rod diameter as three angstroms. Since
the molecular rods are as thick as they are wide, consider the
memory chip's active circuits to be three microns thick, for
comparative purposes. Allowing a chip area of one square
centimeter, the total equivalent active volume of the chip is
300,000,000 cubic microns; this scales to 300,000,000 cubic
angstroms for the molecular mechanical memory. Assuming that the
chip stores 64K bits, the volume per bit is 4,700 cubic
angstroms, or 4.7 cubic nanometers. This result, based on the
first set of numbers that came to mind, is closer to 5 cubic
nanometers (the result of the design exercise above) than one has
any reason to expect.

This method, scaling from integrated circuits, can be used
to estimate the volume of mechanical random-logic "circuits". If
anything, the move from two to three dimensions should give a
relative advantage in "wiring" molecular machines, so this method
may tend to overestimate the volume of molecular random logic.
Since simple CPU's have for years occupied a single chip, this
rough scaling relation (10,000-to-one in linear dimension,
one-trillion-to-one in volume) suggests that the molecular
mechanical equivalent of a simple CPU should fit in roughly
0.0003 cubic microns, a volume less than 0.07 microns on a side.

This result may be surprising, but contradicts no calculations of which I am aware; nevertheless, it deserves a closer look. In particular, can estimates be made of the effects of thermal fluctuations on the computational elements, and can estimates be made of computational speed?

SPEED AND NOISE

The single bonds of carbyne are shorter and stronger than those of diamond. Scaling their properties from those of diamond bonds (14) (based on their greater strength and lesser length, and assuming a similar potential energy curve as a function of nuclear separation), a carbyne chain should have a strength of about 10 nanonewtons, and a linear modulus (analogous to bulk modulus) of about 150 nanonewtons.

To limit errors arising from thermal fluctuations and mechanical noise, a substantial energy must be required to deform a rod excessively. For a suitably designed system, "excessive" deformation (the deformation required to move a bulky group out of the way of another moving part) might be about one nanometer; if the elastic energy for this process is 1 eV or more, thermal errors (at least) will be rare. A carbyne rod can approach 0.5 microns in length before the elastic energy required to stretch it by a nanometer falls below 1 eV. This gives an estimate of maximum rod length; reliable signal transmission over greater distances could be accomplished through use of relays or thicker rods. In RAM arrays of the sort described above, however, a variety of factors (such as friction) may limit the size of a block to less than 0.5 microns.

Signal transmission over these distances can be relatively swift. The speed of sound in carbyne is roughly 30 km/s; an acoustic signal would thus cross 0.5 microns in less than 20 picoseconds. Application of a maximal accelerating force (10 nanonewtons) to a 0.5 micron rod would move it a nanometer in about 4 picoseconds if it could respond as a rigid body.

Allowing 0.2 nanoseconds for the motion (thus simplifying
dynamics by allowing acceleration over 10 times the acoustic
signal time) would require an accelerating force of only 4
piconewtons (about 0.00025 of the breaking stress) and a peak
speed of 10 m/s (implying a kinetic energy of about kT at room
temperature). Thus, a properly designed mechanical system could
apparently yield subnanosecond gate delays.

These estimates omit much mechanical detail, including
supply of power, triggering of one action by the next, resetting
of spring-loaded parts, and so forth. It seems clear, however,
that a sliding framework of rods could transmit mechanical power
and synchronization into the volume of a computational device.
Even if the volume of such a power transmission system
substantially exceeded that of the computational elements, this
would make little practical difference; a factor of ten expansion
in volume would place active elements less than a factor of 2.2
farther apart; a tenfold delay in signal propagation would
require a thousandfold increase in volume. Most conclusions one
might draw are insensitive to assumptions regarding such
specifics.

The carbyne-rod based approach to building a RAM array (and
random logic) may be considered somewhat speculative. Though a
variety of calculations support its plausibility, it is a
specific enough concept that some specific problem could
conceivably invalidate it. Like most molecular mechanical
schemes, however, it is simply a molecular-scale implementation
of a concept that can clearly work on a macroscopic scale. Thus,
by making thick enough rods, using large enough forces, and so
forth, one could modify the design to overcome any problems that
might emerge. The plausibility of this approach at the scale
described, however, suggests that such modifications need not
change the scale dramatically, and that they may well be
unnecessary.

This illustrates one of the strengths of engineering
projection compared to scientific speculation: engineers can

specify conservative lower bounds on performance and begin with tentative conceptual designs. As they firm up a design, engineers can correct their mistakes and produce a working system that does what they projected to be possible. Science, in contrast, generally demands falsifiable statements about specific systems and hence cannot use conservatism and redesign to make speculations come true. Thus, statements in the realms of science and engineering can differ fundamentally.

ENERGY, COMPUTATION, AND MOLECULAR MECHANICS

Molecular machinery provides a particularly clear model of the thermodynamics of computation. For example, the requirement that $\ln(2)kT$ of free energy be dissipated in setting a bit may be illustrated by calculating the mechanical energy needed to confine a gas particle in a specified half of an initial volume. The lower bound on the energy that must be dissipated in computation is of concern because large computational systems, whether mechanical or electronic, will eventually be limited by heat dissipation.

In the most straightforward approach to computation, a certain free energy would be expended per step--perhaps a certain multiple of kT per bit per gate, chosen to ensure a certain reliability against reversal by thermodynamic fluctuations. This would make the energy cost of a computation proportional to its complexity.

It is possible, however, to devise constraints such that a mechanical system will always represent a state in a computation (15); such a computation could be reversible in the way that diffusion of a gas molecule from one end of a tube to the other is reversible. In a system of this sort, a substantial fluctuation would be required to disrupt the computation, but otherwise the process could move forward and backwards by random walk. The energy cost of a computation would thus depend on both its complexity and speed; free energy would be expended only to bias the movement of the computation in the desired direction, to force the computation forward at the desired speed.

Finally, it has been demonstrated that, with Newtonian mechanics, perfect initial conditions, and complete decoupling of thermal and mechanical modes, a computation of arbitrary complexity can proceed at a constant speed with a net input of free energy depending only on the number of bits in the input and output (16). Though this behavior almost certainly cannot be achieved in practice, its consistency with the laws of thermodynamics suggests that energy dissipation in computation could be far less than a naive look at NAND gates (which are not reversible elements) might suggest. Until practical implementations of reversible computation are suggested, however, conservative estimates should probably assume the dissipation of energies in excess of kT per gate per operation; in the scheme outlined above, for example, substantial amounts of energy will be dissipated in friction.

NEW FIELDS OF APPLICATION

Small scale promises to reduce the power requirements and speed the operation of MED's; reduced material requirements (and the use of replicating assemblers) promises to reduce their costs. All these factors point toward compact supercomputers with abilities far beyond those available in the largest, most expensive machines today. The ability to build powerful computers in extremely small volumes, however, promises to open new applications for computation.

Shrinkage of computers from the room-sized and cabinet sized machines of the 1950's and 60's opened an era of personal and pocket computers. Shrinkage of computers with a gigabyte of secondary memory to less than a cubic micron (as indicated by the above calculations) will bring still greater changes--it would be absurd to think of their applications solely in terms of devices in cabinets with keyboards, or even in terms of packages bolted to cars. An obvious area of application for molecular electronics (and perhaps molecular mechanical computers) will be in providing local control of complex processes on a small scale, such as control of molecular machinery interacting with biological systems. This seems an appropriate result; the path outlined above begins from biotechnology, which itself has been funded largely because of its application to medicine.

The last section used calculations based on Newtonian mechanics to explore the abilities of molecular machinery; biological applications lend themselves to another line of argument, to the use of existing biological systems as feasibility proofs. Projections in this area are of interest, both as an exercise in reasoning about the limits of future technology, and for what they may show about the incentives for developing molecular technology.

Consider biochemical and biological analysis. The ability of antibodies to distinguish among proteins, and of cellular molecular machinery to distinguish among all functionally distinct molecular species (a virtual tautology), shows that molecular machinery can sense molecular structures. Further, if it can report its own position during sensing, it can report the spatial distribution of different molecules. This line of reasoning leads to the conclusion that suitably designed molecular machinery, if interfaced to molecular computers and data storage devices, could characterize the molecular structure of cells in virtually complete detail. Procedures ranging from molecule-by-molecule disassembly of preserved cells to the insertion of molecular-scale probes into functioning cells could provide a general, direct solution to the data collection problem in cell (and hence tissue and organ) biology.

Consider biochemical and biological synthesis. The ability of molecular machines in cells to build all the molecular components of cells is well established--they do this every time a cell divides. Again, since existing molecular machinery sets a lower bound on the capabilities of artificial molecular machinery, it is clear that the latter could build or rebuild all or part of a cell.

Consider biochemical and biological repair. The ability to sense a structure, to disassemble it, and to reassemble it implies the ability to repair it, provided that the desired structure is known. Examples of sensing and reassembly were described above; disassembly is demonstrated by digestive enzymes (and by harsher chemicals). Knowledge of correct cellular structures can be obtained through study of healthy tissue. Molecular tools are clearly adequate to repair cells, since all of the elementary operations required parallel those observed in nature; the major remaining issue is computation.

The volume of a typical mammalian cell is roughly 1,000 cubic microns. The calculations above indicate that a simple CPU would occupy less than 0.001 cubic microns, and that 0.1% of the volume of a cell could contain both a powerful computer and secondary storage holding more information than the cell's own DNA. Thus, considerable on-site computational power could be brought to bear. The combination of a computer with a variety of molecular sensors and tools, suitably programmed, could be called a cell repair machine.

Cell repair machines seem a natural outgrowth of the development of molecular machinery, molecular electronics, and improved biological instrumentation. Indeed, pharmaceutical design already provides a driving force behind molecular engineering (17). Cell repair machines will nonetheless bring a profound change in medicine. Neither drug therapy, nor radiation, nor surgery can heal tissue: medicine today can only bring about conditions under which tissue can heal itself. Physicians today lack the tools needed to repair molecular machinery; attempting to repair a cell with a scalpel and an injection would be like trying to repair a watch with an axe and

a drop of oil. Accordingly, medicine must now at all costs
attempt to preserve tissue function, since without it, healing is
impossible. With cell repair machines to perform healing,
however, the emphasis in emergencies will shift to preserving
structure, particularly in diseases (e.g., stroke) that destroy
unique tissue patterns important to the individual. Ultimately,
there seem no barriers to repairing tissue so long as
characteristic cellular structures remain intact; those that do
not could still be replaced. The engineering challenges along
the way are, of course, numerous.

CONCLUSION

By avoiding speculation about unknown scientific facts, one may
project future engineering developments, at least to the extent
of setting some lower bounds on what will be possible. Aided by
chemical and biochemical examples, Newtonian mechanics allows
projection of molecular machinery with a wide range of
capabilities, including the ability to handle and assemble
individual atoms and molecules.

   As one might expect, this ability will lead to broad
synthetic abilities, making possible assembly of virtually any
molecular structure with chemically reasonable bonding.
Assemblers will make possible construction of both molecular
circuits and novel molecular replicators (as one might expect
from the existence of bacteria). Calculations indicate that
molecular mechanical computers with subnanosecond gate delays
could fit in a fraction of a cubic micron. Biological analogies
indicate that molecular devices (perhaps directed by such
computers) could be used in cell repair machines able to
recognize, disassemble, and rebuild cellular structures, thus
effecting repairs.

   In considering the future of MED, we must consider a broad
range of molecular technologies that will mature with it.

Molecular machinery can build MED's and be directed by them; it can even perform traditionally electronic functions, such as computing. It will, however, bring revolutions of its own: replicators could revolutionize the world economy, or could be agents of destruction; cell repair machines could revolutionize medical care, or could be abused. To serve our common future interests, the emerging molecular technology research community can serve best by examining the possibilities carefully, judging them dispassionately, and transmitting its understanding to the public clearly. By distinguishing speculation from projection, considerable foresight seems possible. It may be necessary (18).

REFERENCES

1. J. Angell, S. Terry, and P. Barth, Sci. Am., <u>248</u>, 44 (1983).

2. E. Drexler, Proc. Nat. Acad. Sci. (USA), <u>78</u>, 5275 (1981).

3. K. Ulmer, Science, <u>219</u>, 666 (1983).

4. C. Pabo, Nature, <u>301</u>, 200 (1983).

5. J. Sutcliffe, T. Shinnick, N. Green, and R. Lerner, <u>Science</u>, <u>219</u>, 660 (1983).

6. K. Ulmer, in <u>Molecular Electronic Devices</u>, Marcel Dekker, New York, 1982, p. 213.

7. R. Feynman, in <u>Miniaturization</u>, Reinhold, New York, 1961, p. 282.

8. H. Gutfreund and W. Little, in <u>Highly Conducting</u> One-Dimensional Solids, Plenum Press, New York, 1979, p. 153.

9. F. Carter, in <u>Molecular Electronic Devices</u>, Marcel Dekker, New York, 1982, p. 121.

10. R. Haddon and F. Stillinger, in <u>Molecular Electronic Devices</u>, Marcel Dekker, New York, 1982, p. 19.

11. F. Carter, in <u>Molecular Electronic Devices</u>, Marcel Dekker, New York, 1982, p. 51.

12. C. Guenzer, in Molecular Electronic Devices, Marcel Dekker, New York, 1982, p. 273.

13. J. Knox, Molecular Thermodynamics, John Wiley and Sons, London, 1971, p. 179.

14. A. Kelly, Strong Solids, Clarendon Press, Oxford, 1973, p. 12.

15. T. Toffoli, Math. Systems Theory, 14, 13 (1981).

16. E. Fredkin and T. Toffoli, Conservative Logic, MIT/LCS/TM-197 (1981).

17. P. Gund, J. Andose, J. Rhodes and G. Smith, Science, 208, 1425 (1980).

18. E. Drexler, Smithsonian, 13, 145 (Nov., 1982).

29

Self-Organizing Protein Monolayers as Substrates for Molecular Device Fabrication

Kevin M. Ulmer[1] /Director of Exploratory Research, Genex
Corporation, 16020 Industrial Drive, Gaithersburg, Maryland

I. THE JOURNEY OF A THOUSAND MILES BEGINS WITH THE FIRST STEP

The practical problems of actually fabricating even a very simple
molecular electronic device (MED) are indeed daunting. Two major
fundamental problems must be solved before any significant experimental progress can be made towards this goal. The first of these
concerns the precise nature of the molecular assemblages which will
play the active role as discrete components of a MED, and hence the
physical phenomena which will be employed for signal processing.
The second problem concerns the method for organizing such molecular assemblages into suitable circuits. The ability to experimentally test possible approaches to the first problem is likely to
depend on a satisfactory solution to the second - i.e., it will be
necessary to build actual devices in order to test them. Although

---

1. Present address: Director/ Center for Advanced Research in
    Biotechnology, University of Maryland, 3300 Metzerott Road,
    Adelphi, Maryland 20783.

573

there has been some discussion and debate concerning the physical
phenomena that might be exploited to develop a MED, almost all such
examples explored to date have been either purely theoretical
considerations or analogies drawn from biological systems (1 and
this volume). Such musings may prove useful in providing some
general directions for MED design, but it is unlikely that our
existing theoretical capabilities can be extrapolated to the atomic
dimensions of MEDs with sufficient reliability to provide an
accurate blueprint for device development. Similarly, although the
biological world is full of MED-like models, there is no straightforward method for exploiting them for useful MED fabrication.

    The problem then is really one of materials science. We do
not currently possess either suitable materials or processing
techniques for fabricating MEDs. We simply do not have the tools
that would allow us to arrange appropriate atoms and molecules with
the required accuracy. Furthermore, our theoretical understanding
of solid state phenomena is grossly inadequate for predicting the
properties of MEDs even if we could fabricate them. How do we
begin then to develop this new field in materials science? It
would appear that the most productive approach would be to actually
begin to synthesize new materials and to experimentally study their
properties. Clearly, we are a long way from having the kinds of
materials we will ultimately require and we should therefore choose
our initial research directions with the longer range goals and
requirements in mind. In particular, we should pursue technology
that is scalable. It would be very useful to somehow demonstrate
that it was possible to store and retrieve a single bit of information from a single molecule, but unless there was a path that
would lead from the one molecule case to the N molecule case (where
N is a very big number!), such a demonstration would be of only
limited utility. In addition, we may have to rely on a good bit of
serendipity along the development path to MEDs. As we begin to
develop new materials with ever more complex organization at the
molecular scale, we can anticipate that we will discover new
physical phenomena and properties that might be exploited to

produce useful devices instead of simply interesting materials. We
may not know precisely where we are headed as we depart down the
development path to MEDs, but we should at least try and head in
the right general direction.

II. CRAWL BEFORE YOU WALK, WALK BEFORE YOU RUN, RUN BEFORE...
One of the potential attributes of MEDs that has been most widely
touted is the ability to fully utilize three dimensions, thereby
greatly increasing the device density. Although this is clearly a
desirable long term goal, trying to start off working in three
dimensions only adds unnecessary complications to an already
difficult task. A two dimensional approach will be much simpler,
and there is already some available technology to get us started –
namely the Langmuir/Blodgett (LB) technique for producing molecular
monolayers and multilayers at an air/water interface. As Professor
Kuhn has so elegantly demonstrated at this workshop (see this
volume), it is indeed possible to get down to the molecular world
in at least one out of three dimensions. Using the LB techniques
it has been possible to fabricate heterogeneous multilayers with
very good control of layer composition and film thickness. A
renewed interest in LB films (2,3) stimulated by Gareth Roberts has
led to the fabrication of hybrid solid-state devices in which LB
films are incorporated as functional layers within more conventional planar semiconductor technology. This compatibility of LB
methods with planar processing technology potentially offers at
least a partial solution to another particularly thorny problem
with MEDs, that of connecting them to the outside world.

The next major obstacle to the use of LB films in the fabrication of MEDs is how to achieve control at the molecular scale
over the other two dimensions of a monolayer. How do we put down
the blueprint for a circuit in the other two dimensions? How do we
begin to control the patterning, layout or design of planar
molecular structures? At present we can only control the bulk

composition of a particular monolayer, but we cannot arrange the
molecules within the layer to any appreciable degree. One potential solution which has been suggested is a lithographic approach
on the molecular scale. A direct-write method such as electron
beam lithography potentially could be used to pattern a monolayer
either to directly produce a circuit or to produce a mask for x-ray
lithographic reproduction. Such methods have already been demonstrated to produce features of nanometer dimension (4), but their
principal problem is that they are not scalable in the sense
discussed above. While it might be possible to fabricate a single
MED such as a logic gate or a memory cell using such an approach,
it is very difficult trying to imagine how to scale such an
approach to much greater than 10^{12} elements on a regular production basis. Significant problems arise in beam writing times,
alignment and registration problems, and with the inherent graininess of materials at the atomic scale, to name only a few. The
approach is also inherently limited to planar device fabrication
with no simple means of achieving true three dimensionality. The
fundamental problem with such an approach is that it is still
essentially a bulk processing method and is incapable of dealing
with materials on a molecule-by-molecule basis. The general
outlines of this problem have been discussed previously (5).

III. THE INTEGRATED CIRCUIT JIGSAW PUZZLE

A more attractive approach for generalized MED fabrication would
employ self-organizing or self-assembling systems of molecules. To
illustrate the principle, imagine a jigsaw puzzle made from an
integrated circuit such as a one megabit random access memory (RAM)
chip. For this example we need only to focus on the memory array
portion of the chip, the portion that is most regular in its
layout. The idea would be to cut the chip up into one million tiny
pieces, each of which contains the circuit elements for storage of
a single binary bit of information. We have now reduced the RAM
chip to an extremely fine powder. Next, we take this powder and

sprinkle it on a clean water surface in a dish the same way we might apply talc. We now close our eyes, say a few incantations perhaps, and POOF - the RAM chip magically reassembles in perfect working order! Such spontaneous self-assembly is certainly not a viable engineering process at either a macro- or micro-scale, but such a self-organizing process is in fact the modus operandi at the molecular scale, particularly in biological systems.

To continue with this example, imagine that we now shrink our tiny pieces of the RAM chip to the size of macromolecules such as proteins, which requires a reduction by a factor of about 10^6. Such macromolecular RAM cells might be spread on the surface of a Langmuir trough and under the appropriate conditions, might be induced to crystallize in a two dimensional array to reform the entire megabit RAM chip in an area that was previously required to store a single bit. Such two dimensional crystallization of macromolecules is in fact quite common as shown in Figure 1. This electron micrograph shows a small portion of a two dimensional crystal of the neuraminidase protein of influenza virus (6). Each of the doughnut-shaped objects is in fact a tetramer of four identical protein subunits and the center-to-center distance between doughnuts is 100 Å.

IV. CRYSTALLIZATION AND SELF-ASSEMBLY

Crystallization is certainly an often used processing technique for the materials scientist, and in fact is used to produce the giant single crystals of silicon which are the basis for the entire semiconductor industry today. When we think of crystallization however, we usually think of very homogeneous and well ordered materials made of one or a few kinds of atoms or small molecules and it is therefore difficult to imagine growing a completely crystalline computer. It is also possible though to crystallize larger and more complex molecules such as proteins, supramolecular

FIGURE 1. Electron micrograph of a two-dimensional crystal of influenza neuraminidase. Center-to-center spacing of tetramer units is 100 Å. (The micrography was kindly provided by D.C. Wiley, Biochemistry, Harvard and N. Wrigley and E. Brown, Nat'l Inst. of Med. Research, Millhill, London.)

assemblies such as ribosomes, and even virus particles with atomic masses in excess of 100 million daltons. Co-crystallization of two or more different components is also possible.

The differences between self-assembly and simple crystallization are perhaps best illustrated by the example shown in Figure 2. This shows a small section through a crystal of insulin (7). Each of the four round molecular structures is in fact a hexamer of wedge-shaped insulin molecules which fit together with great precision, mediated by a large number of specific atomic interactions between the individual insulin molecules. The insulin hexamer assembles spontaneously under the appropriate chemical and physical conditions to form a fairly stable, compact, and highly organized structure. Under a much narrower range of physical and chemical conditions it is possible to form macroscopic single crystals with dimensions of 1 millimeter or more. As can be seen from Figure 2, however, the inter-hexamer interactions which stabilize the crystal lattice are both quantitatively and qualitatively different from the intra-hexamer contacts. The crystal lattice interactions are few in number, less specific and often involve ordered solvent molecules (water and ions) as bridges between the protein molecules. The crystal lattice is also a far more open structure with large pores and channels containing disordered bulk solvent. These lattice interactions are also much less stable than the intra-hexamer contacts.

The engineering challenge then is to control the properties of macromolecular crystallization and self-assembly in order to produce highly organized molecular materials. We will need to enhance the interactions stabilizing the crystal lattice so that they become more like the interactions observed in multimerization or other forms of self-assembly. We will also need to tighten up the lattice and eliminate the disordered solvent to provide a more stable and more highly organized molecular material. This will

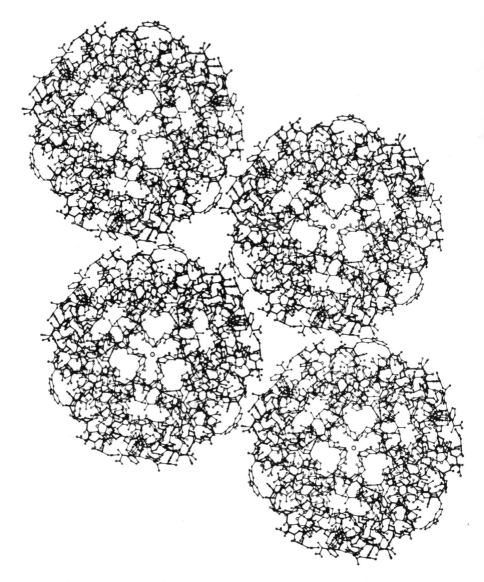

FIGURE 2. A section through a crystal of insulin. Each wedge-
shaped unit is one monomer. These monomers associate into dimers,
which in turn aggregate into hexamers. The hexamers pack into the
crystal. Note the large solvent channels and the relatively few
direct contacts between hexamers. All atoms except hydrogens are
shown. (Reproduced from reference 7.)

require that we understand the atomic interactions which stabilize
the crystal lattice and the macromolecular assemblies in sufficient
detail to permit us to modify them at will. In a sense we must
begin to think in terms of crystal engineering.

V. PROTEIN ENGINEERING: TOOLS OF THE CRYSTAL ENGINEER

Unfortunately, neither influenza neuraminidase nor insulin nor any
other naturally occurring protein crystal is likely to be of any
utility in fabricating MEDs. These natural protein crystals merely
point the way towards self-assemblying molecular materials. How
then do we begin to modify a protein crystal lattice or the
interactions involved in self-assembly of protein subunits? Some
rather gross control of crystal morphology and crystal habit can be
achieved by manipulation of the physical and chemical conditions
during crystallization, but in the case of proteins in particular,
our understanding of crystallization is abysmally poor. The field
is currently more of a black art than a science which is a poor
foundation indeed for developing a new engineering discipline.
Nevertheless, recent advances in molecular biology provide a novel
opportunity to greatly improve our understanding of crystallization
and at the same time will provide the tools necessary to begin to
actually engineer crystals. The newly developed techniques for
site-directed mutagenesis now make it possible to make precise and
facile modifications of the amino acid sequence of any cloned
protein. This provides for unprecedented experimental control of
protein structure and function (including properties like crystallization and self-assembly) and forms the basis for the emerging
field of protein engineering (8).

VI. WHICH PROTEINS TO START WITH?

If we are to begin using protein engineering technology to modify
the self-assembly and crystallization of proteins, which proteins
are the best initial candidates? Many proteins might be considered
if we were only concerned with the structural aspects of the

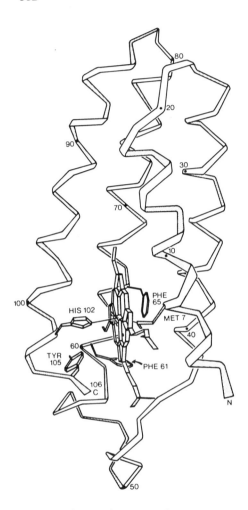

FIGURE 3. Schematic diagram of E coli cytochrome b$_{562}$. The small size, simple structure and easily substituted heme group make this electron transfer protein an interesting candidate for molecular material development. (Reproduced from reference 11.)

problem (9), but if we also want to select proteins which might
provide novel functional properties for MED fabrication we should
focus on electron transport proteins and the immunoglobulins.

Electron transport proteins are a natural choice for molecular
material development work because in many ways they already perform
some of the function we desire for MED fabrication - i.e., electron
storage and transfer at the molecular scale. These electronic
properties arise, not from the protein backbone of these molecules,
but usually from their non-protein prosthetic groups. These
prosthetic groups are inorganic-, organometallic-, or metal atom
cofactors which are integral to the structure of protein. A
particularly interesting protein would be cytochrome b_{562} of E
coli (10,11). This protein is small (12,000 daltons), has a
single polypeptide chain folded into a simple 4-alpha-helical
motif, the x-ray structure is known to 2.5 Å, and most importantly,
the single heme group is non-covalently bound (Figure 3). This
last property allows for the substitution of other porphyrin
analogs with a variety of coordinated metal atoms, greatly increasing the experimental flexibility of the system. One could begin
to engineer crystals of cytochrome b_{562} with an initial goal of
controlling the orientation of the spin axis of the metal atom in
the crystal lattice.

If we are looking for a more general method of incorporating
non-biological molecules into molecularly organized materials, then
the immunoglobulins or antibody molecules offer many attractive
advantages. Using currently available monoclonal antibody technology, it is now possible to generate a specific immunoglobulin
molecule capable of highly specific binding to almost any compound
of interest. The initial goal in this case would be to engineer
crystals of such interesting antibody complexes in which it was
possible to control the arrangement and orientation of the complexed molecules at the molecular scale. There has already been a

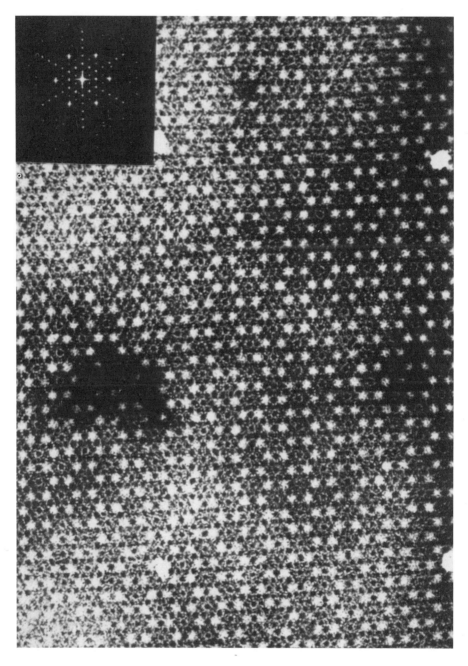

FIGURE 4. Electron micrograph of a two-dimensional crystal of
immunoglobulin produced using Langmuir-Blodgett techniques as
described in reference 11. Each hexagonal unit is composed of six
antibody molecules. (Micrograph courtesy of J.A. Reidler, Genex
Corporation, Gaithersburg, Maryland.)

report (10) of successful application of LB techniques to produce two-dimensional crystals of antibody molecules which clearly holds promise for MED development (Figure 4).

VII. ESCHER, FRACTALS AND PENROSE TILINGS

In keeping with the philosophy of "crawl before you walk...", our first attempts to develop molecular materials should not be too ambitious. At the outset we should focus on attempting to produce two dimensional crystals of a single type of protein molecule. The initial goals should be to achieve control of the type of crystal lattice produced and of the orientation of the protein molecules within the lattice. Other practical concerns will include methods for stabilizing or cross-linking crystals, and for producing macroscopic-size, defect-free crystals. Such homogeneous, crystalline films of proteins may seem a long way off from molecular computers, but they will clearly represent an essential foundation for their development.

The second step might then be to create a checkerboard pattern using two different proteins. This approach could then be extended to produce increasingly complex patterns based on incorporation of ever larger numbers of different protein molecules. The Dutch artist M.C. Escher has provided a wealth of examples which inspire one to consider the possibilities for design of molecular materials (Figure 5). In addition to these regular patterns, Escher also provides us with examples of non-crystalline patterns which may in fact be more useful in MED design. Figure 6 suggests a fractal geometry where a pattern is repeated on an ever smaller scale. Perhaps it will be variations on this theme that will solve the problem of how to connect the macroscopic world of wires and integrated circuits with the atomic world of the MED. Much food for thought on this approach can be found in the works of Mandelbrot (13) who has pioneered the development of the mathematics of fractals.

FIGURE 5. M.C. Escher provides inspiration for self-organizing
macromolecular materials. Imagine a crystalline material made up
of three different types of macromolecules represented by the three
different colored salamanders.

FIGURE 6. Strictly crystalline materials may not be the best
suited for MED fabrication. Another Escher suggests a fractal
geometry which may provide a mechanism to bridge the macro and
molecular worlds.

FIGURE 7. Non-periodic tilings such as those developed by Roger Penrose might also produce novel materials properties for MED development. (Reproduced from reference 15.)

Yet another type of molecular material which might yield very interesting properties could be based on non-periodic tilings of the plane (14,15). Patterns such as the tiling shown in Figure 7 could be generated by designing the protein equivalents of Penrose's chickens!

VIII. ONE PATH TO FOLLOW

Clearly, there are likely to be multiple approaches to the development of MEDs, and this paper hopefully points the way towards at least one very attractive route. We must ground the development of

MEDs in a new field of materials science which is concerned with self-organizing molecular materials, and abandon our traditional methods of bulk processing in favor of technology which allows us to assemble complex structures molecule-by-molecule. We should resist temptations to try and design three dimensional structures at the beginning, and instead should build upon the Langmuir-Blodgett techniques by seeking methods for controlling the two dimensional arrangement of molecules within a monolayer. Our molecular building blocks should be proteins, whose richness of structure and function provide the necessary combination of complexity and flexibility for designing self-assembling struc-

We must employ the tools of the protein engineer to develop the necessary expertise in controlling the structural and functional properties of proteins that will be required for crystal engineering. Finally, we should approach the design problem for MEDs more as an artist creating a mosaic. The overall pattern or layout for the MED should be conceived in such a fashion that it can be reduced to a small number of tiles. In practice the tiles will be individual protein molecules, engineered to spontaneously assemble to form the overall mosaic pattern and also having the necessary functional properties to produce operational MEDs.

REFERENCES

1. F.L. Carter (ed.), Molecular Electronic Devices, Marcel Dekker, New York, 1982.

2. W.A. Barlow (ed.), Thin Solid Films, 68, (1980).

3. G.G. Roberts and C.W. Pitt (eds.), Thin Solid Films, 99, (1983).

4. M. Isaacson and A. Muray, in: F.L. Carter (ed.), Molecular Electronic Devices, Marcel Dekker, New York, 1982, pp. 165-174.

5. K.M. Ulmer, in: F.L. Carter (ed.), Molecular Electronic Devices, Marcel Dekker, New York, 1982, pp. 213-222.

6. J.N. Varghese, W.G. Laver and P.M. Colman, Nature, 303, 35 (1983).

7. T.L. Blundell, et al, Adv. Protein Chem., 26, 279 (1972).

8. K.M. Ulmer, Science 219, 666 (1983).

9. F.C. Bernstein, et al., J. Mol. Biol., 112, 535 (1977).

10. F.S. Mathews, et al., J. Biol. Chem., 254, 1699 (1979).

11. P.C. Weber, et al. J. Biol. Chem., 256, 7702 (1981).

12. E.E. Uzgiris and R.D. Kornberg, Nature, 301, 125 (1983).

13. B.B. Mandelbrot, Fractals: form, chance, and dimension, W.H. Freeman & Co., San Francisco, 1977.

14. M. Gardener, Scientific Amer., 236, 110 (1977).

15. U.S. Patent No. 4,133,152.

30

Stability Consideration in 3-Dimensional Biologically Based Structures

S. R. Quint and R. N. Johnson / Department
of Biomedical Engineering, University of
North Carolina, Chapel Hill, North Carolina

I. INTRODUCTION

Progress in microelectronics is coupled with the ability to
continually place larger numbers of smaller devices on a single
chip. The problems associated with very large scale integration
(VLSI) and continually smaller device structures are extensive.
The separation of device design from system design depends on
being able to isolate individual devices, except for desired
interconnections. In dense device arrays, line-to-line
capacitance and nearest neighbor interactions begin to dominate
the system. Thus, unwanted device or system characteristics may
arise. The 3-dimensional structures utilized in living systems
have the potential for illustrating solutions to the high density
device interaction problem. Moreover, the mammalian central
nervous system (CNS), due to the extreme packing density and high
connectivity of cellular structures, exhibits a variety of
unstable behaviors, clinically described as the epilepsies.

Knowledge of neural circuit architecture could play a significant role in VLSI circuit design at the molecular dimension level. The mammalian central nervous system has approximately 20 billion neurons, surrounded by glial cells in a ratio of 5 to 10 times the number of neurons. The glia provide, among other things, an insulating envelope around the neurons to cut down device-to-device interaction. Total power consumption of the CNS is estimated at 10 watts and the total surface area of the cerebral cortex, including sulci, is over two square feet. Many neurons can have over 100,000 inputs with up to 25,000 outputs. This level of device connectivity exceeds man-made devices with respect to that parameter. Nerve axons may vary from 0.1 microns to 3 microns in diameter and some of the pyramidal neurons in the cerebral cortex can have axons up to a meter long. The active component of the neuron, the cell membrane, is on the order of 50 to 100 angstroms thick and has an electric field gradient of approximately 250,000 volts/inch. Thus, the charge multiplication observed in physiological experiments on the membrane, producing negative resistance characteristics, is not surprising based on the high field intensity across the membrane. Furthermore, information is propagated along axons in a digital form (pulse trains), while at synaptic junctions, analog or integrative signal activity occurs.

It is expected that the minimum feature size in VLSI devices will continue to decrease from 2.5 microns to 0.5 microns by the late 1980's (1). Serious degradation of performance due to such effects as channel capacitance and series resistance can occur. Scaling is detrimental with respect to storage capacitance in MOS dynamic RAM's, reducing the reliability of distinguishing between a logic 0 and 1 as well as making the memory susceptible to soft errors caused by ionizing radiation, such as alpha particles. Scaling also creates serious problems in interconnects, where lead resistance can degrade signal propagation times. Ultimately, multilevel interconnections will be required on VLSI devices to maintain device speed. We believe that studies of 3-dimensional neuronal systems may suggest solutions to some problems of VLSI device scaling and suggest alternative strategies for VLSI system circuit designs.

II. SLOW SPEED SYSTEMS

Biological elements, the neuron and all its appendages, when assembled into a system tend to be extremely slow relative to even the slowest of microprocessors. Where we expect instruction times in microseconds and memory access times in nanoseconds in the microcomputer, when compared to the nervous system, these times are several orders of magnitude faster. The basic action potential itself is a pulse of at least a millisecond long with perhaps the best conduction velocities running on the order of 100 meters per second.

The elegant studies of Stark (2) and colleagues in the 60's relating to response times of the human motor control system document the slowness of the living control system. Their studies have shown that the time delay through the human motor control system from the point of visualizing a change, to the time when movement begins at the effector, is on the order of 300 milliseconds. These extensive time delays are such that the human hand control system can be subject to severe oscillations in the range of about three per second. The studies of Stark and coworkers also illustrated the method by which the nervous system overcomes its slow response in a world where one must respond faster than a third of a second. This is accomplished by a feedforward or predictive type control system. Under these conditions, the biological system can reduce the delay time, provided the desired response can follow a predictable path. Thus, the eye-hand response to random events will always be in

the 300 millisecond range, while the response delay to a
predictable event can be such that a person with some training
can greatly minimize the delay, and in many cases have hand
movements which can slightly lead the anticipated response.
However, the system pays a price for this activity, as under
conditions where a number of repetitive and predictive hand
movements have been made, when a random event is introduced, the
time delay is greater than the normal anticipated 300
milliseconds in order to effect a different response.

Studies of this feedforward controller were first undertaken
by Ito (3), who considered the cerebellum as one of the key
elements of the predictive control system. Thus, clinically, one
finds individuals with cerebellar disease who have lost the
ability to compensate for system delays and tend to be subject to
severe oscillations in the 3 hertz range when required to make
rapid hand movements. Fig. 1 is a simplified diagram of the

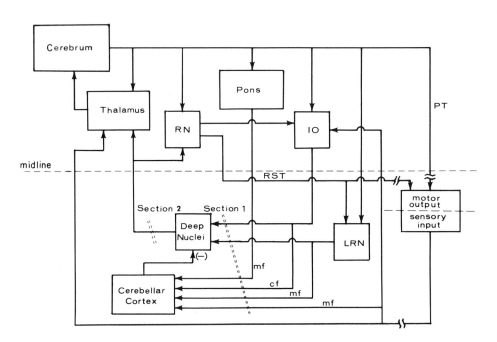

FIG. 1. Human motor control system.

human motor control system emphasizing the complicated pathways and the critical nuclei within the system. Not intending to make this a discussion of anatomy, the three major elements that we would briefly mention are labeled the cerebellar cortex, thalamus and cerebrum. The outer loops on the diagram traverse these nuclei and pass out to the external world providing motor output and resulting sensory input. This outside loop from the cerebrum down the pyramidal track results in a motor output with sensory feedback coming back through the thalamus and cerebellum and forms a system with significant time delays. Other nuclei provide input to the cerebellar cortical and deep nuclei, which project through the thalamus to provide the predictive control necessary to eliminate the time delays within the system. Thus, a slow system which could have a poor response and suffer oscillatory instabilities due to relatively long time delays within the system, requires a predictive computer system, if you will, which can estimate positions, and thus try to override the basic penalty paid by the relatively slow performing elements within the system.

According to Eccles (4), this predictive system is also adaptive, as anyone who has practiced hitting a baseball or golf ball will attest to, since the system must estimate system parameters, make judgments and corrections and continually try to refine the desired response. The boxes of Fig. 1 are thus properly considered as processors, linked in a complicated multiprocessor system designed for real time applications.

III. SYSTEM INTERACTIONS

The extensive complexity of the system brought on by the example listed above, in terms of a predictive controller to mitigate the slow speed elements in the system, complicates the system and increases the connectivity, which subjects the system to global instabilities. As mentioned in the first section above on device parallels, we discussed the high number of inputs that can occur on neurons and in some cases the extensive bifurcation or number of outputs from each neuron. In addition, the need for redundancy occurs since the devices (neurons) exhibit probablistic behavior. Thus, we must expect the potential for widespread instabilities to occur in this system. Early studies by Ashby (5) and colleagues originally pointed to models which show that at certain levels of connectance (i.e., percentage of elements in a system which have a direct effect on each other) in large systems, one might expect the sudden appearance of unstable behavior. Subsequent work by others (6) has suggested that this may not be as clear cut a situation. Nevertheless, it is quite clear that in systems which are stable in isolation from each other, increasing the connectivity increases the probability of unstable behavior. The clinical problem of epilepsy can be compared to the potential for a riot or a mob scene where a large number of people can congregate, become excited, influence each other and ultimately move to a condition of undesirable behavior.

One of the interesting aspects of epilepsy has been the early recognition by Dow (7), Snider and other workers (8) that the mammalian central nervous system has, in fact, found it necessary to develop a control mechanism to prevent widespread epileptic activity. Most work pointed to the cerebellum as one of the key devices within the nervous system that could control epileptic behavior. This experimental work ultimately lead to the beginning of the utilization of cerebellar stimulators (8) in patients in the mid 70's for the partial control of epileptic seizures.

Beginning in the late 60's, we conducted a number of experiments on the thalamo-cortical motor system in animals, all geared towards attempting to understand instability within the system, and particularly emphasizing the possibilities for epileptic behavior and the effects of such stabilizing influences

as anticonvulsant drugs and cerebellar stimulation. Fig. 2 shows
schematically some of the general concepts of the response
profiles that we have utilized as guidelines for studies of this
system. Without elaborating on the details further, which have
been extensively published (9-12), we utilized electrical
stimulation of the ventro-lateral thalamus with recordings
obtained from the sensorimotor cortex, coupled with peripheral
stimulation, and utilization of drugs and cerebellar surface
stimulation, all in an attempt to understand the effects on the
system from these various conditions. Evoked potentials were
quantified using an on-line digital computer for stimulus control
and data collection and basically what might be called in the
most simplistic sense, a series of excitability curves, were

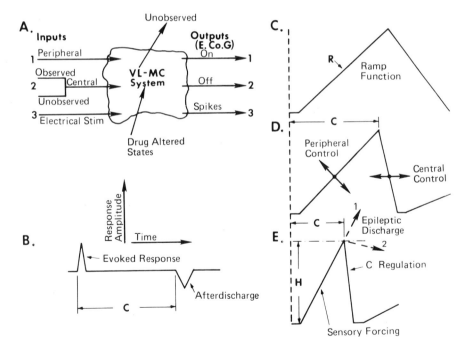

FIG. 2. A. Schematic representation of general concepts used
in study; B. Evoked Response; C.D. & E. Excitability Curves.

generated as shown in Fig. 2, C, D, and E. Basically, one can
visualize periods of increasing excitability within the system,
which are shown by the ramp-like triangular function in C. The
slope of the curve can be changed by sensory input. The duration
of the curve can be reduced by a central controller, since the
undesirable effect is the maximum amplitude or excitability as
diagrammed in E. Thus, if a particular ramp slope were to go
above a certain threshold, shown by the arrow labeled 1 in E, the
potential for extreme excitability within the neural elements and
thus epileptic discharge increases. Thus, it would be necessary
to either reduce the amplitude or find some way to reduce the
period of the ramp function and this is shown by the reduction of
the ramp duration labeled C. This schema is shown in Fig. 3,
where a whole series of ramp functions are depicted by the three
dimensional plots. This family of curves was obtained by varying

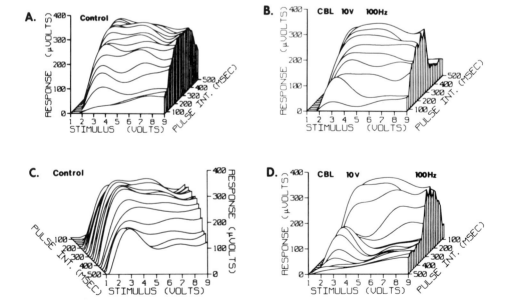

FIG. 3. Thalamus stimulation. A. Control; B. Truncation by
cerebellar stimulation; C&D. Backside of A. and B., respectively.

the potential of stimuli to the thalamus and also varying the
interval between stimulus pulses. These two variables are
plotted on the X and Z axes, respectively, with the amplitude of
the response from the cortex plotted on the Y axis. A control
example is shown in Fig. 3A, while in Fig. 3B, the ramp-like
projection has been truncated by cerebellar stimulation. This is
obvious from the projection on the Y-Z plane of the image
following cerebellar stimulation. The backside of the images are
shown in Fig. 3C and D illustrating that the ramp function has
been truncated by cerebellar stimulation. We have documented
this in many other situations and paralleled this to a situation
using anticonvulsant drugs. Fig. 4 illustrates similar effects

Anticonvulsant Effects

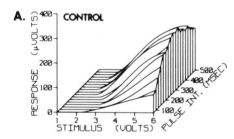
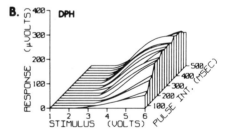

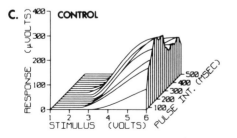
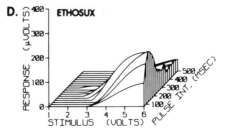

FIG. 4. Drug modification on cerebellar stimulation. A&B.
Phenyltoin (5,5Diphenylhydantoin); C&D. Ethosux.

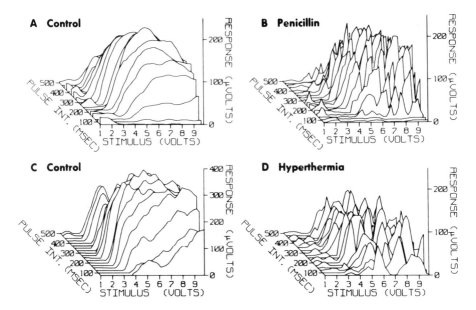

FIG. 5. Effects of penicillin and hyperthermia on system excitability.

to cerebellar stimulation with the utilization of the anticonvulsant drugs, Phenytoin and Ethosuximide. Thus, we have suggested that the simple ramp-like concept, with control of either the amplitude or the duration of the ramp is an example of dynamic regulation by the cerebellum. This is graphically illustrated in Figs. 5A and B, where a family of responses are shown for the control case and then a situation where penicillin has been introduced on the cortex. Penicillin is a known epileptogenic agent. The peculiar jagged response of Fig. 5B is an example of continuous correction by a "regulator" within the nervous system. What actually happens is that at each peak of the response point, a small amount of epileptic activity occurs in the EEG and immediate compensation occurs, reducing the excitability level of the system and controlling the induced epileptiform activity. Thus, we see a picture of a dynamic controller in action, attempting to compensate for system runaway. This is further illustrated in Fig. 5D in a case under hyperthermia. The effect of elevated temperature, particularly in children, and the danger of convulsions from such an elevation is well known. This example is illustrated from animal studies similar to the penicillin study showing a similar type of effect where it would appear that under elevated temperature, the system becomes less stable, and a regulator steps in to control and prevent epileptiform activity.

It is quite obvious in a system such as this that there may be limits to the power of the regulator. Ashby (13) has so aptly pointed out that for the regulator to have 100% control, the variety or power of the regulator must be completely equal to the variety or power of the system being regulated. This would require, of course, that the complexity of the regulator equal that of the system. Although it appears that the nervous system has evolved to use active regulators to control unstable behavior, these systems have limits. They are not of full power and at many times under extreme conditions of external or internal stimuli, the regulator cannot prevent undesirable behavior.

IV. BIOLOGICAL COMPUTERS

It is quite clear that biological computing systems, utilizing neurons and neuronal connections, have developed and become more sophisticated in higher mammals and man. The positive and negative aspects of the features of the system have become evident. Attributes, such as self-constructing, self-organizing, self-repairing, self-programming, adaptive and learning aspects, all give biological systems exceptional power. Biological computers, if you will, are certainly driven by the external environment as well as being potentially victims of their own energy sources. Their great sophistication requires extremely high connectivity with a multitude of signal pathways. Moreover,

they tend to be slow with large delays and use non-deterministic
elements, requiring redundant signal pathways.

The desire to imitate nature in the development of future
computers is, of course, one of the reasons for looking at
biological signal processes. It is important to look to
biological systems for clues to designing future systems of
intelligence. In particular, we need to search for new
architectures. Nature's awesome power, evolved over billions of
years, lies in its replication and interconnection of many
specialized processors. These specialized processors are
interconnected in networks that provide for a range and variety
of activities that no single computer, regardless of its speed,
can currently provide.

REFERENCES

1. M.E. Jones, W.C. Holton and R. Stratton, Proc. IEEE, 70,
   1380 (1982).

2. L. Stark Neurological Control Systems, Plenum, New York,
   1968, pp. 348-403.

3. M. Ito, Brain Res., 40, 81 (1972).

4. J.C. Eccles, M. Ito, and J. Szentagothal, The Cerebellum as
   a Neuronal Machine, Springer, New York, 1967, pp. 300-315.

5. M.R. Gardner and W.R. Ashby, Nature (Lond.), 228, 784
   (1970).

6. J. Daniels and A.L. Mackay, Nature (Lond.), 251, 49 (1974).

7. R.S. Dow, Epilepsia, 6, 122 (1965).

8. I.S. Cooper, M. Riklan and R.S. Snider, The Cerebellum,
   Epilepsy and Behavior, Plenum, New York, 1974, p. 119.

9. R.N. Johnson, G.R. Hanna, and R.F. Munzer, Biol. Cybernet,
   18, 91 (1975).

10. R.N. Johnson, R.N. Englander, S.R. Quint and G.R. Hanna,
    Brain Res., 103, 568, (1976).

11. R.N. Englander, R.N. Johnson, J.J. Brickley and G.R. Hanna, Neurology, $\underline{27}$, 1134 (1977).

12. R.N. Johnson, J.D. Charlton, R.N. Englander, J.J. Brickley, W.J. Nowack and G.R. Hanna, Epilepsia, $\underline{20}$, 247 (1979).

13. W.R. Ashby, An Introduction to Cybernetics, Wiley, New York, 1963, p. 207.

31

Cooperative Effects in Interconnected Device Arrays

R. O. Grondin, W. Porod, C. M. Loeffler and D. K. Ferry/
Department of Electrical Engineering, Colorado State
University, Fort Collins, Colorado

I. INTRODUCTION

Many people have examined the limit to which semiconductor devices
can be scaled downward, and while small devices in the range 0.1-
0.3 μm gate length have been made, problems such as interconnections, electro-migration and thermo-migration of metallization, and
power density within the device strongly affect the packing density
and device size that can be achieved [1]. Yet, no one has evaluated the operation and performance of arrays of very small semiconductor devices, i.e.-those that can be conceived in the sub-0.1 μm
size range, and the interactions within arrays of such devices.
The possible device-device coupling mechanisms are numerous and
include such effects as capacitive coupling, of which line-to-line
parasitic capacitance is one example, and wave-function penetration
(tunneling and charge spillover) from one device to another. The
former is significant as it occurs at distances greater than the
scale of a carrier wavelength of 100Å. Indeed, for device sizes
below 0.3 μm, the line-to-line parasitic capacitance begins to

dominate the direct line capacitance in determining the total capacitance of a logic gate [2]. This parasitic capacitance leads to a
*direct device-device interaction outside the normal circuit or
architectural design.* The constraints imposed by device-device
interactions will have to be included in the design of dense VLSI
systems, and this will most easily be accomplished if these constraints are reflected in the system theory description of the
architecture itself.

Here we consider several approaches to this problem, treating
in particular the first order device-device interaction. First
some general features are established by an approach analogous to a
quantum statistical mechanical transport theory. A semiclassical
theory of a small memory device is used to establish some limits
on RAM density. To gain other general knowledge we study networks
of threshold elements. While the actual physical nature of a
threshold element is not specified, such elements have been widely
used to study various information processing systems.

II. THE STATISTICAL MECHANICAL APPROACH

At the total circuit (or system) level, control over the array of
devices that compose the VLSI structure is exercised by means of
control fields (or voltages) and input or feedback currents (or
voltages). These make up a set of generalized forces $F_{i,ext}$,
i=1,...,N, where N is the total number of devices and exceed 10^5
in modern VLSI. The applied generalized forces are screened by a
variety of interactions, but lead to a set of local applied forces
F_i, which must be found by a total self-consistent method, just as
is done in present two- (and three-) dimensional device modeling.
In the coupled N-device system, we can write the total Hamiltonian
as

$$H = \sum_i H(x_i, F_i) + \sum_{i<j} H_E(x_i, x_j, F_k) \qquad (1)$$

where

$$H_E = H_e + H_{ed}$$

includes the environment plus device-environment terms and thus represents the coupling between devices i and j (assumed to be an instantaneous pair-wise interaction). The variables x_i, x_j refer to complete sets of dynamical variables for the ith and jth devices, respectively. In the absence of inter-device coupling, the second term on the RHS of (1) is zero and we recover the Liouville equation result for each individual device treated separately. The terms in $H(x_i, F_i)$ are assumed moreover to be time-dependent through the coupling to the generalized, time-dependent forces F_i. We note finally that the set of generalized parameters x_i must also include any local structure such as contacts and local interconnects.

In the following, we adopt an approach similar to that of a BBGKY heirarchy, which illustrates the various correlations between devices as they become important. We assume that

$$H(x_i, F_i) \equiv p_i^2/2m + \Phi_F(x_i) \quad , \tag{3}$$

where all external fields are reflected into the generalized local potential Φ_F. Moreover, we assert that H_E is a pair-wise potential interaction. We have separated the generalized parameters into generalized coordinates x_i and conjugate momenta p_i.

It can be shown [3] that the reduced density matrix $\rho_1(x_1)$ for a single device, in the presence of the other devices, can be written as the solution to the Liouville equation

$$ih \frac{\partial \rho_1}{\partial t} = \hat{H}_1 \rho_1(x_1) +$$
$$+ (\frac{N-1}{v}) Tr_2 \{\hat{H}_E(|x_1 - x_2|) \rho_1(x_2)\} \rho_1(x_1) -$$
$$- (\frac{N-1}{v}) Tr_2 \{\hat{H}_E(|x_1 - x_2|) g_2(x_1, x_2)\} \tag{4}$$

where

$$g_2(x, x_2) = \rho_1(x_1)\rho_1(x_2) - \rho_2(x_1, x_2) \tag{5}$$

is the two-device correlation function [4]. In (4), the last term represents high (higher than zero) order effects that arise due to correlated effects of one device upon another. On the other hand,

the second term or the right represents an effect due to the average
presence of all other devices. Thus, this term is analogous to the
Hartree potential for electrons. If the correlation term g_2 is
set to zero, the first two terms on the RHS lead to a Hamiltonian
for an equivalent Hartree equation for the one-device performance,
in which each individual device can be treated in isolation if an
average mean-field background potential is evaluated. Clearly, the
deviation from this simpler approach arises when correlation between
devices begins to become important, a case that already arises when
line-to-line capacitance begins to significantly affect circuit
behavior. *There are in fact two effects here. The first is a
global modification that occurs when the Hartree potential differs
from zero, for example when a substrate current begins to produce
global shifts in substrate bias levels. The second effect is the
correlated device behavior.*

      The possible device-device coupling mechanisms are numerous.
Formally, however, one may describe these effects by assuming the
simplest form of coupling. Arrays of devices, interacting in this
manner, may form a lateral surface superlattice [5]. Lateral
surface superlattices, in which the superstructure lies in a
surface of heterostructure layer, offer considerable advantages for
obtaining superlattice effects in planar technology. While a surface MOS structure is formally similar to an array of CCD devices,
superlattices can also be fabricated through the use of electron
and ion beam lithography and selective area epitaxial growth [6].
If the coupling is capacitive, then the limitation to a spacing
less than the de Broglie wavelength is removed. Monte Carlo simulations of a prototype lateral surface superlattice [5] have clearly
shown how a device-device coupling can cause individual devices to
behave differently in the array than they do in isolation.

      We can further illustrate first order effects and their
destructive behavior for capacitive coupling with a simple example.
In Fig. 1, a single memory cell is shown embedded in an array of
read/write transfer lines. The passage of charge along one of

CHARGE PACKET

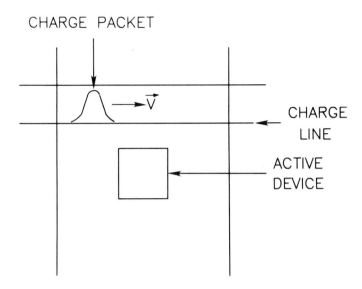

FIGURE 1. Concept of an isolated RAM cell localized within an array of read/write control lines.

these lines "rocks" the potential well of the memory cell, by Coulombic interacting with the charge in the well. If this rocking is sufficiently strong, the memory charge will be de-localized. We first consider the memory cell by itself.

If the potential barriers are slowly varying on the scale of the wave packet, the trajectories are largely those of the classical motion. Even if this is not the case, as we expect for the very-small device, nearly semi-classical trajectories can be expected if the variation of the action is limited to a few low-order derivatives [8]. In semi-classical systems, the phase space of the classical motion forms a natural framework in which to examine problems such as these. While classical one-dimensional transport appears to be basically simple, there exists recent work that suggests that even this simple problem contains a number of unexpected subtleties [8]. The types of subtleties to which we are referring are best illustrated by an area-preserving mapping of particle position and momentum within a device active region bounded by two Gaussian potentials $V(x)$ (which have a weak overlap). In addition, an electric field has been applied. (Here we are considering a collisionless, conservative system and looking at the ballistic transport.) The equations of motion are

$$\dot{p} = -\frac{\partial V(x)}{\partial x}$$

$$\dot{x} = p/m \tag{6}$$

which lead to the discrete mapping

$$p_{n+1} = p_n - T \frac{\partial V}{\partial x}\bigg|_{x_{n+1}}$$

$$x_{n+1} = x_n + T p_n/m \tag{7}$$

For small values of the total particle energy, closed orbits are found inside the cell. More energetic particles are swept out of the well by the field. An energy dissipating collision can drop a high energy particle into the resonant region, thus trapping it within the structure, and this is the basis of charge storage. We expect these particles to contribute a diffusive component of current. Large angle scattering, however, can move a particle to a back-flowing orbit, which effectively causes a reflection of particles from the device input.

We can now examine the role of the additional charge of the read/write lines. This produces an additional force, of the form

$$\frac{eC_L V_L(t)}{2\pi\varepsilon d} \tag{8}$$

where C_L is the capacitance per unit length of the line, $V_L(t)$ is the line potential, and d is the distance to the line. Since $V_L(t)$ is time varying, we can affect its potential pumping, with dissipation, by replacing the first equation of (7) with

$$P_{n+1} = p_n - T \frac{\partial V_0(x_{n+1})}{\partial x} - T\gamma \, p_n + TF_0 \, \sin(\Omega t) \tag{9}$$

where V_0 is the set of Gaussian potentials, $F_0 = eC_L V_{L0}/2\pi\varepsilon d$ is the amplitude of the force in (9), and Ω is the clock frequency of

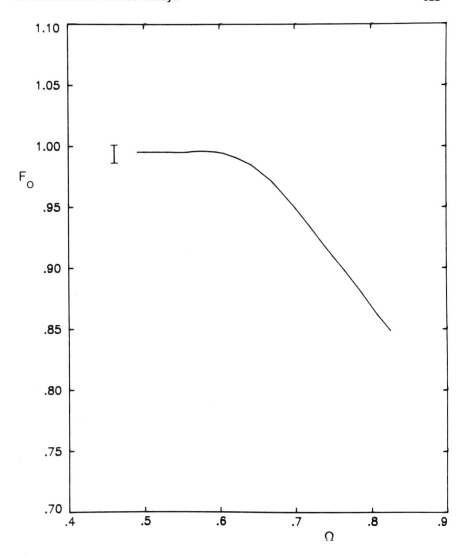

FIGURE 2. Parameter space of stability for the single potential well when subjected to a periodic pumping potential. The axes are normalized quantities (see text).

the system (read/write cycle time). The factor γ is an effective damping factor. In Fig. 2, we plot the normalized (F_0', Ω') plane results. The curve is a separatrix below which a stable device results. Above the curve, the device is unstable. No period-doubling bifurcations within a single device are found. The relation between F_0' and the real parameters is

$$F_0' = 0.0586 \, \frac{C_L}{4\varepsilon} \, \left(\frac{V_L}{V_0}\right) \tag{10}$$

where V_0 is the well depth and the numerical factor incorporates the constants and numerical constants from the simulation. The frequency Ω' scales to very-high ranges and $\Omega'=1$ corresponds to $\Omega=3.3$ THz in a 1.0 μm Si well with $V_0=1V$. For all practical purposes, only the low frequency is of interest, where $F_0'=1$ is the limit. Thus, for SiO_2 insulation, $V_L=V_0=1V$, we find that

$$C_L < 68.4\varepsilon = 23 \text{ pF/cm} \tag{11}$$

or 2.3fF/μm. Integrated circuits today have values of line capacitances well below this value by a considerable amount.

The limit (11) tells us that if we faithfully scale our circuits [9], the line capacitance will not change. While this will have a detrimental effect upon RC delays, the effects observed here will not occur. However, there are trends to move away from such faithful scaling, either to maintain higher potentials or to reduce RC delays and losses. In this case, the usual move is to reduce the line width while maintaining the field oxide thickness [2,10]. If this occurs, line-to-line capacitance begins to dominate C_L, and it is this effect which leads to the loss of charge localization discussed here. Indeed, if we examine the data of [10] for a field oxide thickness of 0.35μm, we find that (11) is violated at a 0.1μm design rule, a size that can be expected before the turn of the century.

III.   NETWORKS OF THRESHOLD ELEMENTS

Another connection to information processing structures is made in this section. One example of such structures, neural networks, is

commonly modeled by representing individual neurons or groups of
neurons by threshold elements. In binary models, the ith such
element has the setting V_i = 0 or 1, where the value is selected
by applying the threshold test:

$$\sum_{j \neq i} T_{ij} V_j \underset{<}{>} U_i \to \begin{array}{c} V_i = 1 \\ V_j = 0 \end{array} \qquad (12)$$

Here U_i is the threshold of the ith element and T_{ij} is an element
of an interconnection matrix. Often [11-13] a neural network is
modeled by a synchronous application of this test to every element
simultaneously. However this simultaneous clocking or synchronism
is not a realistic representation of many systems, in particular
biological neural networks, although it is found for some VLSI
systems. An alternative is an asynchronous application of (12) to
one element at a time [14].

In a more general context, (12) is a realization of a general
cellular automaton [15] which has been applied in diverse areas.
If the range of j in (12) is limited to a near neighborhood, such
as the 4 or 8 nearest neighbors, the operations defined by T_{ij} and
V_i specify cellular logic operations [16] and have been extensively
applied to image processing [17]. On the other hand, when j is
allowed to range over the full spectrum of i, (12) describes
operations that have been applied to simulate distributed, associative memories [14].

We now focus on the differences between synchronous and
asynchronous maps as representations of system dynamics. A map is
the transition from one system state vector to the next produced
by the application of the threshold test (12). Consider the result
of testing one element. Such a test will change only that particular element's setting and therefore can map one initial system
state onto only one system output state, i.e. the mapping is unique
although not necessarily 1:1. This leads to the first effect of
synchronism. If all elements are tested simultaneously, or in
synchrony, then the mapping assigns one unique output state to any
input state. However if only individual elements are tested, then

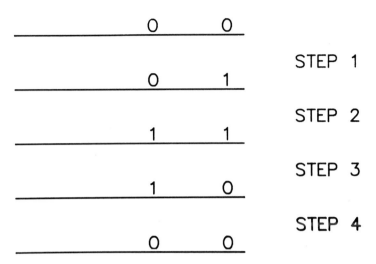

FIGURE 3. An example of the shortest possible limit cycle in an asynchronous system.

there can be many possible output states for any given input state. In these asynchronous systems, the dynamics can be a function of how one selects the next element to be tested [18].

An important feature of threshold test is its unidirectionality. If this test is applied to some element, the resulting setting will not be changed by a future test unless the settings of some other elements are changed first. This logical irreversibility of the threshold test means that all asynchronous and many synchronous maps do not possess an inverse. This non-invertability can have profound implications, the most important of which is that the corresponding logical operations may themselves therefore be thermodynamically irreversible as well [19]. Here we note that this unidirectionality imposes some important constraints on limit cycles in asynchronous systems.

In Fig. 3, an example of the shortest possible limit cycle in an asynchronous system is illustrated. In the first step, one element changes its setting. (In all of these steps it is assumed that only one element will change its setting when tested. This is a necessary condition for a single frequency oscillation in an

asynchronous system.) Since asynchronous maps have no inverse, in
order to return this element to its original setting, some other
element must first change its setting. Thus the second step of
Fig. 3 occurs. In steps 3 and 4, these two elements are returned
to their original settings in the order imposed by the unidirectionality. Thus, the shortest possible limit cycle in an asynchronous system is a 4-cycle.

Some additional constraints also exist for the asynchronous
systems. First an even number of mapping steps are needed to
periodically vary any element's setting and therefore only even
length limit cycles are possible. Secondly, as shown in the example
of Fig. 3, at some point in any limit cycle the following consecutive steps are needed: element i changes its setting to 1 (or 0);
element j changes its setting to 0 (or 1). For these consecutive
steps to occur the interconnection matrix element T_{ji} must be negative. Examination of the threshold tests for the 4 cycle reveals
that while T_{ji} is negative, T_{ij} must be positive. *Therefore a
symmetric interconnection matrix will have no 4-cycles.* One can,
by construction, demonstrate that a symmetric system of 3 elements
cannot have any limit cycles and that symmetric systems of any size
cannot have a 6 cycle. These statements rely on the observation
that a 2m-cycle cannot involve more than m elements but must
involve at least n elements where n is the smallest integer
$\geq \log_2(2m)$. This last criterion ensures the existence of at least
2m states.

The unidirectionality of these maps also allows several special
classes of states to exist. If, for all elements, the associated
asynchronous map does not change that element's setting (i.e. is
directed toward the state) then the system is in a stable state.
If, for all elements, the associated asynchronous map does change
that element's setting (i.e. is directed away from the state), then
the system is in a "garden-of-Eden" state which cannot be reached
by the system in the course of its normal evolution. The existence
of garden-of-Eden states is well known in cellular automata theory
[16]. Although these definitions have been expressed in terms of

asynchronous maps, synchronous and asynchronous systems with an
identical interconnection matrix and identical thresholds will have
identical stable states even though the dynamics of the two systems
may be very different.

These dynamical properties directly appear in a transition
matrix formulation. If the system states are numbered by the
binary sequence $V_N V_{N-1}, \ldots, V_1$, then a vector ρ can be defined whose
ith element is the probability that the system is in state i. The
mapping corresponding to a test of threshold element j can be
written as

$$\rho_{k+1} = M_j \rho_k$$

where k counts the mapping step and M_j is the transition matrix.
Since an asynchronous mapping has 1 output state for every input
state, every column of M_j contains only one non-zero entry and that
has value unity. The unidirectionality property we have discussed
arises from a splitting of the system states into pairs of states,
where the two states differ from each other in only the setting of
element j. Since half of the states cannot be output states of
M_j, half of its rows contain only zero entries. Each remaining row
contains a pair of unity entries with its other entries being zero.
This pair of entries corresponds to one of the pairs of states.
One entry falls on the main diagonal (row c, column c) and it
corresponds to a test of an input state in which the setting of
threshold element j does not change when tested. The other unity
entry corresponds to an input state in which this setting does
change. It is in column $c + 2^{j-1}$ if element j is set to 0 in state
c. If element j is set to 1 in state c, this second entry falls
in column $c - 2^{j-1}$. If the probability that element j is tested
is p_j, then an overall transition matrix M can be defined as

$$M = \sum_j p_j M_j \tag{14}$$

Stable states correspond to a row whose only nonzero entry is on
the main diagonal. Garden-of-Eden states correspond to a row in M
which has only zero entries. Limit cycles of length 2m correspond
to eigenfunctions of M^{2m}.

The directed nature of the threshold test leads to bifurcations when the thresholds are varied. The general behavior will be illustrated here for a simple three-element asynchronous system in which all elements have the same threshold U. The interconnection matrix is taken to be:

$$T = \begin{matrix} 0.0 & -0.5 & 0.2 \\ 0.4 & 0.0 & -0.3 \\ -0.7 & 0.4 & 0.0 \end{matrix} \qquad (15)$$

The diagonal elements are zero as they do not appear in the threshold test. The behavior of this system will be explored for a range of threshold values.

For $U < -0.7$ the various states and mapping are shown in Fig. 4a. Here the state (000) is a garden-of-Eden state and the state (111) is a stable state. Note that the threshold is lower than the sum of any set of elements from the same row of the interconnection matrix. For any size system, either synchronous or asynchronous, when the threshold satisfies the above summation condition, the state of all 0's is a garden-of-Eden state and the state of all 1's is a stable state. In Fig. 4b, the transitions are shown for thresholds in the range $-0.7 < U < 0.0$. Note that now there are no stable states and some looping structures also exist. These looping structures represent oscillating cycles. For $U > 0$, the state (000) ceases to be a garden-of-Eden state and becomes a stable state. For $U > 0.4$ the system becomes essentially reversed from the original system shown in Fig. 4a. This reversed system is shown in Fig. 4c. Again, for any size system, either asynchronous or synchronous, when the threshold is larger than the sum of any set of elements lying in the same row of the interconnection matrix, the state of all 1's is a garden-of-Eden state while the state of all 0's is a stable state which serves as an attractor in the system state space.

The results discussed above explain one difficulty found when networks of this sort are utilized as associative memories [14,20, 21]. An algorithm proposed for storing a set of states with state vectors V^s, s=1,2,...,n, consists of the interconnection matrix

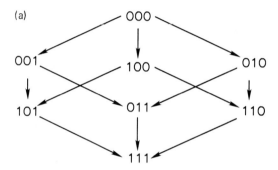

FIGURE 4. (a) Activity topology for the T matrix of equation 6 when the thresholds are all less than -0.7. (b) The activity topology of equation 6 when the threshold is $-0.3 < U < 0.0$. (c) The activity topology of equation 6 for $U > 0.4$.

(b)

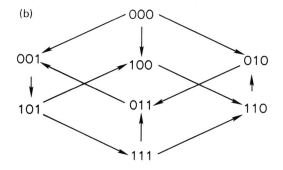

(c)

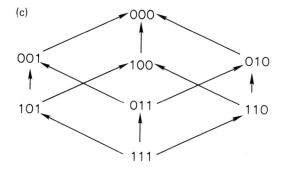

$$T_{ij} = (2V_i^s - 1)(2V_j^s - 1) \tag{16}$$

The contribution to T_{ij} made by a given vector s is shown in
Table I.   This algorithm works well for randomly generated vectors
V^s. However, it works poorly for systems where the stored states
are largely 0's with only a few scattered 1's (or the converse).
These latter states are those found in character recognition, where
a character consists of a pattern of 1's written into a background
of 0's. (A physical two-dimensional array has been used and transformed into the formalism discussed here by assigning a numerical
identifier to each element, i.e. counting the elements). For such
patterns, as Table I shows, one finds mostly positive contributions
are made to the interconnection matrix from each individual pattern,
with the result being a total interconnection matrix whose elements
are almost all positive. Therefore, the system has only one stable
state for a threshold which is non-positive. This is the state
$\{V_i = 1,$ for all i$\}$. Since the memory algorithm works by associ-
ating some stable attracting state with each memory state, it
certainly fails for pattern recognition unless threshold weighting
is employed.

     The above problem is clearly understood from the above analogy
with a random spin glass. Equation (16) is just one model of the
Hamiltonian of such a spin glass. Randomly filling the matrix T
corresponds to a Monte Carlo simulation of this system [22]. In
systems, such as those of the preceeding paragraph, T_{ij} has mostly
+1 elements, so that the threshold lies below the sum of any row or
column, and the system approaches the global stable state.
In analogous magnetic terms, the net magnetization is far above the
phase transition, while for multiple states we need to remain near
the phase transition. Thus, random V^s with nearly equal numbers of
0's and 1's must be used. One should remark, however, that care
must be exercised in pursuing this analogy as spin glasses are
usually simulated by nearest-neighbor coupling only.

IV. SUMMARY

We have discussed the device-device interaction problem from
several vantage points. Such interactions cannot be avoided in

Table I.

V_i^s	V_j^s	sth Contribution to T_{ij}
0	0	+1
0	1	-1
1	0	-1
1	1	+1

dense arrays, such as might be envisioned in molecular electronics. Typically, they are approached in a piecemeal fashion. Here we have tried to formulate a more general theory. In one approach, drawn from statistical mechanics, the dynamics of a collection of interacting devices is described by a Hamiltonian. The device-device interaction arises in two distinct fashions. The first is an average, zero-order effect similar to a Hartree potential. The second is a first-order effect directly related to a two-device correlation function. The existence of minimum device sizes leads us to seek a second approach, which involves a study of model systems constructed from threshold elements. Such systems exhibit a rich variety of dynamical structures including stable states, simple limit cycles, multi-frequency cycles and apparent "chaotic-like" behaviors. While most workers have studied synchronous systems, we have found that important changes occur when an asynchronous evolution is allowed. These differences arise from a multiplicity of possible system state changes, all of which are unidirectional or non-invertible.

This work was supported by the Office of Naval Research.

REFERENCES

1. R. W. Keyes, IEEE Trans. Electron Devices, ED-26, 271, (1979).

2. D. K. Ferry, in Advances in Electronics and Electron Physics, C. Marton, Ed., Academic Press, New York, 1982, vol. 58, p. 311.

3. D. K. Ferry, in Physics of Submicron Semiconductor Devices, H. L. Grubin, D. K. Ferry and C. Jacoboni, Eds., Plenum Press, New York, in preparation.

4. J. R. Barker and D. K. Ferry, Proc. 1979 Intern. Conf. Cybernetics and Society, IEEE Press, New York, 1979, p. 762.

5. R. Reich, R. O. Grondin and D. K. Ferry, Phys. Rev B., in press.

6. G. J. Iafrate, D. K. Ferry and R. K. Reich, Surface Science, 113, 485 (1982).

7. L. S. Schulman, Techniques and Applications of Path Integration, Wiley, New York, 1981.

8. S. Jorna, Topics in Nonlinear Dynamics, American Inst. of Physics, No. 46, New York, 1978.

9. R. H. Dennard, F. H. Gaensslen, H. N. Yu, V. L. Rideout, E. Bassous and A. R. Leblanc, IEEE J. Solid-State Circuits, SC-9, 256 (1974).

10. A. K. Sinha, J. A. Cooper, Jr., and H. J. Levinstein, IEEE Electron Device Letters, EDL-3, 90 (1982).

11. S. Amari, Proc. IEEE, 59, 35 (1971).

12. S. Amari, IEEE Trans. on Computers, C-21, 119 (1972).

13. S. Amari, Kybernetik, 14, 201 (1974).

14. J. J. Hopfield, Proc. Natl. Acad. Sci., USA, 79, 2554 (1982).

15. J. Von Neumann in Cerebral Mechanisms in Behaviour-The Hixon Symposium, L. A. Jeffries, Ed., Wiley, New York, 1951.

16. A. W. Burks, Ed., Essays on Cellular Automata, University of Illinois Press, Urbana, 1970.

17. K. Preston, Jr., M. J. B. Duff, S. Levialdi, P. Norgren and J.-I. Toriwaki, Proc. IEEE, 67, 251 (1979).

18. R. O. Grondin, W. Porod, C. M. Loeffler and D. K. Ferry, submitted for publication.

19. R. Landauer, IBM J. Res. Dev., 5, 183 (1961).

20. L. N. Cooper, in Proc. of the Nobel Symposium on Collective Properties of Physical Systems, Eds. B. Lundqvist and S. Lundqvist, Academic Press, New York, 1973.

21. L. N. Cooper, F. Liberman and E. Oja, Biol. Cybernetics, 33, 9 (1979).

22. R. H. Swendsen, in Real-Space Renormalization, T. W. Burkhardt and J. M. J. van Leewuen, Eds., Springer-Verlag, Berlin, 1982.

32

The Biochip: Now, 2,000 A.D., and Beyond

James H. McAlear, Ph.D and John M. Wehrung / Gentronix
Laboratories, Inc.

It seems that everytime that science should have run out of
wonders still another one comes along. Nowadays there is so much
on TV and in various publications about computers, robots and the
like, it is very hard to tell these wonders or real possibilities
from Star Trek and fiction. Everyone will agree that computers
are changing the way we do things and that they have become a big
business as well as the most critical edge in our nation's
defense. There is less agreement on the future because everything is changing so fast.

A lot has been written in the past couple of years about
biochips, some good information but much of it pure hype. Since
the basic idea for biochips started here over 10 years ago it is
about time we spoke out about what we mean, what they mean and
what we intend to do about it.

The first thing to keep in mind is that the kind of electronic circuits used in computers today have just about gone as far as they can and that unless there are some brave new ideas the computer revolution is going to peter out sometime in the next 10-20 years.

The biochip idea is now pretty well recognized as the best new idea to come along the pike since the last one (the integrated circuit (IC)) almost 25 years ago.

Actually, it is a very old idea, perhaps over four billion years old, when somehow complicated molecular chains became what we call the "genetic code" and life became organized out of the primordial "soup".

Today, virus may have a similar organization to those early "life" forms. The virus attaches to a cell, enters it and releases its DNA material which attaches to the victim's DNA and forces it to make more virus DNA as well as the protein coat that surrounds the virus DNA. The important part is that these virus coat molecules are composed of protein which is programmed in the virus genetic code and which assembles itself into a complete virus around some new DNA.

It just so happens that this same general idea is the way that the genetic code determines all structure, some directly, other indirectly or dynamically.

A protein molecule is a chain made out of about 17 different kinds of links which are just like letters in an alphabet. These letters are "written" by a kind of morse code in the DNA, which can divide itself into more of the same DNA and make more of the kind of protein its code dictates.

Modern bioengineering makes it possible to build DNA to order which can make just about any kind of protein you might ask for whether it occurs in nature or not.

These proteins that are programmed to fit together to form a specific structure like a virus coat, are formed of chains; these chains tend to fold around themselves to make identical three

dimensional units. The surfaces of these proteins have various
patterns of atoms much like the patterns of the faces of a rubics
cube. When the patterns of two proteins are complementary they
stick together like a tinker toy set made of locks and keys,
instead of holes and dowels. This is how proteins assemble
themselves into specific structures called virus, man and
biochips.

The big question in most peoples minds is, so what, you can
make structures out of molecules, how can you make these into
computers. The answer is not simple because it is a long way
from the structure of molecules to the plug in the wall.

We can connect down to the best IC's today which have wires
about 1μ wide (about 1/7th the width of a red blood cell). If we
were to have a micro-man stretch out across one of these wires
our objective would be to eventually use molecules which would be
about the thickness of his fingernail at that scale as the
equivalent of wires, but before we can do that it is necessary
to be able to build things the thickness of his arm and then his
little finger before we hook up fingernails (molecules).

The reason that small size of these little transistors in an
IC is so important is that this is what makes a computer
powerful. As you cut the size of wire by 10-1 you can put 100
times more of them into the same area, this moves them 10 times
closer together, and which makes everything at least 10 times
faster. One big difference between the IC's and the biochip is
that IC's are flat and biochips are cubic, so if you reduce the
size 10-1 you get 1,000 times more on the same area and the thing
begins to look more like a brain than anything built before. But
let's not get carried away with the brain analogy. That's what
most of the hype has been about.

The other aspect of life besides structure is function and
this is most apt to be directed by proteins which are classified
as enzymes. An enzyme is usually more like a work station on an
assembly line, either building something or tearing it down. For

example, an assembly line of enzyme proteins takes apart the food
we eat and another reassembles them into us. As these enzymes go
about their business the different chemical reactions which they
cause to occur can be directed into laying down different metals
and plastics.

This is where we come in, because we recognized that this
was a way to deposit metals which could make up wires used in
computer circuits.

Our first invention was something like the first airplane.
The Wright brothers put an engine on a glider. We covered a
layer of protein with a plastic, plexiglass, which we exposed to
a stream of electrons in a parallel pattern just like in a TV
set. The electrons chopped up the long chains of the plexiglass
molecules (polymethylmethacrylate) so that they could be more
easily dissolved with alcohol, thus uncovering trenches down to
the layer of protein.

We put the metal on by using biological stain for protein,
silver nitrate, and not too surprisingly, it worked. While this
was no big deal in itself, it opens the way to use a long list of
biochemical methods for depositing all kinds of materials, metal
conductors, and semiconductors and insulators, as well as
plastics, which can be useful not just for ordinary
microelectronics applications but for a major potential
breakthrough, three dimensional circuits. If we could make a
three dimensional IC it would be 10,000 times more powerful than
present devices.

This is our goal in what we call Stage I. It is the first
of four steps, each one of which could result in at least 10,000
times improvement over the previous stage until the level of a
single chain of atoms is reached. There is a long way to go. It
will take a tremendous effort to get there, but we are well on
the way. The stakes are of ultimate significance.

Back to Stage I after that brief burst of futuristic euphoria. Our first invention gives us some new tools. It is the only way so far to write a pattern and deposit material directly on it without having to vapor deposit and etch back.

What this means is that it would be fairly simple to make a layer cake of plastic films with conducting wires in it and to connect these layers. The problem is that this doesn't make a computer, you need a semiconductor material. Fortunately, in recent times, there has been a good deal of attention paid to alternatives to silicon chips which includes certain plastics and very small repeating conductors or superlattices which can behave like a semiconductor. These fit well into the three dimensional biochip concept.

The microelectronic industry is a bucket of high tech worms. It is so highly developed and competitive that any radically new technology must meet the most exacting conditions. This is not a garage shop enterprise.

There is another way to make three dimensional structures besides the layer cake approach. This refers to the virus coat and protein self assembly idea. If you have enough different proteins which will naturally self assemble and if you can make an accurate three dimensional model of each of them you may have a tinker toy set at the molecular level. Relative to the ultimate (a fingernail) molecular computer these molecules would be of "finger" size. If you put enough fingers together each bent at the exact angle you (and your computer) decides that you need you could build some pretty elaborate structures at the next level, "arm" up.

This idea has been given some help lately by a new technique called "monoclonal" antibodies which is a big part of the current biotech boom. Monoclonals are being used in all kinds of interesting ways. Let me explain what they are and how we can use them to make three dimensional structures.

You may recall the old fable about the three blind men and the elephant. Each of the blind men variously described the elephant according to the part he touched. This is what happens to a foreign protein like those comprising the coat of a virus when they get inside of the body. Each cell from the antibody forming system in the thymus "sees" a different side of the virus protein and makes an antibody against that side so that these antibodies in the blood will attach to all sides of a virus protein. Using some tricks developed a number of years ago you can cause each one of these antibody producing cells to "light up" so that they can be isolated. Usually these antibody cells will divide a few times, then die, but if you use another trick you can fuse them with a cancer cell which will grow indefinitely in a tissue culture and the resulting cross will go on dividing like the cancer cell but cranking out this one kind of antibody indefinitely. These are the monoclonal antibodies.

Antibody molecules look like a little body with two legs which are almost alike. You can cut the legs off with an enzyme which breaks down protein just like meat tenderizer and these legs, called Fab's, can be used as building blocks or tinker toys which can be produced in endless variety against practically any kind of other protein including Fab's from other animals.

One of the big jobs ahead is to produce a great number of these Fab antibody pairs called "idiotypes" and to analyze their structure using three dimensional graphics methods and to store the information in a data bank so that you and your computer can build any kind of structure that you want. For example, if you want to build a connection between two points in three dimensional space you must start out with an oriented pattern of molecules in a layer one molecule thick. These oriented monolayers are made by floating them onto the surface of a water bath using a trick developed years ago by Irving Langmuir at G.E. These films are compressed into a monolayer using a cursor which moves along the surface until all of the molecules press against

each other. Generally protein molecules can be selected for this
which have one side which is more attracted to the water surface
so that they will all orient that way. Also it is important to
line them up like elephant soldiers on parade by treating them to
an electric field. Such oriented monolayers can be picked up
from the water with a smooth surface like a glass slide, a
plastic film or a silicon wafer. Generally, the water facing
side of the film is the one which makes the best elephant
(antigen) so we could pick up the monolayer so that this is the
upside. This monolayer should be covered by a thin layer of a
plastic like the plexiglass as mentioned earlier and a hole
drilled in it with an electron beam which makes that plexiglass
more easily dissolved in alcohol. Now let's select the first
riders on the elephants so that they lean forward just at the
right angle. The next rider will sit on the shoulders of the
first and so on until they form a stack of the desired height.

Now this is a pretty unstable structure, some of the riders
slip and slide, etc. so in order to stiffen things up we cast the
entire structure in plastic like the same plexiglass in the film.
This can be done by giving each of the antibody riders a plastic
making enzyme and supplying the ingredients for the reaction.
Now I realize that enzymes don't ordinarily make plastics but one
like the zip in horseradish, peroxidase, is a tough little
plastic maker. You give him a chance and he will take his normal
diet, hydrogen peroxide, and chew off the extra oxygen to make
water. The oxygen then takes methacrylate molecules and links
them into plexiglass right on the spot. Before you know it
everything is covered with plexiglass like branches during a
freezing rain.

We now have a fixed structure which goes from A-B at the
angle we want, but plexiglass or the proteins doesn't conduct
electricity very well. We can make these conductors by using the
same enzymes to make another plastic out of DAB (diazobenzidene)

in the same way it makes plexiglass and at the same time. DAB
will react with some metals to form a complex which does conduct
electrons.

So we've done what we have set out to do and while it seems
complicated like most complicated tricks, it's easy when you know
how.

With this method, it becomes possible to make practically
any kind of conducting, semiconductor or insulating structures in
3D right down to the size of elements ("fingers") which can
organize fingernails.

By using combinations of layering techniques and growing
structures stabilizing them, and metalizing them in oriented
monolayers, it becomes technically possible to organize
dimensions starting at the level of present microelectronics
conductors, the body, and to go in four stages to the level of
the fingernail.

Before we go charging off after the molecular computer let's
realize that it will take industry 10-20 years just to absorb
this potential 10,000 time increase in computer power from the 3D
biochip.

The military, however, does have several big reasons for
going ahead. Computers are the single most critical technology
covering everything from smarter missiles and ABM's to
intelligence, mapping data bases and communication. This latter
category is the Achilles heel of our defense (and the Russians),
because the magnetic field (EMP) created by a nuclear explosion
in space could wipe out virtually all of our communications in
satellites and on land, perhaps preventing the launch of a
retalliatory strike. However, if we use light instead of
electrons, EMP has no effect.

Light can be used to interact with molecules to switch
electrons like a transistor, but light can also be used to switch
other light conducted in plexiglass waveguides, like fiber

optics. Furthermore, we can make light waveguides that will work which are smaller than electrically conducting wires.

Furthermore, light travels three times faster than electrons and doesn't create as much heat or interact with neighboring light conductors. This is a new technology but all of the ingredients exist for the development of photo molecular devices which would be about 10,000 times more powerful as computer elements than 3D wire devices. I don't think that the free world can anymore afford to lose a race to build these kinds of EMP immune computers than it could have given the A bomb to the Nazi's in WWII.

There are two more stages, the next involves the transmission of chemically trapped light called excitons along "finger" dimension structures which can also be fabricated by molecular assembly, but by this time we will be designing the molecules from scratch by computer modeling and producing them automatically by genetic engineering.

The final stage is the molecular computer where the signal is a solitary wave or soliton which doesn't require power.

Most researchers are working way down on this end and it must be certainly a fascinating area for those who can understand it. Whenever it is ready and the world is ready for it, I'm sure we can organize it and interface it with Stage III which will be interfaced with Stage II, etc., all the way to that plug in the wall. Altogether, these few stages constitute a combined increase in density and speed which is almost unimaginable with each stage at least 10,000 times more powerful than the previous; compared to these the change from the vacuum tube to the most advanced integrated circuit is more like a family squabble than a revolution.

Like the nuclear bomb and the ballistic missiles which brought us nuclear power and medicine and the space program and communications satellites, the biochip has some major spin-offs.

One is the area of sensors; biosensors are already becoming a major business for medical diagnosis and for process control in industry.

The most potentially important application, however, is in the use of protein polymer composites to produce better prostheses for replacing parts of the body and to make systems for restoring hearing, muscle movement and potentially even sight.   These kinds of impairments cause tremendous suffering in millions of people and cost billions each year in care costs and lost productivity.

There is little doubt now that biochips research will become a rapidly developing area in the national defense and in the microelectronic industry, in the next few decades.

It is impossible to predict the long term consequences. However, the impact of microelectronics on our society in the past 25 years (which has yet to be felt to its full potential) may be a relatively small one.   These changes will be confronted by generations somewhat better prepared in their orientation than our generation and such things as computer-human combinations or symbiotes which are popular in science fiction because of the revulsions they can create in us may be far less forbidding than we presently regard them.

A biologist accustomed to studying the super megatrends of the evolution of living things over billions of years cannot help but speculate where all of this might lead in the future in the same context.

One thing is certain; to evolve by R&D is a lot faster than by variation and natural selection.   To be sure, research and development is something of a trial and error process but to a much lesser degree.   The evolution of the automobile and airplane as well as the computer chip, are examples.   Now computers are designing themselves.   Biochips will certainly be doing this with greater competence at each stage in development.

In a sense, a biochip is a potential new form of living thing which as a computer could be vastly superior to our own brain. It could also make itself from all of the elements with properties which could allow vastly longer functional life spans, broader communication, and travel over far greater distances at higher rates of speed, and existence and development in far more rigorous environments.

33

Molecular Electronic Devices — A System User's Perspective of Opportunities and Challenges

Bernard A. Zempolich / Naval Air Systems Command Headquarters, Washington, DC

In 1963, specifications were written for a new weapon system for use in the naval aviation operational environment. One obscure subelement "hidden" deep within the system design specifications stated that integrated circuits are to be used "wherever practical". Thus, began the advent of major, general (multiple source) procurement of Transistor-Transistor Logic (TTL) integrated circuits for use in modern, high performance weapon systems.

The 15 active devices/chip (circa 1963) now seems archaic - two decades later - when compared with the 100,000 gate/sq. cm. chip and more now available. This manyfold increase in active device density over the past twenty years is a testimonial to the innovative efforts of the myriad number of engineers and scientists who have contributed to this monumental advancement in the state-of-the-art. However, what is even more dramatic from a "forward progress" viewpoint is the fact that the increases in

density were not achieved in a linear fashion over the past 20
plus years. Rather, they occurred with increasing rapidity
during the past five years or so.

In a rush to take advantage of the inherent increases in performance (such as speed, throughput, decreased weight, volume, and
power), users incorporated solid-state devices into system
applications to increase operational performance and decrease
life-cycle maintenance considerations. Unfortunately, very
little management and technical considerations were given to the
ramifications of utilizing the newly developed integrated circuit
- "silicon blue"! It is easily proven that beginning with such
problems with integrated circuits as the nefarious "purple
plague" (the gold aluminum dissimilar metal bonding problem) and
continuing to the current need for manufacturing obsolescence
programs, users were, and still are ill-equipped to effectively
and efficiently make the quantum jump from the discrete transistor to the highly dense integrated circuit. Perhaps this
situation is best summed up in a statement made recently that
"...we have just begun to handle LSI, and now 'they' give us
VLSI"!

This dichotomy of positions - uncontrollable introduction of
highly dense solid-state devices juxtaposed with the uncontrolled
increases in management and technical problems - simultaneously
offers the system user both opportunities and challenges. On the
one hand, it offers opportunities to increase the performance of
the system, while on the other hand, it presents new challenges
which, if not correctly addressed in a timely fashion will impede
the effective introduction of the new technology and add an
unacceptable overhead burden throughout the life-cycle of the
system.

From a user's perspective numerous opportunities and challenges
must be faced in incorporating molecular electronic devices into
an integrated system (generic) design. Specifically, it is the
author's premise that the introduction of molecular electronic

devices will be such a dramatic leap forward in technology usage
that the normal introductory problems will be insurmountable if a
priori systems management, design, and introductory integration
factors are NOT made an inherent, everyday part of the molecular
electronic device innovative process and its subsequent "reduc-
tion-to-practice" phase. Based on this premise, the following
major topics that relate multilevel system factors with micro-
level molecular electronic device characteristics will be
addressed:

1.   Technology transfer and insertion.

2.   Potential systems applications (in both the civil and
military sectors as well as the private and public sectors).

3.   Multilevel system interconnectivity.

4.   Multilevel storage requirements and distribution.

5.   Information transfer and fusion.

6.   Device physical packaging and multilevel systems packaging.

7.   Fault-tolerance, reconfigurability, and adaptation to
changing user needs.

8.   The impact of electromagnetic interference (that is,
radiation, electrical storms (lightning), pulsed energy, r-f
noise, etc.).

9.   Software impact and use (applications packages, support
systems, Computer-Aided-Design, Computer-Aided-Manufacturing, and
Computer-Aided-Test).

10.   Logistical impact (how much, how many).

11.   Economic impact (economy of scale and multiple sources).

12.   Technology obsolescence.

A new system conceptual design definition can also be addressed as "TECHWARE". Techware is defined as the amalgamation of technology, hardware, and software. The concept is predicated on the premise that these three major design aspects of any future system design are no longer individually divisible, and must be henceforth addressed as an integrated whole regardless of the user's end application.

The paper concludes with the identification of the author's opinion on the major system issues that must be examined by those individuals and organizations engaged in the definition, design, and test of molecular electronic devices.

VI
Electron Tunnel Switches

34

Complex Networks in Molecular Electronics and Semiconductor Systems

John R. Barker / Department of Physics, University of Warwick,
Coventry CV4 7AL, England

I.   INTRODUCTION

Recent progress in both solid state materials technology and
synthetic chemistry has independently enhanced interest in a
future molecular-scale electronics which might involve gate
densities of 10^9 cm^{-2} (solid state) to 10^{15} cm^{-3} (molecular
media) (1-3). Inevitably, the focus has been on the development
of novel devices, but which conceptually are traditional circuit
components: communication lines, storage elements, switches and
sensors (4,5,2). Unlike the situation in VLSI semiconductor
electronics, system considerations have largely been ignored in
deference to the formidable problems of how to communicate with
molecular switching groups or moieties, and how these groups are
to be synthesized (2). There are evidently a great number of
possible chemical routes (7), and to a lesser extent, solid-state
routes, to molecular-scale electronics; but as Pascal, a pioneer
of automatic computing, so aptly remarked, "The last thing one
knows in constructing a work is what to put first".

The present article argues that the enormous complexity of systems implied by large gate densities is the first and overriding problem: it runs through problems of design, fabrication, fault tolerance, testing, input-output, temporal and spatial management of the intended information processing structures. The special issues raised by complexity are already perceived in VLSI electronics (reviewed in refs. 6,9,13 and summarized in section 2), where opinion is gaining ground for a holistic structured approach to designing complex systems (8,10,12). The holistic approach is cross-disciplinary. It seeks to reduce the complexity problems by asking what is the system for; what are the constraints; what jointly are the most efficient structured means for balancing logic management and design with its representation and fabrication within physical limitations as a system of suitable devices. This approach is followed here and extends our earlier studies (4,5,10) to consideration of a general conceptual framework for the design of large molecular electronic arrays comprising identical simple logic modules but which might support arbitrary computational processes.

Molecular scale electronics would be advantageous: (a) dense, possibly three-dimensional memory arrays; (b) in compact, ultra-fast computer systems; (c) bio-sensors/bio-technological control/bio-electronic prosthetics. We shall concentrate on area (b), where the combinatorial explosions and structures of real-time control in artificial intelligence schemes such as computer vision, computer reasoning, advanced robotics, machine translation and dictation, require ultra-fast, and in many instances, parallel-processing capabilities. Many of these applications have already been developed in nature, where biological organisms not only display complex logical and computational functions, but in addition have important mechanisms for fault tolerance, self-maintenance and self-repair, self-organization, adaptation and reproduction (13). It is interesting to note that many of the hierarchical modular

structures currently being investigated in VLSI (8,14) have
natural counterparts. Are there other natural principles which
we might exploit?

   The view taken here is that <u>self-organization</u> should be a
key feature of molecular-scale electronics. There are already
promising techniques under development (2,15) which suggest that
the fabrication of large regular molecular assemblies could be
managed by self-organizing chemical processes (7). But the
question of how complex an assembly is required depends on what
computer architecture and its components are considered. In
traditional design a given architecture is achieved by <u>directly</u>
representing the required logic gates, memories, buffers,
registers, clocks, interconnects by fabricating them as devices.
Even with highly regular parallel designs the range of requisite
components and their distribution is large, which poses difficult
problems for molecular fabrication schemes. However, there is an
alternative way of realizing computer architectures of arbitrary
complexity which was first pointed out by von Neumann (16) and
developed by Thatcher (17), Holland (18), Codd (19) and Arib
(20), but abandoned because it required extremely large numbers
of logic gates. This approach is based on the concept of
<u>cellular automata</u> (section 8). A cellular automaton is a regular
(tessellating) array of identical, interconnected finite automata. With the appropriate choice of the cellular space, the
cell neighborhood, and the local transition function for each
automaton in the space, it is possible to <u>embed</u> arbitrary
computer architectures in the cellular space. The program, data,
logic gates, memories and interconnects of the embedded computer
are here identified with <u>spatial configurations</u> or <u>patterns</u> of
the <u>states</u> of the individual finite automata. As an abstract
concept cellular automata provide a useful framework for
examining the structure of computational schemes and as such have
been widely used for developing programming concepts particularly
parallel programming. Automata models have also been used for
analyzing processes in developmental biology (21-23).

The hardware requirements for cellular automata are relatively simple: they are the identical combinational logic modules
plus registers which represent the individual finite automata and
a generally short-range interconnection mechanism for passing
signals between a small number of neighboring modules. In
general, a set of intersecting configurations of cellular states
will interact to produce sequences of new configurations. This
forms the basis for computations. It is not necessary to
directly write in the computer configurations, because cellular
automata exist which can support constructor configurations which
can be programmed to self-construct (and indeed erase) an
arbitrary computer configuration (section 8,9). Our principle
conclusion is that the prime target for molecular scale electronics should be the identification and fabrication of the basic
combinational logic modules for the formation of two-or three-
dimensional molecular cellular automata. The complexity of the
hardware/fabrication problem is thus considerably reduced,
computer design/fabrication becomes a software problem, and there
are added advantages in that the problems of devising fault-
tolerant systems and input-output are more easily analyzed.
Cellular automata satisfy most of the criteria for structured
design and system management that have been put forward for VLSI
systems.

It should be pointed out that cellular automata represent
essentially discrete processes in a quantized effective space and
time, whereas real physical systems are necessarily supported by
underlying continuous processes. Self-organizing processes using
linear and non-linear feedback paths for control have been
extensively studied for analogue systems within cybernetics
theory (24). However, the emphasis has been on achieving home-
ostasis: the self-relaxation of a system to a stable mode of
operation. The recent development of synergetics (25) has gone
considerably further in providing analytical tools and concepts
for describing sequences of discrete structural transitions in
general distributed systems which are governed by continuous

non-linear processes. Such transitions may be modeled in many
cases by cellular automata (if we include stochastic elements)
where the transition rules are derived from the underlying
continuous physical processes. Thus, if we are to follow the
holistic approach it is necessary to use the natural language of
cellular automata theory for the design of molecular electronics
but to temper the design with synergetic analysis of the detailed
physical behavior of the proposed devices and device groups
intended to support the cellular automaton. Without detailed
knowledge of the physical mechanisms for interconnection (both
intended and parasitic) one cannot guarantee that the appropriate
transition functions will actually be stable and as intended. To
elaborate these points we discuss the adverse effects of synergetic processes in VLSI coupled device arrays (sections 2,3).
General tutorial examples of self-organizing effects arising from
non-linear parametric coupling between devices are given in
sections 4-7. A detailed treatment is given in references
(26,27,10). The abstraction to useful cellular automata is
developed in section 8. Extensions to relocatable or floating
computer architectures (10) are sketched in section 9. Analogies
with neural networks are given in reference (10). The relevance
to fault-tolerant systems and schemes for managing the
input/output problem are outlined in sections 10-11. Finally,
section 12 examines the implications for the selection and
fabrication of appropriate materials configurations (the solid
state case is more thoroughly described in reference 10).

II. COMPLEXITY ISSUES

As high density circuits become practicable the prime cost of
production is beginning to shift from fabrication to design.
Thus, there is a significant trend from unstructured, hand-coded
designs to more structured forms which are amenable to computer
aided design (8). Designs which are more regular and simple at
the conceptual level cannot be set up in isolation from physical

constraints. In semiconductor technology the first constraints
come from the planar geometry and fabrication restrictions. To
these must be added physical limitations which are connected with
power dissipation, delays, synchronization and signal distortion
and attenuation (8,10). The latter are exacerbated in traditional sequential logic circuits, where high gate density leads
to long bus and interconnect paths and cross-over problems, with
the interconnect area to gate area ratio increasing at high
complexity levels. In dense systems there are also communication
bottlenecks involved when one considers (a) off chip communication and finite pin-out problems; (b) bandwidth problems of
coupling a central processor to a large memory array. As a
consequence of these problems more systematic holistic design
methodologies are emerging for VLSI with the trend to hierarchical modular structures and concurrent structures in which
logic and memory are regularly distributed throughout the system.
Particular attention is being paid to systolic structures (28,8)
which have regular cellular structures of simple processor/memory
elements connected by short local interconnection nets with
extensive pipelining and multi-processing. Fig. 1 shows some
examples. These have mostly been considered for specialized
computational functions such as signal and image processing.

The timing of complex systems is also problematic. Synchronous timing is not easy to achieve because: (a) delays introduced on the long interconnect lines which distribute the clock
signal inside the system eventually become of similar magnitude
to the gate switching times; (b) synchronization is easily lost
in large designs which require rigid lock step control, and this
includes many systolic designs. Hierarchical timing schemes have
been suggested as intermediate solutions and use sequences of
isochronous zones which are separated by slower timed interfaces
(8,9). For very dense designs it is being argued that self-
timing schemes should be used for which only the sequence of
processing events is important not the length of time.

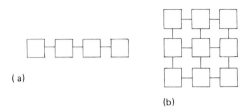

(a)

(b)

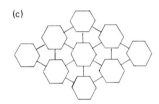

(c)

(d)

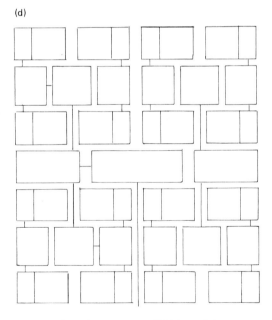

FIGURE 1.   Concurrent VLSI architectures. (a) pipeline; (b) array; (c) hexagonal; (d) binary tree.

Reliability is already a serious issue in VLSI circuits and may well prove the most serious problem for molecular scale electronics. System damage can occur due to intrinsic fluctuations, defects, fabrication error (real and statistical), cross-talk effects, electromigrative disconnection or reconnection, and perhaps most seriously from soft and hard radiation damage due to natural radiation from the host or packaging material and cosmic rays (2,10). Sequential designs are particularly vulnerable. Although redundancy and error detecting/correcting coding and circuitry can (and are) introduced these complicate the design and at sufficient complexity levels become counterproductive. Testing of complex systems is a particularly severe problem and self-testing schemes for high density have yet to be devised. To these problems we may add that the stability of regular concurrent circuits is barely understood particularly with respect to parametric parasitic device-device interactions (29,10).

From this very brief summary of some complexity issues perceived for VLSI we may draw two principal conclusions: complex systems require designs which involve highly concurrent regular processor arrays; fault-tolerant design will be a priority. If we accept the former conclusion it becomes important to develop a general theory of parallel processing structures or find concurrent processing structures which can support arbitrary computational processes as an embedding. The latter approach is described in section 8. For the fault-tolerant problem we might require that self-organizing properties be built into complex designs: as a guide we know that cooperative processes in say laser action can act to stabilize the system behavior and function even though individual elements are defective; the remarkable stability of natural biological information-processing systems against damage shows that solutions are possible if not understood. We might go further and seek a self-organizing computer structure which could detect defects and reallocate its component functions or relocate part or the whole of its struc-

ture to avoid damaged regions. First, let us establish some of
the adverse and advantageous properties of self-organizing
circuits with reference to solid-state systems of devices.

III. SELF-ORGANIZING PROCESSES IN COMPLEX CIRCUITS

A central assumption in the design of LSI circuits is that
each device or interconnect is assumed to behave in the same
manner within a system as it does when it is isolated. The full
function of the system is then determined by the interconnection
matrix which specifies how the devices are joined together. A
different function can only be prescribed by reconnecting the
interconnection net: a practical impossibility for most systems
although microprogramming can assist. Thus, device design is
traditionally separated from system design. However, in VLSI
systems, line-to-line capacitance, a parasitic capacitance, is
beginning to dominate the interconnection capacitance at particular nodes. Some parasitic effects can be allowed for by
equivalent circuit techniques and renormalization of the isolated
device parameters. However, a number of parametric, parasitic
interactions between devices (by which we include "isolation"
areas and interconnects) are possible; where by parametric we
mean dependent on the pattern of charges, currents, voltages,
thermal gradients, etc., in the system, i.e. on the information
state. These include effects of wave-function penetration
(tunneling and charge spill-over), remote energy-loss (especially
near interfaces), electromagnetic cross-talk (already known in
READ/WRITE operations in dense memories), and thermal cross-talk.
In biological media there are similar effects, for example, ion/
electron flow through cell membranes where evidence exists for
parametric transport and switching associated with cooperative
effects between neighboring protein molecules. One also may
anticipate similar effects due to polarization forces, conformational changes and tunneling in molecular electronic device
arrays. Thus, in polymer conductors such as chains of trans-

$(CH)_x$, soliton propagation down a particular chain may be
affected by electron hopping from one soliton to a soliton in a
neighboring chain. One may conclude that true circuit isolation
is likely to be very difficult to achieve in any dense systems.
The stability of large dense circuits is therefore expected to be
very different from situations where isolation of devices is
perfect (3,10,11), and may lead to adverse self-organizing
transitions in the effective system architecture and therefore to
logic errors.

The understanding of self-organizing processes in complex
systems has progressed substantially in recent years due to
developments in non-linear applied mathematics which are grouped
under the heading synergetics (25). Synergetics is concerned
with the non-equilibrium creation of macroscopic structure or
patterns of behavior in systems composed of many competing
sub-units. These sub-units may be, for example, atoms, molecules, biological cells, populations, or devices. Synergetic
phenomena include order transformations which lead to the
formation of (a) spatial structure, e.g. Benard cells in liquids;
(b) temporal structure, e.g. circadian rhythms in biological
systems; coherent modes in lasing media; (c) regular pulse
formation, e.g. solitons, ultra-short laser pulses, nerve pulses;
(d) spiral formations and concentric waves, e.g. in chemically
reacting mixtures; (e) chaos, the sudden appearance of apparently
random processes in a deterministic system, e.g. in the laser in
chemical reactions (25), in neural systems (30). All these
phenomena are, in a sense, global switching phenomena which occur
due to the cooperation of a few order parameters arising from the
many degrees of freedom associated with the large number of
system sub-units. It is interesting to note that one of the
first documented examples of non-trivial synergetic phenomena,
including chaotic behavior, concerned the circuit performance of
thermionic valve "universal circuits" (31), where the self-
organizing effects were traced to a parametric parasitic coupling
between the valves.

In previous studies we used system density matrix techniques
to argue that all the necessary pre-requisites for synergetic
behavior exist in highly concurrent, dense circuits (3,5,10). In
circuit language, the effects are expected whenever there exist
non-linear variable feedback paths in the system network, and are
most favored in systems with large dimensionality or equivalently, large concurrency. To illustrate some of the concepts
let us choose a tutorial example of how self-organization can
occur in the function of a circuit due to order-transformations
in the effective system architecture.

IV. DEPENDENCE OF CIRCUIT FUNCTION ON ARCHITECTURE

Given a set of isolated devices it is obviously possible to
construct a variety of circuit functions by defining an appropriate set of interconnections. We call a specific network the
connection architecture. The influence of the architecture on
the system behavior is not always obvious; some changes to the
architecture may have no effect, others may produce dramatic
changes. To demonstrate let us follow the arguments given in
ref. (29) and consider an isolated integrator with input/output
conditioning and comprising an array of N devices. In the
following we use an analogue approach, but the technique of
control theory which we use may be extended to linear, sequential
digital circuits through the description of the system by an
abstract extension field. The time-dependent state equations for
the isolated system are taken as

$$-\frac{d}{dt} u_i = A_{ij} u_j + B_{ij} y_j \tag{1}$$

$$z_i = D_{ij} u_j \tag{2}$$

(summation convention, $i,j = 1 \ldots N$). Here, u_i, y_i, z_i are the
state variable, input, output to device i, respectively. The
internal device dynamics are determined by the diagonal matrices
$\underset{\sim}{A}$, $\underset{\sim}{B}$, $\underset{\sim}{D}$. Suppose now that an interconnection architecture

is established by introducing a connection matrix $\underset{\sim}{C}$. We relate
the system input and output vectors $\underset{\sim}{g}$ and $\underset{\sim}{h}$ to the individual
device input/outputs by:

$$y = C \cdot z + L \cdot g \qquad (3)$$

$$h = M \cdot z \qquad (4)$$

where the I/O transformation matrices $\underset{\sim}{L}$, $\underset{\sim}{M}$ are added for
generality. From eqns. (1-4) we take Laplace transforms on the
time-dependence to eliminate the state variables and relate the
system input to the output by the transfer relation (Laplace
domain):

$$h = M \cdot D \cdot B (sI - A - C \cdot D \cdot B)^{-1} \cdot L \cdot g \qquad (5)$$

where $\underset{\sim}{I}$ is the unit matrix, s (dimension 1/t) is the transform
variable. Evidently, the zeros of det $\underset{\sim}{S} \equiv$ det $(s\underset{\sim}{I} - \underset{\sim}{A}$
$-\underset{\sim}{C} \cdot \underset{\sim}{D} \cdot \underset{\sim}{B})$ determine the various functional modes of the
system. Since $\underset{\sim}{D}$, $\underset{\sim}{B}$ are diagonal any deviation of the system
response from that of the unconnected system (defined by $\underset{\sim}{A}$)
must arise from elements of the connection matrix $\underset{\sim}{C}$.

As an example, consider the system to be logically con-
nected, i.e. y_i is connected only to z_j, where $j < i$. Then $\underset{\sim}{C}$
has non-zero elements only in the lower triangular sub-diagonal
region. But since $\underset{\sim}{A}, \underset{\sim}{B}, \underset{\sim}{D}$ are diagonal, this form for $\underset{\sim}{C}$
cannot modify the modes determined by $\underset{\sim}{A}$ alone, since det $\underset{\sim}{S} \to$
det $(s\underset{\sim}{I} -\underset{\sim}{A})$. Only if $\underset{\sim}{C}$ has entries above the main diagonal
is det $\underset{\sim}{S} \neq$ det $(s\underset{\sim}{I} - \underset{\sim}{A})$. Thus, not all changes in $\underset{\sim}{C}$ lead
to new modes. But we may generate one new mode by making a
single non-zero entry in $\underset{\sim}{C}$ in the upper right hand corner. This
corresponds to connecting the last output stage to the first
stage and generates the collective ring oscillator mode. An
entire continuum of new modes may be generated by coupling each
device to its nearest neighbor: $\underset{\sim}{C}$ is then a tri-diagonal band
matrix with zeros on the main diagonal, and generates the

equivalent of the Kronig-Penney lattice model for band structures.

It is important to note that the connections need not be
wired -in connections but may also include the parametric
device-device interactions that occur in partially isolated
device arrays: $C_{total} = C + C_p$ where C_p is the parasitic
connection matrix and might include negative entries to allow for
disconnection or reconnection due to interconnect damage or
electro-migrative failure. If the parasitic connection matrix
C_p is functionally dependent on the state variables u_i or on
the control/data signals contained in g, then a non-linear
interaction becomes possible which may lead to the "spontaneous"
restructuring or self-organization of the collective modes of the
system. Generally C_p will be time-dependent and affect the
reliability and noise properties of the system.

V.   SYNERGETIC RESTRUCTURING OF A SYSTEM ARCHITECTURE

Let us consider the spontaneous appearance of non-zero entries
into C_p due to non-linear parasitic device-device coupling.
Call these entries ω_1, ω_2 which we assume can appear in the upper
triangular region of C_p. Suppose that ω_1 and ω_2 represent
contiguous parasitic routes in the system and which are themselves contiguous to devices or interconnects of the hard-wired
system. Let us assume that ω_1 is controlled directly by a state
variable u_1 for one of the nearby devices and also depends on the
parasitic variable ω_2 which is assumed to be influence directly
by a second device variable u_2. ω_1, ω_2 represent an indirect
coupling between devices. We choose the generic form for the
dynamics of ω_1, ω_2 to be:

$$-\frac{d}{dt} \omega_1 = -(\alpha + u_1)\omega_1 - \beta\omega_1\omega_2 \qquad (6)$$

$$-\frac{d}{dt} \omega_2 = -(\delta + u_2)\omega_2 + \delta\omega_1^2 \qquad (7)$$

Here it is assumed that in the absence of inter-device coupling
($\beta=\delta=0$) the parasitic connections are damped for small changes
around $u_1 = u_2 = 0$, i.e. α, $\delta > 0$. Let us further assume $\delta \gg \alpha$;
$\delta \gg u_2$; $\delta \lesssim (\delta + u_2) \gg \alpha' = (\alpha + u_1)$. Physically this means that ω_2
is heavily damped, i.e. fast-relaxing compared to ω_1. This
inequality implies that u_1 is bounded but u_1 may still be
positive or negative. Since ω_2 is fast relaxing compared to ω_1
we may approximate $d\omega_2/dt \approx 0$ on the time scale of variation of ω_1.
Thus, solving for ω_2

$$\omega_2(t) = [\delta']^{-1} \delta \omega_1^2(t) \tag{8}$$

which represents an asymptotic approximation known as adiabatic
slaving in synergetic theory; the fast sub-system (2) is slaved
by the slow sub-system (1), i.e.

$$\frac{d\omega_1}{dt} = -\alpha' \omega_1^1 - (\beta\delta/\delta')\omega_1^3 \tag{9}$$

Equation (9) is closed and is evidently controlled by a generalized potential $V(\omega_1)$, i.e.

$$\frac{d\omega_1}{dt} = \frac{- dV(\omega_1)}{d\omega_1} \tag{10}$$

where (see Fig. 2)

$$V(\omega_1) = (1/2 \ \alpha')\omega_1^2 + (\beta\delta/4\delta')\omega_1^4 \tag{11}$$

The potential V is quartic with topologically different forms
which depend on whether $\alpha' \equiv \alpha + u_1$ is positive or negative
(Figs. 2a,b). If α' is positive (Fig. 2a) there is just one
stable point which corresponds to zero action: $\omega_1\omega_2 = 0$. But if
$\alpha + u_1$ becomes negative, the system bifurcates (Figs. 2b,3), the
previously stable mode becomes unstable and two new stable modes
appear: $\omega_1 = \bar{\omega}_1$; $\omega_2 = \bar{\omega}_2$. A symmetry breaking fluctuation

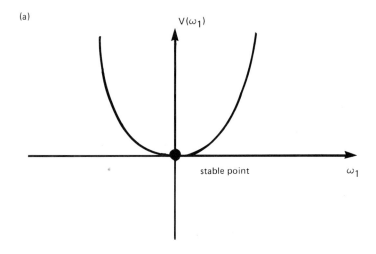

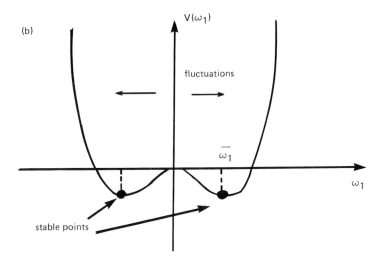

FIGURE 2.   The effective system potential.
            (a) α' positive
            (b) α' negative

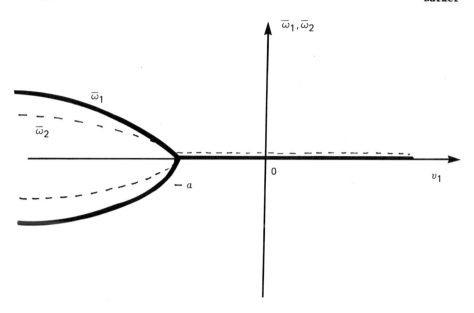

FIGURE 3. Bifurcation diagram.

will thus drive the system to a stable position of non-zero
action $\omega_1, \omega_2 \neq 0$ (Fig. 2b). Evidently, when the slow, master
mode is destabilized by a continuous change in u_1, the inter-
device coupling ω_1, and hence all slaved modes like ω_2 jump
suddenly from zero to finite values and the connection matrix
$\underset{\sim}{C}_p$ and therefore $\underset{\sim}{C}_{total}$ gain new entries. The mode ω_1 can
thus switch large portions of the architecture of the system via
the slaved modes ω_2, ... (10).

        This tutorial example shows that global switching may occur
from a few order parameters (ω_1 in the example) by knock-on
slaving effects.

VI. SELF-ORGANIZATION IN REGULAR DEVICE ARRAYS

If synergetic effects are to be exploited it becomes necessary to
develop a conceptual framework for designing an appropriate
device-device coupling which can lead to useful behavior. To
narrow the problem down we recognize the design/fabrication

constraints discussed in section 2 and consider a planar car-
tesian array of identical processor elements (PEs). Each PE is
located at lattice vectors $\underline{X} = (X,Y)$ and may represent a single
device or an assembly of devices (Fig. 4). The hardwired
control/data lines are not shown. Let us now demonstrate that
cooperative behavior in such arrays can proceed through a
non-linear diffusive process.

Consider first the ideal array for which no parasitic
device-device coupling occurs. Each PE is a generalized switch,
and we describe its state by the variable $\phi(\underline{x})$, and note that
for real digital systems ϕ will be a continuous variable. For
the PE to be a switch the variable ϕ must obey a non-linear
equation of evolution:

$$\frac{\partial}{\partial t} \phi(\underline{x}) = g[\phi(\underline{x}),\phi(\underline{x}')] + \delta\phi(\underline{x}) \qquad (12)$$

$$(\underline{x}' \ \varepsilon \ N(\underline{x}))$$

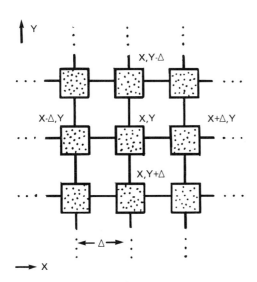

FIGURE 4. A coupled device array.

Here, g is a non-linear function of $\phi(\underset{\sim}{x})$ and will also depend
on the states $\phi(\underset{\sim}{x}')$ of neighboring devices via the hardwired
interconnection net. This wired neighborhood is denoted by
$N(\underset{\sim}{x})$. The linear term $\delta\phi$ may generally be taken as having a
negative coefficient, $\delta<0$ (damping), since the intended changes
in ϕ correspond to rapid switching between stable (damped)
equilibrium states.

Now suppose that each device is parasitically coupled to its
nearest neighbors (the set $G(\underset{\sim}{x})$, Fig. 4). The minimal non-
trivial modification to the equation of state (12) occurs for a
device-device coupling which modifies the rate of change of state
ϕ by terms proportional to the linear difference in states
between neighboring devices:

$$\frac{\partial}{\partial t}\phi(\underset{\sim}{x}) = g + \delta\phi + k \{[\phi(x+\Delta,y)-\phi(x,y)]+[\phi(x-\Delta,y)-\phi(x,y)]$$

$$\qquad (13)$$

$$+ [\phi(x,y+\Delta) - \phi(x,y)] + [\phi(x,y-\Delta) - \phi(x,y)]\}$$

In the limit of very large numbers of devices the above difference equation approximates to the partial differential equation:

$$-\frac{\partial\phi}{\partial t} = g + D\nabla^2\phi + \delta\phi \qquad (14)$$

where ∇^2 is the Laplacian operator in 2-D and $D \equiv k\Delta^2$ is an
effective diffusion coefficient where Δ is the lattice spacing.

Equation (14) expresses the distributed states ϕ as a field
over space-time. It is formally equivalent to the non-linear
reaction-diffusion equations much studied in chemical reaction
kinetics (32) and cellular biology, where inter-cellular coupling
proceeds by diffusion. Reaction-diffusion equations are known to
be capable of displaying both stationary and evolving spatial
patterns, including solitary waves. They also may support
chaotic behavior and turbulence. Similar forms are also known
for discrete lattices of cells provided the cell array is
sufficiently large. Larger numbers of cells generally correspond

to a wider range of synergetic behavior. Variants have been
studied in which the number of degrees of freedom in each cell is
enlarged, or the cell neighborhood is expanded, or noise sources
are added to each cell, or modifications are made to the cell
array boundaries.

For small amplitudes in PE states ϕ it is possible to secure
uniform homogeneous solutions to the evolution equation.
However, at sufficiently large amplitudes the homogeneous
solutions may spontaneously break into an inhomogeneous spatial
or indeed temporal pattern. Similar qualitatively discrete
changes may be induced by raising or lowering the background
noise level or by smoothly changing one of the control parameters
within a particular cell.

We have recently investigated more general models which are
closer to device networks by allowing ϕ to change at a rate
determined by higher powers of $\phi(\underset{\sim}{x}) - \phi(\underset{\sim}{x}')$ to allow for
saturation and overshoot effects. If ϕ is differentiable, the
procedure may be formally encapsulated within catastrophe theory.
These more general models lead to evolution equations of the
form:

$$-\frac{\partial \phi}{\partial t} = g[\phi] + \sum_n [\delta_n \phi + D_n \nabla^2 \phi]^n \tag{15}$$

which involve non-linear terms in the diffusion operator. In all
cases there are either modeling or actual circuit designs.

VII. EXAMPLE OF A COOPERATIVE DEVICE ARRAY

As an example let us consider a model which was originally
derived by us for analyzing the cooperative properties of
radiatively coupled molecules which exhibit proton tautomers
(unpublished). The individual "device" behavior was identified
with intramolecular hydrogen atom shifts between alternative
binding sites (compare reference), but the model has much wider
applicability. Each device in the square lattice can exist in

one of two equilibria which we call "0" and "1". The isolated
devices satisfy

$$-\frac{\partial \phi}{\partial t} (\underset{\sim}{x}) = - \nabla_\phi V[\phi(\underset{\sim}{x})] - |\delta|\phi(\underset{\sim}{x}) \qquad (16)$$

where the asymmetric "potential energy" V is shown in Fig. 5. In
the presence of nearest neighbor coupling we suppose that V is
modulated by a functional of the net state of the neighbors. It
is assumed that the neighbors only strongly influence the PE at
$\underset{\sim}{x}$ when they are in, or close to, state "1". The modulation
function M is shown in Fig. 6, and is oscillatory depending on
the linear sum of the states of the neighboring devices. When M
is positive the device state ϕ = "o" is stabilized; when M is
negative the state ϕ = "1" is stabilized (Fig. 7). Variations in
the pattern of neighboring devices in state "1" thus trigger
switching in the device at x. For example, when the sum of the
neighboring states $\Sigma\phi(\underset{\sim}{x}')$ approximates to "1" or "3" the device
at x switches to state "1". This new state subsequently diffuses

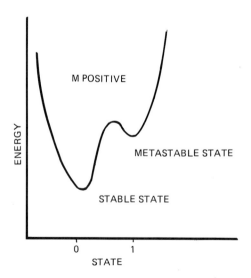

FIGURE 5.   Energy diagram for the isolated device.

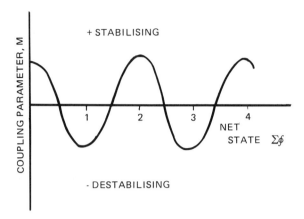

FIGURE 6. Modulation function.

information back to its neighbors, so any initial configuration
of states will evolve in time.

If τ is the time scale on which fluctuations drive the local
switching, and T>>τ is the diffusion delay between devices, we
obtain asymptotically a transition rule (coarse-grained over
time-scale τ):

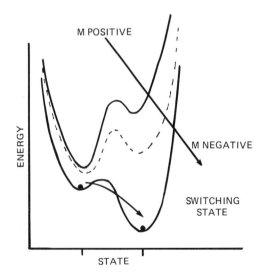

FIGURE 7. Distortion of energy as function of modulation.

$$\phi(\underset{\sim}{x},(N+1)T) = \sum_{\underset{\sim}{x}'\varepsilon G(\underset{\sim}{x})} \phi(\underset{\sim}{x}',nT) \quad MOD\ 2 \qquad\qquad (17)$$

where ϕ is now a binary variable having states "0" or "1". This
rule computes the next state of a device from the parity of its
neighbors at the previous time step. If the transition rule is
applied iteratively to any finite initial configuration of
devices in state "1" it is simple to prove that at a later time
T_r (determined by the spatial extent of the initial configuration), there will result four replicas of the original
configuration. The result generalizes to more general neighbor-
hoods; and Fig. 8 illustrates an iterated sequence based on the

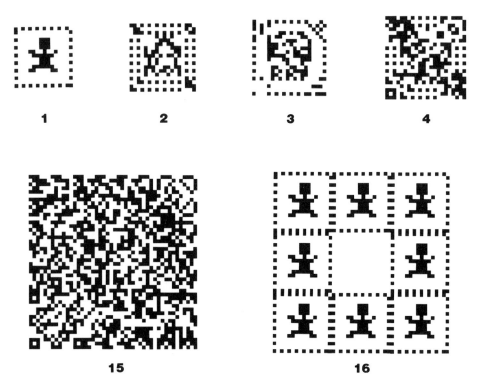

FIGURE 8. Evolution of signal pattern in replicator network. The
dark squares represent cells in state 1.

same rule but using nearest and next nearest neighbor coupling.
Note the apparently random configuration which occurs immediately
before replication is achieved. The continuum version of this
model belongs to the class of equations given by eqn (15) where
up to quadratic terms in the diffusion operator are used. We
observe that the effect of such a network is easily simulated by
an array of parity gates constructed from conventional logic
devices.

VIII.   CELLULAR AUTOMATA

We have made detailed studies of the self-organizing pattern
forming device arrays discussed in sections VI and VII (to be
published). It is found that provided a suitable range of slow
(slaving) and fast (slaved) time scales occur for the non-linear
coupling there generally exist asymptotic time scales on which
the networks behave like the replicator example: the device
states are multistable; initial configurations of device states
evolve or interact to generate a wide range of potentially useful
new configurations at approximately discrete time steps. Of
course, the crucial fluctuations which drive the transitions have
to be chosen in suitable ranges. An important feature is that
although the non-linear interaction between devices may be
short-ranged the order transformations may be on a long range in
the net. Asymptotically, the discrete state approximation is
equivalent to having a regular array of multiple state logic
modules in each of which the next state in time is determined by
the previous state and the states of a finite number of connected
neighbors at the previous time step. In computer science what we
have just described is a regular network of identical finite
automata (Fig. 9); the whole system is termed a cellular automaton. The language and mathematics of cellular automata theory
thus provide a natural framework for designing/understanding the
above classes of coupled device arrays in the asymptotic discrete
approximation. A review of salient properties of abstract
cellular automata is therefore useful.

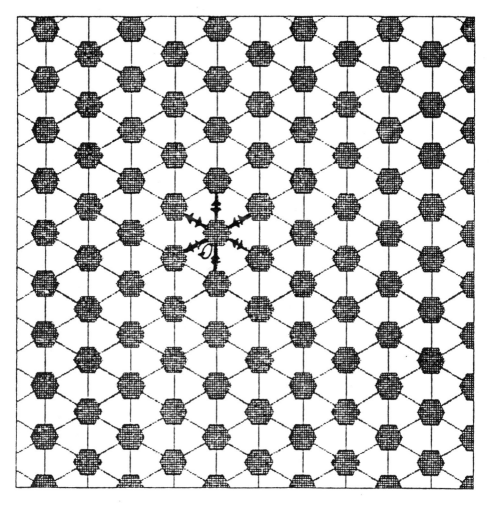

FIGURE 9.   Schematics of a cellular automaton: arrowed lines show
neighborhood connections for a typical cell; hex-connection.

Consider a N X N tessellation of identical finite automata
A, each characterized by a set of m possible states
$\phi E\{\phi_o, \phi_1, \ldots \phi_{m-1}\}$ where ϕ_i is a distinguished quiescent state.
Each A is connected to n neighboring automata (possibly including
itself, which is assumed from now on) located at lattice points
$\{\underset{\sim}{x} + \underset{\sim}{d}_n\}$ where $\underset{\sim}{d}_1 = 0$, $\underset{\sim}{x}$ and $\{\underset{\sim}{d}_n\}$ are lattice vectors.
Each A is described by the same transition function f which maps
the neighborhood states of $A(\underset{\sim}{x})$, $\{d(1), \ldots, \phi(n)\}$ into the next
state of A, with the restriction $f(\phi_o, \phi_o, \ldots \phi_o) = \phi_o$. Thus, if

$h_t(\underset{\sim}{x}) \equiv (\phi^t(\underset{\sim}{x}), \phi^t(\underset{\sim}{x} + \underset{\sim}{d}_2), \ldots \phi^r(\underset{\sim}{x} + \underset{\sim}{d}_n))$ describes the
state of the neighborhood of $A(\underset{\sim}{X})$ at time t, then at the next
time step t+1, A will be in state $\phi^{t+1}(\underset{\sim}{x}) = f(h_t(\underset{\sim}{x}))$ A configuration c is an allowable assignment of non-quiescent states to
automata in the cellular space: c is a function from the cellular
space to the set of cellular states and has finite support
relative to ϕ_0. The global transit on function F is a mapping
from the class C of all configurations into C and is defined by
$F(c)(\underset{\sim}{x}) = f(h(\underset{\sim}{x}))$ for all $\underset{\sim}{x}$ belonging to the lattice.
Starting from any initial configuration c_0, the global transition
function F determines a sequence of configurations $c_0, c_1, \ldots c_t$
where $c_{t+1} = F(c_t)$. A configuration c which satisfies $c = F(c)$ is
evidently passive, and is a fixed point for the transition
function. A configuration c for which $c = F^t(c)$ is periodic with
period t. In general, an initial configuration c_0 will lead to a
sequence which may terminate in the quiescent state, or in a
passive state, or be periodic, or propagate (grow without limit,
reconfigure but remain bounded in size, reconfigure remaining
bounded in size and in spatial location). Spatially separated
configurations may eventually interact. Thus, if c and c´are
initially separated $(\sup(c) \Omega \sup(c) = \phi$ where ϕ is the null set and
$\sup(c) \equiv \{x \ \varepsilon \ \text{Lattice}: c(\underset{\sim}{x}) \neq \phi_0)$, then c and c' are said to
interact such that c passes information to c´if there exists a
time t such that t successive applications of F to the union of c
and c' over the support $\sup(c')$ is not equal to t successive
applications of F to c' over the support $\sup(c')$. Strictly
speaking, this definition needs modification to exclude cases
where c has no effect on c' other than to displace it a distance
δ.

By establishing a correspondence between Turing machines and
cellular configurations it has been proved that two-dimensional
cellular automata exist which admit configurations that can
compute any computable function (20). In naive terms such
automata allow configurations which represent a computer and will
interact with other configurations which represent data and

programs to generate disjoint configurations which represent the
computer and its output. A cellular automaton which can support
a universal computer configuration is called computation-
universal. Computer configurations may be constructed from other
configurations. A configuration c is said to construct c' if
there exists a time t such that c' is a sub-configuration of
$F^t(c)$ disjoint from c; and $F^t(c)-c'$ does not pass information to
c' (again excluding the case of translating in space). Cellular
automata may be proved to exist which allow configurations which
presented with suitable instruction configurations can construct
an arbitrary computer configuration: universal computer-
constructor automata. Von Neumann and Thatcher have proved that
a 29 state, 5 neighbor cellular automaton existed which was a
universal computer-constructor. Codd (19) has demonstrated the
existence of a 8-state, 5 neighbor cellular automaton with the
same property. Codd also demonstrated the possibility of a
universal computer configuration which could be programmed to
reproduce copies of itself. More sophisticated cellular automata
have been discussed by Holland (18) and Arib (20), but involve
much more complex modules than we have described. Codd's design
is at first sight very attractive because it uses few states and
is strongly rotation symmetric: the neighborhood is 90° rotation
symmetric and the local transition function is invariant under a
90° rotation of the neighboring states. However, the transition
table which describes F contains 515 entries coded as 6 digit
numbers: far too complex for the very simple modules that might
be fabricatable in molecular-scale electronics. Part of the
difficulty arises because Codd's choice of states and transition
function was designed to lay down configurations to represent
paths for the propagation of signals which could be coded to turn
signals in the highly symmetric structure; an example is shown in
Fig. (10). A much simpler approach is to devise signal configurations which can propagate in an unsupported fashion which
we now describe.

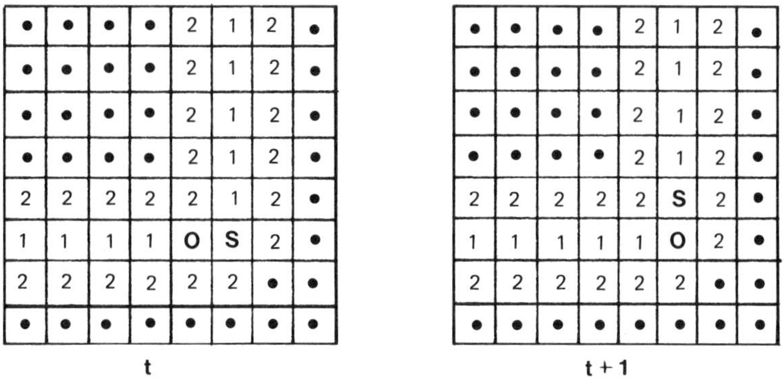

•	•	•	•	2	1	2	•
•	•	•	•	2	1	2	•
•	•	•	•	2	1	2	•
•	•	•	•	2	1	2	•
2	2	2	2	2	1	2	•
1	1	1	1	O	S	2	•
2	2	2	2	2	2	•	•
•	•	•	•	•	•	•	•

t

•	•	•	•	2	1	2	•
•	•	•	•	2	1	2	•
•	•	•	•	2	1	2	•
•	•	•	•	2	1	2	•
2	2	2	2	2	S	2	•
1	1	1	1	1	O	2	•
2	2	2	2	2	2	•	•
•	•	•	•	•	•	•	•

t + 1

FIGURE 10. Signals cornering in the Codd cellular automaton.

IX. FLOATING ARCHITECTURE

One of the problems with the computation-construction universal
cellular automata so far discussed is that they are designed to
admit configurations which represent static signal paths and
logic gates. This makes the transition functions very complicated, but also leads to vulnerability of the paths to physical
damage in the underlying hardware of devices that constitute the
abstract automata. One way out is to devise cellular automata
which permit logic gate configurations which can be moved over
the cellular space; and more importantly to admit self-propa-
gating configurations which do not require path configurations
for their support. One might imagine configurations which: (a)
might be aimed and launched in different directions in the
lattice and be re-routed by collisions with gate configurations
or other signals; (b) necessarily more complex signal configurations which could detect and circumvent "illegal" configurations associated with underlying damage and seek out and
interact with only those configurations with the correct morphology (some scavenger propagating configurations which could
identify and seal off damage regions would be even better). By

arranging for configurations which could relocate the computer
configurations it would be possible to devise "floating computer
architectures" which could be relocated anywhere in the cellular
space. There is in fact one cellular automaton at least which
admits this possibility.

The requisite properties are satisfied by the 2-dimensional
2-state, 9-neighbor (nearest and next nearest neighbor) square
lattice cellular automaton described by the simple transition
function

$$\phi(t+1) = \phi(t)\delta_{\sigma,2} + \delta_{\sigma,3} \tag{18}$$

which is better known as the basis of a mathematical game due to
Conway (33), but which we have found to have considerable power
as a computation universal cellular automaton. Here δ is the
Kronecker delta function, and ϕ is a binary state variable, and σ
is the decimal sum of the 8 neighbor states to a particular cell.
Even with conventional transistor based gates this cell module
requires very little complexity in the combinational logic
design.

The configuration outlined in Fig. (11) is a periodic
configuration of period 30 which at each period emits a self-
propagating configuration which we label "pulse" that moves at
fixed velocity over any quiescent region of the cellular space.
By arranging different angles and timing for "collisions" between
different pulses the transition rule (eqn. 18) leads to the
construction of new stable, unstable, propagating or periodic
configurations or indeed to the erasure of incident pulses. From
these one may select classes of collision which can be used to
re-route pulses or to construct the equivalent of AND, OR, NOT
gates with respect to binarily coded sequences of pulses (no
pulses), where data/program and output can result as pulse
streams or as passive configurations. Alternatively, one may
achieve a particularly simple set of computer primitives by
defining pulse sequences which act as the simple fredkin gates of
conservative logic (34). Thus, one may demonstrate the necessary

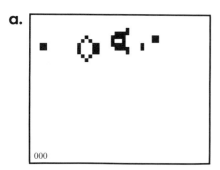

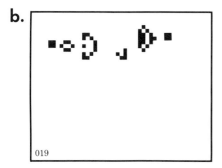

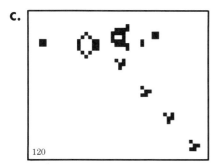

FIGURE 11.   The pulse generator configuration.
             (a) initial
             (b) generation 19
             (c) generation 120 -- showing propogating pulses.

features for "floating architectures" of considerable flexibility. To be implemented directly this cellular automaton would
require well in excess of 1000 x 1000 systolic array of processor
elements with the correct combinational logic if useful abstract
computer configurations were to be achieved. This imposes a
considerable problem for present day VLSI but may well be a
useful target for molecular electronics.

The real power of floating architecture does not concern the
ability to recreate traditional sequentially structed computers,
but instead is the ability to support co-existing families of
co-moving and intercommunicating specialized processor configurations (even more so in 3-D or layered 2-D cellular automata).
In this context we refer to (a) Minsky's "society (35) theory of
thinking" which views the human mind as an organized society of
intercommunicating simple agents; (b) Fahlman's (36) theory of
the parallel organization of knowledge. By providing a general
purpose medium for exploring novel computer architectures,
cellular automata of very large lattice size would be ideal
development systems for problems in artificial intelligence.

X.   FAULT-TOLERANCE

Although the system described by eqn. (18) has, in principle, the
facility to support fault-finding and fault-isolating configurations, there are many technical details to be resolved. It
was also assumed that such faults (and the possible resultant
spreading of illegal configurations) occur in the communication
zones of the lattice. However, the network is prone to rapidly
expanding configurations, which might well be created by a fault
in the support hardware underlying a floating logic region and
thus cause considerable damage to the floating architecture. We
have made some progress in devising more stable cellular automata
including an intercalated system, one sub-system of which acts as
a monitor and isolator for damage in the other computer supporting system but itself is intrinsically stable. But there is
clearly much further work required.

XI.   THE INPUT-OUTPUT PROBLEM

Any computational process is ultimately limited by the rate at
which data is input or output. We have already mentioned the
problem in VLSI where a significant bottleneck is caused by the
small number of edge pins compared with the large internal
address space. Interim solutions, such as enlarging the I/O
bandwidth by using multiple rather than binary logic have been
suggested, but for molecular scale systems more radical
approaches are required. A variety of possible procedures have
been recently reviewed (2,7) including the transducer problem of
mechanisms for communicating from the macroscopic scale to the
micro scale by light or electron beams. Here we discuss the
problem of inputting a large data array into a distinguished
region of a molecular cellular automaton (Fig. 12). Suppose for
example the data array is a 1024 x 1024 X N bits, such as a 2^N
grey scale image, which is required to be input to 8 contiguous
regions (2-9) surrounding the transducer region. This might
occur if it was required to have 8 different parallel operations
on the data such as feature extraction. The transducing cleared
ready for the next input by time $t_n + \tau_c$. The aim is to obtain 8
copies of the data in the regions (2-9) by a time $t_n + \tau_c$ after
which the main processing commences and is concluded by time t_n
$+ \tau_c + \tau_p$. An efficient way to do this is to endow the cells
with time-periodic subunits which act as N parallel replicator
automata of the type discussed in section VIII: binary state, 8
neighbor version. The subunits are active at times $t_n < t < t_n +$
τ_c. By analogy with Fig. (8) it is seen that the 8 replicas are
obtained in the correct locations in a time $\tau_c = 1024\tau_m$, where τ_m
is the inter-cell transfer time, and region (I) is automatically
clear for the next entry. This demonstrates that cellular
automata have potentially optimal capabilities for the widescale
distribution of very large amounts of data. The procedure could
be modified to transport output data to regions reserved for
external sensing.

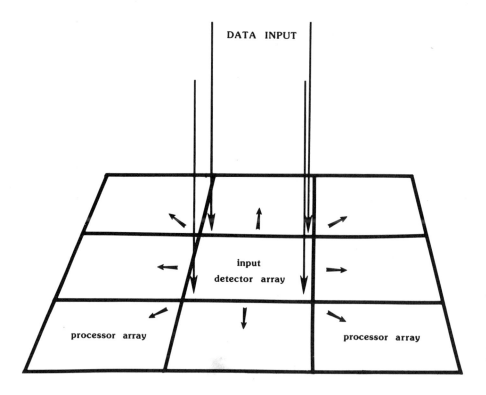

FIGURE 12. The I/O problem.

XII. MATERIALS CONFIGURATIONS

Even if cellular automata are desirable there remains the problem
of choosing suitable materials configurations for physically
realizing the combinational logic cells and their interconnects.
As noted previously, this requires full synergetic modeling of
the underlying processes. Solid-state possibilities have been
discussed indetail elsewhere (9) where the one- and two-dimensional semiconductor quantum-well heterostructures including
superlattices were proposed as useful prototypes. The simplifying assumption was made that it is better to customize a single
multi-stable multiple terminal device to represent the logic cell
rather than use a layout of standard AND/OR/NOT gates. The same

approach is suggested for molecular media where the range of switching possibilities is very large and includes proton tunneling, conformational changes, soliton valving, etc. Lithographic and biotechnological methods of molecular device synthesis have been discussed in (7). The mosting promising route to synthesizing cellular automata however, would appear to be the type of self-organizing process discussed by Stillinger and Wasserman (15) which leads to regular arrays of self-aligning molecular groups. The technique has been suggested for the family of partially fluorinated fused-ring hydrocarbons such as perhydrocoronene. The carbon atoms in fully saturated perhydrocoronene have a pleated sheet structure such that 12 C-H bonds project up, 12 C-H bonds project down, and the remaining C-H bonds are horizontal. By substituting up to 12 fluorines for hydrogens on a particular side, up to 1376 different patterns are possible on each side. Given a side with a hydrogen and fluorine pattern A, we may define a conjugate side to be one in which the hydrogen and fluorine atoms are transposed: pattern A. Consideration of the pair potentials for adjacent sides of two molecules shows that strong bonding only occurs for molecules with conjugate sides. Since each molecule has twosides, one might develop a crystalline series like (F/A) (A/B) (B/C) (C/D)... where each face A,B,C,.. is attracted only by its conjugate form A,B,C,... If a substrate surface is prepared to be attractive to faces D, a single solution which contains the mixed species (F/A), (A/B), (B/C), (C/D) could self-organize into a regular insulating layer four molecules deep, over the substrate.

Interconnections may be vertical or horizontal, and many candidates have been suggested (7) including conducting polymer chains such as strands of $(SN)_x$ or trans-$(CH)_x$. The possibility of linking pairs of fluorinated perhydrocoronene molecules with oxymethylene bridges has been suggested as a technique for making microtubules (7,15). These may be useful for providing insulating sheaths for conducting polymer interconnects to prevent leakage and cross-talk.

The above suggestions are, of course, highly speculative at the present time. But the ideas should indicate the range of possibilities for future attempts to fabricate a molecular cellular automaton. It is hoped that the arguments given might provoke useful discussions between engineers, systems designers and chemists which might narrow down the search for a practical molecular electronic system.

REFERENCES

1. R. Dingle, Ed., "Microelectronics, Structures and Complexity", Plenum Press, in press.

2. F.L. Carter, "Molecular Electronic Devices", Marcel Dekker, Inc., Ed. New York, NY, 1982.

3. J.R. Barker, in "Non-linear Electron Transport in Semiconductors", Eds. D.K. Ferry, J.R. Barker and C. Jacoboni, Ch. 23, Plenum Press (1980).

4. J.R. Barker and D.K. Ferry, Sol. St. Electronics 23 519 (1980).

5. J.R. Barker and D.K. Ferry, Sol. St. Electronics 23 531 (1980).

6. N.G. Einspruch (editor), "VLSI Electronics", Volumes 1-6.

7. F.L. Carter, in reference 1 and in The Chemistry of Future Molecular Computers, in "Computer Applications in Chemistry", Eds. S.K. Heller and R. Potenzone, Jr., Elsevier Science Publishers, BV, Amsterdam, 1983.

8. C.A. Mead and L. Conway, "Introduction to VLSI Systems", Addison Wesley, (1980).

9. J.R. Barker, in reference 1.

10. J.R. Barker, in reference 1.

11. J.R. Barker, Int. Conf. New Trends in Integrated Circuits, Syndicate des Industries et Tubes Electroniques et Semiconducteurs, Paris, 240 (1981).

12. C.A. Mead, in "VLSI 81", ed. J.P. Gray, Academic Press (1981).

13. B. Ransom, "Computers and Embryos, Models in Developmental Biology", Wiley, (1981).

14. C.V. Ramanoorthy and Y.W. Ma, "VLSI Electronics", $\underline{3}$, Academic Press, 2 (1982).

15. F.H. Stillinger and Z. Wasserman, J. Phys. Chem. $\underline{82}$ 929 (1978).

16. J. von Neumann, in "Theory of Self-Reproducing Automata", Ed. A.W. Burks, University of Illinois Press, Urbana (1966).

17. J.W. Thatcher, "Universality in the von Neumann Cellular Model", Technical Report 03105-30-T-ORA, University of Michigan (1964).

18. J.H. Holland, in "Essays on Cellular Automata", Ed. A.W. Burks, University of Illinois Press, Urbana (1968).

19. E.F. Codd, "Cellular Automata", Academic Press (1968).

20. M.A. Arib, "Theories of Abstract Automata", Prentice-Hall (1969).

21. W.R. Stahl, J. Theor. Biology $\underline{14}$ 187 (1967).

22. A. Lindenmayer, "Cellular Automata, Formal Languages and Developmental Systems", in Proc. IV Int. Congress for Logic, Methodology and Philosophy of Science, Bucharest, Rumania (1971).

23. R. Rosen, Foundations of Mathematical Biology, $\underline{2}$ 1 (1972).

24. R.M. Glorioso and F.C.C. Osorio, "Engineering Intelligent Systems", Digital Press, (1980).

25. H. Haken, "Synergetics", 2nd Edition, Springer-Verlag (1979).

26. J.R. Barker, "Physics of Non-linear Transport in Semi-conductors", Eds, D.K. Ferry, J.R. Barker and C. Jacoboni, Plenum Press, Ch. 23 (1981).

27. D.K. Ferry, H.L. Grubin and J.R. Barker, in reference 2.

28. H.T. Kung, Proc. Caltech. Conf. VLSI, 65 (1979).

29. H.L. Grubin, D.K. Ferry, G.J. Iafrate and J.R. Barker, in "VLSI Electronics", $\underline{3}$, Academic Press Chapter 6, 198 (1982).

30. P. Rapp, in reference 1.

31. A. Andronov, A.A. Vitt and S.E. Khaitkin, "Theory of
    Oscillators", Pergammon Press (1966).

32. A. Babloyantz, "Dynamics of Synergetic Systems", Ed. H.
    Haken, Springer-Verlag 180 (1980).

33. M. Gardner, Scientific American 223 120 (1970).

34. T. Toffoli, J. Comp. Sys. Sc., 15 213 (1977).

35. M. Minsky, Artificial Intelligence, Vol. 1, Eds. P.H. Winston
    and R.H. Brown, MIT Press, 421, (1982).

36. S.E. Fahlman, Artificial Intelligence, Vol. 1, Eds. P.H.
    Winston and R.H. Brown, MIT Press, 453 (1982).

35

Quantum Ballistic Transport and Tunneling in Molecular Scale Devices

John R. Barker / Department of Physics, University of Warwick, Coventry CV4 7AL, United Kingdom

I. INTRODUCTION

The understanding and exploitation of the device physics of
sub-micron semiconductor devices and heterostructures will
require an extension of the present quantum theory of transport
which accounts for transient current response, tunneling and
spatial inhomogeneities which result from the built-in potential
barrier profiles and self-consistent driving electric fields (and
we might include special problems such as quantum image forces).
Techniques of molecular beams epitaxy have already provided an
exciting range of devices and structures: HMFETs, planar-doped
barriers, superlattices, real-space transfer devices and lateral
quantum well superlattices (1). Similar theoretical developments
are also perceived for true molecular switches formed from short
periodic barrier arrays (2) and for analyzing device-device
coupling in dense circuits whether they be semiconductor or
molecular media (3). It is likely that concepts appropriate to

molecular electronic devices may be evaluated within semi-conductor media; so in the following we adopt a general quantum mechanical picture which should readily work for any molecular scale device structures.

In previous studies, we and other (4-8) have advocated the use of the Wigner distribution function $f(q,p,t)$ (a particular phase-space representation of the density matrix) as the best fundamental representation for quantum transport and yet which makes close contact with the classical phase-space distributions of traditional Boltzmann-Bloch transport theory (9). The collision-dominated regime has been previously described (6,7) on the assumptions that the electric fields and barrier-profiles were slowly varying. The picture has recently been extended to quantum ballistic transport (3, 10-12) for which collisions on phonons/impurities are substantially reduced by a combination of very high mobility material and short channels, and where the typical de Broglie wavelengths are comparable to feature-sizes within the device. Quantum ballistic transport describes non-dissipative flows, and as such it is an idealization. However, the collision-free propagators are an important pre-requisite for building a theory of dissipative processes: non-local scattering in rapidly varying spatial structures, collision-assisted tunneling, image force effects, dielectric response. The present paper discusses two specific problems: the general problem of pulse propagation in quantum well structures; and, the formal theoretical framework for the evaluation of transport and tunneling. In the following we shall restrict our attention to systems which may be modeled by a simple Hamiltonian of the form: $H = p^2/2m + V(q,t)$; where m is an effective mass, and V is a generic device potential (contacts plus potential barriers and self consistent driving electric fields). V may be time-dependent, spatially smooth or contain surface roughness variations.

II.   THE MOLECULAR SWITCH

A rigorous theory of steady-state wave propagation through a
finite periodic system of inhomogeneous films, plates or slabs
has been developed by Heading (13-15). A typical arrangement is
sketched in Fig. 1. Heading's theory provides a rigorous basis
for Pschenischnov's observation (16) that perfect transmission
may result if the energy E of the incident plane wave matches a
virtual level E_v of the wells between the barriers. Perfect
transmission may occur in an n-barrier array even if the
individual barriers are virtually opaque (15). Carter (2) has
proposed that short periodic barriers might be used as molecular
switches for which waves tuned to a perfectly transmitting state
could be turned off by applying a small perturbation to the
barrier. This is an interesting idea and might well be achieved
in suitable semiconductor superlattices as well.

Let us demonstrate the existence of the perfectly transmitting states by applying the continued-fraction method of
Vigneron and Lambin (17) to an array of four identical barriers
(fig. 1, fig. 2). The method assumes that the potential V is
constant outside the range (q_0, q_{n+1}), and employs a simple

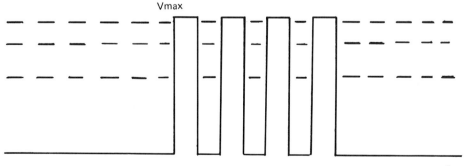

FIGURE 1.   Perfectly transmitting states for a periodic barrier.

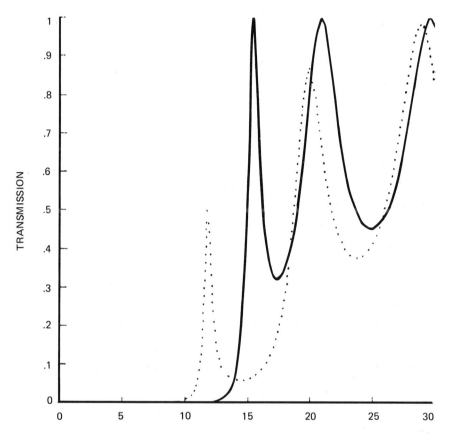

FIGURE 2. Asymptotic calculation of the transmission coefficient
for a four-barrier system (details in text). _____ perfect
barrier; ...... perturbed barrier.

discretization of the Schrodinger equation which leads to a
three-term recursion relation for the wavefunction. The latter
may be exactly solved by continued fractions. Introducing
$v(q_j)=2mV(q_J)/h^2$, $\varepsilon = 2mE/h^2$ and mesh constant $\Delta = (q_{n+1} = q_0)/(n+1)$, the Schrodinger equation becomes:

$$R_{j-1} = b_j(\varepsilon) - 1/R_j \qquad (1)$$

where

$$R_j \equiv \psi(q_j)/t(q_{j+1}) \tag{2}$$

$$b_j \equiv 2 + \Delta^2 [v/q_j) - \varepsilon] \tag{3}$$

If a plane wave of energy E is incident from q < q_0, the reflection coefficient R, and transmission coefficient T are given by (17):

$$R = \frac{|R_{-1} - R^{(-)}|^2}{|R_{-1} - R^{(+)}|^2}; \quad T = 1 - R \tag{4}$$

$$R^{(\pm)} = 1/2 \, \beta \pm i \, (1 - \beta^2/4)^{1/2} \tag{5}$$

Here $\beta \equiv 2 + \Delta^2(v_I - \varepsilon)$, and v_I is the constant value of v for q < q_0. Eqn (1) may be solved by continued-fraction expansion using forward recursion, and rapid convergence requires $\Delta^2(v - \varepsilon)$ << 1. The method is highly accurate, particularly for energies close to the barrier maxima where standard WKB becomes inaccurate. Resonances and virtual levels (17) are easily accommodated.

Figure 2 illustrates the technique for four rectangular barriers (or three wells) of equal widths and spacing: v_{max} = 30, total array-width = 2, in reduced units. This choice gives three perfectly transmitting energies. The dotted curve shows the effect of reducing the second barrier's height by 25%; there is a significant shift and reduction in the transmission peaks which is required if the Carter model is to be effective. We observe that the single barrier transmission is monotonous for E < V_{max}.

III. TRAVERSAL TIMES AND PULSE SHAPES

The simple picture of the tunneling barrier switch breaks down when we consider the practical implementation. Conventional

semiconductor devices are characterized for circuit purposes by
the transit-time τ_d which describes the average time for a
carrier to pass down the switched-on channel from source to
drain. τ_d is the fastest circuit-time, and practical circuits
require that digital pulses have a duration and spacing which are
generally larger than τ_d. The maintenance of pulse shape and
coherence (voltage or charge) is critical for circuit performance. So far no attention has been given to what might
correspond to the transit-time or pulse shapes (for assemblies of
carriers and for individual particle wavepackets) for multiple
tunneling barrier switches. The simple theory of transmission is
asymptotic and utilizes infinitely-extended plane waves and
cannot address what is essentially a problem of transients and
spatially inhomogeneous pulse propagation.

A single carrier must be described by a wavepacket and the
Gaussian form has been the most favored. MacColl showed that for
a simple single barrier (rectangular) an incident Gaussian pulse
with mean energy less than V_{max} splits into a reflected and
transmitted pulse; the latter having a peak which traverses the
barrier at a speed which is dependent on the barrier thickness;
for a wide range of thicknesses the time to traverse the barrier
is independent of the thickness. This surprising result is
discussed in detail by Stevens (19), using a method due to
Brillouin, where it is compared with the equivalent analysis for
a uniform-amplitude pulse. In the latter case the velocity of
the transmitted pulse is associated with the front of the main
part of the pulse (i.e., excluding the characteristic forerunners
which are weak disturbances of indefinite frequencies). The
uniform-amplitude pulse travels with constant velocity $v \sim$
$[2h(V-E)/m]^{1/2}$ and thus takes a time proportional to the barrier
thickness to traverse the barrier. Considerable controversy
surrounds this problem and has recently been critically discussed
by Landauer and Buttiker (20). The latter provide some support
for a traversal time $\tau \sim \int_{X1}^{X2} dx \ \{m/2[v(x)-E]\}^{1/2}$ from consider-

ation of a time-modulated simple barrier where x_1, x_2 are
classical turning points.

However, the multiple-barrier problem is considerably more
complex because of the complicated multiple internal reflections,
resonances and virtual states. Indeed our computer calculations
show that an incident pulse splits into a large number of slowly
varying coherent fragments within the barrier regions and these
persist for very long times after a main part of the transmitted
distribution has left the barriers. Our results, which we
discuss in more detail later, indicate a strong sensitivity to
the details of the incident pulse shape, thus, posing problems
concerning the mechanisms for pulse-shaping and injection. These
problems require considerably more attention if we are to
understand related questions such as scattering-assisted
tunneling, the spread of image charge in the metallic injection
regions as the pulse leaves the interface, and problems of
dielectric response.

IV. QUANTUM BALLISTIC TRANSPORT THEORY

Semi-classical transport theory is based on the Boltzmann
equation for the phase-space probability distribution for the
carriers. It may be written:

$$[\partial_t + v(p).\partial_q + F(q,t).\partial_p]f(q,p,t) = \partial f/\partial t|_{\text{collisions}} \qquad (6)$$

The carrier and current densities follow as projections of
$f(q,p,t)$:

$$n(q,t) = \int d^3p f(a,p.t) \qquad (7)$$

$$j(q,t) = \int d^3p ev(p)f(q,p,t) \qquad (8)$$

(we ignore spin here). In eqns (6-9) $v(p)$ is the carrier
group-velocity and $F = -\nabla V(q,t)$ is the driving force arising from
applied fields and the potential barriers. The RHS of eqn (6)

represents the rate of change of f due to dissipative processes.
If we set $\partial f/\partial t|_c = 0$ we obtain the classical equation for ballis-
tic transport (generally augmented by Maxwell's equations).

There is an analogous quantum description of transport where
f is now interpreted as the Wigner distribution which may be
constructed from the carrier density matrix ρ:

$$f(q,p,t) = (2\pi t)^{-3} \int d^3 P e^{-iq \cdot P/h} \langle p-P/2|\rho|p+P/2 \rangle \tag{9}$$

The Wigner function is not a classical probability distribution
as it is not positive definite; but it does provide correct
statistical expectation values for dynamical observables which
are expressed in Wigner-Weyl form (11). Until recently the
equation of motion for the Wigner distribution has been obscure,
but in recent communications we have shown that f satisfies a
non-local integral equation (10-12):

$$[\partial_t + v(p) \cdot \partial_q] f(q,p,t) =$$

$$- \int \int \frac{d^3 P d^3 Q}{(2\pi h)^3} \; \omega\partial(Q \cdot P/h) F(q,Q) \cdot \partial_p f(q,p+P,t) \tag{10}$$

where the effective force $F(q,Q)$ is defined by

$$F(q,Q) \cdot Q = V(q-Q/2) - V(q+Q/2) \tag{11}$$

Stationary states cannot be obtained directly from eqn (10), but
may be obtained from a similar integral equation (12):

$$\varepsilon f(q,p) = \int \int d^3 P d^3 Q \; W(q,p;Q,P) f(Q,P) \tag{12}$$

where ε is an energy, and the kernel W is

$$W \equiv \frac{d^3 Q' d^3 P'}{(2\pi h)^6} \exp\left[\frac{i}{h} (P' \cdot (Q-q) - Q' \cdot (P \cdot p)) \right]$$

$$\times \; 1/2 \; \{HCq + Q \; 1/2, \; p+P'/2)+H(q-Q'/2,p-P'/2)\} \tag{13}$$

The eigenfunctions $f_n(q,p)$ and eigen-values ε_n of the integral
kernel correspond to the pure eigenstates of the corresponding

Schrodinger problem. The $\{f_n\}$ form a complete orthogonal set with respect to stationary states including mixtures (12):

$$\iint d^3Qd^3Pf_n(Q,P)f_m(Q,P) = \delta mn/(2\pi h)^3 \qquad (14)$$

and defining a scalar product $\langle f|f'\rangle \equiv \iint d^3Pd^3Qf(Q,P)f'(Q,P)/(2\pi h)^3$ it is easy to prove that a general distribution may be factored into an evolving component f_E and stationary part f_s where $\iint d^3Pd^3Qf_s=1$, $\iint d^3Pd^3Qf_E=$); and $\langle f_E|f_s\rangle=0 \cdot f_E, f_s$ satisfy, respectively, equations (10), (13).

The correspondence with classical ballistic transport may be seen more clearly by rewriting eqn (10) as:

$$[\partial_t + v(p)\cdot\partial_q + F(q,t)\cdot\partial p]f(q,p,t) \qquad (15)$$

$$= \iint \frac{d^3Pd^3Q}{(2\pi h)^3} \cos (P.Q/h)\left[F(q,t)-F(q,Q,t)\right]\partial pf(q,p+P,t)$$

The LHS of eqn (15) is identical with eqn (6); the RHS is the non-local quantum correction. If $V(q)$ is of the general quadratic form $V \sim a+b\cdot q+cq^2$, it is simple to prove that the RHS of eqn (15) vanishes identically, so that in such cases, quantum effects can only appear via the initial state $f(q,p,t_0)$. More generally the potential itself induces quantum effects into f. The non-local evolution equations may be solved numerically by a variety of methods such as path-variable or self-scattering techniques (9). The form of eqn (15) is particularly useful since it provides a smooth handling of near-classical limits which are known to be approached ($h \rightarrow o$) non-uniformly in direct constructions of f (11). This approach is computationally more efficient than solving the Schrodinger equation directly, and is advantageous for considering dissipative forces and many body effects (indeed f can be related directly to re-normalized Green functions in many-body theory (12,6,7)).

The formal solution to the transport equation for a given initial state $f_0(q,p)$ (which may represent a wavepacket or pulse)

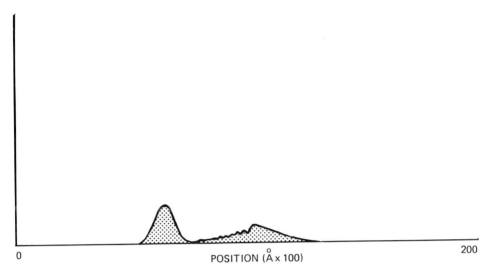

FIGURE 3.   The back–to–back planar doped barrier device.

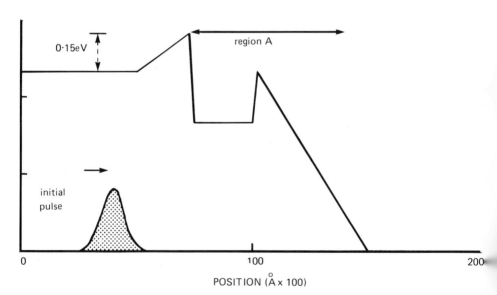

FIGURE 4.   Carrier probability distribution in the back–to–back planar doped barrier.

may be expressed (11) as a linear superposition of <u>classical</u>
distributions $f_c(q,p;Q,P;t)$. Each $f_c(q,p;Q,P;t)$ is parameterized
by phase-space variables Q,P. For a given pair (Q,P), the
corresponding $f_c(q,p;Q,P;t)$ obeys a classical Liouville equation
$(\partial_t f_c + v\partial_f f_c + F(q,Q)\partial_p f_c = 0)$ where the force term is the
effective force $F(q,Q)$ of eqn (11), and the initial condition is
shifted to $f_0(q,p+P)$ rather than $f_0(q,p)$. Each $f_c(q,p;Q,P;t)$ is
weighted by a phase-factor $\cos(Q\cdot P/h)/(2\pi h)^3$ and the correct
Wigner function f is obtained by integrating such terms over all
Q,P. This description provides a useful insight into time-
dependent tunneling phenomena. Contributions to $f(q,p;t)$ in
classically forbidden regions of phase-space occur in two ways:
(a) from trajectories $q(\tau)$, $p(\tau)$ $(\dot{q} = p/m; \, p=F(q,Q)$ belonging
to $f_c(q,p;Q,P;t)$ which enter regions forbidden by $F(q)$ but not by
$F(q,Q)$; (b) from the momentum uncertainty manifest in the
displaced initial condition $f_0(q,p+P)$. The expression of f as a
superposition of classical trajectories is useful for non-
perturbative analytical approximations and for Monte Carlo
simulation which is particularly important for investigating the
statistical fluctuation properties of very small numbers of
carriers anticipated in molecular-scale devices.

V. APPLICATION TO PULSE PROPAGATION IN SUB-MICRON DEVICES AND
    MOLECULAR SWITCHES

    The above formalism is being used to study time-dependent
quantum ballistic transport in semiconductor devices and heterostructures. Figures 4,5 show the carrier density distribution
and Wigner function for a Gaussian wavepacket injected into an
Eastman back-to-back planar doped barrier device (21), where the
incident mean energy is equal to the barrier height V_1 shown in
figure 3. The Wigner function is displayed for the region
between the barrier centre to the collector region and the

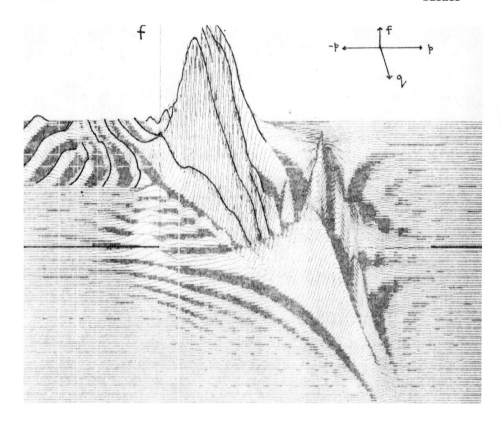

FIGURE 5. Wigner distribution for the region A of Fig. 3, and corresponding to Fig. 4.

accelerating exit pulse is clearly discerned. The region within the main quantum well is complicated and shows flows associated with interference, resonances and tunneling.

A study is also in progress for periodic tunneling barriers of the type discussed in section II. Figures 6,7 show snapshots of the carrier probability distribution $n(q,t)$ at equal time intervals for an initial wavepacket of Gaussian form incident from the left of a four barrier system (figure 1), with mean energy equal to the barrier height. This pulse has momentum components that span the full barrier energy profile. The system

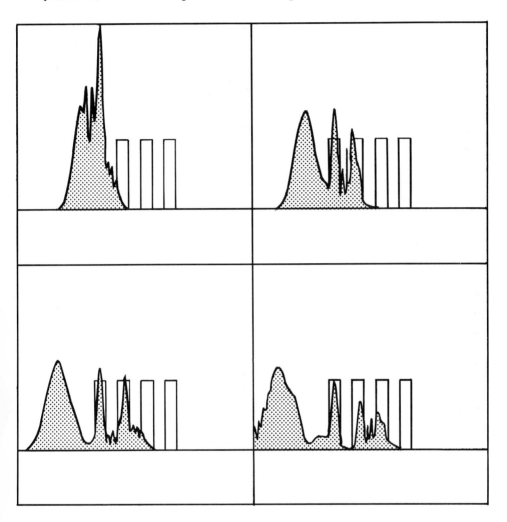

FIGURE 6.   Sequence of pulse evolution in a 4-barrier array.

is bounded on the left by a perfectly reflecting wall to represent a region with a source of carriers.   The subsequent reflected pulse later reflects off the wall as can be seen in
figure 7.   The sharp oscillations are not an artefact but arise
from a condition when the real and imaginary parts of the
wavefunction come into phase.   The illustrated sequence shows

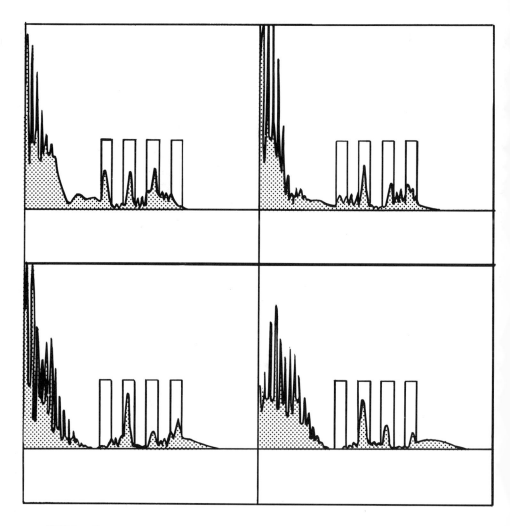

FIGURE 7.   Continuation of Fig. 6.

clearly the problem of assigning a simple traversal time to the
pulse peak; the intra-barrier distribution is a superposition of
different pulses associated with resonances, virtual levels and
multiple reflections/tunneling. Figure 8 shows the momentum
probability distribution for the intra-barrier region and shows
peaks associated with strongly transmitting and reflecting

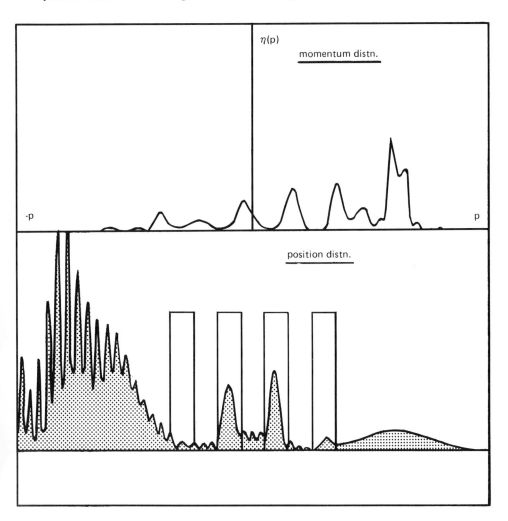

FIGURE 8. Momentum probability distribution for the barrier region of Fig. 1.

states. Although the bulk of the transmitted pulse can escape to over twice the barrier width, there remains a substantial part of the distribution in the barrier which persists for long times. In device operation, this remenant distribution must be allowed to escape before switching or passing a second pulse. The

example illustrates the reality behind naive assumptions of
asymptotic steady-state formation. In a preliminary investigation of noise effects we find that if the barriers are modulated by a static random perturbation with mean amplitude less
than 20% of V_{max} and wavelength components small compared to the
incident wavepacket mean wavelength, then the pulse propagation
is not seriously disturbed.

VI. CONCLUSIONS

We have briefly outlined some of the problems of transport
in molecular scale systems and provided a powerful framework for
calculating the statistical properties of arbitrary charge
distributions. It is hoped that these methods will prove useful
for device design and uncovering exploitable quantum mechanical
transport phenomena.

This work is supported by the Science and Engineering
Research Council.

REFERENCES

1. R. Dingle (Ed), "Microelectronics, Structures and
    Complexity", Plenum Press, in press.

2. F.L. Carter, "Electron Tunneling in Short Periodic Arrays",
    Proc. Mol. Elec. Devices Workshop, NRL Memorandum Report
    4662, 344 (1981).

3. J.R. Barker, "Complexity Issues: Device-device Coupling on
    Chip", in "Microelectronics, Structures and Complexity", Ed.
    R. Dingle.

4. J.R. Barker and D.K. Ferry, "On the Physics and Modeling of
    Small Semiconductor Devices I", Solid State Electronics, 23
    519 (1980).

5. J.R. Barker and D.K. Ferry, "On the Physics and Modeling of
    Small Semiconductor Devices II", Solid State Electronics 23
    531 (1980).

6. J.R. Barker, "Quantum Physics of Retarded Transport", J. de Physique, Supp 10, 42 245 (1981).

7. J.R. Barker and D. Lowe, "Quantum Theory of Hot Electron-Phonon Transport in Inhomogeneous Semiconductors", J. de Physique, Supp 10, 42 293 (1981).

8. J.R. Barker, "Quantum Transport Theory", in "Nonlinear Electron Transport in Semiconductors", Eds. D.K. Ferry, J.R. Barker and C. Jacoboni, Plenum Press, Chapter 5, 23 (1980).

9. D.K. Ferry, J.R. Barker and C. Jacoboni (Eds), "Nonlinear Electron Transport in Semiconductors", Plenum Press, (1980).

10. J.R. Barker, D. Lowe and S. Murray, "A Wigner Function Approach to Transport and Switching in Sub-micron Structures", in "Physics of Submicron Structures", Eds. H.L. Grubin, K. Hess, G.J. Iafrate, and D.K. Ferry, Plenum Press, New York, NY (1982).

11. J.R. Barker and S. Murray, "A Quasi-classical Formulation of the Wigner Function Approach to Quantum Ballistic Transport", Phys. Lett. 93A 271 (1983).

12. J.R. Barker, "A Generalized Wigner Function for the Description of Stationary and Evolving States", in press.

13. J. Heading, "Resonance Effects and Transmission through a System of Periodic Overdense Barriers", J. Atmos. Terr. Phys., 25 519 (1963).

14. J. Heading, "Exact and Approximate Methods for the Investigation of the Propagation of Waves through a System of Barriers", Proc. Camb. Phil. Soc. 74 161 (1973).

15. J. Heading, "Four Parameter Formulae for Wave Propagation through a Periodic System of Inhomogeneous Slabs", Wave Motion 4 127 (1982).

16. E.A. Pshenichnov, "The Tunnel Effect through a System of Identical Periodic Barriers", Sov. Phys. Solid State 4 819 (1962).

17. J.P. Vigneron and P. Lambin, "Transmission Coefficient for One-Dimensional Potential Barriers Using Continued Fractions", J. Phys. A; Math. Gen., 13 1135 (1980).

18. L.A. MacColl, "Note on the Transmission and Reflection of Wavepackets by Potential Barriers", Phys. Rev. 40 621 (1932).

19. K.W.H. Stevens, "A Note on Quantum Mechanical Tunneling",
    Eur. J. Phys. $\underline{1}$ 98 (1980).

20. M. Buttiker and R. Landauer, "Traversal Time for Tunneling",
    Phys. Rev. Lett. $\underline{49}$ 1739 (1982).

21. A. Chandra and L.F. Eastman, J. Appl. Phys. $\underline{53}$ 9165 (1982).

36

Resonant Tunneling in Molecular Beam
Epitaxially-Grown Semiconductor Structures

S. W. Kirchoefer, R. Magno, K. L. Davis,
R. L. Schmidt, and J. Comas / Naval Research
Laboratory, Washington, D.C.

ABSTRACT

Resonant tunneling in various semiconductor heterostructure
geometries is investigated theoretically utilizing
Schrödinger wavefunction techniques. The applicability of
this simple model is discussed, and results of experiments
on MBE-grown single barriers are presented.

Carrier transport across thin multilayer semiconductor hetero-

structures is dependent upon the effect known as resonant tunneling.

Resonant tunneling occurs when the geometry of a potential allows

enhanced coupling through a series of tunnel barriers by means of

the intermediate energy states of well regions between any pair of

barriers. Resonant tunneling enables a carrier wavefunction to

become spatially extended within the heterostructure. An under-

standing of the mechanisms by which resonant tunneling occurs is

essential to the proper understanding of carrier properties within
all types of multilayer semiconductor devices.

Tunnel barriers can be formed in two ways: (1) using an alloy
material ($Al_xGa_{1-x}As$) with wider bandgap than the potential well
material (GaAs), or (2) using a doping discontinuity (p-type barriers and n-type wells). The alloy structures (1) are modeled by
abrupt, one-dimensional, square potential wells and barriers. The
doping-based (2) potentials are derived from the electrostatic field
of fixed ionized impurities and are parabolic rather than square.
This results in the need for larger effective barrier thicknesses
than those of alloy heterostructures to obtain similar barrier
heights.

The heterostructures of interest here are grown by molecular-
beam epitaxy (MBE), and are based upon the GaAs-$Al_xGa_{1-x}As$ ternary
alloy system. MBE allows precise control of layer thicknesses,
alloy compositions, and impurity doping in these heterostructures.
Figure 1 shows the energy gap vs. alloy composition of the Γ, X and
L minima for $Al_xGa_{1-x}As$. For $x \stackrel{<}{\sim} 0.45$, the alloy is a direct bandgap semiconductor, since Γ is the lowest energy minima at these
compositions. The difference in energy gap between GaAs and
$Al_xGa_{1-x}As$ is split with 85% going into the conduction band discontinuity and 15% into the valence band discontinuity. By sandwiching
a thin GaAs layer between two $Al_xGa_{1-x}As$ layers, a double heterostructure (DH) is formed. Carriers within a DH are in their lowest
energy states at the GaAs band edge. However, if the GaAs layer is
made sufficiently thin (<500Å), the carriers begin to exhibit pro-

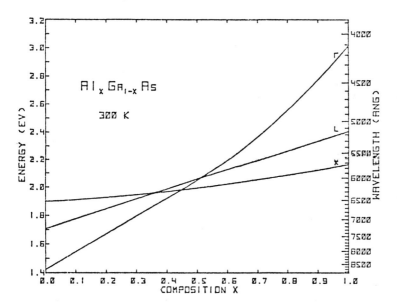

Fig. 1.   Variation of energy gap of the Γ, X, and L minima with alloy composition of $Al_xGa_{1-x}As$.

nounced quantum size effects.   The density of states within the thin GaAs layer is altered by the presence of the quantum states, and the effective bandgap is increased.   It is possible to fabricate many thin GaAs wells separated by equally thin $Al_xGa_{1-x}As$ barriers, and thus form a superlattice.   Carriers in such structures move from well to well by means of tunneling, and resonance tunneling is an important part of the carrier transport in these materials.

The complexity of the expressions for the particle wavefunctions in the coupled potential barriers makes a closed-form analytical solution unwieldy.   However, simple numerical solutions which yield considerably more insight to the nature of the wavefunction

used to describe quantum-well heterostructures [1-3], the effective

mass theorem is applied to reduce the problem to the case of par-

ticles confined to wells-corresponding to the conduction and valence

band discontinuities of bulk alloy heterojunctions. The solution

for Schrödinger's Equation in the various regions of the potential

are:

$$\psi_w = A\cos(kx) + B\sin(kx); \quad k = (2m^*[E-V])^{\frac{1}{2}}/\hbar \qquad (1)$$

$$\psi_B = Ae^{kx} + Be^{-kx}; \qquad k = (2m^*[V-E])^{\frac{1}{2}}/\hbar \qquad (2)$$

The method of solution requires the use of complex arithmetic. The

boundary condition on the transmitted side of the potential requires

that the particle-wave has no component incident on the barrier.

This sets the relative phase of the real and imaginary components of

the transmitted wavefunction. By matching wavefunction amplitude

and first derivative at each interface, the real and imaginary

wavefunctions are calculated from the transmitted side to the inci-

dent side of the potential. Probability densities can be determined

for the wavefunction in all regions of the potential, and the trans-

mission coefficient can be calculated for all particle energies of

interest.

The simplest model of resonant tunneling as applied to semicon-

ductor heterostructures was first outlined by Chang et al. [4].

This model is referred to here as extended-wavefunction (EW) reso-

nant tunneling. In this model, the carriers are considered as

particle-wave incident upon the pontential, and the tunneling probability is modeled as the normalized probability of finding the particle on the transmitted side of the potential. Sample wavefunctions at resonance and off resonance are shown in Fig. 2. In the EW model, the tunneling signal is determined totally by single-particle tunneling completely through the structure via a single quantum state. Although it is generally realized that such resonances are perfectly transmitting, it is easy to show that the full-width half-maximum (FWHM) of these resonances is non-physically small (on the order of 10^{-6} eV), as shown in Fig. 3a. More importantly, such perfect transmission is greatly suppressed by any asymmetries in the potential. Such asymmetries can originate from a number of sources, behavior can be obtained. Consistent with the methods perviously

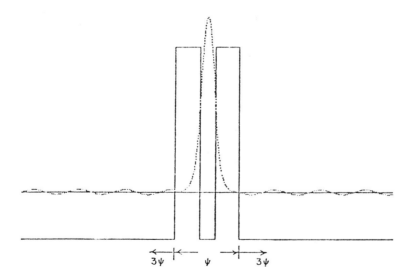

Fig. 2a. Conduction band potential well (double barrier). The barrier layers are Aℓ$_3$Ga$_7$As and all other regions are GaAs. The barriers are 80Å and the well is 50Å thick. Superimposed on the potential plot is the calculated wavefunction for an incident energy equal to the resonant energy in the well.

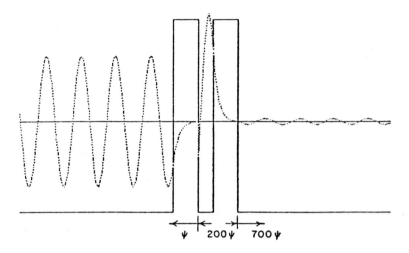

Fig. 2b. Same as Fig. 2a, except the incident energy is off resonance. In this case very little of the wavefunction is transmitted
from left to right through the double barrier.

such as imperfect growth, distortions due to impurities or lattice

phonons, or sloping of the potential by applied electric fields. A

transmission vs applied voltage plot is shown in Fig. 3b. The

sloping of the potential reduces and broadens the resonant peak, and

the resonant signal for the ground state transmission is about nine

orders of magnitude above a 10^{-10} tunneling background for

$Al_{.3}Ga_{.7}As$ 80Å barriers separated by a 50Å GaAs well.

Resonant tunneling in the EW model becomes increasingly

improbable as additional periods are added to multiple barrier

structures. This is due to the fact that the large quantity of

barrier which the carrier must tunnel through becomes non-

transparent with even the slightest asymmetry in the potential. In

addition, resonant tunneling in the EW model fails to take into

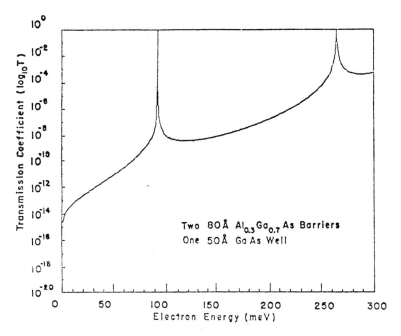

Fig. 3a. Transmission probability vs incident energy for the double barrier potential shown in Fig. 2.

account many physical effects which occur in any real semiconductor device. For distances greater than the mean free path for phonon scattering in these structures, it is meaningless to consider carrier tunneling via a simple unperturbed quantum state [5]. It is clear from the wavefunction plots that a carrier has a high degree of probability of being found in the well regions of the resonant tunnel barrier. Thus it is possible that a large amount of the resonant tunneling measured in an actual device is due to carriers which resonant tunnel into the well regions and then are scattered out of their incoming quantum states. The resulting device current

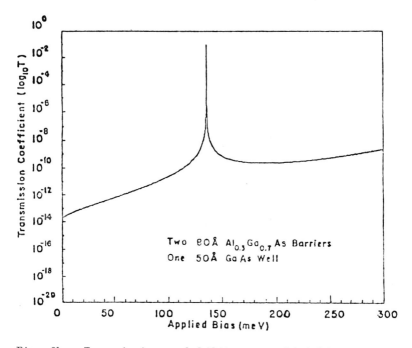

Fig. 3b. Transmission probability vs applied bias for the double barrier potential shown in Fig. 2. The incident energy is fixed at 4 meV above the left-hand GaAs conductiong band edge. The applied bias causes the potentials to slope down from left to right.

would then be produced by simple tunneling from carriers trapped within the wells. Considering the effects of asymmetries on EW tunneling, it is likely that this latter model is, at least qualitatively, a more accurate description of what actually should occur.

An experimental effort is now underway to use quantum well structures fabricated by MBE to test the predictions of the resonance tunneling model. A typical quantum well device consists of alternating layers of undoped $50\overset{\circ}{A}$ thick GaAs and $80\overset{\circ}{A}$ thick $Al_{.3}Ga_{.7}As$ grown on an n-type GaAs substrate and covered by a

0.5 micrometer n-type GaAs layer. A gold-tin metallization is alloyed to the substrate and the top GaAs layer in order to make electrical contact to the sample.

The tunneling experiments are carried out with the samples in liquid helium at 4.2°K in order to limit the amount of thermally stimulated current. While current-voltage measurements are often made, more useful data is frequently obtained by using harmonic detection techniques to measure the voltage dependence of the dynamic resistance $R = dV/dI$ and its derivative d^2V/dI^2.

A first step in this project has been an examination of the characteristics of a single 80Å thick $A\ell_{.3}Ga_{.7}As$ barrier grown between two thick n-type GaAs layers. The voltage dependence of dV/dI and d^2V/dI^2 are shown in Fig. 4 for a 250 micrometer diameter device. The d^2V/dI^2 data have been measured because it is difficult to determine whether tunneling is occurring from the dV/dI alone. The structure near 37 mV in the d^2V/dI^2 is due to the interaction of a tunneling electron with the longitudinal optical (LO) phonon in the GaAs and can be considered proof that tunneling is occurring. The structure near zero bias is also seen in other tunneling experiments on semiconductors and is not well understood. The fact that the dV/dI curve is nearly symmetric about zero bias indicates that the potential barrier is almost constant across the $A\ell_{.3}Ga_{.7}As$. Multiple barrier samples are currently being prepared in order to observe negative resistance phenomena. The area of these devices must be made small in order to increase their resistance at high bias where the negative resistance effects are expected. This will avoid self-heating effects due to large I^2R power dissipation.

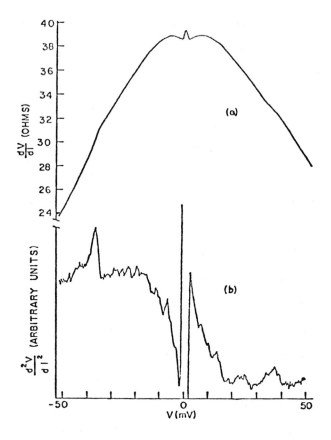

Fig. 4. Experimentally measured voltage dependence of dV/dI and d^2V/dI^2 for a single 80Å Aℓ$_{.3}$Ga$_{.7}$As barrier (no resonant tunneling).

REFERENCES

1.   L. Esaki and R. Tsu, IBM J. Res. Develop. <u>14</u>, 61 (1970).

2.   A. Ya. Shik, Fiz. Tekh. Poluprov. <u>8</u>, 1841 (1974) [Sov. Phys. Semicond. <u>8</u>, 1195 (1975)].

3.   R. Dingle, in "Feskorperprobleme XV (Advances in Solid State Physics), ed. H. V. Queisser (Pergamon, Vieweg, 1975), pp. 21-48.

4.   L. L. Chang, L. Esaki, and R. Tsu, Appl. Phys. Lett. $\underline{24}$, 593 (1974).

5.   R. Tsu and L. Esaki, Appl. Phys. Lett. $\underline{22}$, 562 (1973).

37

Studies of Tunneling in Short Periodic Arrays

Forrest L. Carter / Naval Research Laboratory, Code 6170, Washington, DC

I.   INTRODUCTION

Two promising ideas that could lead to molecular electronic devices (MEDs) include: (1) electron tunnelling in short periodic arrays, and (2) soliton switching devices.   Both have the advantage of being devices which can, to a large measure, operate in largely dissipationless modes.   However, the maximum speed of a soliton switch is limited to the speed of sound in that medium.   This soliton velocity limitation is understandable, considering that it is coupled to the bonding rearrangement.   However, tunnel switching has no such restrictions.

The first suggestion of a short periodic array tunnel switch was made in 1979 (1).   This switch was based on a theoretical quasi-classical proposal of tunnelling by Pschenichnov (2).   The title notes included here, while extending the MED concepts with several minor considerations, do not form a complete picture at this time.

705

Part of the concept of short periodic array tunnelling is illustrated in Fig. la in a square well model. Imagine that an electron beam moving from left to right is impinging on the array of potential barriers. When the energy of the electron, ω, is adjusted to match one of the four pseudo-stationary states, (as shown in the potential wells of Fig. la), then the electron tunnelling through the array barrier is perfect. The tunneling can be turned off or on by changing one or more of the well heights or depths. Most of the mechanisms for changing well characteristics depend on molecular conformation or tautomer changes (Ref. 3, Fig. 6) and are

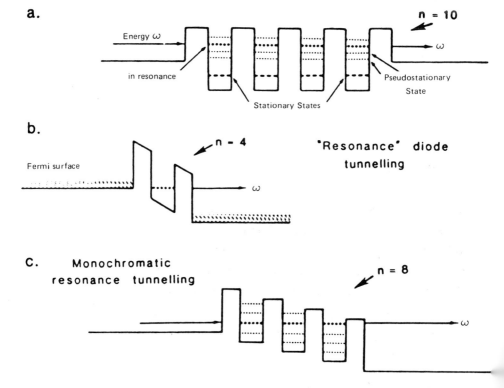

FIG. 1. Three array tunnelling devices are indicated here. The first, Fig. la, combined with Control Groups that change the well depth or barrier heights, provide the mechanism for molecular size switches. Two for of resonant tunnelling are indicated in Fig. lb and lc. The last, a possible monochromator contains n=8 square well potential changes. While all these tunnel forms can be numerically computed, currently only the square well potentials forms for $n \leq 8$ can be given an analytical expression (for n=1 to 6, MED I (6), for n=7 and 8, this paper).

illustrated in Ref. 3, Fig. 4 and Ref. 3, Fig. 5, respectively, for NAND and NOR gates. The tunnel switches contain both operational and dummy tautomers (Ref. 3, Figs. 4, 5 and 6). Numerical calculations (J. Barker, 4) indeed indicate that tunnel switches are theoretical possible.

PERIODIC VARIATIONS IN SHORT ARRAYS

In exploring the possible configurations of tunnel switching, one is not limited to simple short period variations. For example, in Fig. 2 we see two sets of interwoven wells, one with three pseudo-stationary states, the other with only one virtual or pseudo-stationary state. The deeper wells have a near stationary and two pseudo-stationary states, the most energetic of which matches the pseudo-stationary state of the small wells. This arrangement is convenient in that it offers two different types of Control Groups or MED gates (see Fig. 3) attached to the tunnel Body (Ref. 3, Fig. 3). Of course the potential well structure shown in Fig. 2 might be used in molecular electronic, semiconductor, as well as hybrid MED-semiconductor tunnel switches.

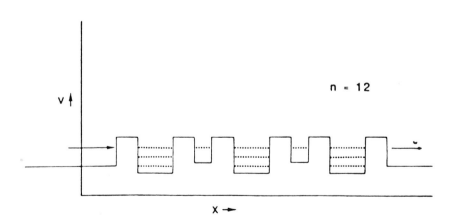

FIG. 2. The configuration for tunnelling need not be simply periodic but might demonstrate a dual or multiple periodicity as suggested here. This structure requires 12 (=n) changes in square well potential, a condition which has not been met yet.

In Fig. 1b, we have illustrated a familiar semiconductor tunnel which is usually in a bias off-resonance mode via the application of a potential difference. By way of contrast in Fig. 1c, we indicate a multiple resonance phenomenon. In this case, the energy of the electron is coupled with the lowest pseudo-stationary state in the first well, the second lowest pseudo-stationary state in the 2nd well and the third lowest pseudo-stationary state in the last well. As suggested in 1979 (1), this resonant tunnelling device has potential use as an electron monochromator. The size of such an individual monochromator makes an array of them possible. One obvious possible use would be in a very thin CRT or video screen application.

SOME MOLECULAR CONSIDERATIONS

Molecular analogues of various semiconductor gates are in the literature (1, 3, 5), including the proceeding of the first MED I workshop (6). This section discusses some molecular aspects of tunnel switches with multiple periodic variations.

The MED tunnel switch is schematically illustrated in Fig. 3 as having (top to bottom) a dummy Control Group, a positively chargeable and a negatively chargeable Control Group. The Control Groups are all separated from the Body of the tunnel switch by σ-bonds to minimize charge leakage. In addition, the pi system of the soliton Control Groups is rotated by 90° from that of the tunnel Body to keep these systems as orthogonal as possible.

Note that the Body of the tunnel switch also indicates a vertical variation at the A' A and the B' B linkages in addition to the charge variation of the Control Groups. These vertical linkages might incorporate a sigma bond or a pi bond character, possibly with a polar nature.

HEURISTIC SQUARE WELL STUDIES

Periodic square-well transmission studies significant enough to be of interest, as in Fig. 1a, 1c and Fig. 2, generally require about 10 or more changes in square potentials.

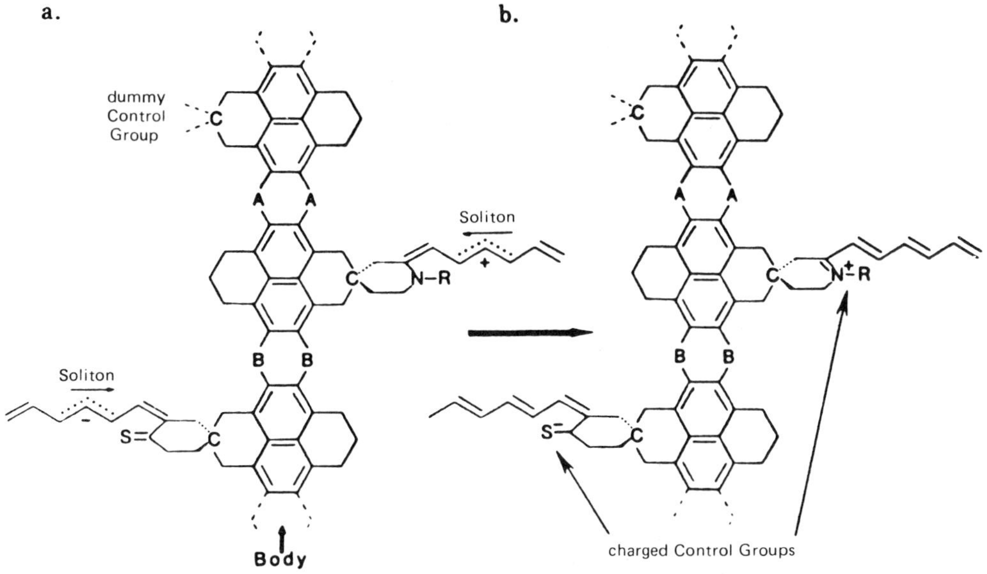

a. **b.**

dummy Control Group

Soliton

Soliton

Body

charged Control Groups

FIG. 3.   Several concepts regarding molecular tunnelling are illustrated with these soliton control groups on the vertical Body.

The reduction of computer or experimental results to understanding and predictive power is highly desirable but rather unlikely without theoretical support.   In the early days of solid state physics, the Kronig-Penny square well potential form provided such guidance.   However, the Kronig-Penny model has not proven very useful in relating chemical concepts to even bond formation or electronegativity.   In 1971, however, it was shown directly by this author (7) that from the chemical bonding and the 3d structure one can directly predict the structure in 3d-reciprocal space.   This leads first to a heuristic picture in unreduced k-space and second to zone descriptions.   At the same time, the application of the square well potential was expanded beyond the Kronig-Penny model to multiple square wells per unit cell (Ref 8, p. 550, 551).

To provide an exact analytical expression for the electron scattering through an array of square well potential changes, one must match both the electron wave equation and the wave equation slope on both sides of each square well step (see Fig. 4b). The electron approaching from the left is given by e^{ikx}, the electron scattered to the left is from the V_0-V_1 interface and is indicated as Be^{-ikx}. From the extreme right no electrons are scattered to the left so only Te^{ikx} is going right (T is the transmission amplitude for the entire process).

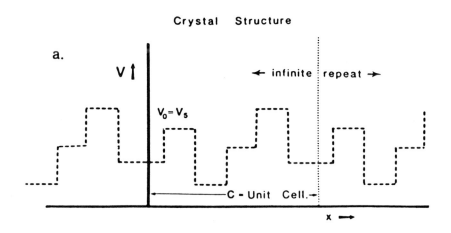

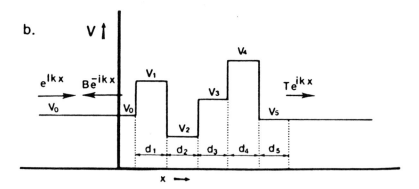

FIG. 4.   Square well potentials in infinite periodic configurations are symboliz
         in Fig. 4a, and short array of square wells and barriers by Fig. 4b.
         The potential wells have heights marked V_i and widths d_i. In Fig.
         the condition for periodicity is that $V_0 = V_5$.

Definitions for the square well potentials are given in Fig. 4b. The flat potential is V_i high and $d_i = a_{i+1} - a_i$ wide with an initial potential of V_0 to the left and V_n to the right. The exact solutions as a function of electron energy, ω, are given in MED I (6, pages 125-129), where the initial and final potentials are V_0 and V_n. While the exact analytical solutions are given in MED I for Case n=0 to n=6 and are tractable, they are not tractable for Case n=7 and Case n=8, see Tables I and II. Terms in Tables I and II are given below:

$$V_i \leq \omega \qquad\qquad V_i > \omega$$
$$\beta_i = \sqrt{\omega - V_i} \qquad\qquad \beta_i = \sqrt{V_i - \omega}$$
$$C_i = \cos \beta_i d_i \qquad\qquad C_i = \cosh \beta_i d_i$$
$$S_i = \sin \beta_i d_i \qquad\qquad S_i = \sinh \beta_i d_i$$

The transmission coefficient and its deviation is given as below (6, 126):

$$T^2 = \frac{\left[(U_{no}^+)^2 + (V_{no}^+)^2 - (U_{no}^-)^2 - (V_{no}^-)^2 \right]^2}{4 \left[(U_{no}^+)^2 + (V_{no}^+)^2 \right]}$$

Note that transmission coefficients can be calculated for any system of dual periodic array barriers, Fig. 2, n=10, given ones patience. The analytical number of terms for the coefficients of $U_{70}^\pm$, $V_{70}^\pm$ and $U_{80}^\pm$, $V_{80}^\pm$ are 32 and 128, respectively. This range of square well changes is indicated in the coefficient Tables I and II by the subscript "nO" (as in $V_{no}^\pm$ for square well 0 to well n). The implication of square well array tunnelling is partially indicated in Fig. 1a, b and c. While short periodic array tunnelling was suggested by the pseudo-classical work of Pschenichov (2), the quantum analytical results presented here and in MED I (6) offer possibilities of considerable utility. For example, resolution of the electron as it tunnels through the barrier n=9, 10 can be computed as its energy ω adjusts so that it is in resonance with the pseudo-stationary states.

Now in situations such as in Fig. 1b and in more bias situations, "resonance" has a somewhat different meaning and a much larger transmission half-width occurs. However, in Fig. 1c, where the bias is small, the meaning of resonance is enhanced and the resolution is increased. This suggests that the non-periodic array might serve as an electron monochromator. This tunnelling use was mentioned in 1979 (1) and might

Table I — Coefficients of $U_{70}^{\pm}$ and $V_{70}^{\pm}$

$$
\begin{aligned}
U_{70}^{\pm} ={}& \left[1 \pm \frac{\beta_0}{\beta_7}\right] C_1 C_2 C_3 C_4 C_5 C_6
- \left[\frac{\beta_5}{\beta_6} \pm \frac{\beta_0\beta_6}{\beta_5\beta_7}\right] C_1 C_2 C_3 C_4 S_5 S_6
- \left[\frac{\beta_4}{\beta_6} \pm \frac{\beta_0\beta_6}{\beta_4\beta_7}\right] C_1 C_2 C_3 S_4 C_5 S_6 \\[4pt]
&- \left[\frac{\beta_3}{\beta_6} \pm \frac{\beta_0\beta_6}{\beta_3\beta_7}\right] C_1 C_2 S_3 C_4 C_5 S_6
- \left[\frac{\beta_2}{\beta_6} \pm \frac{\beta_0\beta_6}{\beta_2\beta_7}\right] C_1 S_2 C_3 C_4 C_5 S_6
- \left[\frac{\beta_1}{\beta_6} \pm \frac{\beta_0\beta_6}{\beta_1\beta_7}\right] S_1 C_2 C_3 C_4 C_5 S_6 \\[4pt]
&- \left[\frac{\beta_4}{\beta_5} \pm \frac{\beta_0\beta_5}{\beta_4\beta_7}\right] C_1 C_2 C_3 S_4 S_5 C_6
- \left[\frac{\beta_3}{\beta_5} \pm \frac{\beta_0\beta_5}{\beta_3\beta_7}\right] C_1 C_2 S_3 C_4 S_5 C_6
- \left[\frac{\beta_2}{\beta_5} \pm \frac{\beta_0\beta_5}{\beta_2\beta_7}\right] C_1 S_2 C_3 C_4 S_5 C_6 \\[4pt]
&- \left[\frac{\beta_1}{\beta_5} \pm \frac{\beta_0\beta_5}{\beta_1\beta_7}\right] S_1 C_2 C_3 C_4 S_5 C_6
- \left[\frac{\beta_3}{\beta_4} \pm \frac{\beta_0\beta_4}{\beta_3\beta_7}\right] C_1 C_2 S_3 S_4 C_5 C_6
- \left[\frac{\beta_2}{\beta_4} \pm \frac{\beta_0\beta_4}{\beta_2\beta_7}\right] C_1 S_2 C_3 S_4 C_5 C_6 \\[4pt]
&- \left[\frac{\beta_1}{\beta_4} \pm \frac{\beta_0\beta_4}{\beta_1\beta_7}\right] S_1 C_2 C_3 S_4 C_5 C_6
- \left[\frac{\beta_2}{\beta_3} \pm \frac{\beta_0\beta_3}{\beta_2\beta_7}\right] C_1 S_2 S_3 C_4 C_5 C_6
- \left[\frac{\beta_1}{\beta_3} \pm \frac{\beta_0\beta_3}{\beta_1\beta_7}\right] S_1 C_2 S_3 C_4 C_5 C_6 \\[4pt]
&- \left[\frac{\beta_1}{\beta_2} \pm \frac{\beta_0\beta_2}{\beta_1\beta_7}\right] S_1 S_2 C_3 C_4 C_5 C_6
+ \left[\frac{\beta_3\beta_5}{\beta_4\beta_6} \pm \frac{\beta_0\beta_4\beta_6}{\beta_3\beta_5\beta_7}\right] C_1 C_2 S_3 S_4 S_5 S_6
+ \left[\frac{\beta_2\beta_5}{\beta_3\beta_6} \pm \frac{\beta_0\beta_3\beta_6}{\beta_2\beta_5\beta_7}\right] C_1 S_2 S_3 C_4 S_5 S_6 \\[4pt]
&+ \left[\frac{\beta_1\beta_5}{\beta_4\beta_6} \pm \frac{\beta_0\beta_3\beta_6}{\beta_1\beta_3\beta_7}\right] S_1 C_2 C_3 S_4 S_5 S_6
+ \left[\frac{\beta_2\beta_5}{\beta_3\beta_6} \pm \frac{\beta_0\beta_3\beta_6}{\beta_2\beta_5\beta_7}\right] C_1 S_2 C_3 S_4 S_5 S_6
+ \left[\frac{\beta_1\beta_5}{\beta_3\beta_6} \pm \frac{\beta_0\beta_3\beta_6}{\beta_1\beta_5\beta_7}\right] S_1 C_2 S_3 C_4 S_5 S_6 \\[4pt]
&+ \left[\frac{\beta_1\beta_5}{\beta_2\beta_6} \pm \frac{\beta_0\beta_2\beta_6}{\beta_1\beta_5\beta_7}\right] S_1 S_2 C_3 C_4 S_5 S_6
+ \left[\frac{\beta_2\beta_4}{\beta_3\beta_6} \pm \frac{\beta_0\beta_3\beta_6}{\beta_2\beta_4\beta_7}\right] C_1 S_2 C_3 S_4 C_5 S_6
+ \left[\frac{\beta_1\beta_4}{\beta_3\beta_6} \pm \frac{\beta_0\beta_3\beta_6}{\beta_1\beta_4\beta_7}\right] S_1 C_2 S_3 S_4 C_5 S_6
\end{aligned}
$$

$$
\begin{aligned}
V_{70}^{\pm} =\;& C_1 C_2 C_3 C_4 C_5 C_6
+ \left[\frac{\beta_0}{\beta_6} \pm \frac{\beta_6}{\beta_7}\right] C_1 C_2 C_3 C_4 C_5 S_6
+ \left[\frac{\beta_0}{\beta_5} \pm \frac{\beta_5}{\beta_7}\right] C_1 C_2 C_3 C_4 S_5 C_6
+ \left[\frac{\beta_0}{\beta_4} \pm \frac{\beta_4}{\beta_7}\right] C_1 C_2 C_3 S_4 C_5 C_6 \\[4pt]
&+ \left[\frac{\beta_0}{\beta_3} \pm \frac{\beta_3}{\beta_7}\right] C_1 C_2 S_3 C_4 C_5 C_6
+ \left[\frac{\beta_0}{\beta_2} \pm \frac{\beta_2}{\beta_7}\right] C_1 S_2 C_3 C_4 C_5 C_6
+ \left[\frac{\beta_0}{\beta_1} \pm \frac{\beta_1}{\beta_7}\right] S_1 C_2 C_3 C_4 C_5 C_6 \\[4pt]
&- \left[\frac{\beta_0 \beta_5}{\beta_3 \beta_6} \pm \frac{\beta_3 \beta_6}{\beta_5 \beta_7}\right] C_1 C_2 S_3 C_4 S_5 S_6
- \left[\frac{\beta_0 \beta_5}{\beta_4 \beta_6} \pm \frac{\beta_3 \beta_6}{\beta_5 \beta_7}\right] C_1 C_2 C_3 S_4 S_5 S_6
- \left[\frac{\beta_0 \beta_5}{\beta_2 \beta_6} \pm \frac{\beta_2 \beta_6}{\beta_5 \beta_7}\right] C_1 S_2 C_3 C_4 S_5 S_6 \\[4pt]
&- \left[\frac{\beta_0 \beta_4}{\beta_1 \beta_6} \pm \frac{\beta_1 \beta_6}{\beta_5 \beta_7}\right] S_1 C_2 C_3 C_4 S_5 S_6
- \left[\frac{\beta_0 \beta_4}{\beta_2 \beta_6} \pm \frac{\beta_2 \beta_6}{\beta_4 \beta_7}\right] C_1 S_2 C_3 S_4 C_5 S_6
- \left[\frac{\beta_0 \beta_4}{\beta_3 \beta_6} \pm \frac{\beta_3 \beta_6}{\beta_4 \beta_7}\right] C_1 C_2 S_3 S_4 C_5 S_6 \\[4pt]
&- \left[\frac{\beta_0 \beta_3}{\beta_2 \beta_6} \pm \frac{\beta_2 \beta_6}{\beta_3 \beta_7}\right] C_1 S_2 S_3 C_4 C_5 S_6
- \left[\frac{\beta_0 \beta_3}{\beta_1 \beta_6} \pm \frac{\beta_1 \beta_6}{\beta_3 \beta_7}\right] S_1 C_2 S_3 C_4 C_5 S_6
- \left[\frac{\beta_0 \beta_3}{\beta_1 \beta_6} \pm \frac{\beta_1 \beta_6}{\beta_4 \beta_7}\right] S_1 C_2 C_3 S_4 C_5 S_6 \\[4pt]
&+ \left[\frac{\beta_1 \beta_4}{\beta_2 \beta_6} \pm \frac{\beta_0 \beta_3 \beta_6}{\beta_1 \beta_4 \beta_7}\right] S_1 S_2 C_3 C_4 C_5 S_6
+ \left[\frac{\beta_1 \beta_3}{\beta_2 \beta_6} \pm \frac{\beta_0 \beta_2 \beta_6}{\beta_1 \beta_3 \beta_7}\right] S_1 S_2 C_3 C_4 C_5 S_6
+ \left[\frac{\beta_2 \beta_4}{\beta_3 \beta_5} \pm \frac{\beta_0 \beta_3 \beta_5}{\beta_2 \beta_4 \beta_7}\right] C_1 S_2 S_3 S_4 S_5 C_6 \\[4pt]
&+ \left[\frac{\beta_1 \beta_4}{\beta_3 \beta_5} \pm \frac{\beta_0 \beta_3 \beta_5}{\beta_1 \beta_4 \beta_7}\right] S_1 C_2 S_3 S_4 S_5 C_6
+ \left[\frac{\beta_1 \beta_4}{\beta_2 \beta_5} \pm \frac{\beta_0 \beta_2 \beta_5}{\beta_1 \beta_4 \beta_7}\right] S_1 S_2 C_3 S_4 S_5 C_6
+ \left[\frac{\beta_1 \beta_3}{\beta_2 \beta_5} \pm \frac{\beta_0 \beta_2 \beta_5}{\beta_1 \beta_3 \beta_7}\right] S_1 S_2 S_3 C_4 S_5 C_6 \\[4pt]
&+ \left[\frac{\beta_1 \beta_3}{\beta_2 \beta_4} \pm \frac{\beta_0 \beta_2 \beta_4}{\beta_1 \beta_3 \beta_7}\right] S_1 S_2 S_3 S_4 C_5 C_6
- \left[\frac{\beta_1 \beta_3 \beta_5}{\beta_2 \beta_4 \beta_6} \pm \frac{\beta_0 \beta_2 \beta_4 \beta_6}{\beta_1 \beta_3 \beta_5 \beta_7}\right] S_1 S_2 S_3 S_4 S_5 S_6
\end{aligned}
$$

Table I — Coefficients of $U_{70}^{\pm}$ and $V_{70}^{\pm}$ (Continued)

$$-\left[\frac{\beta_0\beta_2}{\beta_1\beta_6} \pm \frac{\beta_1\beta_6}{\beta_2\beta_7}\right] S_1S_2C_3C_4C_5S_6 - \left[\frac{\beta_0\beta_2\beta_5}{\beta_1\beta_4\beta_6} \pm \frac{\beta_1\beta_4\beta_6}{\beta_2\beta_5\beta_7}\right] C_1C_2S_3S_4S_5C_6 - \left[\frac{\beta_0\beta_4}{\beta_3\beta_5} \pm \frac{\beta_3\beta_5}{\beta_4\beta_7}\right] C_1S_2C_3S_4S_5C_6$$

$$-\left[\frac{\beta_0\beta_4}{\beta_1\beta_5} \pm \frac{\beta_1\beta_5}{\beta_4\beta_7}\right] S_1C_2C_3S_4S_5C_6 - \left[\frac{\beta_0\beta_3}{\beta_2\beta_5} \pm \frac{\beta_2\beta_5}{\beta_3\beta_7}\right] C_1S_2S_3C_4S_5C_6 - \left[\frac{\beta_0\beta_3}{\beta_1\beta_5} \pm \frac{\beta_1\beta_5}{\beta_3\beta_7}\right] S_1C_2S_3C_4S_5C_6$$

$$-\left[\frac{\beta_0\beta_2}{\beta_1\beta_5} \pm \frac{\beta_1\beta_5}{\beta_2\beta_7}\right] S_1S_2C_3C_4S_5C_6 - \left[\frac{\beta_0\beta_3}{\beta_2\beta_4} \pm \frac{\beta_2\beta_4}{\beta_3\beta_7}\right] C_1S_2S_3S_4C_5C_6 - \left[\frac{\beta_0\beta_3}{\beta_1\beta_4} \pm \frac{\beta_1\beta_4}{\beta_3\beta_7}\right] S_1C_2S_3S_4C_5C_6$$

$$-\left[\frac{\beta_0\beta_2}{\beta_1\beta_4} \pm \frac{\beta_1\beta_4}{\beta_2\beta_7}\right] S_1S_2C_3S_4C_5C_6 - \left[\frac{\beta_0\beta_2}{\beta_1\beta_3} \pm \frac{\beta_1\beta_3}{\beta_2\beta_7}\right] S_1S_2S_3C_4C_5C_6 + \left[\frac{\beta_0\beta_3\beta_5}{\beta_2\beta_4\beta_6} \pm \frac{\beta_2\beta_4\beta_6}{\beta_3\beta_5\beta_7}\right] C_1S_2S_3S_4S_5S_6$$

$$+\left[\frac{\beta_0\beta_3\beta_5}{\beta_1\beta_4\beta_6} \pm \frac{\beta_1\beta_4\beta_6}{\beta_3\beta_5\beta_7}\right] S_1C_2S_3S_4S_5S_6 + \left[\frac{\beta_0\beta_2\beta_5}{\beta_1\beta_4\beta_6} \pm \frac{\beta_1\beta_4\beta_6}{\beta_2\beta_5\beta_7}\right] S_1S_2C_3S_4S_5S_6 + \left[\frac{\beta_0\beta_2\beta_5}{\beta_1\beta_3\beta_6} \pm \frac{\beta_1\beta_3\beta_6}{\beta_2\beta_5\beta_7}\right] S_1S_2S_3C_4S_5S_6$$

$$+\left[\frac{\beta_0\beta_2\beta_4}{\beta_1\beta_3\beta_6} \pm \frac{\beta_1\beta_3\beta_6}{\beta_2\beta_4\beta_7}\right] S_1S_2S_3S_4C_5S_6 + \left[\frac{\beta_0\beta_2\beta_5}{\beta_1\beta_3\beta_6} \pm \frac{\beta_1\beta_3\beta_6}{\beta_2\beta_5\beta_7}\right] S_1S_2S_3S_4S_5C_6 + S_1S_2S_3S_4S_5S_6$$

Table II — Coefficients $U_{80}^{\pm}$ and $V_{80}^{\pm}$

$$
\begin{aligned}
U_{80}^{\pm} =\ & \left[1 \pm \frac{\beta_0}{\beta_8}\right] C_1 C_2 C_3 C_4 C_5 C_6 C_7
- \left[\frac{\beta_6}{\beta_7} \pm \frac{\beta_0\beta_7}{\beta_6\beta_8}\right] C_1 C_2 C_3 C_4 C_5 C_6 S_7
- \left[\frac{\beta_5}{\beta_7} \pm \frac{\beta_0\beta_7}{\beta_5\beta_8}\right] C_1 C_2 C_3 C_4 S_5 C_6 S_7 \\[4pt]
& - \left[\frac{\beta_4}{\beta_7} \pm \frac{\beta_0\beta_7}{\beta_4\beta_8}\right] C_1 C_2 C_3 S_4 C_5 C_6 S_7
- \left[\frac{\beta_3}{\beta_7} \pm \frac{\beta_0\beta_7}{\beta_3\beta_8}\right] C_1 C_2 S_3 C_4 C_5 C_6 S_7
- \left[\frac{\beta_2}{\beta_7} \pm \frac{\beta_0\beta_7}{\beta_2\beta_8}\right] C_1 S_2 C_3 C_4 C_5 C_6 S_7 \\[4pt]
& - \left[\frac{\beta_1}{\beta_7} \pm \frac{\beta_0\beta_7}{\beta_1\beta_8}\right] S_1 C_2 C_3 C_4 C_5 C_6 S_7
- \left[\frac{\beta_5}{\beta_6} \pm \frac{\beta_0\beta_6}{\beta_5\beta_8}\right] C_1 C_2 C_3 C_4 C_5 S_6 C_7
- \left[\frac{\beta_4}{\beta_6} \pm \frac{\beta_0\beta_6}{\beta_4\beta_8}\right] C_1 C_2 C_3 S_4 C_5 S_6 C_7 \\[4pt]
& - \left[\frac{\beta_3}{\beta_6} \pm \frac{\beta_0\beta_6}{\beta_3\beta_8}\right] C_1 C_2 S_3 C_4 C_5 S_6 C_7
- \left[\frac{\beta_2}{\beta_6} \pm \frac{\beta_0\beta_6}{\beta_2\beta_8}\right] C_1 S_2 C_3 C_4 C_5 S_6 C_7
- \left[\frac{\beta_1}{\beta_6} \pm \frac{\beta_0\beta_6}{\beta_1\beta_8}\right] S_1 C_2 C_3 C_4 C_5 S_6 C_7 \\[4pt]
& - \left[\frac{\beta_4}{\beta_5} \pm \frac{\beta_0\beta_5}{\beta_4\beta_8}\right] C_1 C_2 C_3 S_4 S_5 C_6 C_7
- \left[\frac{\beta_3}{\beta_5} \pm \frac{\beta_0\beta_5}{\beta_3\beta_8}\right] C_1 C_2 S_3 C_4 S_5 C_6 C_7
- \left[\frac{\beta_2}{\beta_5} \pm \frac{\beta_0\beta_5}{\beta_2\beta_8}\right] C_1 S_2 C_3 C_4 S_5 C_6 C_7 \\[4pt]
& - \left[\frac{\beta_1}{\beta_5} \pm \frac{\beta_0\beta_5}{\beta_1\beta_8}\right] S_1 C_2 C_3 C_4 S_5 C_6 C_7
- \left[\frac{\beta_3}{\beta_4} \pm \frac{\beta_0\beta_4}{\beta_3\beta_8}\right] C_1 C_2 S_3 S_4 C_5 C_6 C_7
- \left[\frac{\beta_2}{\beta_4} \pm \frac{\beta_0\beta_4}{\beta_2\beta_8}\right] C_1 S_2 C_3 S_4 C_5 C_6 C_7 \\[4pt]
& - \left[\frac{\beta_1}{\beta_4} \pm \frac{\beta_0\beta_4}{\beta_1\beta_8}\right] S_1 C_2 C_3 S_4 C_5 C_6 C_7
- \left[\frac{\beta_2}{\beta_3} \pm \frac{\beta_0\beta_3}{\beta_2\beta_8}\right] C_1 S_2 S_3 C_4 C_5 C_6 C_7
- \left[\frac{\beta_1}{\beta_3} \pm \frac{\beta_0\beta_3}{\beta_1\beta_8}\right] S_1 C_2 S_3 C_4 C_5 C_6 C_7 \\[4pt]
& - \left[\frac{\beta_1}{\beta_2} \pm \frac{\beta_0\beta_1}{\beta_1\beta_8}\right] S_1 S_2 C_3 C_4 C_5 C_6 C_7
+ \left[\frac{\beta_2\beta_6}{\beta_5\beta_7} \pm \frac{\beta_0\beta_5\beta_7}{\beta_2\beta_6\beta_8}\right] C_1 S_2 C_3 C_4 S_5 S_6 S_7
- \left[\frac{\beta_4\beta_6}{\beta_5\beta_7} \pm \frac{\beta_0\beta_5\beta_7}{\beta_4\beta_6\beta_8}\right] C_1 C_2 C_3 S_4 S_5 S_6 S_7 \\[4pt]
& - \left[\frac{\beta_1\beta_6}{\beta_5\beta_7} \pm \frac{\beta_0\beta_5\beta_7}{\beta_1\beta_6\beta_8}\right] S_1 C_2 C_3 C_4 S_5 S_6 S_7
- \left[\frac{\beta_3\beta_6}{\beta_5\beta_7} \pm \frac{\beta_0\beta_5\beta_7}{\beta_3\beta_6\beta_8}\right] C_1 C_2 S_3 C_4 S_5 S_6 S_7
- \left[\frac{\beta_3\beta_6}{\beta_4\beta_7} \pm \frac{\beta_0\beta_4\beta_7}{\beta_3\beta_6\beta_8}\right] C_1 C_2 S_3 S_4 C_5 S_6 S_7
\end{aligned}
$$

Table II — Coefficients $U_{80}^{\pm}$ and $V_{80}^{\pm}$ (Continued)

$$-\left[\frac{\beta_2\beta_6}{\beta_4\beta_7}+\frac{\beta_0\beta_4\beta_7}{\beta_2\beta_6\beta_8}\right]C_1S_2C_3S_4C_5S_6S_7-\left[\frac{\beta_1\beta_6}{\beta_4\beta_7}+\frac{\beta_0\beta_4\beta_7}{\beta_1\beta_6\beta_8}\right]S_1C_2C_3S_4C_5S_6S_7-\left[\frac{\beta_2\beta_6}{\beta_3\beta_7}+\frac{\beta_0\beta_3\beta_7}{\beta_2\beta_6\beta_8}\right]C_1S_2S_3C_4C_5S_6S_7$$

$$-\left[\frac{\beta_1\beta_6}{\beta_3\beta_7}+\frac{\beta_0\beta_3\beta_7}{\beta_1\beta_6\beta_8}\right]S_1C_2S_3C_4C_5S_6S_7-\left[\frac{\beta_1\beta_6}{\beta_2\beta_7}+\frac{\beta_0\beta_2\beta_7}{\beta_1\beta_6\beta_8}\right]S_1S_2C_3C_4C_5S_6S_7+\left[\frac{\beta_3\beta_5}{\beta_4\beta_7}+\frac{\beta_0\beta_4\beta_7}{\beta_3\beta_5\beta_8}\right]C_1C_2S_3S_4S_5C_6S_7$$

$$-\left[\frac{\beta_2\beta_5}{\beta_4\beta_7}+\frac{\beta_0\beta_4\beta_7}{\beta_2\beta_5\beta_8}\right]C_1S_2C_3S_4S_5C_6S_7-\left[\frac{\beta_1\beta_5}{\beta_4\beta_7}+\frac{\beta_0\beta_4\beta_7}{\beta_1\beta_5\beta_8}\right]S_1C_2C_3S_4S_5C_6S_7-\left[\frac{\beta_2\beta_5}{\beta_3\beta_7}+\frac{\beta_0\beta_3\beta_7}{\beta_2\beta_5\beta_8}\right]C_1S_2S_3C_4S_5C_6S_7$$

$$-\left[\frac{\beta_1\beta_4}{\beta_3\beta_7}+\frac{\beta_0\beta_3\beta_7}{\beta_1\beta_4\beta_8}\right]S_1C_2S_3S_4C_5C_6S_7-\left[\frac{\beta_1\beta_5}{\beta_2\beta_7}+\frac{\beta_0\beta_2\beta_7}{\beta_1\beta_5\beta_8}\right]S_1S_2C_3S_4S_5C_6S_7-\left[\frac{\beta_2\beta_4}{\beta_3\beta_7}+\frac{\beta_0\beta_3\beta_7}{\beta_2\beta_4\beta_8}\right]C_1S_2S_3S_4C_5C_6S_7$$

$$-\left[\frac{\beta_1\beta_4}{\beta_3\beta_7}+\frac{\beta_0\beta_3\beta_7}{\beta_1\beta_4\beta_8}\right]S_1C_2S_3S_4C_5C_6S_7-\left[\frac{\beta_1\beta_3}{\beta_2\beta_7}+\frac{\beta_0\beta_2\beta_7}{\beta_1\beta_3\beta_8}\right]S_1S_2C_3S_4C_5C_6S_7-\left[\frac{\beta_1\beta_3}{\beta_2\beta_7}+\frac{\beta_0\beta_2\beta_7}{\beta_1\beta_3\beta_8}\right]S_1S_2S_3C_4C_5C_6S_7$$

$$+\left[\frac{\beta_3\beta_5}{\beta_4\beta_6}+\frac{\beta_0\beta_4\beta_6}{\beta_3\beta_5\beta_8}\right]C_1C_2S_3S_4S_5S_6C_7-\left[\frac{\beta_2\beta_5}{\beta_4\beta_6}+\frac{\beta_0\beta_4\beta_6}{\beta_2\beta_5\beta_8}\right]C_1C_2S_3S_4S_5S_6C_7+\left[\frac{\beta_1\beta_5}{\beta_4\beta_6}+\frac{\beta_0\beta_5\beta_6}{\beta_1\beta_3\beta_8}\right]S_1C_2S_3S_4S_5S_6C_7$$

$$-\left[\frac{\beta_2\beta_5}{\beta_3\beta_6}+\frac{\beta_0\beta_3\beta_6}{\beta_2\beta_5\beta_8}\right]C_1S_2S_3C_4S_5S_6C_7-\left[\frac{\beta_1\beta_5}{\beta_3\beta_6}+\frac{\beta_0\beta_3\beta_6}{\beta_1\beta_5\beta_8}\right]S_1C_2S_3C_4S_5S_6C_7-\left[\frac{\beta_1\beta_5}{\beta_2\beta_6}+\frac{\beta_0\beta_2\beta_6}{\beta_1\beta_5\beta_8}\right]S_1S_2C_3C_4S_5S_6C_7$$

$$-\left[\frac{\beta_2\beta_4}{\beta_3\beta_6} + \frac{\beta_0\beta_3\beta_6}{\beta_2\beta_4\beta_8}\right] C_1 S_2 S_3 S_4 C_5 S_6 C_7 - \left[\frac{\beta_1\beta_4}{\beta_3\beta_6} + \frac{\beta_0\beta_3\beta_6}{\beta_1\beta_4\beta_8}\right] S_1 S_2 C_3 S_4 C_5 S_6 C_7$$

$$-\left[\frac{\beta_1\beta_3}{\beta_2\beta_6} + \frac{\beta_0\beta_2\beta_6}{\beta_1\beta_3\beta_8}\right] S_1 S_2 C_3 S_4 C_5 S_6 C_7 - \left[\frac{\beta_2\beta_4}{\beta_3\beta_5} + \frac{\beta_0\beta_3\beta_5}{\beta_2\beta_4\beta_8}\right] S_1 C_2 S_3 S_4 S_5 C_6 C_7$$

$$-\left[\frac{\beta_1\beta_4}{\beta_2\beta_5} + \frac{\beta_0\beta_2\beta_5}{\beta_1\beta_4\beta_8}\right] S_1 S_2 C_3 S_4 S_5 C_6 C_7 - \left[\frac{\beta_1\beta_3}{\beta_2\beta_4} + \frac{\beta_0\beta_2\beta_4}{\beta_1\beta_3\beta_8}\right] S_1 S_2 S_3 S_4 C_5 C_6 C_7$$

$$+\left[\frac{\beta_2\beta_4\beta_6}{\beta_3\beta_5\beta_7} + \frac{\beta_0\beta_3\beta_5\beta_7}{\beta_2\beta_4\beta_6\beta_8}\right] C_1 S_2 S_3 S_4 S_5 S_6 S_7 + \left[\frac{\beta_1\beta_4\beta_6}{\beta_3\beta_5\beta_7} + \frac{\beta_0\beta_3\beta_5\beta_7}{\beta_1\beta_4\beta_6\beta_8}\right] S_1 C_2 S_3 S_4 S_5 S_6 S_7$$

$$+\left[\frac{\beta_1\beta_4\beta_6}{\beta_2\beta_5\beta_7} + \frac{\beta_0\beta_2\beta_5\beta_7}{\beta_1\beta_4\beta_6\beta_8}\right] S_1 S_2 C_3 S_4 S_5 S_6 S_7 + \left[\frac{\beta_1\beta_3\beta_6}{\beta_2\beta_5\beta_7} + \frac{\beta_0\beta_2\beta_5\beta_7}{\beta_1\beta_3\beta_6\beta_8}\right] S_1 S_2 S_3 C_4 S_5 S_6 S_7$$

$$+\left[\frac{\beta_1\beta_3\beta_6}{\beta_2\beta_4\beta_7} + \frac{\beta_0\beta_2\beta_4\beta_7}{\beta_1\beta_3\beta_6\beta_8}\right] S_1 S_2 S_3 S_4 C_5 S_6 S_7 + \left[\frac{\beta_1\beta_3\beta_5}{\beta_2\beta_4\beta_7} + \frac{\beta_0\beta_2\beta_4\beta_7}{\beta_1\beta_3\beta_5\beta_8}\right] S_1 S_2 S_3 S_4 S_5 C_6 S_7$$

$$+\left[\frac{\beta_1\beta_3\beta_5}{\beta_2\beta_4\beta_6} + \frac{\beta_0\beta_2\beta_4\beta_6}{\beta_1\beta_3\beta_5\beta_8}\right] S_1 S_2 S_3 S_4 S_5 S_6 C_7$$

Table II — Coefficients $U_{80}^{\pm}$ and $V_{80}^{\pm}$ (Continued)

$$
\begin{aligned}
V_{80}^{\pm} =\;& \left[\frac{\beta_0}{\beta_7} \pm \frac{\beta_7}{\beta_8}\right] C_1C_2C_3C_4C_5C_6C_7
+ \left[\frac{\beta_0}{\beta_6} \pm \frac{\beta_6}{\beta_8}\right] C_1C_2C_3C_4C_5S_6C_7
+ \left[\frac{\beta_0}{\beta_5} \pm \frac{\beta_5}{\beta_8}\right] C_1C_2C_3C_4S_5C_6C_7 \\[4pt]
+\;& \left[\frac{\beta_0}{\beta_4} \pm \frac{\beta_4}{\beta_8}\right] C_1C_2C_3S_4C_5C_6C_7
+ \left[\frac{\beta_0}{\beta_3} \pm \frac{\beta_3}{\beta_8}\right] C_1C_2S_3C_4C_5C_6C_7
+ \left[\frac{\beta_0}{\beta_2} \pm \frac{\beta_2}{\beta_8}\right] C_1S_2C_3C_4C_5C_6C_7 \\[4pt]
+\;& \left[\frac{\beta_0}{\beta_1} \pm \frac{\beta_1}{\beta_8}\right] S_1C_2C_3C_4C_5C_6C_7
+ \left[\frac{\beta_0\beta_6}{\beta_5\beta_7} \pm \frac{\beta_5\beta_7}{\beta_6\beta_8}\right] C_1C_2C_3C_4S_5S_6S_7
- \left[\frac{\beta_0\beta_6}{\beta_4\beta_7} \pm \frac{\beta_4\beta_7}{\beta_6\beta_8}\right] C_1C_2C_3S_4C_5S_6S_7 \\[4pt]
-\;& \left[\frac{\beta_0\beta_6}{\beta_3\beta_7} \pm \frac{\beta_3\beta_7}{\beta_6\beta_8}\right] C_1C_2S_3C_4C_5S_6S_7
- \left[\frac{\beta_0\beta_5}{\beta_2\beta_7} \pm \frac{\beta_2\beta_7}{\beta_5\beta_8}\right] C_1S_2C_3C_4S_5C_6S_7
- \left[\frac{\beta_0\beta_6}{\beta_2\beta_7} \pm \frac{\beta_2\beta_7}{\beta_6\beta_8}\right] C_1S_2C_3C_4C_5S_6S_7 \\[4pt]
-\;& \left[\frac{\beta_0\beta_5}{\beta_3\beta_7} \pm \frac{\beta_3\beta_7}{\beta_5\beta_8}\right] C_1C_2C_3S_4S_5C_6S_7
- \left[\frac{\beta_0\beta_5}{\beta_3\beta_7} \pm \frac{\beta_2\beta_7}{\beta_5\beta_8}\right] C_1C_2S_3C_4S_5C_6S_7
- \left[\frac{\beta_0\beta_4}{\beta_2\beta_7} \pm \frac{\beta_2\beta_7}{\beta_4\beta_8}\right] C_1S_2C_3S_4C_5C_6S_7 \\[4pt]
-\;& \left[\frac{\beta_0\beta_5}{\beta_1\beta_7} \pm \frac{\beta_1\beta_7}{\beta_5\beta_8}\right] S_1C_2C_3C_4S_5C_6S_7
- \left[\frac{\beta_0\beta_4}{\beta_3\beta_7} \pm \frac{\beta_3\beta_7}{\beta_4\beta_8}\right] C_1C_2S_3S_4C_5C_6S_7
- \left[\frac{\beta_0\beta_4}{\beta_2\beta_7} \pm \frac{\beta_2\beta_7}{\beta_4\beta_8}\right] S_1C_2S_3C_4C_5C_6S_7 \\[4pt]
-\;& \left[\frac{\beta_0\beta_4}{\beta_1\beta_7} \pm \frac{\beta_1\beta_7}{\beta_4\beta_8}\right] S_1C_2C_3S_4C_5C_6S_7
- \left[\frac{\beta_0\beta_3}{\beta_2\beta_7} \pm \frac{\beta_2\beta_7}{\beta_3\beta_8}\right] S_1C_2S_3C_4C_5C_6S_7
\pm \left[\frac{\beta_0\beta_3}{\beta_1\beta_7} \pm \frac{\beta_1\beta_7}{\beta_3\beta_8}\right] C_1S_2C_3S_4C_5C_6S_7 \\[4pt]
i\;& \left[\frac{\beta_0\beta_2}{\beta_1\beta_7} \pm \frac{\beta_1\beta_7}{\beta_2\beta_8}\right] S_1S_2C_3C_4C_5C_6S_7
- \left[\frac{\beta_0\beta_5}{\beta_4\beta_6} \pm \frac{\beta_4\beta_6}{\beta_5\beta_8}\right] C_1C_2C_3S_4S_5S_6C_7
- \left[\frac{\beta_0\beta_3}{\beta_3\beta_6} \pm \frac{\beta_3\beta_6}{\beta_5\beta_8}\right] S_1C_2S_3C_4C_5S_6C_7 \\[4pt]
-\;& \left[\frac{\beta_0\beta_5}{\beta_2\beta_6} \pm \frac{\beta_2\beta_6}{\beta_5\beta_8}\right] C_1S_2C_3C_4S_5S_6C_7
- \left[\frac{\beta_0\beta_5}{\beta_1\beta_6} \pm \frac{\beta_1\beta_6}{\beta_5\beta_8}\right] S_1C_2C_3C_4S_5S_6C_7
- \left[\frac{\beta_0\beta_4}{\beta_3\beta_6} \pm \frac{\beta_3\beta_6}{\beta_4\beta_8}\right] C_1C_2S_3S_4C_5S_6C_7 \\[4pt]
-\;& \left[\frac{\beta_0\beta_5}{\beta_2\beta_6} \pm \frac{\beta_2\beta_6}{\beta_5\beta_8}\right] C_1S_2C_3C_4S_5S_6C_7
- \left[\frac{\beta_0\beta_5}{\beta_1\beta_6} \pm \frac{\beta_1\beta_6}{\beta_5\beta_8}\right] S_1C_2C_3C_4S_5S_6C_7
- \left[\frac{\beta_0\beta_4}{\beta_3\beta_6} \pm \frac{\beta_3\beta_6}{\beta_4\beta_8}\right] C_1C_2S_3S_4C_5S_6C_7
\end{aligned}
$$

$$-\left[\frac{\beta_0\beta_4}{\beta_2\beta_6}+\frac{\beta_2\beta_6}{\beta_4\beta_8}\right]C_1S_2C_3S_4C_5S_6C_7$$

$$-\left[\frac{\beta_0\beta_3}{\beta_1\beta_6}+\frac{\beta_1\beta_6}{\beta_3\beta_8}\right]S_1C_2S_3C_4C_5S_6C_7$$

$$-\left[\frac{\beta_0\beta_4}{\beta_2\beta_5}+\frac{\beta_2\beta_5}{\beta_4\beta_8}\right]C_1S_2C_3S_4S_5C_6C_7$$

$$-\left[\frac{\beta_0\beta_3}{\beta_1\beta_5}+\frac{\beta_1\beta_5}{\beta_3\beta_8}\right]S_1C_2S_3C_4S_5C_6C_7$$

$$-\left[\frac{\beta_0\beta_3}{\beta_1\beta_4}+\frac{\beta_1\beta_4}{\beta_3\beta_8}\right]S_1C_2S_3S_4C_5C_6C_7$$

$$-\left[\frac{\beta_0\beta_4}{\beta_1\beta_6}+\frac{\beta_1\beta_6}{\beta_4\beta_8}\right]S_1C_2C_3S_4C_5S_6C_7$$

$$-\left[\frac{\beta_0\beta_2}{\beta_1\beta_6}+\frac{\beta_1\beta_6}{\beta_2\beta_8}\right]S_1S_2C_3C_4C_5S_6C_7$$

$$-\left[\frac{\beta_0\beta_4}{\beta_1\beta_5}+\frac{\beta_1\beta_5}{\beta_4\beta_8}\right]C_1C_2C_3S_4S_5C_6C_7$$

$$-\left[\frac{\beta_0\beta_7}{\beta_1\beta_5}+\frac{\beta_1\beta_5}{\beta_2\beta_8}\right]S_1S_2C_3C_4S_5C_6C_7$$

$$-\left[\frac{\beta_0\beta_2}{\beta_1\beta_4}+\frac{\beta_1\beta_4}{\beta_2\beta_8}\right]S_1S_2C_3S_4C_5C_6C_7$$

$$\left[\frac{\beta_0\beta_3}{\beta_2\beta_6}+\frac{\beta_2\beta_6}{\beta_3\beta_8}\right]C_1S_2S_3C_4C_5S_6C_7$$

$$\left[\frac{\beta_0\beta_4}{\beta_3\beta_5}+\frac{\beta_3\beta_5}{\beta_4\beta_8}\right]C_1C_2S_3S_4S_5C_6C_7$$

$$\left[\frac{\beta_0\beta_3}{\beta_2\beta_5}+\frac{\beta_2\beta_5}{\beta_3\beta_8}\right]C_1S_2S_3C_4S_5C_6C_7$$

$$\left[\frac{\beta_0\beta_3}{\beta_2\beta_4}+\frac{\beta_2\beta_4}{\beta_3\beta_8}\right]C_1S_2S_3S_4C_5C_6C_7$$

$$\left[\frac{\beta_0\beta_2}{\beta_1\beta_3}+\frac{\beta_1\beta_3}{\beta_2\beta_8}\right]S_1S_2S_3C_4C_5C_6C_7$$

$$+\left[\frac{\beta_0\beta_4\beta_6}{\beta_3\beta_5\beta_7}+\frac{\beta_3\beta_5\beta_7}{\beta_4\beta_6\beta_8}\right]C_1C_2S_3S_4S_5S_6S_7$$

$$+\left[\frac{\beta_0\beta_4\beta_6}{\beta_1\beta_5\beta_7}+\frac{\beta_1\beta_5\beta_7}{\beta_4\beta_6\beta_8}\right]S_1C_2C_3S_4S_5S_6S_7$$

$$+\left[\frac{\beta_0\beta_3\beta_6}{\beta_1\beta_5\beta_7}+\frac{\beta_1\beta_5\beta_7}{\beta_3\beta_6\beta_8}\right]S_1C_2S_3C_4S_5S_6S_7$$

$$+\left[\frac{\beta_0\beta_3\beta_6}{\beta_2\beta_4\beta_7}+\frac{\beta_2\beta_4\beta_7}{\beta_3\beta_6\beta_8}\right]C_1S_2S_3S_4C_5S_6S_7$$

$$+\left[\frac{\beta_0\beta_4\beta_6}{\beta_2\beta_5\beta_7}+\frac{\beta_2\beta_5\beta_7}{\beta_4\beta_6\beta_8}\right]C_1S_2C_3S_4S_5S_6S_7$$

$$+\left[\frac{\beta_0\beta_3\beta_6}{\beta_2\beta_5\beta_7}+\frac{\beta_2\beta_5\beta_7}{\beta_3\beta_6\beta_8}\right]C_1C_2S_3S_4S_5S_6S_7$$

$$+\left[\frac{\beta_0\beta_2\beta_6}{\beta_1\beta_5\beta_7}+\frac{\beta_1\beta_5\beta_7}{\beta_2\beta_6\beta_8}\right]S_1S_2S_3C_4S_5S_6S_7$$

$$+\left[\frac{\beta_0\beta_3\beta_6}{\beta_1\beta_4\beta_7}+\frac{\beta_1\beta_4\beta_7}{\beta_3\beta_6\beta_8}\right]S_1C_2S_3S_4C_5S_6S_7$$

Table II — Coefficients $U_{80}^{\pm}$ and $V_{80}^{\pm}$ (Continued)

$$+ \left[\frac{\beta_0\beta_2\beta_6}{\beta_1\beta_4\beta_7} + \frac{\beta_1\beta_4\beta_7}{\beta_2\beta_6\beta_8}\right] S_1S_2C_3S_4C_5S_6S_7 + \left[\frac{\beta_0\beta_2\beta_6}{\beta_1\beta_3\beta_7} + \frac{\beta_1\beta_3\beta_7}{\beta_2\beta_6\beta_8}\right] S_1S_2S_3C_4C_5S_6S_7$$

$$+ \left[\frac{\beta_0\beta_3\beta_5}{\beta_2\beta_4\beta_7} + \frac{\beta_2\beta_4\beta_7}{\beta_3\beta_5\beta_8}\right] C_1S_2S_3S_4S_5C_6S_7 + \left[\frac{\beta_0\beta_3\beta_5}{\beta_1\beta_4\beta_7} + \frac{\beta_1\beta_4\beta_7}{\beta_3\beta_5\beta_8}\right] S_1C_2S_3S_4S_5C_6S_7$$

$$+ \left[\frac{\beta_0\beta_2\beta_5}{\beta_1\beta_4\beta_7} + \frac{\beta_1\beta_4\beta_7}{\beta_2\beta_5\beta_8}\right] S_1S_2C_3S_4S_5C_6S_7 + \left[\frac{\beta_0\beta_2\beta_5}{\beta_1\beta_3\beta_7} + \frac{\beta_1\beta_3\beta_7}{\beta_2\beta_5\beta_8}\right] S_1S_2S_3C_4S_5C_6S_7$$

$$+ \left[\frac{\beta_0\beta_3\beta_4}{\beta_1\beta_3\beta_7} + \frac{\beta_1\beta_4\beta_6}{\beta_2\beta_4\beta_8}\right] S_1S_2S_3S_4C_5C_6S_7 + \left[\frac{\beta_0\beta_3\beta_5}{\beta_2\beta_4\beta_6} + \frac{\beta_2\beta_4\beta_6}{\beta_3\beta_5\beta_8}\right] C_1S_2S_3S_4S_5S_6C_7$$

$$+ \left[\frac{\beta_0\beta_2\beta_5}{\beta_1\beta_3\beta_6} + \frac{\beta_1\beta_4\beta_6}{\beta_2\beta_5\beta_8}\right] S_1C_2S_3S_4S_5S_6C_7 + \left[\frac{\beta_0\beta_2\beta_5}{\beta_1\beta_4\beta_6} + \frac{\beta_1\beta_4\beta_6}{\beta_2\beta_5\beta_8}\right] S_1S_2C_3S_4S_5S_6C_7$$

$$+ \left[\frac{\beta_0\beta_2\beta_4}{\beta_1\beta_3\beta_5} + \frac{\beta_1\beta_3\beta_6}{\beta_2\beta_5\beta_8}\right] S_1S_2S_3C_4S_5S_6C_7 + \left[\frac{\beta_0\beta_2\beta_4}{\beta_1\beta_3\beta_6} + \frac{\beta_1\beta_3\beta_6}{\beta_2\beta_4\beta_8}\right] S_1S_2S_3S_4C_5S_6C_7$$

$$+ \left[\frac{\beta_0\beta_2\beta_4}{\beta_1\beta_3\beta_5} + \frac{\beta_1\beta_3\beta_5}{\beta_2\beta_4\beta_8}\right] S_1S_2S_3S_4S_5C_6C_7 \pm \left[\frac{\beta_0\beta_2\beta_4\beta_6}{\beta_1\beta_3\beta_5\beta_7} + \frac{\beta_1\beta_3\beta_5\beta_7}{\beta_2\beta_4\beta_6\beta_8}\right] S_1S_2S_3S_4S_5S_6S_7$$

achieve an even larger energy resolution monochromatic effect operating in the reverse direction.

Earlier, we had indicated that the driving force to link bond concepts to the Fermi surface had resulted in the chemical bond interpretation of the filling of the unreduced k-band scheme. It had also led to the analytical solution of the square well potential for periodic structures (Kronig-Penny like). However, it has not as yet led to the factorization of the analytical wave equations into terms which would help better understand, for example, electronegativity. Ultimately, a better understanding of chemical concepts and why they work as well as they do will be synthesized.

However, we noted in MED I (6) that the square well potential can link the unit cell of the periodic structure in Fig. 4a with the scattering square well potential in Fig. 4b. Given that c is the unit cell length and that k is the wave vector, then

$$\cos \ kc \ = \ U_{no}^{+}/2$$

for all n given that the potential equations $V_0 = V_n$ are satisfied. The interesting point is that for calculation of the transmission tunnelling effect the terms U_{no}^{-} and $U_{no}^{\pm}$ and $V_{no}^{\pm}$ are also required. Is it possible that the U_{no}^{-}, $V_{no}^{\pm}$ terms needed in the electron scattering in Fig. 4b are mutually canceling by the successive unit cell terms in Fig. 4a?

SUMMARY

The analytic solution for solving the wave equation for electron tunnelling through square well potential changes has been expanded from 1 to 6 (6) changes to 1 to 8 changes. In essence this solves the related problem for 8 changes per unit cell in a one dimensional crystal. A variety of new periodic tunnel switches were discussed in addition to forms of tunnel monochromators. Molecular soliton control gates for tunnel switches were discussed.

REFERENCES

1. F.L. Carter in the NRL Program on Electroactive Polymers; a. First Annual Report, Ed. Luther B. Lockhart, Jr., NRL Memo Report 3960, p. 121; b. Second Annual Report, Ed. Robert B. Fox, NRL Memo Report 4335, p. 35.

2. E.A. Pschenichnov, Soviet Physics - Solid State, 4, 819 (1962).

3. F.L. Carter, "From Electroactive Polymers to the Molecular Electronic Device Computer", in "VLSI-Through the 80's and Beyond", Eds. D. McGreevy and K. Pickar, IEEE Computer Soc. Wash. DC, Cat. EHO192-5, 1982.

4. J.R. Barker, "Quantum Ballistic Transport and Tunnelling in Molecular Scale Devices", in Molecular Electronic Devices II, Ed. F.L. Carter, Marcel Dekker, Inc., 1987.

5. F.L. Carter, Physica 10D(1984) 175-194.

6. F.L. Carter, in "Electron Tunnelling in Short Periodic Arrays", in Molecular Electronic Devices, Ed. F.L. Carter, Marcel Dekker, Inc., New York, 1982.

7. F.L. Carter, "On Deriving Density of State Information from Chemical Bond Considerations", in Electronic Density of States, Ed. L.H. Bennett, N.B.S. Spec. Publ. 323, 1971, p. 85.

8. F.L. Carter, "Valence Bonding in Some Refractory Transition Metal Compounds with High Coordination", in Proc. of 5th Materials Research Symposium, Eds. R.S. Roth and S.J. Schneiner, Jr., N.B.S. Spec. Publ. 364, 1972, p. 515.

38

Closing Remarks: New Directions for Chemists Toward Modular Chemistry and Molecular Lithography

Forrest L. Carter / Chemistry Division, Code 6170 Naval Research
Laboratory, Washington, DC

The future of Molecular Electronics necessarily lies in the hands
of those who will fabricate the ultimate devices as well as the
components for concept testing. While the fabrication skills
required for the development of this Molecular Technology (1)
will be those of physicists (as in molecular beam epitaxy),
chemists, molecular biologists, it is the object of this paper to
stress the need for the development of a new chemistry. In
short, we need to learn the principles of, and develop a practice
for the sharpest tools available, namely the atomic and molecular forces. The skillful control of these forces will produce
a revolution in chemistry and its applications.

After a short listing of MED problems we will make modest
suggestions of how some fabrication problems might be approached
via "modular" chemistry. "Molecular wires" prepared by modular
chemistry is then used as an approach to a discussion of mole-

cular communication, in this case bridging the gap between
microscopic leads and molecular functional groups. As an example
of the importance and use of molecular forces in the self-organization of composite nanostructures we will extend the ideas of
Stillinger and Wasserman (2). Another suggestion demonstrating
the use of molecular forces will concern itself with the development of lithography at the molecular size scale. The final
section will take note of other new developing chemical techniques of potential importance to the fabrication of molecular
electronic devices.

THE CHALLENGE OF MED PROBLEMS

It is difficult not to be overwhelmed by the magnitude and
complexity of the problems (Table I) facing molecular electronics
when one recalls that the development of the modern semiconductor
computer required more than 100K manyears. On the other hand,
the viability and existence proof for much more complicated
molecular electronic and biochemical device systems is
demonstrated and provided by every sentient being with whom we
interact.

That the challenge of the fabrication of molecular electronic systems can be met in innumerable ways is celebrated by
our cultural recognition and semantic designation of such
biological systems (persons) by the name <u>individuals</u>. (It is
marvelous to think that these individual units can interact to
form a much more complex system of unending variations called
societies.)

Table 1 provides a very abbreviated list of MED problems.
The courage to comtemplate the simultaneous solution of all these
major problems is drawn from the above recognition that there
exists not just one but an infinite variety of solutions. In
listing only self-organization and self-synthesis under Fabrication heading of Table 1 emphasizes again that standard chemical

TABLE 1.   MED Problems

1. Fabrication

    a.   Self-organization

    b.   Self-synthesis

2. Communication

    a.   Molecular addressability

    b.   Redundancy (e.g., Single or Multiple Gates)

    c.   Local or Global Interconnects

3. Reliability

    a.   Statistics

    b.   Quantum Effects

    c.   Radiation

4. Architecture

    a.   Molecular Structure

    b.   Logic Structure

    c.   Programmatic

        Self-organization

        Self-correction

5. Heat Build-up

6. I/O

---

approaches are not going to be practical in the preparation of
MED systems. Instead we must ultimately learn from the biological world how all the problems of Table 1 can be dealt with
simultaneously. In the long run it is inappropriate to isolate
the synthesis prblem from those of structural organization. That
is, morphology is seen to be important, even critical, at every

level of organization, (see Table 1). However, in order to make
a start one must step from the present with current tools.

MODULAR CHEMISTRY

The concept of modular chemistry takes the Merrifield synthesis
of polypeptides (3) as its source of inspiration. As in Merrifield technique in the modular approach complex structures are
built up step by step adding single or modular units one at a
time. In the Merrifield synthesis this is achieved through a
slow and painstaking series of blocking reactions to exactly
control the course of reaction. Before suggesting a modular
chemical approach to the synthesis of "molecular wires" in
inorganic and organic systems it is interesting to look even
further in the future. Ultimately, it will be better to model
the fabrication after self-synthesis in nature where polypeptides
are synthesized at a great rate from a soup of components.
Currently, our understanding is insufficient to seriously
contemplate this, however, self-organization of an insulating
layer from a soup is discussed below. In passing we note that
protein synthesis is not only much, much faster in nature but it
is also in the opposite direction to that of the Merrifield
synthesis in which the reactive end of the growing protein is a
-COOH group.

The modular synthesis of the conducting inorganic polymer
polysulfur nitride $(SN)_x$ borrows from the oligomeric synthesis of
$(SN)_x$ by J. Milliken (4). This reaction involves the reaction of
sulfur diimide (I), that is, $(CH_3)_3SiNSNSi(CH_3)_3$ with sulfur
dichloride in a stepwise fashion as indicated in Fig. 1. The
bulk silicon surface in Fig. 1a is presumed to have undergone a
surface modification reaction, perhaps by bombardment with SCl_2,
leaving behind a coating of $\equiv Si-SCl$ on the silicon surface. The
coating is then reacted with the diimide I, Fig. 1a, to split out
gaseous $ClSi(CH_3)$, which is removed from the system by pumping.

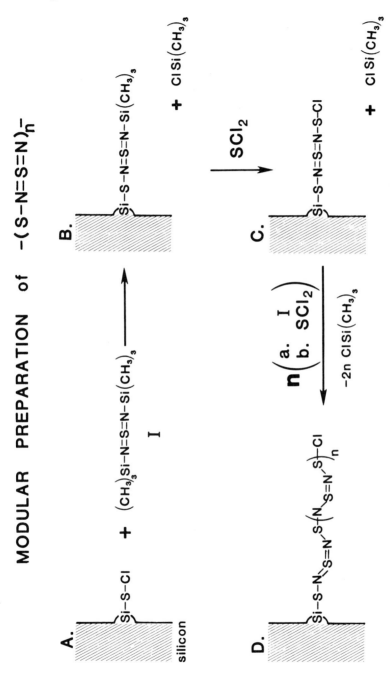

FIG. 1. A principal advantage of this proposed synthesis of $(SN)_x$ is that the maximum length of the "molecular wire" can be controlled exactly, in this case, to n+1 $(SN)_2$ units.

Now the second important step is the reaction of this modified
surface (Fig. 1b) with SCl_2 to give a short chain (S_2N_2) terminated by the -SCl group (Fig. 1c). But this termination is the
same as on the initial silicon surface, Fig. 1a. In other words,
the pair of reactions, surface treatment with the diimide
followed by SCl_2, provided one $-S_2N_2-$ modular unit. By repeating
this sequence, first treatment with the diimide followed by
separate treatment with SCl_2, n times provides n modular units of
$-(N_2S_2)-$. In this way the exact maximum length of the 'molecular
wire' can be precisely controlled. (However, this says nothing
about the configuration in space or its morphology.)

The chemistry of sulfur-nitrogen compounds is strikingly
versatile: it would be surprising if other examples of modular
chemistry could not be devised. In fact, the presence of ionic
compounds among them suggests that the electrochemical technique
suggested earlier (5) might be applicable. In that proposal, a
single modular ionic reaction was controlled by a positive
electric potential and the second ionic reaction was initiated by
reversing the potential such that a series of alternating
reversals resulted in a structure such as -ABABABAB-. In this
case the length of the nanostructure was controlled by the number
of potential reversals. A crude morphological structuring might
also be achieved in a chemical cell via the use of electric field
to stretch out a charged polymer chain (5).

MICRO-TO-MOLECULAR SCALE COMMUNICATION

A challenge inherent in the question of molecular addressability
is that of the communication between the microscopic world and
the molecular size world. This topic can be discussed with the
aid of Fig. 2a and b where the $(SN)_x$ chain attached to the
macroscopic silicon lead has been terminated with chromophore.
This chromophore is an electron donor and in a electrochemical
cell under a potential, might be expected to lose an electron to

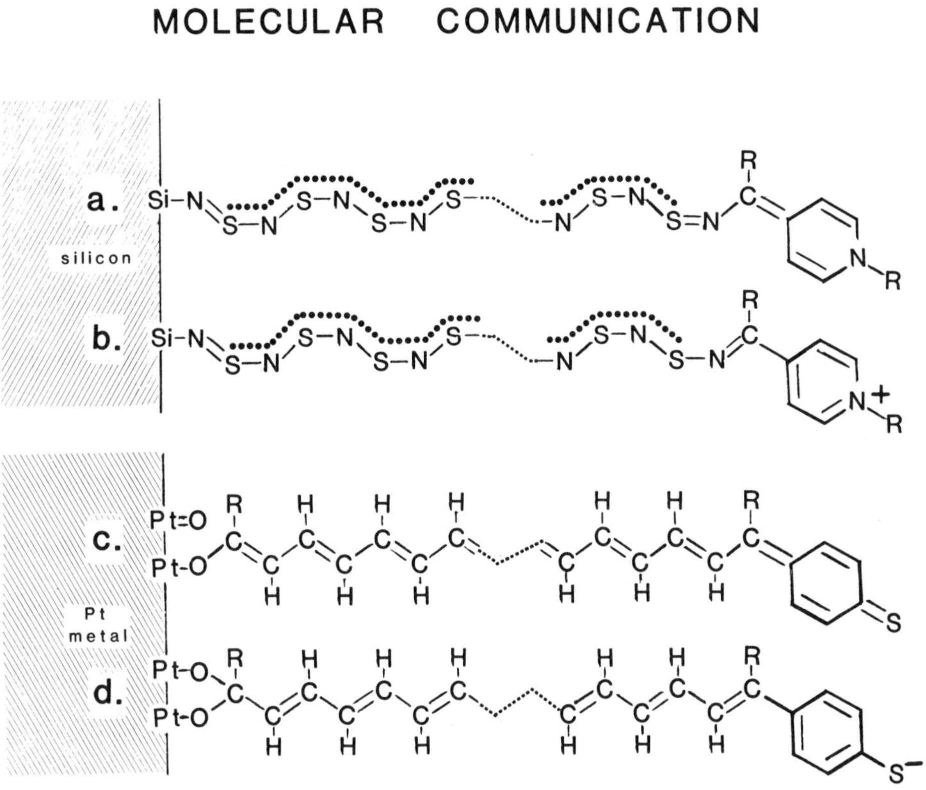

MOLECULAR   COMMUNICATION

FIG. 2. Molecular wires of polysulfur nitride and trans-polyacetylene are seen as communication links between microscopic
leads and chargeable molecular groups or chromophores.

the silicon substrate as in Fig. 2b. One may establish that the
molecular-size chromophore has in fact communicated with the
silicon substrate by observing the adsorption spectrum change of
the chromophore between the two cases Fig. 2a and Fig. 2b. Since
this proposed experiment can be performed on microscale (two-
dimensional) such an adsorption spectrum change would confirm
microscopic to molecular scale communication. However, this
would not, of course, demonstrate molecular addressability. In
order to achieve addressability molecular switches must be
inserted between the macroscopic world and an array of attached

chromophores. Such switches, for example employing electron
tunneling or soliton phenomena, are discussed elsewhere, (6) or
(9), respectively.

The first modular reactions in the growth of 'molecular
wires' on microscopic leads as in Fig. 2a and Fig. 2c could be
discussed as part of the new developing field of surface modification chemistry. The covalent bonds shown in Figs. 2c and 2d
linking Pt metal to the conjugated transpolyacetylene system is a
simple representative of how such an interconnect might be
accomplished. Another route might be to go through a sulfur-
containing chelate. Signal (i.e., charge) transport in the
'molecular wires' of Figs. 2c and 2d is rather different than in
the case above. In Figs. 2a and 2b the signal travels at a
fraction of the velocity of light whereas in the polyacetylene
case, Figs. 2c and 2d, the signal travels at a fraction of the
velocity of sound. Modular chemistry can be used to build up
polyacetylene molecular wires (8), however such conjugated
systems are likely to have some aliphatic R groups periodically
replacing some of the hydrogens in their $-(CH=CH)_x-$chains.

SELF-ASSEMBLY IN A NON-BIOLOGICAL SYSTEM

The fabrication of sophisticated molecular electronic devices
(9,10) will require degrees of self-synthesis and self-organization appreciably beyond current capabilities. While these
advanced fabrication techniques might best be learned by the
study of biological systems, it is rather doubtful if the
application of such techniques in the assembly of molecular
computers will be limited to the biological regime. Accordingly,
in the assembly of nanostructures, it is appropriate to consider
other starting points as well. What might be achieved is
suggested by the pair potential calculations of dimer formation
in a paper entitled, "Molecular Recognition and Self-Organization
in Fluorinated Hydrocarbons", by Stillinger and Wasserman (2).

Their pair potential calculations were carried out fluorinated derivatives of fused ring systems such as perhydrocoronene. In the fluorine, free prototype, Fig. 3a, 12 of 36
hydrogens project out in the plane of the carbon ring, 12
hydrogens project up perpendicular to the plane (small solid
dots) and 12 project down (unmarked). For each side fluorine can
replace hydrogen in 1376 different ways (corresponding from 0 to
12 fluorines). Fig. 3b illustrates a fluorinated perhydrocoronene (FPHC) containing five fluorines (heavy solid dots).
This isomer is labeled A while its conjugate in the words of
Stillinger and Wasserman (S&W) is labeled $\bar{A}$. Pair calculations
show that an FPHC with an A side fits one with an $\bar{A}$ face
perfectly to yield a deep pair potential. However, an FPHC with
some other fluorine count or distribution give with a shallow
potential minimum if one exists at all. Now since each FPHC has
two faces, S&W pointed out that a series like $(F/A)(\bar{A}/B)(\bar{B}/C)$
$(\bar{C}/D)$ is self organizing into a crystalline like stack.
Here each face A, B or C is assumed to be attached only by its
conjugate face $\bar{A}$, $\bar{B}$ or $\bar{C}$.

Using these concepts of self-organization we had extended
the ideas of S&W and indicated how a prescribed thickness of
insulation or a definite length of insulating "spaghetti" can be
obtained from a single solution (10) as below.

Assume there is a substrate surface (either organic or
inorganic) upon which an FPHC isomer $\bar{A}$ deposits on epitaxially.
Such a substrate surface labeled A' must be similar in some ways
to the A face of a FPHC isomer. Then one could make up a single
solution of four FPHC molecules with the faces indicated by
$(\bar{A}/B)$, $(\bar{B}/C)$, $(\bar{C}/D)$, and $(\bar{D}/E)$. Then using such a
solution an insulating layer of exactly four molecules thick and
no more could be put on the A' substrate. Since the top layer is
specified by E, for which no conjugates exists in solution, four
and four only layers would be formed even if excess components
were available.

CONJUGATE PAIRING and SELF-ASSEMBLY

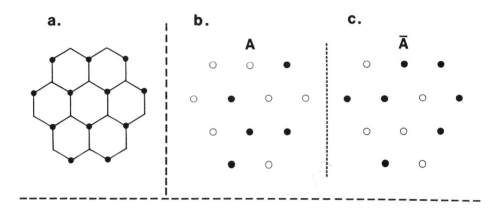

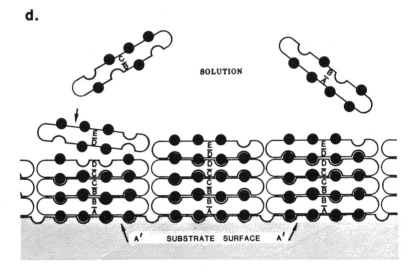

FIG 3. Of the 36 C-H bonds in perhydrocoronene (Fig. 3a), 12 project up (solid dots), 12 project down, and 12 are in the plane. In Fig. 3b, 5 of the 12 hydrogens on the top side have been replaced by fluorine to form one member, A, of a conjugate pair; while Fig. 3c illustrates the fluorine arrangement, $\overline{A}$ in the other half of the conjugate pair. Fig. 3d suggests the use of conjugate pairs to form a self-organizing layer exactly four molecules thick.

The fabrication of microtubular with pairs of FPHC molecules linked by an oxymethylene bridge has been described by S&W (2). Such an insulating "molecular spaghetti" (see Fig. 4) might be useful for isolating strands of conducting $(SN)_X$. In the same way as was described for the insulating layer above a "pre-organized" solution can be prepared of bridged FPHC pairs for assembling a predetermined length of insulating "molecular spaghetti".

While the two examples above are very simple they should open our eyes as to what is possible through molecular engineering, that is by using rather weak forces in the design and assembly of nanocomposite structures. Paying close attention to packing considerations provides a useful step of fabrication beyond the Merrifield approach of molecule-by-molecule buildup and begins to approach what must happen in some biological systems.

MOLECULAR   SPAGHETTI

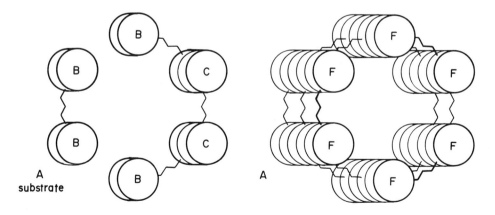

FIG. 4. Conjugate pairing can be used with bridged fluorinated perhydrocoronene pairs to form controlled lengths of insulating "molecular spaghetti." Adapted from Ref. (2).

LITHOGRAPHY FROM THE ATOM UP

A new approach to the fabrication of fine scale material architectures is clearly needed not only for the near term improvement
of semiconductor lithography but certainly for the future
development of molecular electronics. The technique proposed
here is to again use atomic and molecular scale forces to build
up devices rather than extend the semiconductor technological
approach of sculpturing out of crystalling material a pattern
reduced from macroscopic dimensions. In short, we propose that
lithography be accomplished from the atom up and in this section
we propose a modest point of departure along this new path.

The following approach leads to an epitaxial superstructure
on a semiconductor like crystalline silicon or gallium arsenide.
The idea suggested at a 1982 NATO Institute on Microelectronics-
Structures and Complexity (11) was that a large protein molecule
B could be found or prepared by recombinant DNA techniques that
would fit epitaxially on a (100) surface of silicon but not cover
the surface completely as illustrated in Fig. 5a and 5b. Since
the B molecules are not completely area filling they will leave a
systematic array of vacant areas, labeled A in Fig. 5, which are
not covered or masked out by the interlocking B molecules.
Various techniques might be used to fill the A areas such as
evaporation, sputtering, electrochemical deposition or cyto-
chemical staining. After the A areas are filled, weak bonds in
the B molecules are photolytically destroyed allowing the B
molecule fragments to be washed away. The resulting superstructure is illustrated in Fig. 5c. The A areas are accurately
centered within atomic dimensions.

The naive conception of the B molecule is pictured in Fig.
6c which suggest that the B molecule forms an interlocking
network with its neighbors and is composed of four parts joined
by photolysisable bonds. Recent advances in recombinant DNA
techniques suggests that if such a material is not currently

MOLECULAR EPITAXIAL SUPERSTRUCTURING

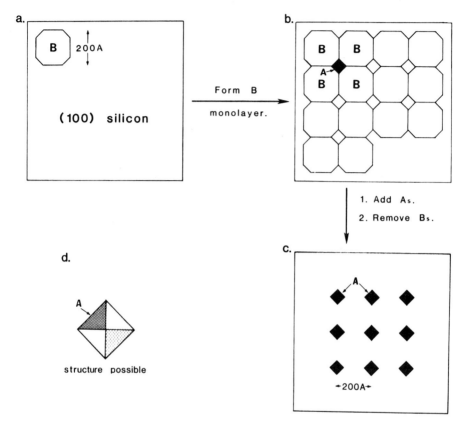

FIG. 5. Lithography at the molecular size level can be obtained with atomic precision via the epitaxial deposition of
large B molecules which form a mask on a single crystal as in
Fig. 5a. The pattern superstructure shown in Fig. 5c is obtained by filling in the holes of Fig. 5b and then removing
the epitaxial B layer.

available one could be bioengineered (12). As if to substantiate
the above 1982 speculation (11) K. Ulmer has discussed at this
yearshop (13) self-organizing protein monolayers drawing particular attention to the influenza virus neuraminidase. This
compound is very interesting in that it: i) forms a two-dimen-

sional monolayer, ii) which monolayer has an appearance very much like Fig. 5b but with dimensions 100Å center to center rather than 200Å as in Fig. 5, iii) which monolayer is composed of box shaped spore heads easily decomposed into four parts as suggested by Fig. 6.

Until the above approach to molecular lithography is demonstrated more directly, Ulmer's support via the neuraminidase monolayer is encouraging enough to make this approach worthy of serious consideration.

MOLECULAR   LITHOGRAPHIC
DETAILS

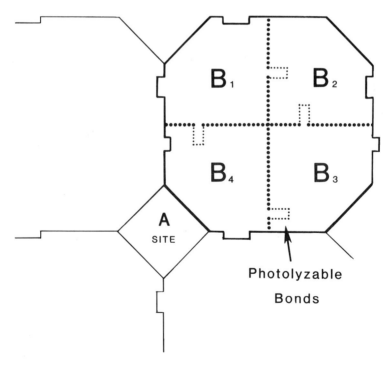

FIG. 6. The large B molecule with interlocking edges might be prepared via recombinant DNA techniques. It should be designed for easy removal after the A sites have been filled in or decorated.

NEW CHEMISTRY

In a strong measure, the future of molecular electronics is in new chemical techniques currently being developed or to be imagined in the future and developed. Table 2 summarizes some promising techniques and areas but Table 2 is not meant to be exclusive. In general, only a few words will be added in consideration.

The importance of learning the principles of self-organization and self-synthesis is difficult to exaggerate. However, it seems optimistic to think that an appropriate long term basic program will be initiated in this current era of directed research and zero sum budgeting. In this regard, the interesting and more immediate calculations as per Stillinger and Wasserman

Table 2.   New Chemistry to be Developed
_____

* Learn principles of self-organization and self-synthesis from biological systems.

* Extend self-organization computations as per Stillinger and Wasserman.

* Surface Modification Chemistry, principles and practice.

* Very small electrochemical cells.

* Laser-catalyzed synthesis.

* Recombinant DNA techniques.

* Langmuir-Blodgett film techniques.

* Surface Acoustic Wave (SAW) techniques.
_____

will both fare better and still be most instructive. Surface
modification chemistry either in its own right or viewed as a
component of modular chemistry is of great importance to molecular electronics as well as a variety of more macro chemical
phenomenon.

The demonstrated release of dopamine by an electric field in
Miller's and Lau's surface modification study (15) has important
implications linking molecular electronics and biological
(neural) systems.

While very small electrochemical cells are currently
employed primarily in analysis (15) their adaptation to fabrication of nanocomposites might be associated with the following
advantage: i) rapid exchange of solution; ii) use of a potential
to drive electric cell reactions to completion; iii) the alteration of potential in order to perform modular chemistry using
reagents of different charges as per ref. (5); and iv) use of the
potential to achieve a coarse morphological structuring.

Examples of the next entry Laser-Catalyzed Synthesis are now
common in the literature and the importance of recombinant DNA
techniques (12,13) and Langmuir-Blodgett film techniques (16,17)
have been stressed at this proceedings.

The last entry in Table 2, i.e. surface acoustic wave (SAW)
techniques in chemical sensors has been briefly discussed by H.
Wohltjen at this proceedings (18), as well as more thoroughly in
the literature (see ref. 19). SAW techniques are important to
molecular electronics: first, because a feasible extension of the
technique would permit one to monitor the progress of a reaction
at the monolayer level (19), and second, surface acoustic waves
could be employed to pump solutions through small electrochemical
cells.

While all of the above chemical techniques are still in the
developing stages, they, and others to come should prove to be of
considerable use in the development of molecular electronics.

CONCLUSIONS

In bringing to a close this paper and hence the 2nd International
Workshop on Molecular Electronic Devices, let me extend a
challenge to the varied imaginative scientists whose surprising
synthetic skills have brought us so far in the organic,
biomolecular, and semiconductor world in the last 35 years. Can
we devise a new chemistry or new fabrication techniques using
molecular forces and self-organization and self-synthesis
effects to achieve in a small way in the organic and inorganic
regimes what nature achieves with wonderful complexity in a
massive scale in the biological world? Synthetic chemists have
now shown (often with much effort) that virtually any compound
which can be drawn on paper can be synthesized in the laboratory
(at least with a poor yield). If this remarkable feat could be
duplicated with near perfect yields, a giant move to closure with
nature would have been achieved and a viable molecular
electronics would be within our grasp.

REFERENCES

1. K.E. Drexler, "Molecular Machinery and Molecular Electronic
   Devices", in these proceedings.

2. F.H. Stillinger and Z. Wasserman, J. Phys. Chem. 82 (1978),
   929.

3. C. Birr, Aspects of the Merrifield Peptide Synthesis,
   Springer-Verlag, Berlin, Heidelberg, New York, 1978.

4. J.W. Milliken, "Metallic Covalent Polymers: Derivatives of
   $(SN)_x$ and $(CH)_x$", Chemistry, Ph.D Dissertation, Univ. of PA,
   1980.

5. F.L. Carter, "The Chemistry of Future Molelcular Computers",
   in Computer Applications in Chemistry, Eds. S.R. Heller and
   R. Potenzone, Jr., Elsevier Science Publishers B.V.,
   Amsterdam, 1983, p.225.

6. F.L. Carter, "Electron Tunnelling in Short Periodic Arrays", in Molecular Electronc Devices, Ed. F.L. Carter, Marcel Dekker, Inc., New York, 1982, p. 121.

7. F.L. Carter, Physica 10D (1984) 175-194.

8. Long chain conjugated systems have been made prepared by condensation reactions or via the Wittig reaction, see H.R. Brakmana, K. Katsuyama, J. Inamaga, T. Katsuki, and M. Yamaguchi, Tetrahedral Letts. 22 18, 1696 (1981).

9. F.L. Carter, "From Electroactive Polymers to the Molecular Electronic Device Computer", in VLSI Technologies: Through the 80's and Beyond, Eds. D.V. Greivy and K.A. Pickar, IEEE Computer Society Press, No. EHO192-5, p. 328.

10. F.L. Carter, "Further Considerations on 'Molecular' Electronic Devices", in Second Annual Report, NRL Program on Electroactive Polymers, Ed. R.B. Fox, NRL Memorandum Report 4335.

11. F.L. Carter, "Toward Computing at the Molecular Level", NATO Advanced Research Institute at Les Deux Alpes, France, 22-26 March 1982, in Microcircuitry-Structure and Complexity, Ed. K. Dingle, in press; see also J. Vac. Sci. Technol. B1 (4), Oct.-Dec. 1983.

12. K.M. Ulmer, Science, 219, 666, 1983.

13. K.M. Ulmer, "Self-Organizing Protein Monolayers as Substrates for Molecular Device Fabrication", in this proceedings.

14. L.L. Miller and A.N.K. Lau, "Chemical Communication Involving Electrically Stimulated Release of Chemicals from a Surface", in this proceedings.

15. C.E. Lante, P.T. Kissinger, and R.E. Stroup, Anal. Chem. 57 (8), 1541 (1985).

16. H. Kuhn, "Self-Organizing Molecular Electronic Devices?", in this proceedings.

17. F.L. Carter, A. Schultz and D. Duckworth, "Soliton Switching and Its Implications for Molecular Electronics", in this proceedings.

18. H. Wohltjen, "Some Thoughts on Molecular Electronic Devices and Chemical Sensors", in this proceedings.

19. A. Snow and H. Wohltjen, Anal. Chem. (1984) 1411.

39

**Reliability in High Density Hierarchical Devices: Possible
Lessons from Neutral Systems**

P.E. Rapp / Department of Physiology and Biochemistry, The
Medical College of Pennsylvania, 3300 Henry Avenue, Philadelphia,
PA

ABSTRACT

Increased use of advanced technologies will greatly accelerate
the transition to electronic control systems that will be
characterized by greater density and increased speed. Perhaps of
even greater importance from the point of view of dynamical
stability considerations will be the placement of these devices
in increasingly complex, hierarchical, parallel processing
networks. Simply stated, these systems are becoming more
biological.

It is therefore possible to ask if future generations of
control networks will be vulnerable to forms of failure
previously only observed in biological systems, such as the
mammalian central nervous system.

This paper considers the possibility that complex electronic
networks may display a failure mode analogous to a convulsion.
It is explicitly recognized that present understanding of

dynamical failure in large systems, both biological and
technological, is very incomplete. Available evidence is
indirect and inconclusive. However, the consequences of a
convulsive failure in a military control system could be grave.
For this reason, the possibility, even if seemingly remote,
merits examination.

I.   INTRODUCTION

Very large scale integration offers the prospect of ultra-high
density, parallel processing systems. A cautionary question
should be considered: are there reliability problems unique to
ultra-dense systems? The only functioning systems of this type
presently available for examination are biological. The
mammalian central nervous system provides a unique laboratory for
assessing the dynamical reliability of the engineering
technologies that will become available during the next ten
years. Thus, the original question can be restated: do neural
systems present examples of failure modes that may become more
prominent in engineering systems as they approach biological
complexity? A corollary question that can also be asked is, does
the examination of neural systems suggest design protocols that
could improve the robustness of engineering devices?

     Classically, the studies of failure in neural systems
followed a pattern established by clinical practice. Disorders
were classified according to their presentation in the clinic as
movement disorders, auditory disturbances, seizure disorders and
the like. Recently however, efforts have been made to establish
classifications of neural function based on the underlying
topology of the dynamical behavior. The introduction of the
methods of dynamical systems theory to neurophysiology is
beginning to produce some encouraging results. From the medical
perspective, this program has the virtue of indicating that
seemingly very different disorders are dynamically similar and
are the consequence of equivalent control failures. From a more

general perspective, mathematical investigations have the virtue of producing results that are at least potentially generalizable to engineering systems.

As a specific example, a highly irregular form of behavior that is termed chaos in the topological literature will be described. Evidence suggesting that some convulsive disorders could be the result of transitions to this behavior will be summarized, and the possible parallels to engineering systems will be considered.

II. FAILURE IN THE HUMAN CENTRAL NERVOUS SYSTEM

IIa. The Brain as a Control Device

Viewed as an information processing and signal transmitting device, the human brain is extraordinary. The basic component of the system is the neuron. There are about 10^{11} neurons per brain. This is roughly equal to the number of stars in our galaxy (1). Even when considered individually, neurons are dynamically complex, highly nonlinear devices. Some of this complexity is suggested in Fig. 1.

Not only is the behavior of the individual neurons complex, they form highly interconnected circuits. In the human, a typical central nervous system neuron makes synaptic connections with 10^3 to 10^4 other neurons (2). This geometric complexity is reflected in the associated behavior. In engineering terms this behavior is characterized by the extensive use of parallel processing and by hierarchical networks displaying a large measure of autonomy at each level. It is therefore not surprising to learn that the human brain has the highest energy requirement per unit weight of any biological tissue.

To summarize, the human brain has the following systems characteristics: a) Individual elements are highly nonlinear, b) Elements are densely packed; c) The network has a highly complex, hierarchical, parallel processing circuit geometry; (d) Large amounts of energy per unit volume move through the system.

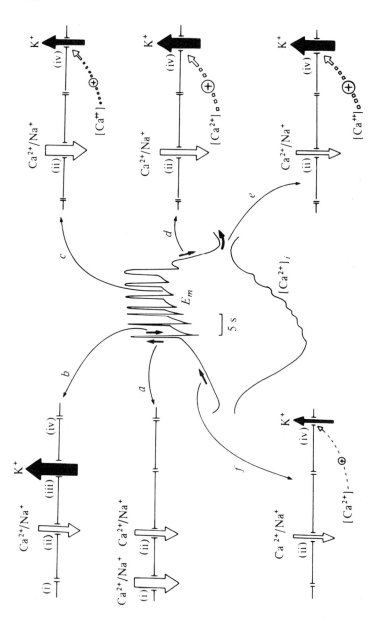

FIGURE 1. The burst of a molluscan neuron results from the interaction of several coupled and nonlinear processes. Downward arrows denote ionic flows into the cell, and upward arrows denote movements out of the cell. An approxiamte indication of the magnitude of the current is given by the thickness of the arrow (70).

Thus, from an engineering point of view, the human brain would seem to be a prime candidate for displaying some form of dynamical instability. The fact that the brain works as well as it does indicates that given enough time, evolutionary selection is an incredibly effective engineering design procedure. However, neural systems do fail. Indeed, they do so with a high frequency. An important distinguishing feature of biological control systems is that, usually, they fail soft; typically the consequences of a failure are confined to a local hierarchical level and do not have a dramatic effect on overall system performance. This is in marked contrast to engineering systems. However, the efficient implementation of hierarchical structures in engineering systems evidently presents formidable technical difficulties (3,4). Biological systems have solved these problems. An understanding of how hierarchical biological networks are structured and why they are so effective in confining most failures to soft failure modes could provide valuable engineering design protocols.

Hard failures do occur in the human central nervous system. They are far more frequent than is commonly supposed. Usually these failures are sufficiently dramatic that they lead to immediate medical intervention. It is one of the theses of this paper that viewing these pathologies as control system failures provides a useful paradigm for developing strategies for responding to these defects. A specific class of hard failure, the epilepsies, will be considered.

IIb. The Epilepsies: Hard Failure in a Dynamical System

It is now recognized that epilepsy is not strictly a disease but rather a collection of symptoms characteristic of a class of convulsive disorders which is observed throughout the world in about 1% of the population. The two principal subcategories are generalized epilepsy and partial (or focal) epilepsy (5).

Generalized epileptic convulsions can be subdivided into grand mal and petit mal types. A grand mal convulsion (abrupt loss of consciousness and generalized convulsion) is often preceded by a warning sensation or hallucination called an aura. The nature of the aura corresponds to the affected region of the brain where the disturbance responsible for the seizure is initiated. The subsequent convulsion has two distinct phases. In the first, the tonic stage, muscles contract continuously, facial features are distorted, arms are flexed and the lower limbs are rigidly extended. The clonic stage is marked by alternating contraction and relaxation of muscles. Petit mal seizures are far less dramatic and consist of brief periods in which a patient seemingly stares into space (transient absence). There are no generalized convulsions.

Partial epileptic seizures include Jacksonian convulsions and the various forms of psychomotor epilepsy. A typical convulsion of this class begins as a muscular twitch in a specific part of the body. This location varies from patient to patient, but for any given individual it is almost invariably in the same location. The region of the brain that controls these muscles is the site where the disturbance begins. The disturbance may remain confined to that local focus or it may spread to adjacent areas of the brain in which case there is a corresponding spread in the muscular contractions. Provided that the electrical disturbance is confined to one hemisphere, the patient retains consciousness. Complex partial seizure disorders include psychomotor epilepsy. The patient experiences a feeling of unreality (often forms of conscious amnesia such as deja vu or deja entendu). During a confused period the patient may perform complex semi-purposive acts but will have no recollection of them. There is no generalized convulsion.

There is a large body of literature describing convulsions of various types. However, though we know a great deal about what a convulsion looks like, we know far less about what causes them. The cause of the disorder is suggested (strictly it is never proved) only about 50% of the time. Just as the cause of the disease is often unknown, the immediate trigger of a given convulsive episode is often undetermined. However, some fairly common triggering stimuli can be identified (5). As one would expect, there is considerable variation in sensitivity to different triggers from patient to patient. Triggers encountered in everyday environments include visual stimuli (flickering lights, viewing certain geometrical patterns and, for some individuals, television) and psychological factors. Triggering should be distinguished from activation. Activation procedures are specific methods used to elicit abnormal behavior during EEG recording sessions (6,7). Activation procedures include hyperventilation, photic stimulation, auditory activation, skin stimulation, sleep deprivation, hypoglycemia, administration of convusants, withdrawal of anticonvulsants and psychological activators specific to the individual.

As previously stated most epileptic seizures are the consequence of an electrical disturbance that begins at some discrete location in the brain. As suggested in Fig. 2, the subsequent course of the convulsion depends on whether or not the disturbance spreads to adjacent areas. If the disturbance spreads it can do so in one of two ways. Propagation across the cortex (Path 1 in the figure) results in a Jacksonian march. Spread by projections into the interior of the cerebrum (Path 2) can result in generalized convulsions.

The spread of an epileptic electrical disturbance can be observed in an EEG record. In Fig. 3, the experimental subject was an adult male Rhesus monkey. A chronic epileptogenic focus was artificially created by subpial (beneath the pia matter, a membrane that envelops the brain and spinal cord) injection of aluminum hyroxide. The records of the first four channels are normal sleep signals. The spikes in channels 5 to 8 are epileptogenic bursts (8).

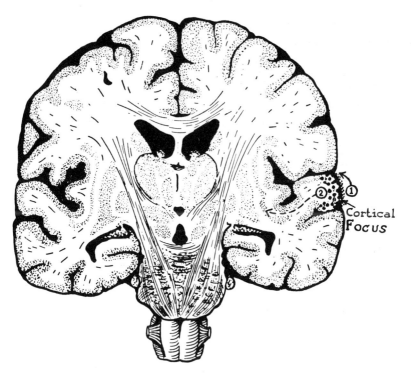

FIGURE 2. An electrical disturbance in a cortical epileptogenic
focus can either propagate into adjacent areas of the cortex
(Path 1) or subcortically (Path 2). Each form of propagation
results in different clinical observations (71, reproduced
courtesy Charles C Thomas, Publisher, Springfield, IL).

        Both the intracellular and extracellular electrical activity

within the epileptogenic lesion is correlated with the surface

EEG.   Fig. 4 shows the potential records in an artifically

created focus in an adult cat.   The time axis is divided into two

regions: the interictal period (between seizures, ictus = stroke)

and the ictal period (during and immediately following a

seizure).   The interictal period is further subdivided into an

interictal EEG paroxysm (paroxysm = convulsion) and a silent

period.   Spikes of this type are seen in the EEG's of epileptics

between seizures.   At the cellular level a depolarization termed

the paroxysmal depolarization shift is responsible for the

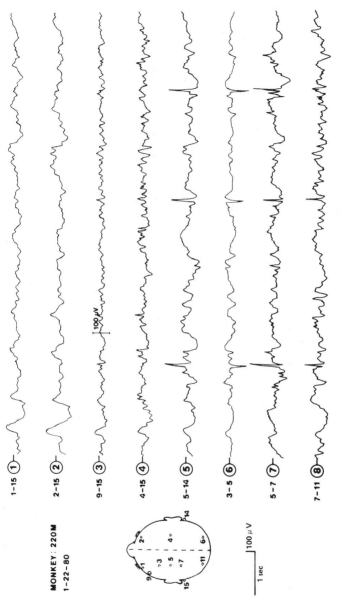

FIGURE 3. An EEG recording from a male rhesus monkey made chronically epileptic by subpial injection of aluminum hydroxide. Channels 1–4 show a sleep recording. Channels
5–8 include epileptiform bursts (8, reproduced courtesy Raven Press, New York, NY).

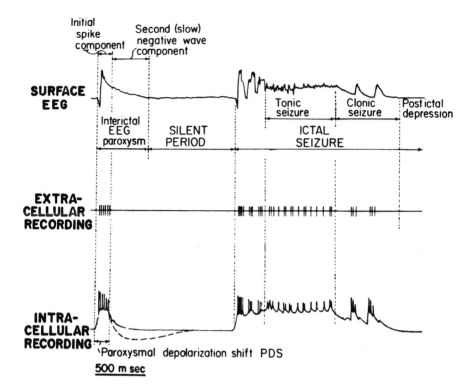

FIGURE 4. The relationship between surface EEG, extracellular
potential and intracellular membrane potential during seizure
activity produced in an adult cat by topical application of
penicillin (10).

electrical activity that is reflected in EEG spikes. It is
generally held that an understanding of this depolarization is
central to a understanding of epileptogenesis.

Theories of epileptogenesis can be roughly classified into
two groups. The first emphasizes processes intrinsic to
individual neutrons (9) and suggests that seizures are the
consequence of the activity of aberrant neurons. The second
group of models emphasizes the collective multicellular aspects
of neural behavior. In these models it is proposed that the
electrical disturbance is the result of activity in a network
(10,11). These views are not mutually exclusive, and it is

possible to construct a synthesis in which aberrant neurons
acting collectively in networks with abnormal connectivity and
transmission properties produce a propagating disturbance that
results in the observed seizure. Both views and their synthesis
are amenable to analysis by the mathematical methods described in
the next section.

III. TOPOLOGICAL ANALYSIS OF DYNAMICAL SYSTEMS

III.A. Flows and Attractors Defined by Differential Equations
Consider an n-dimensional autonomous nonlinear ordinary
differential equation

$$dx_1/dt = f_1(x_1 \ldots x_n, p_1 \ldots p_m)$$
$$dx_n/dt = f_n(x_1 \ldots x_n, p_1 \ldots p_m)$$

(The results presented here can be stated in more general terms.
This differential equation is general enough to display the
behaviors that will be of importance in the subsequent
discussion). Time, t, is the independent variable. The x's are
the dependent variables. In neural models x's would represent
the electrical potential across cell membranes, membrane
currents, and some chemical concentrations. All of these
quantities have their analogs in physical systems. The
parameters $p_1 \ldots p_m$ assume a constant value for any given
case but vary from case to case. Often some of the parameters
are subject to experimental control. They can include quantities
such as temperature, the composition of extracellular fluid and
drug concentrations.

It is possible to think of a differential equation as a
purely numerical device. For a given set of input values
$x_1(o) \ldots x_n(o)$ and a given value of time, the equation generates
another set of numbers $x_1(t) \ldots x_n(t)$. An important alternative
view is contained in the geometric theory of differential

equations (12,13). The three dimensional example shown in Fig. 5
would correspond to a three dimensional differential equation (an
equation of the type stated above where n = 3). Each axis
corresponds to one of the dependent variables. A solution to the
equation is a parametric curve in time called a trajectory,
beginning at the point defined by the initial values and moving
through the three space as time increases. Each point in this
space (which is often referred to as the phase space or the
behavior space) is a possible initial point. The differential

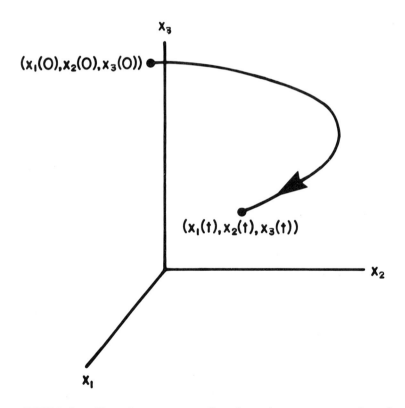

FIGURE 5. The time course of a dynamic system can be viewed as
the movement of a trajectory through a behavior space with time.
In this example, a single trajectory of a three-dimensional system is shown.

equation determines the movement of all of these points. Because
the concerted movement of solutions in behavior space is analogous to the flow of water through a region, the functions
defined by a differential equation that describe this motion are
called flows.

On first encountering this interpretation of a differential
equation it might seem that little insight is likely to come from
it since the trajectories could twist about in an unenlightening
arbitrarily complicated way. Typically, however, this is not the
case. The trajectories form smooth, well ordered flows precisely
because they must obey the law of motion which is expressed by a
differential equation.

The treatment of the solutions of differential equations as
geometric objects rather than as numerical processes, allows the
introduction of very powerful methods of topological analysis.
It is important to note that geometric analyses of modern
topology are not limited to the two and three dimensional
geometries of human experience. Though they are difficult to
obtain, many important topological results can be generalized to
arbitrary dimension. Even in higher dimensional systems the
geometric behaviors of flows are constrained because they must
satisfy the differential equation. This results in a
surprisingly limited number of behaviors. The most important is
the convergence of trajectories onto objects called attractors.

The most important and topologically simplest attractor is a
fixed point. A point in the behavior space is said to be an
attractor if all trajectories that begin close to the attractor
converge on it with increasing time. Dynamically, systems
governed by fixed point attractors display steady state behavior.
This includes control systems that converge on an equilibrium
after receiving a transient displacement.

The second classical attractor is the periodic attractor. A
famous two dimensional example (14) is shown in Fig. 6. Nearby
trajectories spiral to the closed loop attractor as time

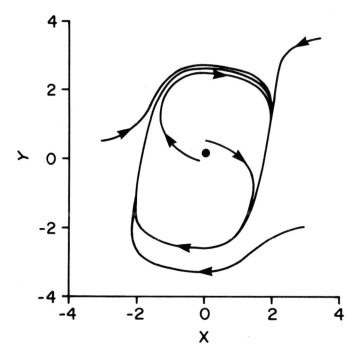

FIGURE 6. The periodic attractor of the van der Pol equation
(14) for μ=2.

$$dx/dt = y + \mu \ (x-x^3/3)$$
$$dy/dt=-x$$

increases. If the variables were plotted independently against
time, oscillations result (Fig. 7). The closed ring of the
attractor has the same character if it is drawn on a plane,
placed in a three dimensional space or embedded in a two
hundred-dimensional behavior space.

    The classical (early twentieth century) geometric theory of
differential equations was constructed almost entirely on the
topology of fixed point and periodic attractors. Subsequently, a
third class of attractor was identified among the solutions of
comparatively simple equations. These attractors have such
thoroughly disagreeable properties that they have been termed
strange attractors.

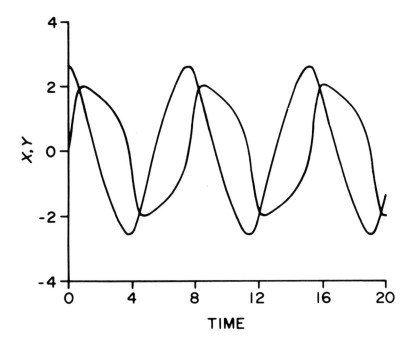

FIGURE 7. The periodic attractor of the van der Pol equation
plotted against time. The equations are those in Fig. 7.

A strange attractor is an attractor; all nearby solutions
converge on it. Further, it is not a fixed point, and it is not
periodic. The geometrical structure of strange attractors is
incompletely understood. Examination of some examples
established that these attractors have an exceptionally complex
form (15,16). However, simply having a very unusual shape does
not immediately establish the dynamical importance of strange
attractors. From the point of view of behavior, their
significance lies not simply in their shape, but in the motion of
trajectories as they approach the attractor. These solutions
display a sensitive dependence on initial conditions; that is, an
infinitesimal change in initial conditions results in solutions
that diverge exponentially. This is dramatically different from
the behavior of solutions approaching fixed point or periodic
attractors where solutions initially close remain close and

approach each other as they move to the attractor. Sensitive
dependence on initial conditions as the phrase is used here does
not mean that two specific solutions diverge; rather, all
solutions separate. This has been termed the blunderbus effect
by Mees and Sparrow (17), and results in a highly irregular
quasirandom motion which is called chaos in the mathematical
literature. This is an appropriate term. The degree of disorder
exhibited by chaotic solutions can be profound. For example, it
has been suggested that turbulent fluid motion is due to a
strange attractor (18,19,20). It should be stressed that this
irregular behavior occurs in the solutions of completely
deterministic differential equations. These systems do not
contain stochastic terms. It should also be noted that there are
other forms of deterministic chaos. Not all chaotic systems
contain strange attractors. Indeed, as the consequence of
continued research, the term is losing rather than gaining
specificity as the range of disordered systems, both mathematical
and experimental, increases.

III.B. Parameter Dependent Changes of Attractor Topology
Recall that the previously stated equations contain constant
parameters $p_1 \ldots p_m$. In electronic circuits the parameters
specify the resistances and capacitances. In fluid systems the
Reynolds number appears as a parameter, and in biological systems
the concentration of administered drugs is a parameter. The
behavior of the solutions to a differential equation is parameter
dependent. The number and topological class of a system's
attractors can change in response to small, theoretically
infinitesimal, changes in the numerical values of specific
parameters. Dramatic changes in dynamical behavior called
bifurcations result when a parameter value moves through a
critical value (a bifurcation value).

The simplest examples of bifurcations are the changes in the
stability characteristics of fixed point attractors. An example

modified from the ecological literature is shown in Fig. 8. As
the result of a change in the value of a parameter, solutions no
longer converge on the attracting fixed point at $(x_1, x_2) =$
$(1,1)$. That point ceases to be an attractor when a parameter is
increased through a critical threshold value. In this example,
all attractors both before and after bifurcation are fixed point
attractors. This need not always be the case. Bifurcations can
result in the appearance of periodic and strange attractors.

Just as a steady state system can undergo transitions to
oscillatory behavior, systems can also undergo bifurcations to
chaos. There is no single pattern of behavior in the transition
to chaotic motion. A number of known scenarios have been
collected by Eckmann (21). A common pattern is one in which
attracting periodic solutions become progressively more complex
as a parameter approaches a critical value. At that value the
periodic solution ceases to be attracting and chaotic solutions
are observed. This is reflected in the solution's spectrum. A
broadband background component abruptly appears in the spectrum
when the chaotic parameter domain is entered.

In summary, two important results should be retained from
this brief discussion of dynamical systems theory. (i). Fully
deterministic systems of ordinary differential equations can
display highly disordered dynamical behavior called chaos. (ii).
Differential equations can display parameter dependent
transitions to and from chaotic behavior. There is an additional
important observation that should be added to the discussion that
is not a mathematical result. It is a result of experimental
studies of both physical and biological systems. (iii). Dramatic
parameter dependent transitions to disordered behavior are not an
exclusively theoretical phenomenon observed only in the solutions
of unusual differential equations. Transitions of this type can
be reproducibly observed experimentally. The implications of
this statement will be explored in the next section.

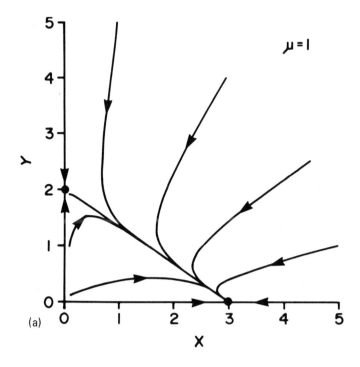

(a)

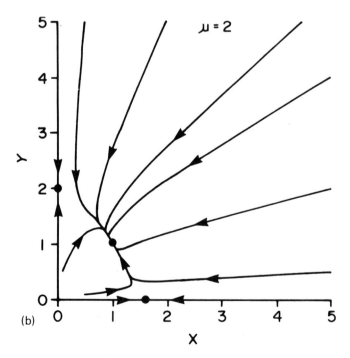

(b)

IV. BIFURACTION PHENOMENA IN BIOLOGICAL AND PHYSICAL SYSTEMS

IV.A. Speculations Concerning Epileptogenesis

When considering the dynamic etiology of an epileptic seizure, it is convenient to restate the problem as two sequential questions. First, is an epileptic episode the consequence of a bifurcation in a dynamical system? Second, is an epileptic episode the consequence of a bifurcation to chaotic behavior? It is explicitly recognized that neither of these questions can now be answered with certainty.

The defining characteristic of a bifurcation is an abrupt and dramatic change in the qualitative characteristics of behavior obtained reproducibly in response to a small change in the value of a system parameter. If epileptiform activity is bifurcation phenomenon, it should be possible to identify these bifurcation parameters and their critical values. A large experimental literature demonstrates that this is in fact possible. For example, Pedley and Traub (11) have listed some of the bifurcation parameters leading to epileptiform electrical activity in hippocampal slice preparations. (The hippocampus is on the inner surface of the temporal lobe of the cerebral cortex. Epilepsy in which the primary epileptogenic focus is in the temporal lobe is referred to as psychomotor epilepsy.) The bifurcation parameters include: (i) convulsant drugs such as

---

FIGURE 8. The effect of changing the value of a parameter on the geometry of a flow is illustrated by this system of differential equations modified from the ecology literature (72).

$$dx/dt = x(3 - \mu x - y)$$
$$dy/dt = y(2 - x - y)$$

In Case (a), $\mu=1$, there are three singular points. (Singular points are the solutions of the steady-state equations $dx/dt=0$ and $dy/dt=0$.) The points $(0,0)$ and $(o,2)$ are nonattracting. The singular point $(3,0)$ is a fixed point attractor. In Case (b), $\mu=2$, there are four singular points. Three, $(0,0)$, $(0,2)$, and $(3/2,0)$ are nonattracting. The point $(1,1)$ is a fixed point attractor.

penicillin and bicuculline that decrease the effectiveness of inhibitory neurotransmitters; (ii) drugs such as kainic acid that augment excitatory activity; (iii) lowering extracellular chloride (22); (iv) raising extracellular potassium (23) and (v) depolarization of cells in the CA_2-CA_3 region of the hippocampus in the presence of convulsants (24).

The dramatic parameter sensitive appearance of epileptiform electrical activity suggests that it results from a bifurcation in a nonlinear system, but this does not demonstrate that this is a chaotic bifurcation. Indeed, though preliminary theoretical and experimental results suggest that neural systems are capable of generating chaotic signals (25,26; and references in the following discussion), the implicit preliminary question "are neuronal systems capable of autonomous chaotic behavior?" has not been resolved in a definitive manner. The involvement of chaotic dynamics in epileptogenesis is even less certain.

Holden (27,28) has reported a drug dependent transition to chaotic behavior analogous to the convulsant dependent transitions to epileptiform episodes. He has conjectured that the procedure forces the system controlling neuron membrane potential to undergo multiple bifurcations that result in the appearance of a strange attractor and deterministic chaos. However, demonstrating that an irregular signal is chaotic introduces a number of complex technical problems (29). While Holden's results are highly suggestive, the signal has not been proved to be chaotic in the strict sense of the term. The present author believes that Holden's result is the first experimental evidence explicitly supporting a connection between a known convulsant agent and chaotic neural activity. However, the mechanistic relation of irregular, possibly chaotic, behavior in a molluscan neuron to seizures in the mammalian central nervous sytem has not been established.

The difficulties of experimental work have motivated several preliminary theoretical studies. Carpenter's investigations

consider behavior at the level of a single cell. Studies by
Kaczmarek, Mackey, Traub and their respective colleagues consider
the behavior of small networks. Results of these studies that
are relevant to an analysis of the etiology of epileptic seizures
will now be summarized.

Carpenter (30,31) in an analysis of Hodgkin-Huxley models of
the neuron using singular perturbation theory demonstrated that
these systems can have chaotic solutions. The theoretical system
of Kaczmarek (32) consists of populations of excitatory and
inhibitory cells. The excitatory cells provide input to the
inhibitory cells and positive feedback to themselves. These
functional differential equations can undergo parameter dependent
bifurcations to firing patterns analogous to tonic and clonic
stages of some convulsions. No observations of chaos in the
first Kaczmarek model were reported.

In the second model (33) the cell properties are those of
the first model. The revised model, however, has a defined
spatial structure. Three network geometries were examined. In
the third sparsely connected network the strength of connection
decreases as a function of distance. In this system a parameter
dependent loss of regularity in the firing pattern resulted in
disordered behavior. The authors refer to this as chaotic, but
an explicit characterizatio of the system's attractor was not
attempted. Kaczmarek and Babloyantz speculated that seizures
might begin as chaotic disturbances and that the regular rhythmic
pattern of epileptic discharges results from the subsequent
recruitment of other neurons.

Mackey and an der Heiden's model (34,35) considers firing
stability in region CA_3 of the hippocampus. In the model
presynaptic fibers provide excitatory input to postsynaptic
pyramidal cells which in turn excite basket cell interneurons.
The basket cells send inhibitory signals to the pyramidal cells.
The Mackey-an der Heiden equations model a specific process, the
blockade of gamma-aminobutyric acid (GABA) receptors by

convulsants. Erratic, possibly chaotic, solutions appear as the number of GABA receptors decreases, but these calculations do not establish a causative link between chaotic behavior and seizure activity.

Traub (36,37) has constructed a very different mathematical model of hippocampal epileptogenesis. The basic component of the network is a model of a hippocampal pyramidal cell (38). These neural elements are used to form a 10x10 array. An epileptiform event is initiated by modeling the application of a constant depolarizing current (1.5 nA) to four adjacent cells. Computations resulted in an impressive degree of agreement between simulated cellular membrane potentials and extracellular field potentials and experimentally obtained records from penicillin treated guinea pig transverse hippocampal slices recorded in the CA_2-CA_3 region. The Traub system differs in an important aspect from the Mackey-an der Heiden system in that epileptiform behavior is not endogenously generated in response to penicillin treatment alone. A depolarizing stimulus is also required. It is possible to speculate that this spontaneous synchronized discharge in a small group of cells could be the result of a bifurcation in a local network.

Irregular behavior can occur as the result of periodic input signals. When an autonomous oscillator is subjected to a periodic input, the common response is N:M integer ratio phase locking in which N input cycles result in M oscillations of the driven oscillator. This behavior can be observed experimentally in biological systems subject to periodic stimulation. The most systematic investigations have been those of Glass and his colleagues using cultured embryonic chick heart cells (39,40). Period doubling bifurcations leading to chaos were predicted by their mathematical model and observed experimentally. Similar experiments have been performed in neural preparations (41,42,43).

On the basis of the preceding discussion, the following provisional conclusions can be made.

1. Transition from normal to convulsive behavior is
probably the result of a nonlinear bifurcation.

2. Neurons and neural networks are probably dynamically
capable of entering domains of chaotic behavior.

3. Though indirect theoretical and experimental evidence is

suggestive, there is at present no hard evidence indicating that
the bifurcation that results in a convulsion is a chaotic
bifurcation.

The evidence suggesting that chaotic behavior is possible in
physical systems is considered in the next section.

IV.B. Bifurcations to Chaos in Chemical and Physical Systems

Theoretical studies suggest that comparatively simple chemical
systems can display chaotic concentration functions (44,45).
Irregular concentration variation, that is suspected to be
chaotic, has been observed experimentally in a biochemical
reaction in which total enzyme concentration is the bifurcation
parameter (46). A more systematic experimental investigation of
chaotic chemical systems has been performed with the
Belousov-Zhabotinskii reaction. The constant flow rate through
the reaction vessel was used as a bifurcation parameter. Passage
through a region of periodic behavior to seemingly chaotic
dynamics is indicated by the appearances of a broadband component
in the spectrum of a concentration function (47,48). Similarly,
it has been proposed that the transition to turbulence in a fluid
system may be the result of bifurcations to chaos (49,50,51,52).

According to theoretical studies chaotic dynamics may be
possible in laser discharge (53) and pulse width modulated
voltage converters (54). Combined theoretical and experimental
studies have indicated that even comparatively simple electronic
circuits can be chaotic (55,56,57,58). The appearance of
broadband background noise in Josephson junction oscillators used
as parametric amplifiers has been attributed to the appearance of
a strange attractor and a transition to chaos (59,60).

The term solid state turbulence has been coined to include a
number of nonlinear phenomena in solid state systems. Huberman
and Cruthfield (61) have published calculations which suggest
that conduction instabilities in charge density wave systems
could be driven into turbulent behavior by the application of
microwave or infrared fields of small magnitude. In support of
their analysis they cite experimental results of Fleming and
Grimes (62) of conduction noise in niobium triselenide. Huberman
and Boyce (63) have also proposed that superionic conductors
could be driven into chaos by an infrared laser.

It is impossible to draw detailed conclusions from this
cursory listing of proposed or observed chaotic physical systems.
However, the breadth of the phenomena listed supports the
previously stated contention that chaotic behavior can occur in
many different kinds of systems (64). The most important
cnclusion of this section from the perspective of this report is
certainly established: chaotic behavior is not limited to
biological systems, and in particular it is observed in the kinds
of physical devices and networks that are used to construct the
hardware of electronic and optical control systems.

V.   CONCLUSIONS

V.A. Counterarguments and Refutations

In constructing conclusions a procedure will be employed that has
proved useful in many areas of reliability analysis. It is often
instructive to make an extrapolation to the worst case; state the
worst case as a thesis, and then produce counterarguments to this
worst case. In attempts to assess the validity of the
counterarguments, reference will be made to the mammalian central
nervous system. This is done not simply because of the intrinsic
interest of these biological systems, but rather because they are
the only ultra-high density parallel processing devices presently
available for examination. However, comparison to biological,

instead of physical, examples inevitably results in a decrease in
the significance that can be assigned to any conclusions obtained
by this procedure.

Thesis: (1) Engineering systems, particularly future
generations of hardware employing very large scale integration
technologies and parallel processing networks, could undergo
transitions to destabilized behavior dynamically analogous to
convulsive disorders. (2) Dynamical disturbances of this type
could propagate through a system and result in major failures.

Counterargument: The central nervous system is very reliable;
convulsive disorders are rare. Therefore, analogous defects in
engineering systems will be correspondingly rare.

The premise of this counterargument is false. Epilepsy, for
example, is not a rare disorder. It is present in approximately
1% of the population and shows little variation from country to
country. In the United States the disease costs approximately $3
billion a year. (This includes the direct cost of health care as
well as indirect costs of income lost due to illness (65).)

Counterargument: The density of excitatory connectivity in the
central nervous system is enormous. No forseeable electronic
system would even approach this level of excitability, and
therefore it is impossible for electronic systems to display
convulsive failures.

Again the premise of this counterargument is false. High
levels of mutual excitation can appear in regions of the central
nervous system, but evidently this is not essential to
epileptogenesis. The hippocampus provides an example of this.
Two pyramidal cells of region CA_3 of the hippocampus were
simultaneously impaled with microelectrodes. One cell was
stimulated to fire. Membrane potential was measured in the
second cell. In 5 cases of the 88 pairs of cells, current evoked
action potentials in the first cell resulted in excitatory
postsynaptic potentials and/or spikes in the second cell (66).
Thus, the pyramidal cell-pyramidal cell excitatory connectivity
is only about 6%.

Counterargument: A disturbance, if it should appear, would be
confined to a discrete area of the network. It would not
propagate to large areas, thus a major system failure would not
occur.

This argument can be restated as a question: What are the
intrinsic excitation and connectivity characteristics necessary
to prevent the propagation of a disturbance to adjacent regions
of the network? The counterargument assumes an answer to this
question. Namely, it assumes that a disturbance will propagate
through a region only if it has a high intrinsic excitability.

The neurological evidence suggests that this is an unwar-
rantably sanguine assumption. Experimental evidence indicates
that epileptiform activity in region CA_3 of the hippocampus can
drive burst behavior in area CA_1. Bursts in CA_3 always precede
CA_1 bursts, and a cut between CA_1 and CA_3 abolishes epileptiform
electrical activity in CA_1 but not in CA_3 (67,68). However, CA_1
appears to have a low intrinsic excitability. Certainly it is
lower than in CA_3. "In recording for 101 pairs of pyramidal
cells (of region CA_1 of the hippocampus) no interactions were
detected in 87% of the pairs. In 13% of the pyramidal pairs,
spike trains in one cell caused inhibitory postsynaptic potentials in the second cell. No excitatory interactions were
detected," (69). Though pyramidal cells do form excitatory
connections with interneurons in 28% of the interneuron-pyramidal
cell pairs, these interneurons do not in turn excite pyramidal
cells of CA_1. It seems that this is a region of the brain with
low intrinsic excitability. For example, CA_1 is apparently
incapable of endogenous generation of epileptiform activity.
Nonetheless CA_1 is able to transmit CA_3-generated disturbances
resulting in major psychomotor seizures.

V.B. Summarized Conclusions
This paper began with an examination of the human brain and
progressed to a consideration of electronic networks. However,

in many important respects they are not that different. They are both dynamical systems. They are both subject to instabilities. They are both amenable to analysis using techniques of topological dynamics.

This study has resulted in the following provisional conclusions. Parameter dependent transitions of attractor topology may provide a common dynamical mechanism for understanding nervous system failures, particularly convulsive disorders, and major control failures in electronic systems. Presently very little is known about what properties predispose an electronic control system to a major convulsive failure. Argument by analogy with the mammalian central nervous system suggests that the predisposing factors include: (i) a high degree of nonlinearity in component input/output relations, (ii) complex network geometry (that is, a large degree of connectivity, hierarchical structures with local autonomy and parallel processing), (iii) high density, and (iv) high energy/unit volume transfer rates.

If these are indeed the properties that predispose a system to disordered behavior, then as technological systems become faster, smaller and are incorporated into increasingly complex distributed networks a failure mode analogous to a central nervous system convulsions becomes more probable. At present dynamical theorists are confronted with a terra incognita. Present ignorance makes it impossible to confidently identify which systems are, or are not, robust against parameter dependent failures. Further, it seems that these failures may become increasingly probable as electronic hardware becomes more sophisticated.

The author wishes to take care not to overstate the speculations incorporated into these conclusions. Present evidence is very incomplete, and it is inconclusive. However, the consequence of a convulsive failure in a military control system could be grave. For this reason, the possibility, even if seemingly remote, almost certainly warrants investigation.

ACKNOWLEDGMENTS

The author wishes to thank Richard Latta for reading the
manuscript, for the computations in the figures, and for
preparing the glossary. This research has been supported by NIH
grant NS 19716 of the National Institute of Neurological and
Communicative Disorders and Stroke.

REFERENCES

1. C.F. Stevens, Sci. Amer., 241(3), 48-59 (1979).

2. D.H. Hubel, Sci. Amer., 241(3), 38-47 (1979).

3. M.D. Mesarovic', D. Macko and Y. Takahara, Theory of Hier-
    archical Multilevel Systems, Academic Press, New York, 1970.

4. M.G. Singh, Dynamic Hierarchical Control, Revised Edition,
    Elsevier, Amsterdam, 1980.

5. E. Niedermeyer, Epileptic seizure disorders, In: Electro-
    encephalography, E. Niedermeyer and F. Lopes da Silva, Eds.,
    Urban and Schwarzenberg, Baltimore, 1982, pp. 339-428.

6. R.G. Bickford, Activation procedures and special electrodes,
    In: Current Practice of Clinical Electroencephalography, D.W.
    Klass and D.D. Daly, Eds., Raven Press, New York, 1979, pp.
    269-305.

7. T. Takahashi, Activation methods, In: Electroencephalography,
    E. Niedermeyer and F. Lopes da Silva, Eds., Urban and
    Schwarzenberg, Baltimore, 1982, pp. 179-195.

8. J.S. Lockard, A primate model of clinical epilepsy: Mechanism
    of action through quantification of therapeutic effects. In:
    Epilepsy, A Window to Brain Function, J.S. Lockard and A.A.
    Ward, Eds., Raven Press, New York, 1980, pp. 11-50.

9. D.A. Prince, Annu. Rev. Neurosci., 1, 395-415 (1978).

10. G.F. Ayala, M. Dichter, R.J. Gumnit, H. Matsumoto, and W.A.
    Spencer, Brain Res., 52, 1-17 (1973).

11. T.A. Pedley and R.D. Traub, Physiology of epilepsy, In:
    Scientific Basis of Clinical Neurology, M. Swash and C.
    Kennard, Eds., Churchill Livingstone, London, (1985).

12. L.E. El'gsol'c, Quantitative Methods in Mathematical Analysis, American Mathematical Society, Providence, Rhode Island, 1964.

13. V.I. Arnold, Geometric Theory of Differential Equations, Springer-Verlag, New York, 1983.

14. B. van der Pol, Philos. Mag., 43, 700-719 (1922).

15. D. Ruelle, Math. Intelligencer, 3, 126-137 (1980).

16. E. Ott, Rev. Modn. Phys., 53, 655-671 (1981).

17. A.I. Mees and C.T. Sparrow, IEE Proc., 128D, 201-205 (1981).

18. E.N. Lorenz, J. Atmo. Sci., 20, 130-141 (1963).

19. D. Ruelle and F. Takens, Commun. Math. Phys., 20, 167-192 (1971).

20. J.B. McLaughlin and P.C. Martin, Phys. Rev. 12A, 186-203 (1975).

21. J.P. Eckmann, Rev. Modn. Phys., 53, 643-654 (1981).

22. N. Ogata, Experientia, 34, 1035-1036 (1978).

23. N. Ogata, Exptl. Neurol., 46, 147-155 (1975).

24. R.K.S. Wong and R.D. Traub, J. Neurophysiol., 49, 442-458 (1983).

25. P.E. Rapp, Chaotic neural dynamics: Turbulent behavior in a biophysical control system, Proceedings: NATO Advanced Research Institute, R.A. Dingle, Ed., Les deux Alpes, France, in press.

26. M.R. Guevara, L. Glass, M.C. Mackey and A. Shrier, Chaos in Neurobiology, IEEE Trans. Systems, Man and Cybernetics, 13, 790-798 (1983).

27. A.V. Holden, W. Winlow and P.G. Heydon, Biol. Cybernetics, 43, 163-173 (1982).

28. A.V. Holden and M.A. Muhamed, J. Electrophysiol. Technique, 11, 135-147 (1984).

29. H.S. Greenside, A. Wolf, J. Swift and T. Pignataro, Phys. Rev. 25A, 3453-3456 (1982).

30. G.A. Carpenter, SIAM J. Appl. Maths., 36, 334-372 (1979).

31. G.A. Carpenter, Normal and abnormal signal patterns in nerve cells, In: Mathematical Psychology and Psychophysiology, S. Grossberg, Ed., SIAM-AMS Proceedings, American Mathematical Society, Providence, Rhode Island 13, 49-90, 1981.

32. L.K. Kaczmarek, Biol. Cybernetics, 22, 229-334 (1976).

33. L.K. Kaczmarek and A. Babloyantz, Biol. Cybernetics, 26, 199-208 (1977).

34. M.C. Mackey and U. an der Heiden, Funkt. Biol. Med., 1, 156-164 (1982).

35. M.C. Mackey and U. an der Heiden, Dynamics of Recurrent Inhibition, McGill University, Preprint, 1983.

36. R.D. Traub and R.K.S. Wong, Science, Wash., 216, 745-747 (1982).

37. R.D. Traub and R.K.S. Wong, J. Neurophysiol., 49, 459-471 (1983).

38. R.D. Traub, Neurosci., 7, 1233-1242 (1982).

39. M.R. Guevara, L. Glass and A. Shier, Science, Wash., 214, 1350-1353 (1981).

40. M.R. Guevara and L. Glass, J. Math. Biol., 14, 1-24 (1982).

41. A.V. Holden and S.M. Ramadan, Biol. Cybernetics, 41, 157-163 (1981).

42. H. Hayashi, S. Ishizuka, M. Ohta and H. Hirakawa, Phys. Lett., 88A, 435-438 (1982).

43. G. Matsumoto, K. Aihara, M. Ichikawa and A. Tasaki, J. Theor. Neurobiol., in press (1983).

44. O.E. Rossler, Chaos and Strange Attractors in Chemical Kinetics, In: Synergetics: Far from Equilibrium, Springer Series in Synergetics, Vol. 3, A. Pacault and C. Vidal, Eds., Springer-Verlag, Berlin, 1979, pp. 107-113.

45. R.H. Simoyi, A. Wolf, and H.L. Swinney, Phys. Rev. Lett., 49, 245-248 (1982).

46. H. Degn, L.F. Olsen and J.W. Perram, Ann. N.Y. Acad. Sci. 316, 623-637 (1979).

47. J.C. Roux, A. Rossi, A. Bachelart and C. Vidal, Physica, 2D, 395-403 (1981).

48. J.S. Turner, J.C. Roux, W.D. McCormick, and H.L. Swinney, Phys. Lett. A., 85A, 9-12 (1981).

49. G. Ahlers and R.P. Behringer, Phys. Rev. Lett., 40, 712-716 (1978).

50. G. Ahlers and R.P. Behringer, Suppl. Prog. Theor. Phys., 64, 186-201 (1978).

51. R.J. Donnelly, K. Park, R. Shaw and R.W. Walden, Phys. Rev. Lett., 44, 987-989 (1980).

52. M. Gorman and H.L. Swinney, Phys. Rev. Lett., 46, 992-995 (1981).

53. H. Haken, Phys. Rev. Lett., 53A, 77 (1975).

54. J. Ballieul, R.W. Brockett and R.B. Washburn, IEEE Trans., CAS-27, 990-997 (1980).

55. J.P. Gollub, T.O. Brunner and B.G. Danly, Science, Wash., 200, 48-50 (1978).

56. J.P. Gollub, E.G. Romer and J.E. Socolar, J. Stat. Phys., 23, 321-333 (1980).

57. J. Testa, J. Percy and C. Jeffries, Phys. Rev. Lett., 48, 714-717 (1982).

58. H. Ikezi, J.S. deGrassie and T.H. Jensen, Observation of multiple-valued attractors and crises in a driven nonlinear circuit. Preprint, GA Technologies, 1983.

59. B.A. Huberman, J.P. Crutchfield and N.H. Packard, Appl. Phys. Lett., 37, 750-752 (1980).

60. M.J. Feldman and M.T. Levison, IEEE Magnet., 17, 834-837 (1981).

61. B.A. Huberman and J.P. Crutchfield, Phys. Rev. Lett., 43, 1743-1747 (1979).

62. R.M. Fleming and C.C. Grimes, Phys. Rev. Lett., 42, 1423-1426 (1979).

63. B.A. Huberman and J.B. Boyce, Solid State Commun., 25, 759-762 (1978).

64. N. MacDonald, Nature, Lond., <u>271</u>, 305-306 (1978).

65. U.S. Dept. of Health and Human Services, National Institute of Neurological and Communicative Disorders and Stroke, NIH (National Institutes of Health) Publication No. 82-1683, Wash., DC, 1981.

66. B.A. MacVicar and F.E. Kudek, Brain Res., <u>184</u>, 220-223 (1980).

67. P.A. Schwartzkroin and D.A. Prince, Ann. Neurol., <u>1</u>, 463-469 (1977).

68. P.A. Schwartzkroin and D.A. Prince, Brain Res., <u>147</u>, 117-130 (1978).

69. W.D. Knowles and P.A. Schwartzkroin, J. Neurosci. <u>1</u>, 318-322 (1981).

70. M.J. Berridge and P.E. Rapp, J. Exptl. Biol., <u>81</u>, 217-279 (1979).

71. W. Penfield and K. Kristiansen, <u>Epileptic Seizure Patterns</u>, Charles Thomas, Springfield, 1951.

72. V. Volterra, Acta. Biotheoretica, <u>3</u>, 1-16 (1937).

Glossary

BIOLOGICAL AND MOLECULAR ELECTRONIC SWITCHING

Acceptor - A molecule, group, or moiety that accepts one or more electrons (like TCNQ or $-NO_2$).

Action potential - Transient local reversal of membrane potential that propagates itself along a neuron.

Actuator - A device producing mechanical motion; molecular examples include the actin/myosin system in muscle.

Acridine - $C_{13}H_9N$

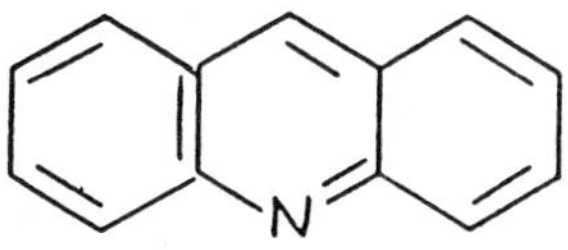

AEM - Analytical Electron Microscopy, a variety of EPMA carried out in a transmission electron microscope and suitable for thin film specimens.

Allosteric Interaction - A process which is regulated by a species (enzyme) at another site.

Amino Acids - The building blocks of proteins. There are 20 common amino acids having the same basic structure both with different side groups (R):

$$R-\underset{\underset{NH^+_3}{|}}{C}HCOO^- \quad = \quad R-\underset{\underset{NH_2}{|}}{C}HCOOH$$

Amino Acid Sequence - The linear order of the amino acids in a protein. Also known as the primary structure of a protein.

Amplification (biotechnical) - Replication of a gene library in bulk.

Amphiphilic - Description of a molecule that has both hydrophilic and hydrophobic properties.

Anthracene - $C_{14}H_{10}$

Antiferromagnet - A substance in which the magnetic moments of the atomic sublattice align themselves so as to make the net magnetization zero.

Arrhenius Plot - Graph of the logarithm of a reaction rate vs. the inverse of the absolute temperature. According to transition state theory, the slope of the line is proportional to the activation energy of the reaction.

Axons - The core of a nerve fiber which conducts impulses away from the nerve cell.

Bacteria - Unicellular microorganisms containing a single very large DNA molecule per cell (the chromosome).

Bacteriophage - A virus that grows and reproduces in bacteria.

1,2 - Benzanthracene $C_{18}H_{12}$

Bilayer Phase Transition - A gel to liquid crystalline phase transition that occurs in the acyl side chain region of phospholipid bilayers as they are heated.

Bistable Switching - A regime in which there exist two separately identifiable states (e.g. 1 and 0, ON and OFF, or two molecular tautomers).

Canting - Certain substances that are primarily antiferromagnetic exhibit a weak ferromagnetism due to the canting or imprecise alignment of antiparallelism in the spins.

Carrier - A special molecule in the cell membrane that specifically binds to an ion or molecule and transports it across the membrane.

CCD - Charge Coupled Device - An electronic device which is capable of precisely transferring and manipulating small amounts ("buckets") of electronic charge. Used for computer memory, optical and IR imaging.

Centrioles - A tiny cylindrical organelle located at the center of a centrosome; used during mitosis (cell division).

Channel - A pore for the passage of ions through a membrane. The pore is formed by a special membrane protein.

Charge Transfer Interaction - The interaction by charge transfer of electronic levels in disparate electronic species.

Charge Transfer or Ion-Radical Salt - When the ionization potential of a donor is low and the electron affinity of an acceptor is high, electron transfer may occur to give a phase markedly different from the weak pi-complex. In a 1:1 stoichiometry, the crystal structures may consist of segregated linear chains of each molecular species.

Chemical Excitation - Process by which an action potential is propagated chemically, such as at the synapse by way of the chemical transmitter acetylcholine.

Chlorophyll - The intensely colored pigment used for light harvesting and primary charge separation in photosynthetic organisms.

Cilia - Microscopic hair-like appendages which extend from the cell surface.

Clamp potential - The membrane potential kept constant by a voltage clamp technique while measuring current.

Clone - A group of genetically identical cells or organisms reproduced asexually from the same cell.

Codon - A sequence of three adjacent nucleotides (coding triplet) in messenger RNA that code for an amino acid (or chain termination).

Colony - A group of contiguous cells, usually derived from a single ancestor, growing on a solid surface.

Complementary Base Sequences - Polynucleotide sequences that are related by the base-pairing rules.

Contractile ATPase - An enzyme which exerts catalytic influence on ATP (adenosine triphosphate) in contractile muscle.

Contrast Curve - In lithography, a plot of resist dissolution rate versus the incident radiation dose.

Critical Behavior - A phase change characterized by critical points such as an order parameter, coexistence curve, etc.

Crystal Field Theory - The theory which treats a metal complex as if the only interaction between the central metal atom and its set of nearest neighbor molecules or ions is a purely electrostatic one.

Curie Temperature - T_c, the temperature for ferromagnetic ordering.

Curie-Weiss Law - The presence of interactions between atomic magnets modifies Curie's law such that $\chi = C/(T-O)$, where O is called the Weiss constant. Susceptibility data is often displayed as "Curie-Weiss" plots of $1/\chi$ vs. T to exhibit the intercept on the T axis, O. If this is at $T > 0$, then $O > 0$ and the ferromagnetic interaction is dominant. If the intercept is at $T < 0$, then $O < 0$ and the antiferromagnetic interaction is dominant. (If $O = 0$ there is no interaction, the material is a paramagnet and Curie's law is recovered.)

Cytochemistry - The chemistry of plant and animal cells.

Cytoplasmic Movement - Movement of the protoplasm outside of the cell nucleus.

Dendrites - The branched parts of nerve cells which transmit impulses toward the cell body.

Depolarization - Reduction or reversal of the membrane potential as during an action potential. Also referred to as hypopolarization.

N_1N - Diethylaniline $C_6H_5N(C_2H_5)_2$

N,N'-Dioctadecyl-4,4'-bipyridinium

$R = C_{18}H_{37}$

N,N'-Dioctadecyloxacarbocyanine

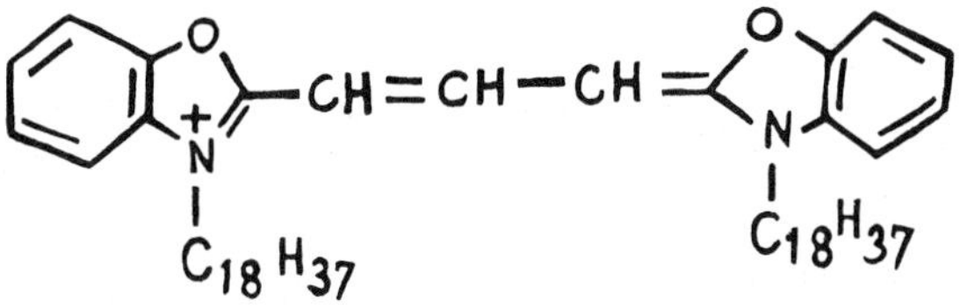

N,N'-Dioctadecyloxacyanine

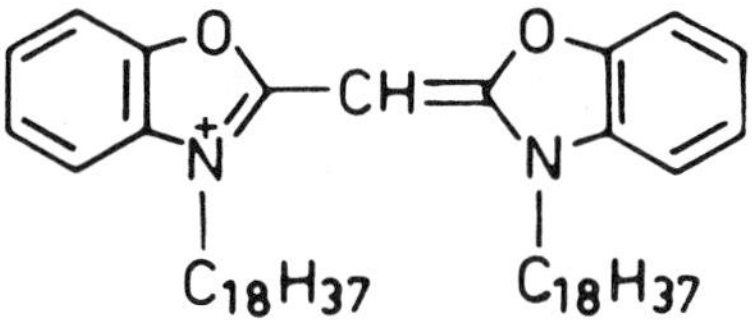

N,N'-Dioctadecylthiacarbocyanine

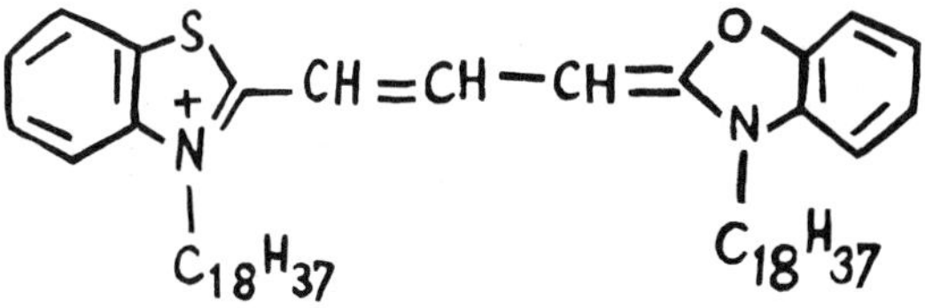

N,N'-Dioctadecylthiacyanine

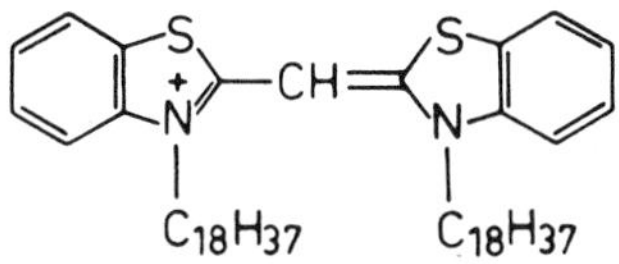

Dioleolylphosphatidylcholine (DOPC) - A phospholipid similar to those that comprise the bulk of the cell membrane. In aqueous solutions, DOPC forms phospholipid bilayers that are model systems for cell membranes.

Dipolar or Solvent Relaxation - A phenomenon wherein solvent (or other surrounding) molecules reorient themselves around the new electron distribution of an excited fluorophor. If the relaxation occurs before the fluorophor emits, the additional stability of the excited state appears as a red shift.

DNA Deoxyribonucleic acid - The genetic material of all cells and many viruses. Most DNA molecules consist of two interwound strands of four basic units called nucleotides or bases, A, C, G, T. Their structure is such that A in one strand is always found opposite T in the other strand, and vice versa. Similarly, C and G always pair with each other. DNA molecules can be either linear or circular (for micro-organisms, i.e. bacteria and viruses).

DNA Polymerase - An enzyme that can fill in single-stranded gaps in double-stranded DNA by inserting the proper complementary bases opposite the bases in the intact strand.

DNA Replication - The process by which the two complementary strands of a DNA molecule separate and a new complementary strand is synthesized by DNA polymerase on each of the separated strands. This process gives rise to two daughter DNA molecules identical to the parental molecule.

Donor - A molecule or group that gives up one or more electrons (like TTF or $-NH_3$).

Donor Acceptor Complex - (See Weak PI Complex).

Dynamic Memories - Digital electronic devices which store binary information in the form of electric charge on a capacitor. Due to leakage of this capacitor, it is necessary to "refresh" the memory by recharging it at frequent intervals typically 10 -> 100 times per second.

Dynein - A large protein which catalyzes ATP hydrolysis in the presence of Mg^{+2}.

Dzyaloshinsky-Moriya Interaction - An interaction Hamiltonian used to describe spin-spin interaction.

E-Beam - Electron beam.

E. Coli, Escherichia coli - The most extensively studied species of bacteria, isolated from the gastrointestinal tract of most mammals including man. The present degree of understanding of E. coli is probably an order of magnitude greater than that of any other bacterium, as is the capability for manipulating it in precise ways by genetic techniques.

EDS - Energy Dispersive X-ray Spectrometry.

EELS - Electron Energy Loss Spectrometry.

Electrical excitation - Process by which an action potential is propagated electrically, such as along an axon.

Electron Microprobe - Method for quantitative chemical analysis. An electron beam impinging on a sample excites characteristic X-rays which are analyzed.

Electron Spin Resonance (ESR) - A magnetic sample is placed in a uniform magnetic field. An electric field of (usually microwave) frequency ν is then applied. Energy is absorbed by the sample when $h\nu$ = energy level splitting induced by the magnetic field plus the internal field.

Electron tunneling - The penetration of a potential energy barrier by an electron. The effect is due to the wave nature of the electron.

Enzyme - A protein molecule that catalyzes (increases the rate of) a chemical reaction.

Episome - A circular gene fragment.

EPMA - Electron probe microanalysis, an analysis technique based on spectrometry of characteristic x-rays generated by electron bombardment of solid.

N-Ethylcarbazole - $C_{14}H_{13}N$

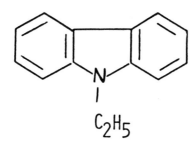

Extrinsic Phase - An ion-radical solid whose formation is accounted for by invoking chemical species other than those implied by the name of the solid.

Femtosecond - 10^{-15} sec.

Ferromagnetism - An example of cooperative behavior characterized by a spontaneous macroscopic magnetization in the absence of an applied magnetic field and resulting from an exchange interaction between electrons such that their spins are parallel.

Flagella - A long filamentous appendage of certain cells which frequently assists in locomotion.

Flagellar motor - A reversible, variable speed motor, driven by proton flow, that turns bacterial flagella.

Four Point Probe - A device for measuring the electrical resistivity of materials by passing a known constant current between two precisely spaced outer electrodes and measuring the voltage produced between two precisely spaced inner electrodes.

Gate - That part of a channel allowing for control of flow of ions through the channel. Notice that this term stems from a particular model that visualizes the site of passive transport.

Gated - A channel is said to be gated if the conductance is controlled by the voltage across the membrane (electrically gated, voltage-gated) or by special chemical transmitters (chemically gated).

Gating - Opening and closing of a membrane channel.

Gating current - Current due to the opening and closing of membrane channels, as opposed to the ionic current which is due to the flow of ions through the channels.

Gene - A sequence of DNA base pairs that codes for a single species of protein, which is itself a sequence of amino acid units in a chain. Proteins (enzymes) are thus gene products.

Genetic code - Dictionary of codons or code words consisting of three adjacent nucleotides which specify the twenty amino acids and stop signals in protein synthesis.

Gene Expression - A gene is expressed in a cell when it is transcribed into RNA and the RNA is translated into a functional protein.

Hard Errors - Data processing errors encountered in digital electronic hardware which result from the failure of an electronic device and not just the loss of information.

Heisenberg Exchange interaction - The most commonly assumed form of strong interaction between atomic magnets. It is written $JS_i \cdot S$, where $\vec{S_i}$ and $\vec{S_j}$ are spins on sites i and j. It is <u>symmetrical</u> in that the three components of the vector S are equivalent.

Heisenberg Model - A model, in which for a description of magnetic and other systems, the magnetic moments are related to a quantum mechanical three component spin operator, and the assumption is made that the energy is proportional to the scaler product of these operators. There is no exact solution for a three-dimensional lattice.

Heme - A complex between a protoporphyrin and iron (II). Protoporphyrin is a derivative of a tetrapyrrole compound.

Hemoglobin - A complex four chain molecule, one chain of which has Fe^{2+} surrounded by a porphyrin ring with either an alpha or beta protein attached, and, if in the oxy form, an oxygen molecule in the sixth position of the metal on. The heme of hemoglobin is Fe-photoporphyrin IX.

Host-Vector - A system designed for preparation of recombinant DNAs, consisting of a vector or molecular vehicle to carry the recombinants and host cells for growth of the vector.

Hydrogen bonds - An important but relatively weak chemical bond as in - $OH \cdots NH_2$-, -$OH \cdots Cl$-, or water. It is responsible for the α-helix structure of proteins and holds complementary base pairs (A and T, C and G) together in DNA, etc.

Hydrophilic - Water-soluble.

Hydrophobic - water-insoluble, synonymous with lipophilic, lipid-soluble.

Inactivation - Refers to the decrease of the sodium conductance in response to a sudden and maintained depolarization under voltage clamp.

Inducible Enzymes - enzymes whose rate of production can be increased by the presence of inducers in the cell.

Interaction volume - the region of a sample in which a primary excitation beam deposits energy and generates secondary radiation.

Intrinsic Phase - Ion-radical solid where initiation may be discussed by involving only the donor and acceptor, their redox states, and their mutual interactions.

Inverse Faraday Effect - The resonance enhancement of internal magnetic moments.

Ionic current - Current due to the passive flow of ions through the membrane.

Ion microprobe - A SIMS technique which utilizes a focused primary ion beam for microanalysis.

Ion microscope - A SIMS technique which utilizes ion optical microscopy to form an image of a sample from a particular sputtered ion species.

Ion Radical Salt - (See Charge Transfer Salt) Often involves a donor or acceptor molecule (like TCNQ) that becomes charged and contains an unpaired electron in the valence orbitals (like $TCNQ^{\sigma}$).

Ion specificity - Property of a membrane channel that allows it to prefer the passage of one ion type over other ion types.

Ion selectivity - Property of a transport system in membranes such as carriers or channels.

Ising Model - A model in which the magnetic moments are assumed to be classical, one-dimensional "sticks" capable of two orientations. Solutions for the model exist for the one and two dimensional cases.

J-aggregate - Brick-stone-work-like arrangement of chromophores showing narrow and high absorption bands shifted to larger wavelengths with respect to the monomeric absorption.

Jahn-Teller Instability - This instability theorem states that a complex
which has an even number of electrons and a degenerate ground state
will distort so that the degeneracy is removed, producing a
configuration which is lower in energy.

LAMMA - Laser Microprobe Mass Analyzer, an analysis technique based
upon laser ablation and time-of-flight mass spectrometry.

Langmuir-Blodgett Technique - A process for generating adsorbed films by
repetitively dipping a substrate into a liquid which is covered with a
monolayer of the desired film material. If the material is an aliphatic
organic compound with a polar end, then each dip will result in the
deposition of a film whose thickness is twice the length of the
compound.

Larmor Precession Theorem - In a magnetic field an electron in orbital
motion may be regarded as a vector precessing about the direction of
the magnetic field.

Laser Raman Microprobe - An analysis technique based on inelastic light
scattering.

Ligand Field Theory - Crystal field theory modified in such a way as to
take account of the existence of moderate amounts of delocalization
between the metal and ligand orbitals.

Linker - A small fragment of synthetic DNA that has a restriction site
useful for gene splicing.

Lipid - Water-insoluble compound made of fatty acids and alcohol. There
are many types of which polar lipids are of special interest, because
they form a molecule with a hydrophilic head group and two
hydrophobic hydrocarbon tails.

Lipid bilayer - Two-dimensional structure formed in a solvent, usually
water, by two single layers of parallel lipid molecules, whereby the
hydrocarbon tails of the two layers keep away from the solvent and
touch each other, while the head groups form the two surfaces in
contact with the water.

Lipophilic - Fat-soluble. Synonym for hydrophobic.

Lithography - The process of transferring two dimensional images by
irradiation of radiation sensitive resist materials with either photons
or charged particles. The desired pattern is created by development
and etching.

Macromolecule - A large molecule, usually composed of a sequence of a
limited number of different kinds of basic subunits. DNA, RNA,
proteins, and polysaccharides are all examples of macromolecules.

Magneto-Crystalline Anisotropy - A property of a magnetic material whose magnet moments show a preferred direction with respect to crystal axes.

Manganese Stearate - A soap in which two stearate chains combine with one divalent manganese to form $Mn(C_{18}H_{35}O_2)_2$.

Membrane - Lipid bilayer structure separating the cell interior from the cell exterior (or forming compartments within the cell such as mitochondrion, nucleus).

Membrane Potential - Potential difference between extra- and intracellular fluid, also referred to as transmembrane potential.

MER - A repeat unit, as in polymer, but differing from the monomer due to a condensation reaction plus any bonding rearrangement.

Metamagnetic Transition - The conversion of an antiferromagnet to a ferromagnet by the application of a strong external magnetic field. The ferromagnetic state exists only in the presence of the field.

Microanalysis - The analysis of selected small regions of a sample where the dimensions of the analyzed region are on the order of micrometers.

Microtubles - Cylindrical protein polymers composed of multiple strands of α and β tubulin assembled in a helical fashion.

Mitotic Spindles - Microfilaments that pull chromosomes apart during cell division.

Molecular Machine - A device having moving parts of molecular scale, structured to atomic or nanometer specifications, and organized to accomplish some purpose.

Molecular Technology - A technology based on handling atoms (or molecules) as individuals, combining them to build structures to complex, atomic specifications.

Monolayer Organizate - Rigid structure of specifically designed architecture obtained by depositing monolayers of planned composition on top of each other.

mRNA, Messenger RNA - The kind of RNA that is synthesized from genes and that acts as the intermediate in the conversion of information stored as a base-pair sequence in DNA into an amino acid sequence in a protein.

Multistage Redox System - A molecule system capable of the reversible transfer of n electrons in n steps.

Mutation - A change in the genetic material of an organism. Mutations can be base-pair changes, deletions, additions, or inversions of a series of base pairs.

NAND (Not And) - A logical operation in which a false result is obtained only if all input conditions are true and a true result is obtained if any input condition is false.

2-Naphthol - Fluorescent molecule whose spectrum shifts when the phenolic proton dissociates. The propensity for ionization is much higher when the molecule is excited; i.e. the ground state pK_a is 9.2, and the excited state pK_a is 2.0.

NATA, N-Acetyl-L-Tryptophanamide - Model compound with properties similar to tryptophan, the amino acid principally responsible for protein fluorescence.

Neel Temperature - T_n, is the ordering temperature of antiferromagnets.

Neuron - Cell that can process and propagate nerve signals. It has the structurally distinct regions of cell body, dendrites, and axon. Also called a nerve fiber.

NMP - N-Methyl Phenazinium

NMP^+

NOR   (Not Or) - A logical operation in which a true result is obtained only if all input conditions are false and a false result is obtained if any input condition is true.

Nuclear Backscattering - Method for quantitative chemical analysis. Incident nuclei are scattered from the nuclei in a sample. Depending on the rebound energy, the masses of the elements in the sample can be determined.

Operator - In molecular biology, a region of DNA that interacts with a repressor protein to control the expression of an adjacent gene or group of genes.

Operon - A gene unit consisting of one or more genes that specify a polypeptide and an "operator" that regulates the transcription of the structural gene. The regulator and the coding genes are adjacent on the DNA molecule.

Organelles - The highest level of organization in the hierarchy of cell structure in which various supra molecular complexes are assembled into such things as nuclei, mitochondria and chloroplasts.

Passive transport - Process whereby ions flow "downhill" from the high concentration side of the membrane to the low concentration side.

Palindrome - A self-complementary nucleic acid sequence, that is, a sequence identical to its complementary strand; perfect palindromes (for example, GAATTC) frequently occur as sites of recognition for restriction enzymes.

Paramagnet - A material whose atomic magnets weakly interact and normally do not form an ordered magnetic arrangement but will order in an external magnetic field. The tendency toward alignment is opposed by thermal agitation.

Phase-Modulation Fluorometry - A method for measuring fluorescence lifetimes and other temporal properties of fluorescence emission. The sample is excited with light sinusoidally modulated at a MHz frequency; its sinusoidal emission is phase-delayed and demodulated to an extent determined by the lifetime(s) of the sample.

Phase Sensitive Detection of Fluorescence - A technique of phase-modulation fluorometry wherein the emissions from two ensembles of fluorophors can be resolved if their lifetimes differ.

Phonons - A quantum of acoustic or vibrational energy.

Phosphodiester bond - The type of covalent bonds that link nucleotides together to form the polynucleotide strands of a DNA or RNA molecule. Phosphodiester bonds can be cut by endonucleases and exonucleases, formed by DNA polymerase, and repaired (reformed) by DNA ligase.

Phospholipid - General term for fatty molecules that are the principal constituents of cell membranes. They are derivatives of glycerol-3-phosphate, with fatty acids esterified at the 1 and 2 positions of the glycerol, and usually some charged moiety attached to the phosphate group. Below is illustrated Dimyristoylphos-phatidylcholine:

hydrophobic acyl sidechain   polar moiety

Phospholipid Bilayer - Basic structure of the cell membrane. Phospholipids
spontaneously assemble into bilayers in aqueous solution, because this
configuration isolates the hydrophobic acyl side chains from the polar
solvent.

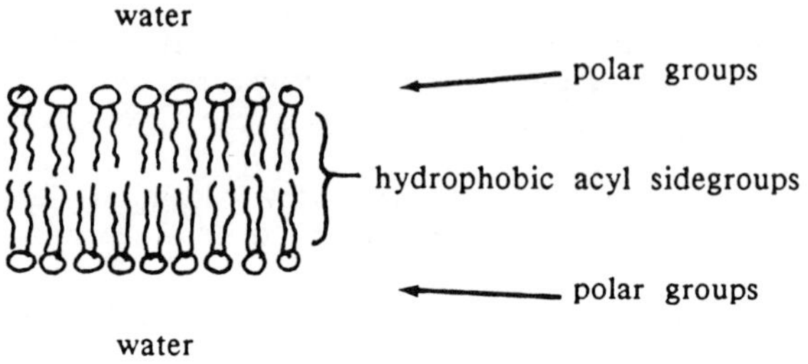

water

polar groups

hydrophobic acyl sidegroups

polar groups

water

Phospholipid Vesicle - A planar phospholipid bilayer prepared so as to
enclose a small volume of water.

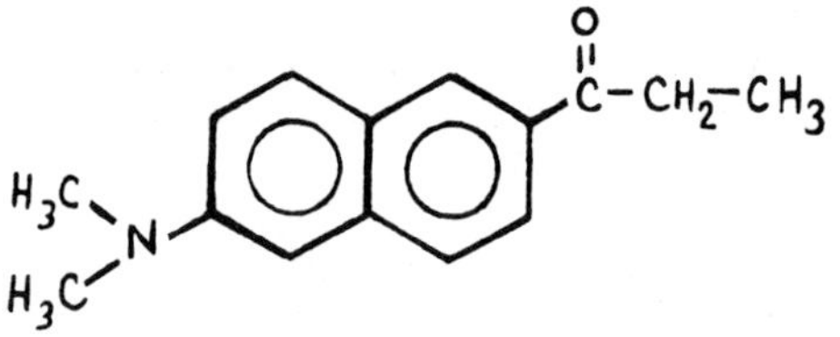

Photosynthesis - The conversion of light energy (photons) into chemical
energy by natural or artificial means.

Photosynthetic Membrane - Natural membranes in photosynthetic organisms
in which all the light reactions of photosynthesis take place.

Phthalocyanine -

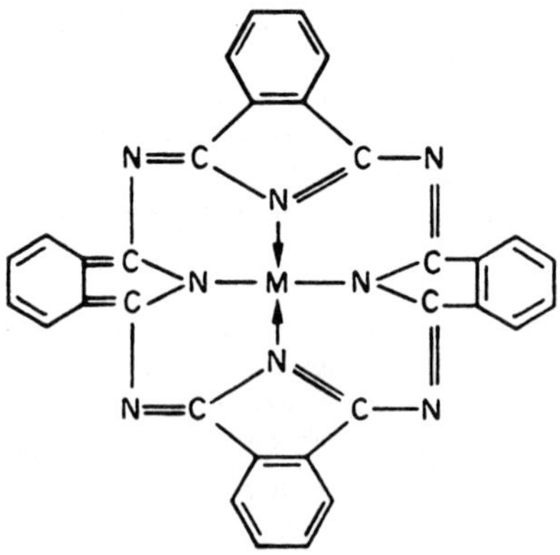

Pi-Amphoteric Molecule - A molecule which behaves as an electron donor or
acceptor, depending on specific experimental conditions.

Picosecond - 10^{-12} sec.

Plasmid - Extrachromosomal, autonomously replicating, circular DNA segment. Associated with drug resistance in bacteria.

Plastoquinone - An electron carrier in photo systems.

$$\text{CH}_3 \left(\!\!\begin{array}{c} O \\ \| \end{array}\!\!\right) H \quad \text{CH}_3$$

$$\text{CH}_3 \bigcirc \text{CH}_3 -(\text{CH}_2-\text{CH}=\overset{\text{CH}_3}{\underset{}{\text{C}}}-\text{CH}_2)_n\,\text{H}$$

PMMA (Polymethymethacrylate) - A popular resist polymeric material in E-beam and X-ray lithography.

$$\left[\begin{array}{c} \overset{\text{C}\,H_3}{\underset{|}{\text{|}}} \\ -\text{C}-C\,H_2- \\ \underset{\text{O}}{\overset{|}{\text{C}}}\text{-O-}C\,H_3 \end{array}\right]_m$$

Polarons - An excited state of an electron that is coupled to a semilocalized hole state (+ charge). The location of the hole is partially trapped by relaxation of the atoms near the hole.

Polymerase - Enzyme that catalyzes the assembly of nucleotides into RNA and of deoxynucleotides into DNA.

Polymerized 1, 11-Dodecadiyne Dimer ——

$$\text{C} \equiv \text{C} \equiv \text{C} \equiv \text{C}$$
$$(\text{CH}_2)_8 \qquad\qquad (\text{CH}_2)_8$$
$$-\!\left(\text{C} = \overset{H}{\underset{}{\text{C}}} - \overset{H}{\underset{}{\text{C}}} = \text{C}\right)_N$$

Polystyrene

$$-\!\left(\text{CH}_2 - \text{CH}\right)_N$$

Porphyrin - A macrocyclic tetrapyrrole compound.

Porphine   **Porphyrin System**

Primary Events of Photosynthesis - The earliest photochemistry in
   photosynthesis.

Prodon, 2-Propionyl-6-dimethylaminonaphthalene - A recently developed
   fluorophor that exhibits a large increase in dipole moment upon
   excitation.

Prokaryotes - Simple, unicellular organisms such as bacteria and blue-green
   algae. Prokaryotes have a single chromosome and lack a defined
   nucleus and nuclear membrane.

Promoter - A DNA sequence at which RNA polymerase binds, and then
   initiates transcription.

Protein - The major structural and catalytic macromolecules in cells.
   Proteins consist of linear chains of 20 different kinds of building
   blocks called amino acids. Each triplet of base pairs in a DNA
   molecule codes for a different amino acid, so that the sequence of
   base pairs is converted into a sequence of amino acids during protein
   synthesis.

Proximity Effect - Loss of resolution in charged particle beam lithography
   caused by scattering of the incident particles in the resist layer and
   substrate.

Push-Pull Olefin - A molecule containing donor and acceptor groups
   separated by one or more double bonds like $H_3N-HC=CH-NO_2$.

Radical Anion - A charged ion or group whose charge includes an unpaired
   electron near or in the valence orbitals.

Raman Spectroscopy - Inelastic light scattering by molecules due to the induction of a dipole moment by the incident, usually laser, light.

Reaction Center - A natural complex between protein, chlorophylls, and other electron transfer components. The primary events occur within the reaction center.

Reading Frame - There are three reading frames in any nucleotide sequence, since there are three nucleotides per codon. Generally only one reading frame of a gene will translate into a functional protein.

Reading - One-way linear process by which nucleotide sequences are decoded, for example, by protein-synthesizing systems.

Recombinant DNA, chimeric DNA, hybrid DNA - DNA molecules of different origin that have been joined together by biochemical techniques to make a single molecule, usually circular and usually capable of some specific biological function, especially self-replication in the appropriate cell.

Registration - In lithography, the process of aligning new patterns to be defined with existing patterns on the same substrate.

Regulatory sequence - A DNA sequence involved in regulating the expression of a gene (for example, promoters, operators).

Relaxation Time - The time required for an excited molecule (or moiety) to decay to a less excited or ground state.

Repressor - The protein that binds to a regulatory sequence (operator) adjacent to a gene and which, when bound, blocks transcription of the gene.

Replicon - An autonomous genetic element that replicates as a single unit, having an origin at which replication commences.

Resonance Raman Spectroscopy - Similar to Raman spectroscopy, except that the irradiating light source resonates with what is usually the electronic absorption band corresponding to an induced electric dipole. Magnetic effects are possible.

Resting potential - In excitable cells the membrane potential when the membrane is not electrically active.

Ribosomal RNA, rRNA - The nucleic acid component of ribosomes, making up two-thirds of the mass of the ribosome in E. coli, and about one-half the mass of mammalian ribosomes.

Ribosomes - Small macromolecular cellular particles (200Å in diameter) made up of rRNA and protein and held together by weak noncovalent forces (e.g. van der Waals, hydrogen bonding, etc.). Ribosomes are the site of protein synthesis.

RNA - Ribonucleic acid - All cells contain both DNA and RNA, and some viruses have RNA as their genetic material. RNA differs from DNA in that it has the sugar ribose instead of deoxyribose, it is usually single-stranded instead of double-stranded, and it has the base U (uracil) in place of the base T (thymine). RNA is used as the molecule that transmits information encoded in the DNA base sequence to the protein-synthesizing apparatus of the cell (see mRNA), as the major structural synthesizing apparatus of the cell (see mRNA) and as the major structural components of ribosomes.

SAM - Scanning Auger Microscope/Microprobe - An analysis technique based on stimulation of Auger electron emission under primary electron bombardment.

Sampling Volume - The region of a sample from which secondary radiation is emitted under primary radiation bombardment.

Short Intermolecular Contacts - Crystallographically observed interatomic distances between molecules which are shorter than the sum of accepted van der Waals radii.

SIMS - Secondary Ion Mass Spectrometry - An analysis technique based upon primary ion beam sputtering and mass spectrometry of the sputtered secondary ions.

Single Event Upset - A soft error caused by an X-ray or cosmic ray.

Sodium pump - Abbreviation for sodium-potassium adenosine triphosphatase pump. A complex of protein subunits of about 275,000 daltons molecular weight that is located in cell membranes and can exchange sodium ions on the cell inside for potassium ions on the cell outside in a ratio of about 3:2.

Soft Error - In digital electronic systems, the loss of information caused by a transient failure of an electronic device (sometimes radiation induced).

Soliton - A solitary wave moving in a non-dissipating or nondispersive mode in a dissipating medium.

Stacked Porphyrin - Metal porphyrin group bridged via the metal atom so that the porphyrin centers are directly over one another like a stack of poker chips.

Stearic Acid - A fatty acid, a long-chain molecule with a carboxylic acid (COOH) group at one end: $H_3C-(CH_2)_{16}-COOH$.

STEM - Scanning Transmission Electron Microscope.

Synapse - Microscopic region of close proximity between the presynaptic membrane of the terminal of one neutron and the receiving postsynaptic membrane of another. Synapses are usually made between axon and dendrite (axo-dendritic), but also between axon and axon (axo-axonal), and axon and cell body (axo-somatic).

Synthetic DNA - DNA of a desired nucleotide sequence that has been synthesized chemically rather than enzymatically.

Synthetic Endonuclease Site - A synthetic DNA that contains the specific base pair sequences recognized by a site specific endonuclease. See Linker.

Tautomerism - Chemical isomerism characterized by relatively easy interconversion of isomeric forms in equilibrium, like

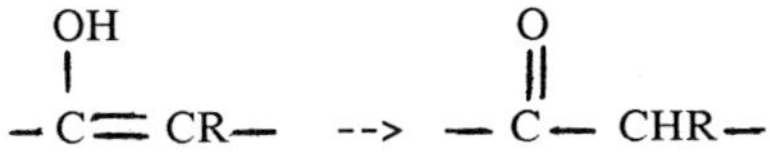

$$
\begin{array}{cc}
\underset{|}{\text{OH}} & \underset{\|}{\text{O}} \\
-\text{C}\equiv\text{CR}- \quad \text{-->} \quad & -\text{C}-\text{CHR}-
\end{array}
$$

Termination codon - A codon that specifies the termination of translation.

Termination sequence - A DNA sequence at the end of a transcriptional unit that signals the end of transcription.

Tetraazaanulene -

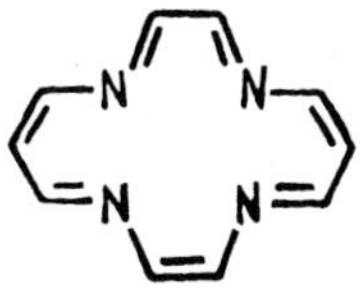

T-4 Bacteriophage - A much-studied virus that infects bacterial cells; it demonstrates several forms of molecular machinery.

TCNE - Tetracyanoethylene

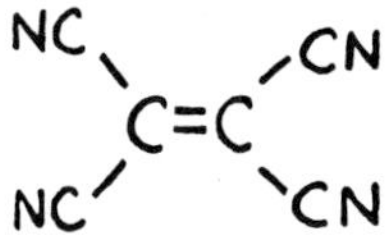

TCNQ - Tetracyanoquinodimethide -

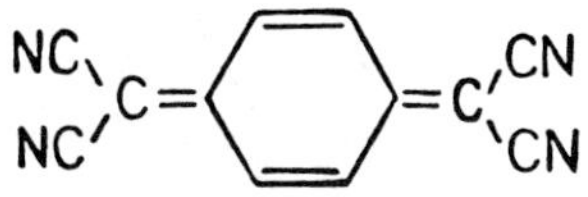

TNAP - Tetracyanonapthoquinodimethene

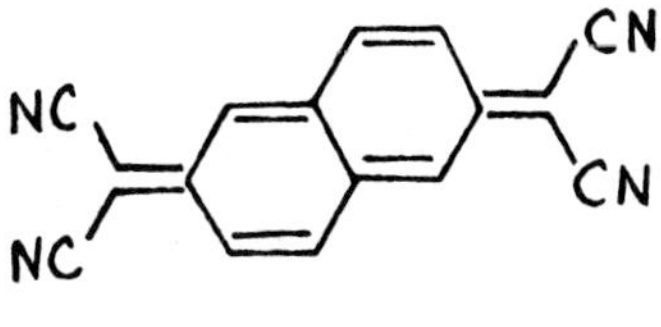

TNS2-(p-Toluidinyl)Naphthalene-6-sulfonic acid - A widely used fluorophor that has a much larger dipole moment in the excited state, and hence exhibits large Dipolar Relaxation effects. It binds specifically to Phospholipid Bilayers near the lipid-water interface.

Trace analysis - The analysis of a constituent which forms a minute fraction of the analyzed volume.

Transcription - The process of synthesizing an RNA chain using a DNA molecule as a template.

Translation - The process in which the genetic code contained in the nucleotide sequences of mRNA directs the order of amino acids in the formation of peptide.

Tricritical Point - A point representing an equilibrium state in which three phases coexist.

p-Tricyanovinylphenyldicyanamide -

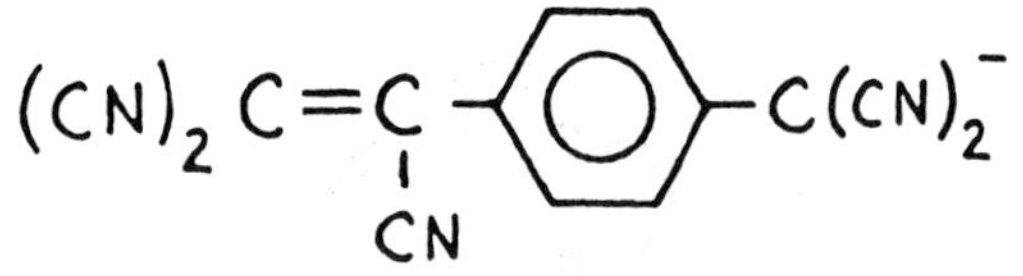

$$(CN)_2 \, C{=}C \underset{CN}{\overset{|}{}} \text{—}\!\!\left\langle\bigcirc\right\rangle\!\!\text{—} C(CN)_2^-$$

Trinitrofluorenone - (TNF)

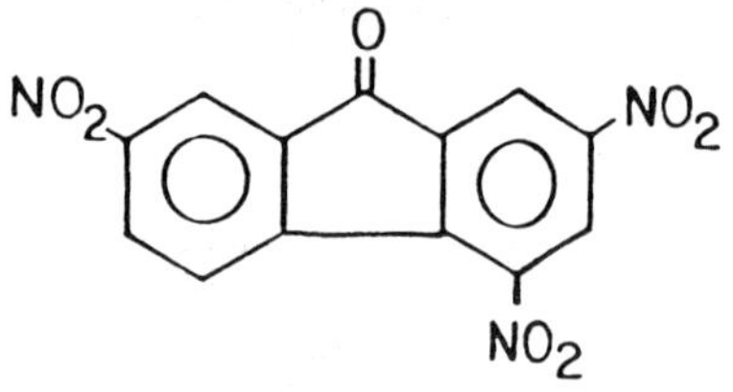

tRNA - Transfer RNA.

TTF - Tetrathiafulvaline

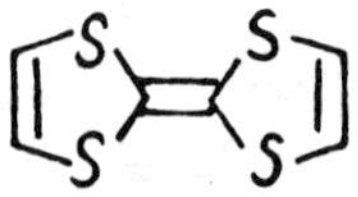

Ubiquinone - An electron carrier which is ubiquitous in cells consisting of a benzoquinone derivative with a long isoprenoid side chain (Coenzyme Q).

Vector - An agent consisting of a DNA molecule known to autonomously replicate in a cell to which another DNA segment may be attached experimentally so as to bring about the replication of the attached segment.

Virus - A DNA or RNA molecule surrounded by a protective protein coat. The viral DNA or RNA is capable of replicating itself inside an appropriate type of cell. Some viral DNA's can be covalently inserted into the chromosomal DNA of the cells they infect, in which case the viral DNA is replicated as part of the cellular DNA and does not kill the cell.

Voltage Clamp - An electronic technique involving a feedback circuit by which the membrane potential is held constant while the membrane current is measured.

WDS - Wavelength Dispersive X-ray Spectrometry.

Weak-ferromagnet - This is an antiferromagnet, whose sublattices are not exactly opposed (i.e. they are canted), so that there is a small net magnetic moment.

Weak Pi-Complex or Donor-Acceptor Complex - A crystallographic phase formed from two or more molecular components and manifesting no charge-transfer in the ground state. The geometries of the components are unperturbed compared to the isolated components. However, charge transfer between the components can readily take place and may lead to enhanced electrical conductivity.

Wild type - The genetic state of an organism isolated from nature.

Zero-field Splitting - The splitting of degenerate energy levels in the absence of an external magnetic field.

Index